Nanotheranostics

Mahendra Rai • Bushra Jamil
Editors

Nanotheranostics

Applications and Limitations

Editors
Mahendra Rai
Department of Biotechnology
Nanobiotechnology Laboratory
Amravati, Maharashtra, India

Department of Chemistry
Federal University of Piauí
Teresina, Piauí, Brazil

Bushra Jamil
Department of DMLS
University of Lahore
Islamabad, Pakistan

ISBN 978-3-030-29770-1 ISBN 978-3-030-29768-8 (eBook)
https://doi.org/10.1007/978-3-030-29768-8

This Springer imprint is published by the registered company Springer Nature Switzerland AG
The registered company address is: Gewerbestrasse 11, 6330 Cham, Switzerland

Preface

Theranostics is an emerging area, where therapeutic and diagnostic platforms are integrated together to perform disease diagnosis and therapy simultaneously. It provides a noninvasive method to determine the targeted delivery of drugs and to evaluate drug efficacy as well. Targeted delivery not only aids in reducing the therapeutic dose but also helps in minimizing dose-related side effects. Theranostic approaches have already been proposed for conditions like cancer, inflammatory diseases, and infections. Nonetheless, the practical form of theranostics can be achieved through the application of nanotechnology. Owing to certain unique properties of nanoparticles like large surface area, miniscule size, enhanced retention capability and minimal off-target accumulation, surface functionalization, and the ability to escape host defenses, they may provide a basis for personalized medicines. Nanotheranostics can consequently encourage stimuli-responsive release and may provide a basis for siRNA co-delivery and oral delivery of peptides. They can also be used for delivery across the blood–brain barrier. They may play a remarkable role in combating multidrug-resistant pathogens.

It is generally believed that early diagnosis promotes cure, and this is the basic philosophy behind nanotheranostics: to diagnose diseases before infected individuals start to show symptoms. However, there are certain limitations and problems to achieve this goal. This book is specifically designed to provide information about various nanocarriers developed under nanotheranostics for a sustained, controlled, and targeted co-delivery of diagnostic and therapeutic agents. In addition, diverse theranostic applications of nanotechnology and their limitations have also been addressed.

This book is highly interdisciplinary and is very useful for a diverse group of readers including pharmacologists, nanotechnologists, microbiologists, biotechnologists, clinicians, cancer specialists, and those who are interested in development of nanoproducts used in therapeutics. Students should find this book useful and reader friendly.

This book has been divided into various chapters as follows: Nanotheranostics: an emerging nanoscience; Nanoparticles in nanotheranostics applications; Nanotheranostics approaches in antimicrobial drug resistance; Theranostic nano platforms as a promising diagnostic and therapeutic tool for *Staphylococcus aureus*; Current status and prospects of chitosan: metal nanoparticles and their applications as theranostic agents; Nanomaterials for selective targeting of intracellular pathogens; Nanoformulations: a valuable tool in the therapy of viral diseases attacking humans and animals; The potential of gold and silver antimicrobials: nanotherapeutic approach and applications; Theranostic potential of aptamers in antimicrobial chemotherapy; Current and future aspects of smart nanotheranostic agents in cancer therapeutics; Biosynthesized metallic nanoparticles as emerging cancer theranostics agents; Superparamagnetic iron oxide nanoparticles for cancer theranostic applications; Theranostic applications of nanobiotechnology in cancer; Magnetic/superparamagnetic hyperthermia as an effective noninvasive alternative method for therapy of malignant tumors; Emerging role of aminolevulinic acid and gold nanoparticles combination in theranostic applications; Gold nanorods as theranostic nanoparticles for cancer therapy. All chapters are written by experts in the field and provide the latest in nanotheranostics in a reader-friendly style. Readers will be enriched by emerging nanotheranostics, and their applications.

We would like to thank the authors, who have made noteworthy contributions to this book: to Carolyn Spence, senior publishing editor, and Priyadharsini, project co-ordinator, Springer Nature for their generous co-operation and patience during the whole process of editing the book. We also express our gratitude to the reviewers for their particular informative comments and suggestions on these book chapters. MR thankfully acknowledges the financial support rendered by CNPq Brazil (process number 403888/2018-2).

Amravati, Maharashtra, India — Mahendra Rai
Islamabad, Pakistan — Bushra Jamil

Contents

1 **Nanotheranostics: An Emerging Nanoscience** 1
Bushra Jamil and Mahendra Rai

2 **Nanoparticles in Nanotheranostics Applications** 19
Nadun H. Madanayake, Ryan Rienzie, and Nadeesh M. Adassooriya

3 **Nanotheranostics Approaches in Antimicrobial Drug Resistance** 41
Juan Bueno

4 **Theranostic Nanoplatforms as a Promising Diagnostic and Therapeutic Tool for *Staphylococcus aureus*** 63
Bushra Uzair, Anum Shaukat, and Safa Mariyam

5 **Current Status and Prospects of Chitosan: Metal Nanoparticles and Their Applications as Nanotheranostic Agents** 79
Dilipkumar Pal and Supriyo Saha

6 **Nanomaterials for Selective Targeting of Intracellular Pathogens** 115
Muhammad Ali Syed and Nayab Ali

7 **Nanoformulations: A Valuable Tool in the Therapy of Viral Diseases Attacking Humans and Animals** 137
Josef Jampílek and Katarína Kráľová

8 **The Potential of Gold and Silver Antimicrobials: Nanotherapeutic Approach and Applications** 179
Heejeong Lee and Dong Gun Lee

9 **Theranostic Potential of Aptamers in Antimicrobial Chemotherapy** 197
Bushra Jamil, Nagina Atlas, Asma Qazi, and Bushra Uzair

10 Current and Future Aspects of Smart Nanotheranostic Agents in Cancer Therapeutics 213
Qurrat Ul Ain

11 Biosynthesized Metallic Nanoparticles as Emerging Cancer Theranostics Agents 229
Muhammad Ovais, Ali Talha Khalil, Muhammad Ayaz, and Irshad Ahmad

12 Superparamagnetic Iron Oxide Nanoparticles for Cancer Theranostic Applications 245
Dipak Maity, Ganeshlenin Kandasamy, and Atul Sudame

13 Theranostic Applications of Nanobiotechnology in Cancer 277
Rabia Javed, Muhammad Arslan Ahmad, and Qiang Ao

14 Magnetic/Superparamagnetic Hyperthermia as an Effective Noninvasive Alternative Method for Therapy of Malignant Tumors 297
Costica Caizer

15 Emerging Role of Aminolevulinic Acid and Gold Nanoparticles Combination in Theranostic Applications 337
Lilia Coronato Courrol, Karina de Oliveira Gonçalves, and Daniel Perez Vieira

16 Gold Nanorods as Theranostic Nanoparticles for Cancer Therapy . . 363
Maria Mendes, Antonella Barone, João Sousa, Alberto Pais, and Carla Vitorino

Index 405

Contributors

Nadeesh M. Adassooriya Department of Food Science & Technology, Wayamba University of Sri Lanka, Makandura, Gonawila, Sri Lanka

Irshad Ahmad Department of Life Sciences, King Fahd University of Petroleum and Minerals (KFUPM), Dhahran, Saudi Arabia

Muhammad Arslan Ahmad Department of Tissue Engineering, China Medical University, Shenyang, China

Key Lab of Eco-restoration of Regional Contaminated Environment, Shenyang University, Ministry of Education, Shenyang, China

Qurrat Ul Ain Department of Molecular Medicine, National University of Medical Sciences, Rawalpindi, Pakistan

Nayab Ali Department of Microbiology, The University of Haripur, Haripur, Pakistan

Qiang Ao Department of Tissue Engineering, China Medical University, Shenyang, China

Nagina Atlas Department of Biological Sciences, International Islamic University, Islamabad, Pakistan

Muhammad Ayaz Department of Pharmacy, University of Malakand, Chakdara, Khyber Pakhtunkhwa, Pakistan

Antonella Barone Department of Health Sciences, University "Magna Græcia" of Catanzaro, Catanzaro, Italy

Juan Bueno Research Center of Bioprospecting and Biotechnology for Biodiversity Foundation (BIOLABB), Armenia, Quindío, Colombia

Costica Caizer Department of Physics, West University of Timisoara, Timisoara, Romania

Lilia Coronato Courrol Laboratory of Applied Biomedical Optics, Physics Department, Federal University of São Paulo, Diadema, São Paulo, Brazil

Karina de Oliveira Gonçalves Laboratory of Applied Biomedical Optics, Physics Department, Federal University of São Paulo, Diadema, São Paulo, Brazil

Bushra Jamil Department of DMLS, University of Lahore, Islamabad, Pakistan

Josef Jampílek Faculty of Natural Sciences, Department of Analytical Chemistry, Comenius University, Bratislava, Slovakia

Rabia Javed Department of Tissue Engineering, China Medical University, Shenyang, China

Ganeshlenin Kandasamy Department of Biomedical Engineering, Vel Tech Rangarajan Dr. Sagunthala R&D Institute of Science and Technology, Chennai, TN, India

Ali Talha Khalil Department of Eastern Medicine and Surgery, Qarshi University, Lahore, Pakistan

Katarína Kráľová Faculty of Natural Sciences, Institute of Chemistry, Comenius University, Bratislava, Slovakia

Dong Gun Lee School of Life Sciences, College of Natural Sciences, Kyungpook National University, Daegu, Republic of Korea

Heejeong Lee School of Life Sciences, College of Natural Sciences, Kyungpook National University, Daegu, Republic of Korea

Nadun H. Madanayake Department of Botany, University of Sri Jayewardenepura, Nugegoda, Sri Lanka

Dipak Maity Department of Chemical Engineering, Institute of Chemical Technology Mumbai, IOC Campus, Bhubaneswar, OD, India

Safa Mariyam Department of Biological Sciences, International Islamic University, Islamabad, Pakistan

Maria Mendes Faculty of Pharmacy, University of Coimbra, Coimbra, Portugal

Centre for Neurosciences and Cell Biology (CNC), University of Coimbra, Coimbra, Portugal

Muhammad Ovais CAS Key Laboratory for Biomedical Effects of Nanomaterials and Nanosafety, CAS Center for Excellence in Nanoscience, National Center for Nanoscience and Technology (NCNST), Beijing, People's Republic of China

University of Chinese Academy of Sciences, Beijing, People's Republic of China

Alberto Pais Coimbra Chemistry Centre, Department of Chemistry, University of Coimbra, Coimbra, Portugal

Dilipkumar Pal Department of Pharmaceutical Sciences, Guru Ghasidas Vishwavidyalaya (Central University), Bilaspur, Chhattisgarh, India

Asma Qazi Department of Biogenetics, National University of Medical Sciences, Rawalpindi, Pakistan

Mahendra Rai Department of Biotechnology, Nanobiotechnology Laboratory, Amravati, Maharashtra, India

Department of Chemistry, Federal University of Piauí, Teresina, Piauí, Brazil

Ryan Rienzie Faculty of Agriculture, University of Peradeniya, Peradeniya, Sri Lanka

Supriyo Saha School of Pharmaceutical Sciences & Technology, Sardar Bhagwan Singh University, Dehradun, Uttarakhand, India

Anum Shaukat Department of Biological Sciences, International Islamic University, Islamabad, Pakistan

João Sousa Faculty of Pharmacy, University of Coimbra, Coimbra, Portugal

Coimbra Chemistry Centre, Department of Chemistry, University of Coimbra, Coimbra, Portugal

Atul Sudame Department of Mechanical Engineering, Shiv Nadar University, Dadri, UP, India

Muhammad Ali Syed Department of Microbiology, The University of Haripur, Haripur, Pakistan

Bushra Uzair Department of Biological Sciences, International Islamic University, Islamabad, Pakistan

Daniel Perez Vieira Radiobiology Laboratory, Nuclear and Research Institute, IPEN/CNEN-SP, São Paulo, Brazil

Carla Vitorino Faculty of Pharmacy, University of Coimbra, Coimbra, Portugal

Coimbra Chemistry Centre, Department of Chemistry, University of Coimbra, Coimbra, Portugal

Chapter 1
Nanotheranostics: An Emerging Nanoscience

Bushra Jamil and Mahendra Rai

Abstract Theranostic approaches have been suggested for various ailments, particularly cancer, microbial diseases, AIDS, and many others. This is a kind of personalized treatment where the treatment is guided according to the individual molecular profile or on the basis of biomarker identification. Combination of diagnostics and therapeutic strategy into a single platform can be made possible with the help of nanotechnology. Usually most of the nanomedicines act by increasing bioavailability of the drug, protection from degradation, and controlled biodistribution in the body system. Nanotheranostics thus encompass all those nano stages that can be used for simultaneous detection and treatment of disease by providing better penetration of drugs within the body systems with reduced risks as compared to other conventional therapies. Theranostics offer new and emerging applications of nanotechnology. Nonetheless, the nanocarrier should have the capacity to accommodate multiple agents such as stabilizer, therapeutics, and targeting and imaging moieties.

The pharmaceutical and healthcare industry is the one that has the most benefits of this new and emerging field of nanotechnology. It can also play a key role in the field of molecular biology by the development of molecular sensors or imaging agents for diagnosis and carriers or vehicle development for therapeutic agents. These innovative carriers and agents can make difference in the treatment of cancer, AIDS, cardiovascular diseases, burn wounds, infections, etc. by the development of nanotheranostic diagnostic systems like immunoassays or colorimetric assays and in therapeutic approaches through gene therapy or by biomarker identification and targeting systems.

Keywords Nanotheranostics · Biomarkers · Molecular imaging · Imaging probes · Image-guided therapy · Aptamer theranostics

B. Jamil
Department of DMLS, University of Lahore, Islamabad, Pakistan

M. Rai (✉)
Department of Biotechnology, Nanobiotechnology Laboratory, Amravati, Maharashtra, India

Department of Chemistry, Federal University of Piauí, Teresina, Piauí, Brazil

M. Rai, B. Jamil (eds.), *Nanotheranostics*,
https://doi.org/10.1007/978-3-030-29768-8_1

Nomenclature

ATP	Adenosine-5′-triphosphate
ADCC	Antibody-dependent cellular cytotoxicity
CDC	Complement-dependent cytotoxicity
cDNA	Complementary single-stranded DNA
CT	Computed tomography
Dox	Doxorubicin
ELISA	Enzyme-linked immunosorbent assay
FA	Folic acid
FDA	Food and Drug Administration
Gd	Gadolinium
GSH	Glutathione
GSSH	Glutathione disulfide
IHC	Immunohistochemistry
ICG	Indocyanine green
MRI	Magnetic resonance imaging
mAbs	Monoclonal antibodies
NPs	Nanoparticles
NIR	Near-infrared spectroscopy
PLA2	Phospholipase A2
PT	Photothermal
PTT	Photothermal therapy
PET	Positron emission tomography
QDs	Quantum dots
ROS	Reactive oxygen species
Si NPs	Silica nanoparticles
SPECT	Single-photon emission computed tomography
ssDNA	Single-stranded DNA
SWNTs	Single-walled carbon nanotubes
SPIO	Superparamagnetic iron oxide
SIPPs	Superparamagnetic iron platinum particles
SPR	Surface plasmon resonance
SELEX	Systematic evolution of ligands by exponential enrichment
USPIO	Ultrasmall superparamagnetic iron oxide
US	Ultrasound

1.1 Introduction

There is a drastic change in the healthcare services in the field of drug delivery, imaging modalities, and diagnosis with the evolution of nanotheranostic, an emerging field of nanotechnology. This new field of nanotechnology aims at combining both diagnostics and therapeutics to bring results that are more prolific in the cure of diseases (Prabhu and Patravale 2012; Sharma et al. 2019). Nonetheless, nanoparticles (NPs) themselves may also act as multifunctional agents due to their unique properties; for instance, gold (Au) has many unique properties like surface functionalization, plasmon resonances, photo thermal ablation, and ease of detection (Yeh et al. 2012; de Melo-Diogo et al. 2017). Surface plasmon resonance (SPR) is an optical and quantitative detection phenomenon (as the incident light is converted into both scattered and absorbed component. The scattered component gives optical properties and the absorbed portion gives the thermal effect). It is used to determine molecular binding kinetics in real time. It is a label free, highly sensitive detection method and requires a very minute quantity of sample. Whereas the other techniques like ELISA gives only the binding affinity, SPR gives binding kinetics or the on-and-off phenomenon depending upon association and dissociation of molecules. The only problem with plasmon resonant NPs is low sensitivity because of background scattering by cells and tissues (Jain et al. 2007; Khlebtsov and Dykman 2010). The image quality can be improved with photothermal (PT) techniques. In photothermal therapy (PTT) electromagnetic radiation, most often near-infrared (NIR) wavelengths, are used. NIR radiation upon absorption generates heat. This heat kills the surrounding cells. This approach has already been used successfully to not only kill local cancerous cells but also the cancer cells that have been metastasized (Zou et al. 2016). PTT is actually an extension of photodynamic therapy. In photodynamic therapy a photosensitizer (for instance, porfimer sodium) is used. This photosensitizer is injected in blood. All body cells take photosensitizers. However, normal body cell releases photosensitizer more quickly as compared to cancer cell. After 24–72 h most of the photosensitizers are therefore retained only by cancerous cells and when the body is irradiated with laser beam of specific wavelength only cancerous cells are exposed to radiations. Photosensitizer produces reactive oxygen species (ROS) that kills the nearby cells (Shirata et al. 2017). Nonetheless, PTT offers more advantages over photodynamic therapy as it does not require oxygen. The penetration of PTT is better and can be used to cure deep cancer and cancer metastasis as well but the problem with PT techniques is that it requires higher laser-induced temperatures that can be detrimental to cells and molecules (Lukianova-Hleb et al. 2010).

Targeted drug delivery systems may be applied to increase the therapeutic index of the drugs and imaging agents at the targeted site. Nonetheless, the convergence of therapeutics and diagnostics in combination with nanotechnology can play a vital role for personalized and precision medicine where the drug release would be on demand (Vinhas et al. 2015; Silva et al. 2019). Nanotheranostics provide an unprecedented opportunity to integrate various components along with customized therapeutic agents, controlled-release mechanisms, targeting strategies, and reporting functionality for therapeutic detection/visualization within a nano-scaled architecture (Wang et al. 2017a; Sonali et al. 2018) (Fig. 1.1).

In this chapter, the general overview of nanotheranostics and its important components have been discussed. In addition, all the creative approaches being developed for these classes of therapies, imaging modalities, and the recent developments in the field have also been examined. It can be said that nanotheranostic is a promising

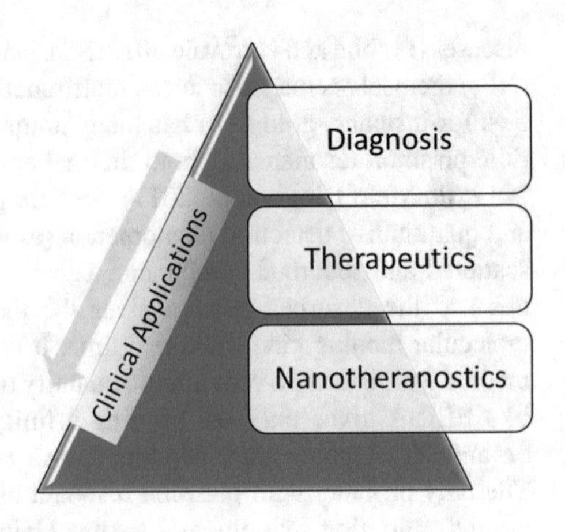

Fig. 1.1 Nanotherapeutics may offer a wide range of clinical and medical applications with less side effects

and emerging field that can offer rapid detection and targeted delivery system, which is rapid and cost-effective with reduced risks. However, there is a need to address some of the limitations related to this field before properly introducing this application in the clinics.

1.2 Nanotheranostic as a Novel Platform

An important component of nanotheranostics is developing a nanocarrier or nano-platform that has the potential to accommodate all requirement of an efficacious theranostic in one system (Kang et al. 2008; Rai and Morris 2019).

Several nanotheranostics platforms have been presented over the past decade. Nonetheless, most frequently used are the traditional ones, namely metal (gold and silver) and silica nanoparticles (Si NPs), liposomes, quantum dots, and composite NPs (Miao et al. 2019; Parchur et al. 2019; Silva et al. 2019; Xu et al. 2019) (Fig. 1.2).

Generally, the physicochemical characteristics (like size, shape, charge, and surface functionalization) of nanoparticles (NPs) determine their fate (Penet et al. 2014). For instance, very small nanoparticles, <20 nm, have rapid body distribution but are also subjected to rapid renal clearance. Whereas larger nanoparticles, >200 nm, are cleared by mononuclear phagocytic system and accumulate in various body organs like liver and spleen (Zhang et al. 2009). The pore size of tight endothelial junction in normal blood vessels is usually <10 nm whereas the size of tight junction in tumor microenvironment is much larger, i.e., >200 nm to 1200 nm. In addition to that, there is no lymphatic drainage in tumor tissues so that nanoparticles cannot escape the tumors; this is enhanced permeability and retention (EPR) effect.

These platforms allow visualization and monitoring of the route taken by the formulation, providing information about delivery kinetics, intra-organ and/or intratumor distribution, and drug efficacy. Using a single NP, it is possible to tune

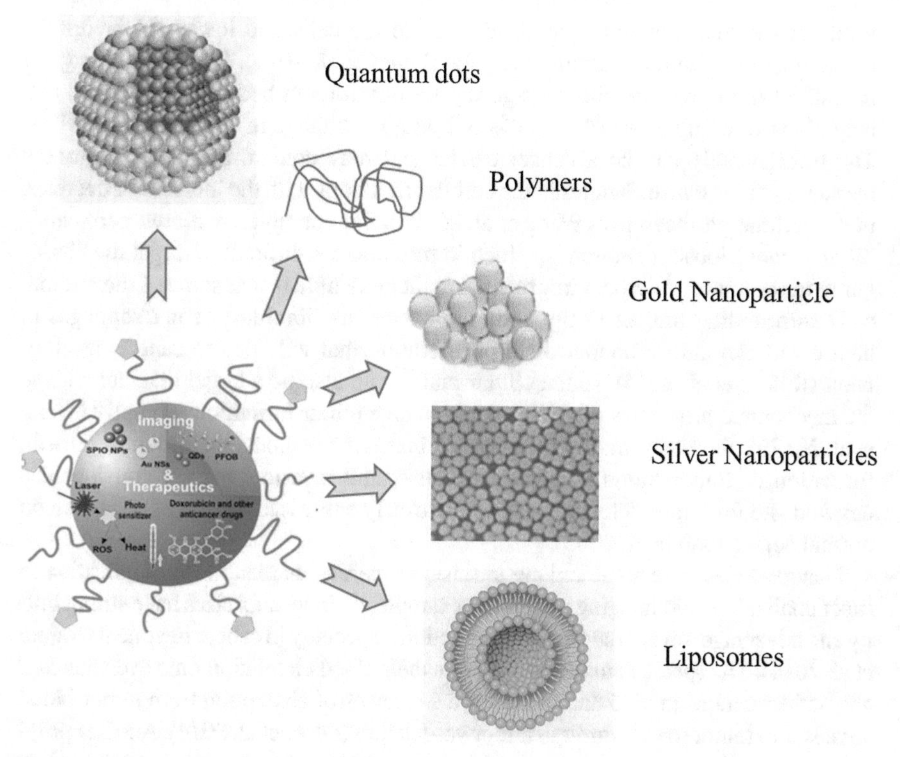

Fig. 1.2 Nano-platforms for developing a successful theranostics

therapy while simultaneously providing for real-time monitoring of disease progression (Vinhas et al. 2015).

Other important components while working with nanoparticles in nanotheranostics are understanding of type of target cells and biomarkers at the target site. Secondly, the route of administration or pathway they will follow to reach their target site (Rajora et al. 2014). Stability of nanoparticles while in the body system (in vivo), i.e., they must resist attack from body's natural immune system till they reach their specific target site and get absorbed. Shape and size of NPs used in theranostics is directly related to their efficacy (Penet et al. 2014). For example, spherical NPs are commonly used in cancer theranostics. Filamentous, rod-shaped, disk-shaped, worm like, or well-shaped, all have different features like drug loading and absorption capacity, circulation time, target uptake, absorption at target site, etc. Similarly, size range is important for the transport and absorption of NPs. Their size may vary from 50 to 200 nm according to Banerjee et al. (2016). They have tested this size range in the intestine. In another study by Loverde et al. (2012), it was found that worm-like NPs with specific probe of PEG-PBD are more effective in cancer treatment.

NPs may target the specific receptor related to the specific cellular microenvironment like hypoxia, pH, and interstitial fluid pressure. For example, lower oxygen levels can induce cancerous characteristics in the cells and lower pH favors the tumor microenvironment around the cells (Penet et al. 2014). In case of change of interstitial fluid pressure, the passageway for any foreign body can be opened that may allow free influx or efflux in the cell with the change in cellular permeability. The foreign body can be a cancer carrier and may contain any other unwanted marker that may cause changes in the cellular functions with the increase or decrease in the cellular permeability (Wang et al. 2017b). For example, increased permeability may cause loose membranes, which in turn make it difficult to target the vasculature in the cancer microenvironment. Another example is that some of the stromal cells surrounding tumors in the body are cancerous fibroblast or macrophages in nature and can induce uncontrolled proliferation that will lead to cancer development (Rajora et al. 2014). Extracellular matrix can also be a target that can change the mechanical properties of cells and can induce tumor formation promptly (Wang et al. 2017b). Similarly, matrix metalloproteinases are responsible for tumor growth, formation of tumor blood vessels, metastasis, and invasion and are considered a targeted site for tumor detection as they are overly expressed on cancer cells than on normal cells (Yoon et al. 2003).

Nanoparticles have some unique intrinsic properties that lead to its application in functionalization and imaging utility. For example, their specific sizes have strong utility and advantage toward the target site of action, especially in cancer treatment (Rajora et al. 2014). The specific small size supports their blood circulation time over standard chemotherapeutics in vivo and increases the chances of absorption from tumor blood vessels into tumor tissues through tumor vasculature (Penet et al. 2014). Another property that NPs have surface-area-to-volume ratio that is high enough to give loading capacity to imaging probe and targeting ligands in case of cancer therapy is another benefit for them to become nanotheranostics (Loverde et al. 2012). All these features of nanoparticles strongly benefit the field of personalized medicine in diagnosis even based on biomarker identification (Mura and Couvreur 2012). These diversified properties of NPs make them a choice in cancer treatment and management for optimizing treatment strategies and their effects along with targeted diagnosis.

It is strongly observed in past studies that nanoparticles are very useful tools that can play a vital role in early diagnosis and in targeted drug delivery. Nonetheless, human body is not a single compartment, rather a combination of multifaceted and multifarious enzymes, organs, and systems, and every individual is genetically and phenotypically different from all others. Every individual responds to treatment differently, and thus the development of bioresponsive nanotheranostics to cater to the requirement of individual need requires a deep understanding of the pathophysiological features of many distinct types of diseases. The tissue microenvironment in disease is different from healthy tissues like in case of infections and tumors blood flow and pH changes. Consequently, there are many other biomarkers (it's an indicator to show the presence or severity of a disease like antibody for a specific antigen indicates infection) that are specifically expressed in disease tissues. These biomarkers basically provide foundation for personalized treatment (treatments individually tailored to specific patients).

1.3 Biomarkers for Theranostics

Biomarkers can be used synonymously to molecular markers. Although biomarkers are in clinical use since ancient times, nowadays more focus is on development and identification of molecular biomarkers that could diagnose disease accurately at early stages and could predict treatment in response to stimuli and can also be able to predict prognosis. Stimulus for nanotheranostics might be external or internal (Raza et al. 2019).

External stimulus is commonly used as imaging biomarkers. Imaging biomarkers offer several advantages like patient convenience, noninvasive procedure, and highly intuitive. By imaging biomarkers one can get both qualitative and quantitative data. It includes computed tomography (CT), magnetic resonance imaging (MRI), positron emission tomography (PET), near-infrared spectroscopy (NIR) that responds to magnetic field, temperature, ultrasound, light, or electric pulses (Wilhelm et al. 2016).

Considering the limitations of external stimuli, numerous internal stimulus such as glucose concentrations, pH differences, redox reactions, and other ions/small biomolecules are more efficient for the designing of smart drug delivery systems (Fig. 1.3).

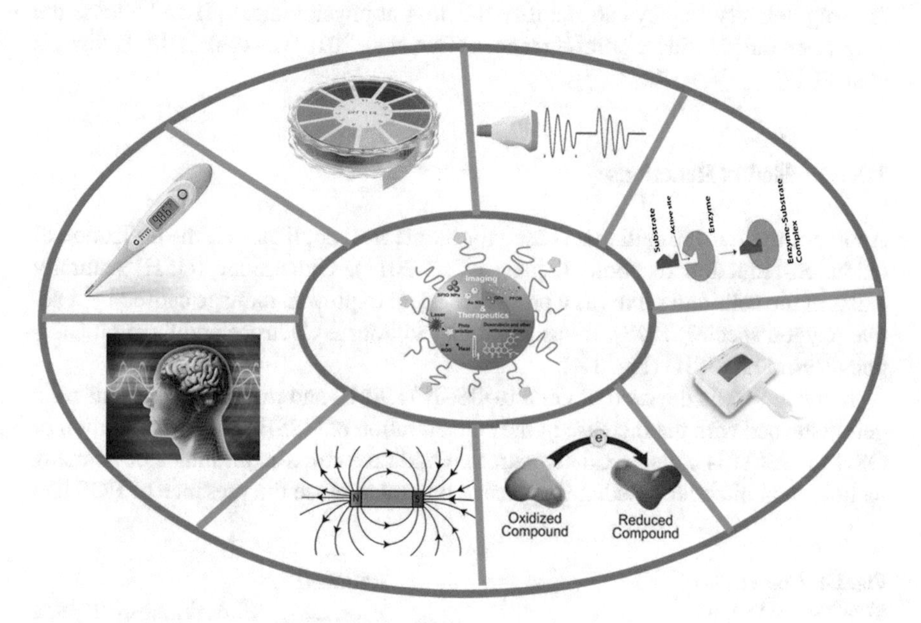

Fig. 1.3 External and internal stimuli for theranostics

1.3.1 Internal Body Signals for Theranostics

1.3.1.1 pH as Stimuli

Among all internal stimuli, pH difference is a most regularly investigated internal body signal for developing targeted theranostics. Tumor microenvironment, infected tissues, ischemia, and rheumatoid arthritis have acidic pH as compared to normal body tissues (Du et al. 2011; Wu et al. 2014; Ju et al. 2016; Fernandez-Piñeiro et al. 2017; Li et al. 2017a). Tumor microenvironment has more energy requirement because of rapid and uncontrolled growth of tumor cells. It results in high lactate and hydrogen ion concentration which reduces the pH to 6.5. It is pertinent to mention that normal body tissues have 7.4 pH. Hence, the difference in pH can serve as a key biomarker for targeting certain disease.

There certainly exist many materials that are sensitive to slight pH differences. Tertiary amines and imidazoles have the property of switching their state after sensing pH difference. They are hydrophobic at pH > 7 whereas hydrophilic at pH < 7. They form agglomerates at higher pH and as the pH drops they release their drug cargo in aqueous environment (Li et al. 2017b). In addition to that, intracellular compartments like endosomes and lysosomes are also acidic in nature having 5.5 and 5.0 pH, respectively. Many endogenous pathogens can therefore be targeted directly by these intracellular vesicles (Wu et al. 2018).

pH-responsive nanomaterials have been used to design sensitive nano-systems for drug delivery as they can stabilize the drug at physiological pH and release the drug when the pH trigger point is reached (Gao et al. 2010; Liu et al. 2014; Eskiizmir et al. 2017).

1.3.1.2 Redox Reactions

Another important stimuli that is used for biomedical appliances is the utilization of oxidation-reduction reactions (Zhang et al. 2017). Glutathione (GSH) naturally exists in the cells and serves as a protective agent to prevent damage caused by reactive oxygen species (ROS). It has the ability to donate electrons and form glutathione disulfide (GSSH) (Fig. 1.4).

In healthy cells the ratio of GSH to GSSH is 90:10 and in disease state this ratio gets disturbed with the increase in the concentration of GSSH. The concentration of GSH to GSSH is called oxidative stress. Oxidative stress is an indicator of many pathological disorders. Redox-responsive polymers sense the presence of ROS and

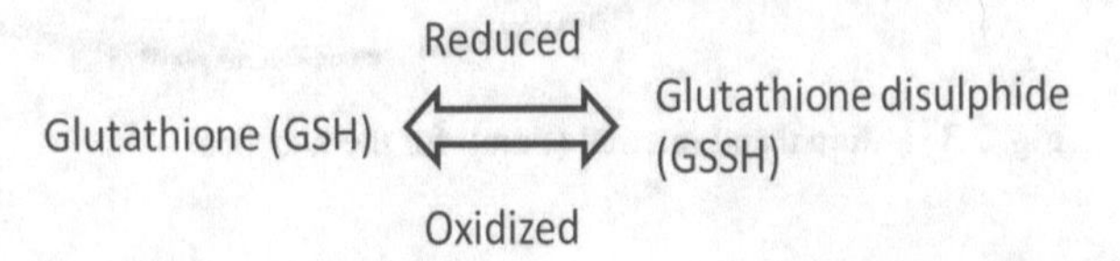

Fig. 1.4 Conversion of glutathione (GSH) to glutathione disulfide (GSSH)

release their cargo where there is high oxidative stress (Iamsaard et al. 2018; Yang et al. 2018). Redox-responsive polymers can be fabricated by incorporating disulfide, diselenide, and boronic ester linkages in the polymers (Huo et al. 2014).

1.3.1.3 Hypoxia

Another important feature of tumor microenvironments is low oxygen pressure, called hypoxia. It is caused by consumption of oxygen by rapidly proliferating cells. The tumor cells adapt to hypoxic environment by their genetic instability. Hypoxia microenvironment is also a marker of angiogenesis, poor prognosis and enhanced tumor aggressiveness and metastasis (Jiang et al. 2018). Likewise, the therapeutic efficiency of radiotherapy also gets reduced because of hypoxia (Hu et al. 2018). However, hypoxia serves as an opportunity to target tumors which have reductive state. Many hypoxia-selective drugs like azobenzene, nitroaromatics, and quinones have been developed to trigger drug release in the absence of oxygen (Wang et al. 2017a).

1.3.1.4 Enzyme-Responsive Nanotheranostic Agents

Enzymes are essential component of all metabolic reactions/processes because of their catalytic properties and their dysregulation leads to many abnormalities and pathological conditions. Enzyme-responsive nanotheranostics can bring more significant and controlled response with small dose by their biocatalytic nature. In addition to that, they also exhibit more specific chemical reactions (Andresen et al. 2010; de la Rica et al. 2012; Popat et al. 2012; Hu et al. 2014). Enzyme-responsive nanotheranostic offers many advantages, such as more selectivity and specificity. Many enzymes like hydrolases, oxidoreductases, proteases, transferases, and phospholipases can be used in nanotheranostics. Phospholipase-based theranostics utilizes the upregulation of phospholipase A2 (PLA2). PLA2 upregulation has been a pathological indicator for multiple kinds of cancers and many other disease processes, including thrombosis, congestive heart failure, inflammation, neurodegeneration, and infectious pathogens (Scott et al. 2010; Hu et al. 2014).

1.3.1.5 Nucleic Acids (i.e., DNAs or RNAs) as Bioresponsive Switches

Nucleic acid sequences (both DNA and RNA) have unique sequences and hence can also be used as reliable biomarkers. These are generally used as fluorophore-labeled single-stranded DNA (ssDNA) in combination with nanoparticles (Xia et al. 2017). In one approach, polymer-coated iron oxide nanoparticles were used and fluorophore-labeled single-stranded DNA (ssDNA) was joined with the nanoparticles (Fig. 1.5). As the system target mRNA inside the cells dye-labeled ssDNA is released from the closed system and produces fluorescence (Lin et al. 2014).

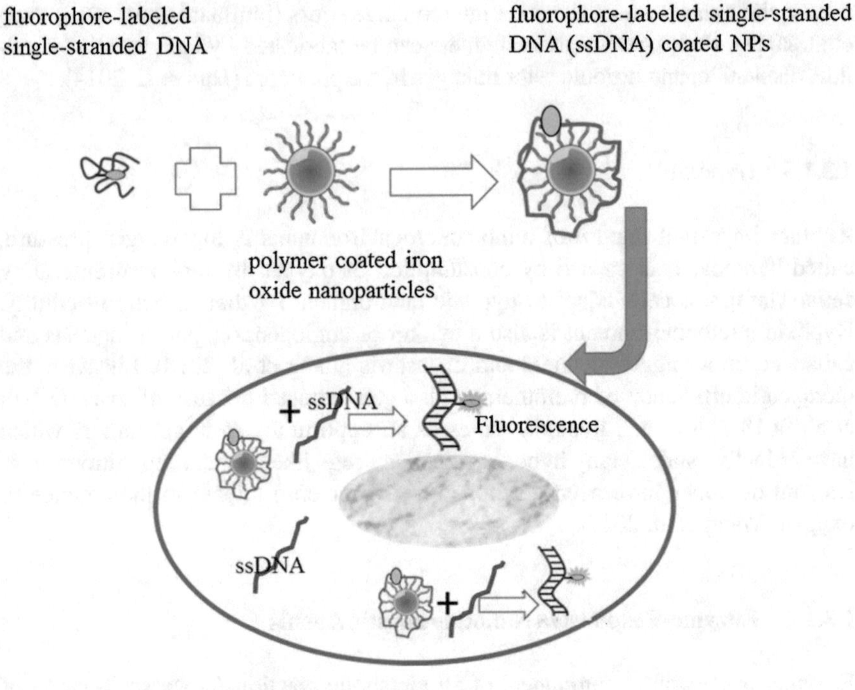

Fig. 1.5 Mechanism of using ssDNA as bioresponsive switch

1.3.1.6 Small Biomolecule-Responsive Theranostics

In addition to specific macromolecules, small biomolecules such as adenosine-5′-triphosphate (ATP) and GSH have also been exploited as biological triggers for stimuli-responsive theranostic applications. ATP is a readily available coenzyme in intracellular environment and is used to provide energy during cellular metabolism. In this regard, Mo et al. (2014) developed a polymeric nano-vehicle comprising an ATP-responsive DNA motif (containing ATP aptamer and its complementary single-stranded DNA (cDNA)) with anti cancerous drug (doxorubicin (Dox)), protein (m-protamine) and hyaluronic acid for "on-demand" release of drugs. The drug was loaded in GC sites of DNA. In the presence of ATP, the doxorubicin was released from DNA motif after conformational changes (Mo et al. 2014; Wang et al. 2017a).

1.3.2 External Body Signals or Molecular Imaging Probes/ Agents Used for Theranostics

Molecular imaging can be defined as "it is the visualization, characterization, and measurement of biological processes at the molecular and cellular levels in humans and other living systems" (Thakur and Lentle 2005; Mankoff 2007). Currently, many techniques like optical imaging (bioluminescence and fluorescence), magnetic resonance imaging (MRI), nuclear imaging, computed tomography (CT), single-photon emission computed tomography (SPECT), positron emission tomography (PET), and ultrasound (US) are being used for molecular imaging (Mankoff 2007; Janib et al. 2010). CT and MRI are not sensitive enough to detect pathological disease markers at early stages; however, PET can do so. PET can efficiently investigate sophisticated processes like receptor binding, DNA synthesis or enzyme activity, oxygen metabolism and blood flow (Vinhas et al. 2015; Wang et al. 2017a).

Nonetheless, for developing smart bioresponsive nanotheranostics that can monitor drug trafficking and therapeutic efficiency and that can provide an on-demand drug release after sensing the disease progression, more sensitive nanoimaging modalities are required (Kelkar and Reineke 2011).

1.3.2.1 Types of Imaging Probes

Near-Infrared (NIR) Imaging

The electromagnetic spectrum of NIR ranges from 780 to 1700 nm. It is further subdivided into NIR I and NIR II regions. NIR I ranges from 780 to 900 nm and NIR II from 900 to 1700 nm (Fig. 1.6).

- NIR I imaging can be done by two type of dyes, namely,
  - Ujoviridin (indocyanine green)
  - Provayblue (methylene blue)

These two dyes emit short wavelength radiation in NIR I region. Although they are better in terms of tissue penetration as compared to visible probes, they are incompatible with CT and MRI because of photon scattering by tissues.

Imaging in NIR II window is characterized by reduced photon scattering and thus aid in improving image quality with lessened background fluorescence. Two types of NIR II imaging techniques are used, namely NIR II-emission imaging and NIR II-excitation imaging.

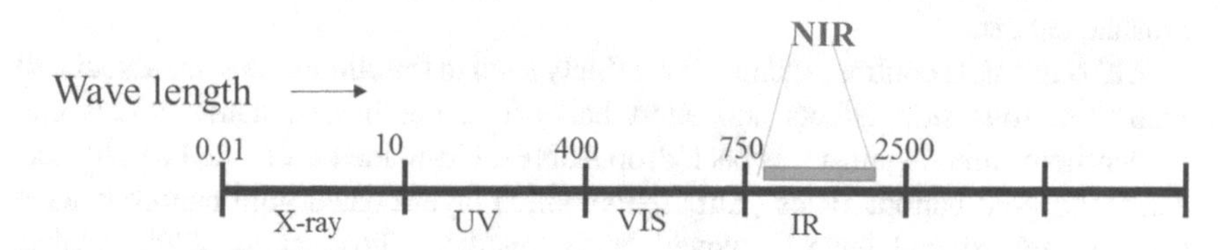

Fig. 1.6 Electromagnetic spectrum

- **NIR II-emission imaging**. Probes used for this type of imaging emit NIR II radiation upon excitation. Four types of probes are used for this purpose, namely,
  - ***Small molecule dyes*** like indocyanine green (ICG) which usually emits luminescence in the NIR I region; however, its liposomal formulation emit radiations in the second near-infrared window (NIR II) (Bhavane et al. 2018).
  - ***Single-walled carbon nanotubes (SWNTs)***
  - ***Quantum dots (QDs)***
  - ***Rare earth-doped nanoprobes***, for example, ytterbium (Yb^{3+}), neodymium (Nd^{3+}), praseodymium (Pr^{3+}), holmium (Ho^{3+}), erbium (Er^{3+}), and thulium ions (Tm^{3+})-doped nanoprobes can emit strong NIR II emission under suitable external excitation sources.
- **NIR II-excitation imaging.** In this type NIR II radiations as excitation source is used. Upon light excitation, the NIR II-excited probes release the luminescence signals in combination with acoustic signal (sound) or thermal heat. Consequently, three types of probes are used in this type of imaging to get luminescence, photoacoustic and thermal image (Liu et al. 2018).

Magnetic Resonance Imaging (MRI)

In MRI contrast agents are widely used to increase the contrast difference between normal and abnormal tissues. The majority of MRI contrast agents are either paramagnetic (having unpaired electron and are attracted toward an applied magnetic field. It includes aluminum, oxygen, titanium, and iron oxide. These are also temporarily magnetized upon application of external magnetic field) or superparamagnetic (iron oxide) magnetite particles (Xiao et al. 2016). However, MRI contrast agents may be divided into lanthanides and transition elements. Lanthanides include gadolinium (Gd), iron, nickel, and cobalt (Faulkner and Blackburn 2014). However, Gd is the only one that can be magnetized at room temperature. Transition element includes manganese and iron (Cortezon-Tamarit et al. 2017). Manganese is used to detect liver lesion and functional brain imaging. Iron is used either as superparamagnetic iron oxide (SPIO) or as ultrasmall superparamagnetic iron oxide (USPIO) and both are used for detection of liver cancer. Superparamagnetic iron platinum particles (SIPPs) after encapsulated in phospholipids are used to detect prostate cancer.

Although MRI contrast agents are routinely used in the clinics, they are associated with numerous side effects and short half-life. Nonetheless, nanoparticles can render them safer. Against this back drop AuNPs, Gd-loaded acetylated dendrimer-entrapped gold nanoparticles (AuDENPs), silica layer-coated gold nanorods have been developed and have displayed better results (Ghosh et al. 2008; Vinhas et al. 2015).

Nuclear Imaging Agents (PET/SPECT Agents)

Nuclear imaging agents are done via gamma and positron emitters. It includes positron emission tomography (PET) and single photon emission computed tomography (SPECT) (Velikyan 2012). Technetium-99m or iodine-123 are gamma emitters and SPECT uses gamma cameras to detect them. Whereas, gallium-68 is positron emitter and SPECT uses it (Yordanova et al. 2017).

1.4 Targeted Therapy or Image-Guided Therapy

Targeted therapy is the foundation of precision therapy and a wide variety of molecules have been utilized for this purpose that may range from hormones to cytokines, antibodies, and peptides. In image-guided therapy digital imaging is utilized to guide therapy or surgery.

1.4.1 Antibody Theranostics

For a successful theranostic agent, targeted therapy is essential that could be achieved through monoclonal antibodies (mAbs). It makes treatment more specific. To date, several mAbs have been approved by the Food and Drug Administration (FDA) for the treatment of cancer. However, for antibodies to be used determination of target expression is necessary and also to understand whether target antigen is associated with disease progression or not.

Target expression can be determined by various techniques, such as:

- Immunohistochemistry (IHC).
- Hematological analysis by ELISA or flow cytometry.
- Immuno-positron emission tomography (PET)/immuno-single photon emission computed tomography (SPECT) (Fleuren et al. 2014).

1.4.2 Photothermal Ablation Agents

Thermal ablation of cancerous cells is a well-known therapy in which nanoparticles absorb energy of the illuminating laser and generate heat. The heat ultimately destroys the surrounding cells that might be diseased or healthy cells as well (Grosges and Barchiesi 2018). Scientists, therefore, are trying to develop target-specific photothermal ablation agents. Gold nanorods and colloidal gold nanospheres have more frequently been used as theranostic agents (Vinhas et al. 2015).

1.4.3 Aptamers Theranostics

Aptamers are peptide molecules or oligonucleotides that bind to specific target only and are selected from pools of oligonucleotides by a process known as systematic evolution of ligands by exponential enrichment (SELEX). They have the ability to recognize a wide variety of targets with high affinity and specificity just like antibodies. However, they are more preferred over antibodies as they are more stable, less immunogenic, and easy to prepare (Xing et al. 2014).

1.5 Limitations

Herceptin® is the first of the humanized antibody capable of simultaneously detecting and treating HER2-positive metastatic breast cancer. Unfortunately, many other scientists tried to follow Herceptin® model but failed to develop successful therapeutic agent. However, the emerging field of nanotechnology has paved the way for the development of theranostics agents for a sustained, controlled, and targeted delivery of therapeutics coupled with the capability to follow in real-time distribution to tissues and organs, thus allowing therapy evaluation. These nanotheranostics have been designed keeping in mind the selectivity for the most appropriate and effective therapy with fewer side effects. Albeit, many nanomedicines like Doxil® Myocet®, DaunoXome®, DepoCyt®, Abraxane®, Resovis®, Genexol-PM®, and Oncaspar® have already been to the market. Thus, smartly designed NPs have already made their way to the clinics, but the development of an equivalent able to combine both strategies is still underway. Indeed, several platforms have already been proposed integrating therapeutic strategies (i.e., chemo-, genetic-, immunotherapy, photothermal ablation) and imaging agents to monitor NP fate, e.g., MRI, CT, and photoacoustic tomography (PAT) (Vinhas et al. 2015).

Potential obstacles to successful nanotheranostics include the discovery and targeting of new biomarkers, the innate toxicity of the nanoparticle components, formulation stability, production costs, and control of intellectual property (Janib et al. 2010; Vinhas et al. 2015).

There is no single therapeutic agent, which has the same effect on all patients, suffering with the same type and stage of cancer while treating cancer. This is the point where precision medicine or personalized medicine are considered as a treatment of choice, as they can personalize a treatment that is best suited for individual patient. The molecular mechanism of action of these medicines is designed on the basis of patient genetic makeup, expressed proteins, and metabolites. Therefore, with the help of genetic sequencing of patient, personalized medicines help in discerning individual patient susceptibility (Kim et al. 2016) toward the developed disease and in turn designing the disease prevention regimes. Understanding and further research in this field specifically will help the pharmaceutical industry in improving the efficacy of these drugs for individual patient to benefit at maximum. The therapy design based on identified molecular information of individual patient may promise patient treatment with lesser side effects (Kim et al. 2016).

1.6 Conclusions

Converging diagnosis and therapeutics based on nanotechnology is opening a new way toward accurate and personalized medicine. Developing smart nanotheranostics that act on the bioresponsive systems have recently evolved and offering promising result with high accuracy and efficiency at targeted point via on-demand drug release. Intelligent nanotheranostics has both therapeutic and diagnostic capabilities. It can be said that nanotheranostics is a promising and emerging field that can offer rapid detection and targeted delivery system that is rapid and cost-effective with reduced risks. However, there is a need to address some of the limitations related to this field before properly introducing this application in the clinics.

Considering the limitation of external stimuli, numerous internal stimuli such as glucose concentrations, pH differences, redox reactions, and other ions/small biomolecules are more efficient for the designing of smart drug delivery systems.

The emergence of nanotechnology has provided an opportunity to promote the development and design of novel nanotheranostics. From this perception, the performance of nanotheranostics should be observed not only before or after but also throughout the therapy and after gaining extensive information about their genotoxicity, cytotoxicity, immunotoxicity, and cost-effectiveness, the nanotheranostics can be introduced into routine healthcare as an important element of personalized and predictive medicine.

References

Andresen TL, Thompson DH, Kaasgaard T. Enzyme-triggered nanomedicine: drug release strategies in cancer therapy (invited review). Mol Membr Biol. 2010;27:353–63.

Banerjee A, Qi J, Gogoi R, Wong J, Mitragotri S. Role of nanoparticle size, shape and surface chemistry in oral drug delivery. J Control Release. 2016;238:176–85.

Bhavane R, Starosolski Z, Stupin I, Ghaghada KB, Annapragada A. NIR-II fluorescence imaging using indocyanine green nanoparticles. Sci Rep. 2018;8:14455.

Cortezon-Tamarit F, Ge H, Mirabello V, Theobald MB, Calatayud DG, Pascu SI. Carbon nanotubes and related nanohybrids incorporating inorganic transition metal compounds and radioactive species as synthetic scaffolds for nanomedicine design. In: Inorganic and organometallic transition metal complexes with biological molecules and living cells. Hong Kong: Elsevier Academic Press Inc.; 2017. p. 245–327.

Du JZ, Du XJ, Mao CQ, Wang J. Tailor-made dual pH-sensitive polymer–doxorubicin nanoparticles for efficient anticancer drug delivery. J Am Chem Soc. 2011;133:17560–3.

Eskiizmir G, Ermertcan AT, Yapici K. Nanomaterials: promising structures for the management of oral cancer. In: Nanostructures for oral medicine. 2017. p. 511–544. Elsevier.

Faulkner S, Blackburn OA. The chemistry of Lanthanide MRI contrast agents. In: The chemistry of molecular imaging. 2014. p. 179–198.

Fernandez-Piñeiro I, Badiola I, Sanchez A. Nanocarriers for microRNA delivery in cancer medicine. Biotechnol Adv. 2017;35:350–60.

Fleuren ED, Versleijen-Jonkers YM, Heskamp S, van Herpen CM, Oyen WJ, van der Graaf WT, Boerman OC. Theranostic applications of antibodies in oncology. Mol Oncol. 2014;8:799–812.

Gao W, Chan JM, Farokhzad OC. pH-responsive nanoparticles for drug delivery. Mol Pharm. 2010;7:1913–20.

Ghosh P, Han G, De M, Kim CK, Rotello VM. Gold nanoparticles in delivery applications. Adv Drug Deliv Rev. 2008;60:1307–15.

Grosges T, Barchiesi D. Gold nanoparticles as a photothermal agent in cancer therapy: the thermal ablation characteristic length. Molecules. 2018;23:1316.

Hu Q, Katti PS, Gu Z. Enzyme-responsive nanomaterials for controlled drug delivery. Nanoscale. 2014;6:12273–86.

Hu M, Yang C, Luo Y, Chen F, Yang F, Yang S, Chen H, Cheng Z, Li K, Xie YA. Hypoxia-specific and mitochondria-targeted anticancer theranostic agent with high selectivity for cancer cells. J Mater Chem B. 2018;6:2413–6.

Huo M, Yuan J, Tao L, Wei Y. Redox-responsive polymers for drug delivery: from molecular design to applications. Polym Chem. 2014;5:1519–28.

Iamsaard S, Seidi F, Dararatana N, Crespy D. Redox-responsive polymer with self-immolative linkers for the release of payloads. Macromol Rapid Commun. 2018;39:e1800071.

Jain PK, Huang X, El-Sayed IH, El-Sayed MA. Review of some interesting surface plasmon resonance-enhanced properties of noble metal nanoparticles and their applications to biosystems. Plasmonics. 2007;2:107–18.

Janib SM, Moses AS, MacKay JA. Imaging and drug delivery using theranostic nanoparticles. Adv Drug Deliv Rev. 2010;62:1052–63.

Jiang X, Wang C, Fitch S, Yang F. Targeting tumor hypoxia using nanoparticle-engineered CXCR4-overexpressing adipose-derived stem cells. Theranostics. 2018;8:1350.

Ju KY, Kang J, Pyo J, Lim J, Chang JH, Lee JK. pH-induced aggregated melanin nanoparticles for photoacoustic signal amplification. Nanoscale. 2016;8:14448–56.

Kang JH, Asai D, Kim JH, Mori T, Toita R, Tomiyama T, Asami Y, Oishi J, Sato YT, Niidome T, Jun B. Design of polymeric carriers for cancer-specific gene targeting: utilization of abnormal protein kinase Cα activation in cancer cells. J Am Chem Soc. 2008;130:14906–7.

Kelkar SS, Reineke TM. Theranostics: combining imaging and therapy. Bioconjug Chem. 2011;22:1879–903.

Khlebtsov NG, Dykman LA. Plasmonic nanoparticles: fabrication, optical properties, and biomedical applications. In: Handbook of photonics for biomedical science, vol. 18. Boca Raton, FL: CRC Press; 2010. p. 37–82.

Kim D, Lee N, Park YI, Hyeon T. Recent advances in inorganic nanoparticle-based NIR luminescence imaging: semiconductor nanoparticles and lanthanide nanoparticles. Bioconjug Chem. 2016;28:115–23.

de La Rica R, Aili D, Stevens MM. Enzyme-responsive nanoparticles for drug release and diagnostics. Adv Drug Deliv Rev. 2012;64:967–78.

Li X, Kim J, Yoon J, Chen X. Cancer-associated, stimuli-driven, turn on theranostics for multimodality imaging and therapy. Adv Mater. 2017a;29:1606857.

Li F, Lu J, Kong X, Hyeon T, Ling D. Dynamic nanoparticle assemblies for biomedical applications. Adv Mater. 2017b;29:1605897.

Lin LS, Cong ZX, Cao JB, Ke KM, Peng QL, Gao J, Yang HH, Liu G, Chen X. Multifunctional Fe_3O_4@ polydopamine core–shell nanocomposites for intracellular mRNA detection and imaging-guided photothermal therapy. ACS Nano. 2014;8:3876–83.

Liu J, Huang Y, Kumar A, Tan A, Jin S, Mozhi A, Liang XJ. pH-sensitive nano-systems for drug delivery in cancer therapy. Biotechnol Adv. 2014;32:693–710.

Liu Y, Jia Q, Zhou J. Recent advance in near-infrared (NIR) imaging probes for cancer theranostics. Adv Ther. 2018; 1(8): 1800055.

Loverde SM, Klein ML, Discher DE. Nanoparticle shape improves delivery: rational coarse grain molecular dynamics (rCG-MD) of taxol in worm-like PEG-PCL micelles. Adv Mater. 2012;24(28):3823–30.

Lukianova-Hleb EY, Hanna EY, Hafner JH, Lapotko DO. Tunable plasmonic nanobubbles for cell theranostics. Nanotechnology. 2010;21:085102.

Mankoff DA. A definition of molecular imaging. J Nucl Med. 2007;48:18N–21N.

de Melo-Diogo D, Pais-Silva C, Dias DR, Moreira AF, Correia IJ. Strategies to improve cancer photothermal therapy mediated by nanomaterials. Adv Healthc Mater. 2017;6:1700073.

Miao T, Oldinski RA, Liu G, Chen X. Nanotheranostics-based imaging for cancer treatment monitoring. In: Nanotheranostics for cancer applications. Cham: Springer; 2019. p. 395–428.

Mo R, Jiang T, DiSanto R, Tai W, Gu Z. ATP-triggered anticancer drug delivery. Nat Commun. 2014;5:3364.

Mura S, Couvreur P. Nanotheranostics for personalized medicine. Adv Drug Deliv Rev. 2012;64:1394–416.

Parchur AK, Jagtap JM, Sharma G, Gogineni V, White SB, Joshi A. Remotely triggered nanotheranostics. In: Nanotheranostics for cancer applications. Cham: Springer; 2019. p. 429–60.

Penet MF, Krishnamachary B, Chen Z, Jin J, Bhujwalla ZM. Molecular imaging of the tumor microenvironment for precision medicine and theranostics. In: Advances in cancer research, vol. 124. London: Academic Press; 2014. p. 235–56.

Popat A, Ross BP, Liu J, Jambhrunkar S, Kleitz F, Qiao SZ. Enzyme-responsive controlled release of covalently bound prodrug from functional mesoporous silica nanospheres. Angew Chem. 2012;124:12654–7.

Prabhu P, Patravale V. The upcoming field of theranostic nanomedicine: an overview. J Biomed Nanotechnol. 2012;8:859–82.

Rai P, Morris SA, editors. Nanotheranostics for cancer applications. Cham: Springer Nature; 2019.

Rajora AK, Ravishankar D, Osborn HM, Greco F. Impact of the enhanced permeability and retention (EPR) effect and cathepsins levels on the activity of polymer-drug conjugates. Polymers. 2014;6:2186–220.

Raza A, Rasheed T, Nabeel F, Hayat U, Bilal M, Iqbal H. Endogenous and exogenous stimuli-responsive drug delivery systems for programmed site-specific release. Molecules. 2019;24(6):1117.

Scott KF, Sajinovic M, Hein J, Nixdorf S, Galettis P, Liauw W, de Souza P, Dong Q, Graham GG, Russell PJ. Emerging roles for phospholipase A2 enzymes in cancer. Biochimie. 2010;92:601–10.

Sharma M, Dube T, Chibh S, Kour A, Mishra J, Panda JJ. Nanotheranostics, a future remedy of neurological disorders. Expert Opin Drug Deliv. 2019;16(2):113–28. https://doi.org/10.1080/17425247.2019.1562443.

Shirata C, Kaneko J, Inagaki Y, Kokudo T, Sato M, Kiritani S, Akamatsu N, Arita J, Sakamoto Y, Hasegawa K, Kokudo N. Near-infrared photothermal/photodynamic therapy with indocyanine green induces apoptosis of hepatocellular carcinoma cells through oxidative stress. Sci Rep. 2017;7:13958.

Silva CO, Pinho JO, Lopes JM, Almeida AJ, Gaspar MM, Reis C. Current trends in cancer nanotheranostics: metallic, polymeric, and lipid-based systems. Pharmaceutics. 2019;11(1):22.

Sonali MK, Singh RP, Agrawal P, Mehata AK, Datta Maroti Pawde N, Sonkar R, Muthu MS. Nanotheranostics: emerging strategies for early diagnosis and therapy of brain cancer. Nano. 2018;2:70.

Thakur ML, Lentle BC. Joint SNM/RSNA molecular imaging summit statement. J Nucl Med. 2005;46:11N–3N.

Velikyan I. Molecular imaging and radiotherapy: theranostics for personalized patient management. Theranostics. 2012;2:424.

Vinhas R, Cordeiro M, Carlos FF, Mendo S, Fernandes A, Figueiredo S, Baptista P. Gold nanoparticle-based theranostics: disease diagnostics and treatment using a single nanomaterial. Nanobiosensors Dis Diagn. 2015;4:11–23.

Wang J, Tao W, Chen X, Farokhzad OC, Liu G. Emerging advances in nanotheranostics with intelligent bioresponsive systems. Theranostics. 2017a;7:3915.

Wang M, Zhao J, Zhang L, Wei F, Lian Y, Wu Y, Gong Z, Zhang S, Zhou J, Cao K, Li X. Role of tumor microenvironment in tumorigenesis. J Cancer. 2017b;8:761.

Wilhelm S, Tavares AJ, Dai Q, Ohta S, Audet J, Dvorak HF, Chan WC. Analysis of nanoparticle delivery to tumours. Nat Rev Mater. 2016;1:16014.

Wu W, Zhang Q, Wang J, Chen M, Li S, Lin Z, Li J. Tumor-targeted aggregation of pH-sensitive nanocarriers for enhanced retention and rapid intracellular drug release. Polym Chem. 2014;5:5668–79.

Wu W, Luo L, Wang Y, Wu Q, Dai HB, Li JS, Durkan C, Wang N, Wang GX. Endogenous pH-responsive nanoparticles with programmable size changes for targeted tumor therapy and imaging applications. Theranostics. 2018;8:3038.

Xia Y, Zhang R, Wang Z, Tian J, Chen X. Recent advances in high-performance fluorescent and bioluminescent RNA imaging probes. Chem Soc Rev. 2017;46:2824–43.

Xiao YD, Paudel R, Liu J, Ma C, Zhang ZS, Zhou SK. MRI contrast agents: classification and application. Int J Mol Med. 2016;38:1319–26.

Xing H, Hwang K, Li J, Torabi SF, Lu Y. DNA aptamer technology for personalized medicine. Curr Opin Chem Eng. 2014;4:79–87.

Xu X, Bayazitoglu Y, Meade A. Evaluation of theranostic perspective of gold-silica nanoshell for cancer nano-medicine: a numerical parametric study. Lasers Med Sci. 2019;34(2):377–88.

Yang X, Shi X, Ji J, Zhai G. Development of redox-responsive theranostic nanoparticles for near-infrared fluorescence imaging-guided photodynamic/chemotherapy of tumor. Drug Deliv. 2018;25(1):780–96.

Yeh YC, Creran B, Rotello VM. Gold nanoparticles: preparation, properties, and applications in bionanotechnology. Nanoscale. 2012;4:1871–80.

Yoon SO, Park SJ, Yun CH, Chung AS. Roles of matrix metalloproteinases in tumor metastasis and angiogenesis. J Biochem Mol Biol. 2003;36:128–37.

Yordanova A, Eppard E, Kürpig S, Bundschuh RA, Schönberger S, Gonzalez-Carmona M, Feldmann G, Ahmadzadehfar H, Essler M. Theranostics in nuclear medicine practice. Onco Targets Ther. 2017;10:4821.

Zhang S, Li J, Lykotrafitis G, Bao G, Suresh S. Size-dependent endocytosis of nanoparticles. Adv Mater. 2009;21:419–24.

Zhang X, Han L, Liu M, Wang K, Tao L, Wan Q, Wei Y. Recent progress and advances in redox-responsive polymers as controlled delivery nanoplatforms. Mat Chem Frontiers. 2017;1:807–22.

Zou L, Wang H, He B, Zeng L, Tan T, Cao H, He X, Zhang Z, Guo S, Li Y. Current approaches of photothermal therapy in treating cancer metastasis with nanotherapeutics. Theranostics. 2016;6:762.

Chapter 2
Nanoparticles in Nanotheranostics Applications

Nadun H. Madanayake, Ryan Rienzie, and Nadeesh M. Adassooriya

Abstract Nanotheranostics, the amalgamation of diagnosis and therapeutic functions with nanotechnology, is a novel approach in personalized medicine. The advancement of nanotechnology offers a greater opportunity to engineer nanoparticles in theranostics applications and it has shown promising results especially in cancer therapy compared to conventional treatments. Since nanoparticles possess enhanced surface properties, they are capable of orienting nanotheranostic agents in specific sites of disease through which it significantly reduces the undesired side effects. In addition, the biocompatibility of those nanotheranostic agents with target cells or tissues provides a greater advantage to apply them in therapeutic functions as well as in imaging. Primarily metallic, magnetic, polymeric nanoparticles and quantum dots are used in nanotheranostics applications and gold-based nanomaterials and superparamagnetic iron oxide nanoparticles have attracted significant attention in recent years. Therefore, the aim of this chapter is to discuss the use of nanoparticles in theranostic applications while made them functionally important in nanotheranostics for personalized medicine.

Keywords Nanoparticles · Nanotheranostics · Personalized medicine · Therapeutics · Cancer treatments

N. H. Madanayake
Department of Botany, University of Sri Jayewardenepura, Nugegoda, Sri Lanka

R. Rienzie
Faculty of Agriculture, University of Peradeniya, Peradeniya, Sri Lanka

N. M. Adassooriya (✉)
Department of Food Science & Technology, Wayamba University of Sri Lanka, Makandura, Gonawila, Sri Lanka
e-mail: nadeesh@wyb.ac.lk

M. Rai, B. Jamil (eds.), *Nanotheranostics*,
https://doi.org/10.1007/978-3-030-29768-8_2

Nomenclature

CT	Computed tomography imaging
DNA	Deoxyribonucleic acid
EPR	Enhanced permeability and retention
FA	Folic acid
Gd_2O_3-AuNCs	gadolinium oxide–gold nanoclusters
IONPs	Iron oxide nanoparticles
MPNs	Magnetic nanoparticles
MRI	Magnetic resonance imaging
NIR	Near infrared
NPs	Nanoparticles
PA	Photoacoustic imaging
PDT	Photodynamic therapy
PEG	Polyethylene glycol
PET	Positron emission tomography
PL	Photoluminescence
PS	Photosensitizers
PTT	Photothermal therapy
PTX	Paclitaxel
ROS	Reactive oxygen species
RT	Radiotherapy
SERS	Surface-enhanced Raman spectroscopy

2.1 Introduction

Nanotechnology is an interdisciplinary technological approach that has been webbed over a range of fields including environmental remediation, energy, medicine, agriculture, engineering, and food industry that offers a greater impact to uplift the well-being of life (Shi et al. 2010; Duncan 2011; Kottegoda et al. 2016; Madusanka et al. 2016; Madusanka et al. 2017a, b; Ashiq et al. 2019). Interestingly, medicine is such a field which experiences its tremendous influence on personalized applications. Thanks to nanotheranostics which has been enabled to combine the diagnosis and therapeutic functions into a single platform especially on cancer treatments and other related diseases (Ma et al. 2016; Yang et al. 2019). Accordingly, an array of applications of nanotheranostic agents exist for applications including diagnosis, target drug delivery, real-time monitoring of the therapy with a minimum level of side effects and toxicity to the patients (Cabral and Baptista 2014). Nanotheranostics have given promising results for the development of personalized medicine at peak level (Jo et al. 2016). Due to the heterogeneity of cancers and inefficient results of common prescriptions, nanotheranostics have upgraded the development of optimized protocols providing remarkable results in personalized medicine applications (Mura and Couvreur 2012).

Inherent properties of nanoparticles due to its high surface-area-to-volume ratio as well as quantum confinement enable them to be successfully modulated in different fields (Madusanka et al. 2014). Currently, the advancement in nanotechnology has contributed to design nanoparticles with variable shapes (e.g., spheres, rods, and cubes) and sizes having enhanced optical, electronic, magnetic, and catalytic properties. Therefore, the use of nanoparticles in theranostics application has granted several opportunities to mitigate the limitations of conventional treatments. In addition, the ability to modify their surface with different functional groups and compounds elevates their value. This approach can improve their biocompatibility toward biological systems to provide further extensions of its utilization. Moreover, ease of localizing different agents along with nanoparticles at the target site also adds additional value for individualized medicine.

Up to date, nanotheranostics has experimented with different nanoparticles, the unique surface plasma resonance of metallic nanoparticles, fluorescent properties of quantum dots and encapsulation properties of polymeric nanoparticles and magnetic properties of magnetic nanoparticles attracted immensely toward personalized medicine (Selvan and Narayanan 2016). Moreover, cost-effectiveness and biocompatibility enable them to be promising candidates for theranostics (Wang et al. 2012). The schematic representation shown in Fig. 2.1 summarizes different types of nanoparticles which are used in theranostic applications while explaining different drug delivery methods and imaging techniques.

This chapter provides a comprehensive overview of the use of nanoparticles in theranostic applications which made them functionally important for personalized medicine while focusing on future concerns in nanotheranostics.

2.2 Metallic Nanoparticles

Most of the nanoformulations used in recent theranostic applications are based on metallic nanoparticles. Thus, metals such as gold (Au), silver (Ag), zinc (Zn), and titanium (Ti) have attracted immensely to apply as theranostics agents due to their unique physical and chemical properties (Cole et al. 2011). Having a larger surface area as well as a high surface-area-to-volume ratio enhance their physiochemical properties making them suitable especially on cancer treatments (Sharma et al. 2015). Furthermore, ease of surface fabrication with theranostic agents transforms them to excellent candidates with greater biocompatibility toward the living system (Jiang et al. 2015).

Surface plasmon resonance (SPR) is a major property shown by metallic nanoparticles compared to their bulk state. Therefore, it has become an x-factor for image-guided drug delivery of metallic nanoparticles as well as a fingerprint feature (Usov et al. 2009; Scholl et al. 2012). Metallic NPs can be appreciated as a lattice of ionic cores having conduction electron moving almost freely inside their lattice structure. This nature enables NPs to get illuminated under electromagnetic radiation of light and this enhances the collective vibrations of conductive electrons leading to trigger surface plasmon resonance. This resonating optical property due

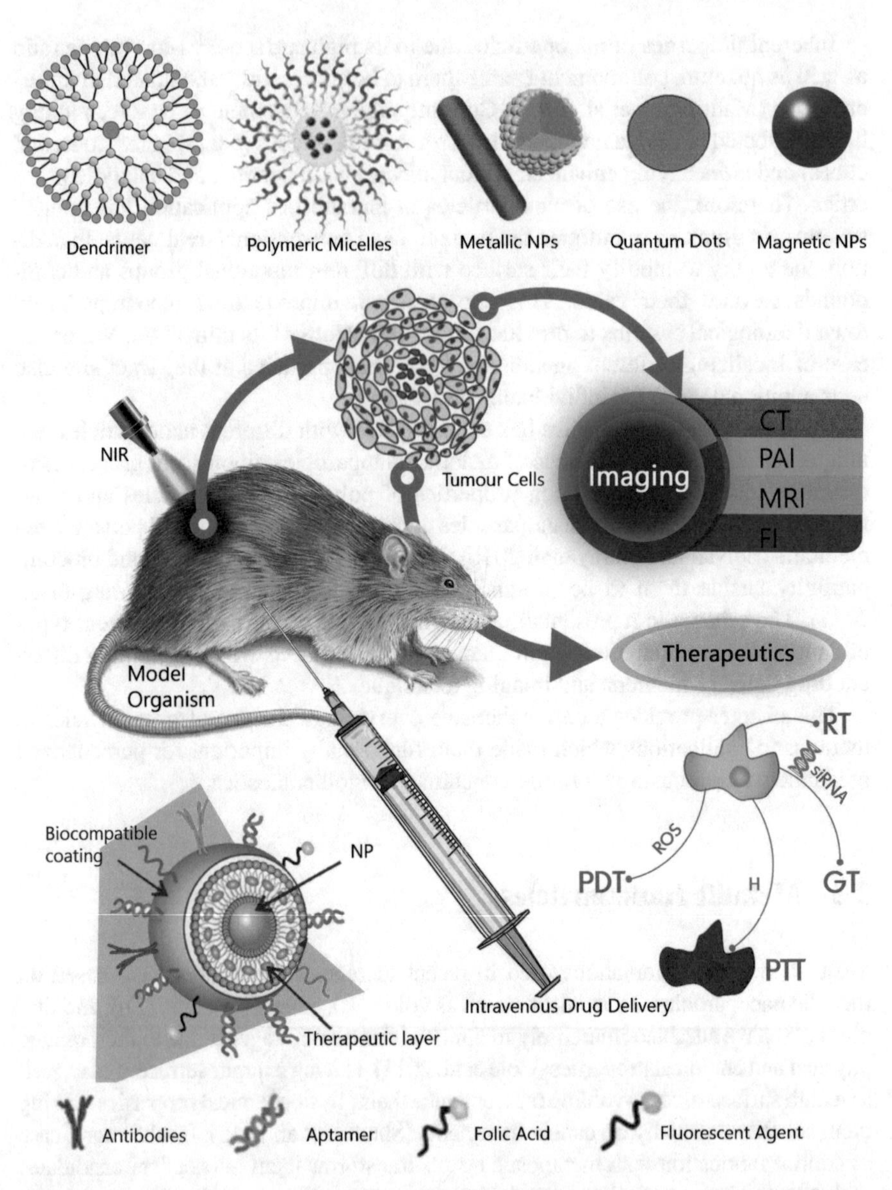

Fig. 2.1 Schematic illustration of nanotheranostic applications on cancer treatments. *FI* fluorescent imaging, *RT* radio therapy, *PTT* photothermal therapy, *PDT* photodynamic therapy, *GT* gene therapy, *CT* computed tomography, *MRI* magnetic resonance imaging, *PAI* photoacaustic imaging, *NIR* near-infrared light, *NP* nanoparticle, *ROS* reactive oxygen species, *siRNA* small interference RNA, *H* hyperthermia

to excitation of metallic NPs will be hardly achieved by other optical materials. SPR utterly depends on the size, shape, surrounding medium, and the charge of the metallic NP. Taking that into advantage, different metallic NPs with specific plasmonic properties are being used in theranostic applications (Templeton et al. 2000; García 2011).

Ease of surface functionalization with a variety of molecules enables to upgrade their biocompatibility toward comprehensive applications on theranostics. For instance, metallic nanoparticles can be coated with multiple ligands such as polyethylene glycol (PEG), biotin, paclitaxel (PTX), folic acid (FA) on their surface exploited to be advantageous as theranostic agents especially for cancer treatments (Bhattacharya et al. 2007; Fernández-López et al. 2009; Heo et al. 2012). Because of this metallic NPs have been extensively used as imaging agents in computed tomography imaging (CT), photoacoustic imaging (PA), and also in image-guided drug delivery (Agarwal et al. 2007; Mallidi et al. 2009; Kim et al. 2010). In addition, they can be utilized as therapeutic agents in photothermal therapy (PTT), photodynamic therapy (PDT), and radiotherapy (RT) (Khan et al. 2012; Hwang et al. 2014). Hence, this section will discuss the use of metallic nanoparticles toward the enhanced performance of personalized medicine.

2.2.1 Gold Nanoparticles

Multifunctional agent for theranostic applications has been increasingly exploited using gold nanoparticles due to its high degree of biocompatibility and ease of surface fabrication. The inherent optical and photothermal characteristics of gold nanoparticles enable scientists to engineer them as sensing modalities as well as therapeutic agents (Wang et al. 2012). Moreover, the localized surface plasmon resonance can be manipulated by tuning their size, shape, structure, and surface chemistry providing more and more chances to apply them as favorable imaging agent (Cobley et al. 2011). Additionally, different forms of gold nanoformulations with varying sizes and shapes claim to be used as successful multifunctional agents in cancer treatments (Guo et al. 2017).

Slow and controlled drug release is an important aspect of drug and chemical delivery. This strategy has been widely applied in the medical field as well as in agriculture (Kottegoda et al. 2011). Gold NPs-thiol bonding triggers the subsequent attachment of drugs and enables to release them by reacting with cytosolic glutathione (Zhang et al. 2010). Furthermore, gold nanoparticles can be modified with multiple ligands including polyethylene glycolate (PEG), biotin, paclitaxel, and rhodamine B-linked β-cyclodextrin on its surface without any cytotoxic effect on nontarget cells (Heo et al. 2012).

Because of the localized surface plasmon resonance, gold nanoparticles exhibit fascinating optical properties enabling them to be used as imaging modalities depending on their size and shape (Cabral and Baptista 2014). For an instance, Jain et al. (2006) have demonstrated that the increase in gold nanosphere size from 20 to 80 nm enhances the magnitude of excitation and absorption cross section compared to conventional absorbing dyes such as indocyanine green and rhodamine-6G. Therefore, this property makes gold nanoparticles as an efficient candidate for diagnosis and imaging of cancer.

At present, computed tomography (CT) has become a genuine and broadly practiced clinical imaging technology for their higher resolution (Peng et al. 2012; Meir et al. 2015). Gold has shown promising results compared to conventional iodine-based agents in CT imaging. Primarily, higher X-ray attenuation coefficient harness high-quality CT imaging compared to conventional iodinated CT imaging agents especially due to the higher atomic number and electron density of gold nanoparticles. Furthermore, higher plasticity to laminate with target moieties along with minimal toxicity makes them better CT imaging nanoplatforms (Popovtzer et al. 2008; Peng et al. 2012; Guo et al. 2017). Moreover, Peng et al. (2012) developed PEGylated dendrimer-entrapped gold nanoparticles for tumor imaging by computed tomography and shown that gold nanoparticles with proper surface tempering enhance and lengthen their blood circulation time by reducing the effects from the immune system.

Photoacoustic or thermoacoustic imaging is a recently developed application which is dependent on size and shape as well as plasmonic properties of metallic nanoparticles (Li and Chen 2015). Different forms of gold-based nanoformulations such as nanospheres, rods, shells, prisms, cages, and stars are commonly employed for PA imaging. Because of their simplicity in synthesis and manipulatable electromagnetic absorption in the near-infrared region, increases their efficiency of heat generation at the imaging site. Thermal variation that results in photoactive sites can generate thermoelastic expansions in surrounding air, forming an acoustic wave (Stevenson and Heffern 2018). Therefore, the optical properties of gold nanoparticles utilized them as contrast agents as well as tumor imaging agents (Li and Chen 2015).

Gold nanoparticles are competent enough to act as enhancers for surface-enhanced Raman spectroscopy (SERS) through its surface electromagnetic field enhancement and chemical contribution for the detection of viruses and tumor cells (Conde et al. 2014). In addition, hybrid dual imaging technologies have drawn their attention toward cancer diagnosis in order to conquer the limitations of individual imaging modes to obtain accurate and comprehensive information during diagnosis (Padmanabhan et al. 2016). Positron emission tomography/computed tomography, magnetic resonance imaging/positron emission tomography, and photoacoustic imaging/CT are currently available techniques for detection (Hoejgaard and Hesse 2011).

Gold nanoparticles cannot be underestimated as imaging agents; it is also an excellent therapeutic agent for cancer treatments. Furthermore, gold nanoshells, nanorods, and nanocages are widely applied in therapeutic applications, especially in photothermal therapy (Melamed et al. 2015; Cheng et al. 2017). These nanomaterials can withdraw tumor cells or tissues by transforming near-infrared light into thermal energy (Gao et al. 2016) that cause hyperthermia on targets followed by cell destruction. In addition, certain studies explain that photothermal therapy can trigger cell death via necrosis and apoptosis under controlled radiation condition (Melamed et al. 2015).

Photosensitizers (PS) can convert surrounding oxygen molecules into toxic reactive oxygen species (singlet oxygen generation) which can destroy tumor cells as well as other malignant cells (Guo et al. 2017). Mostly their activation occurs under ultraviolet or visible light, and this is therapeutically inefficient due to its poor penetration power in biological systems. Other than that, lower molar excitation coefficient results in photobleaching and enzymatic degradation (Schweitzer and Somers 2010). Hence, gold nanoparticles deserve to be an excellent nominee to overcome these restrictions since they have a higher molar excitation coefficient with a greater photostability and resistance to enzymatic degradation (Xu et al. 2019). Therefore, gold nanoparticles can enhance their photodynamic efficiency to a vast range of photosensitizers (Phthalocyanines, Toluidine blue O, ENREF18, indocyanine green, AlPcS4, and hematoporphyrin). Furthermore, gold nanoparticles especially conjugated with PS have shown promising results in photodynamic therapy (Zhang et al. 2015). The selective binding affinity of gold nanoparticles functionalized with multiple targeting moieties and the enhanced permeability and retention (EPR) effect play a crucial role in photothermal therapy (Barreto et al. 2011). To date, the use of gold-based nanoplatforms in radiotherapy has been extensively exploited due to its high X-ray absorption coefficient and ease of synthesis via controlled physicochemical properties (Her et al. 2017).

2.2.2 *Silver Nanoparticles*

Extraordinary properties of silver nanoparticles resulted in a range of applications in medicine as well as in the industrial field (Madusanka et al. 2014; Zhang et al. 2016a, b). Even more, silver nanoparticles represent the size- and shape-dependent physiochemical and biological properties with photonic, electronic, catalytic, and therapeutic properties enabling them to be used as biodiagnostic, imaging, therapeutic, and drug delivery applicants (Austin et al. 2014; Di Pietro et al. 2016). Same as in gold nanoparticles, localized surface plasmon resonance of silver nanoparticles enable them to integrate on theranostic applications.

Attributes of nanomaterials can enhance their properties toward different aspects but the same thing can lead to the toxicity of biological systems. This has been frequently experienced in agricultural practices as well (Rienzie and Adassooriya 2018). Despite silver nanoparticles' potential for theranostic applications, genotoxicity and cytotoxicity at certain concentrations resulted in adverse effects on human cells, thus limiting its application (Bhushan and Gopinath 2015). Therefore, most silver nanoformulations are designed by levering with stabilizing agents to reduce its toxicity. Recently, it has been reported that silver graphene quantum dots tethered with carboxymethyl inulin have mitigated their toxicity and inhibit the development of pancreatic tumor cells (Joshi et al. 2017). Moreover, silver nanorods, silver nanocubes, and silver nanospheres are widely applied forms in nanotheranostic applications (Austin et al. 2014).

Photoacoustic imaging based on silver nanocores or shells has shown promising results compared to gold-based nanoformulations (Wang et al. 2012). Firstly, silver nanoparticles provide much stronger photoacoustic signals than gold because of their better light absorption properties. Secondly, silver is more likely to degrade in the body and its antimicrobial properties provide additional values in its application. In addition, surface functionalization gives a better chance to manipulate as required (Homan et al. 2010).

The ability of silver nanoparticles to stimulate the oxidative DNA destruction and chromosomal aberrations has given opportunities toward gene therapy and cancer therapy. Moreover, the ability to bump up reactive oxygen species generation that triggers mitochondrial-dependent apoptosis has given the attraction toward therapeutics (Bhushan and Gopinath 2015). For instance, silver nanoparticles with 5-fluorouracil can synergistically induce apoptotic pathways in uracil phosphoribosyl transferase expression system which sensitize the cell as a strategy for gene therapy (Gopinath et al. 2008).

Image-guided therapy is another application where silver nanoparticles act as an optical identification code. Wu et al. (2010) developed core shell-structured hybrid nanogel with silver nanoparticle core and pH-responsive gel of poly (*N-isopropylacrylamide-co-acrylic* acid) shells as a carrier of hydrophobic dipyridamole drug. Therefore, silver nanoparticles have gained wider attraction as a theranostic platform in personalized medicine due to its tremendous physiochemical properties.

2.2.3 *Liquid Metal Nanoparticles*

Light-driven liquid metal nanotransformers, which is a strange application toward theranostics, has drawn the attraction of researches, especially on bioimaging. Liquid metals such as gallium, gallium-indium eutectic alloys, and gallium-indium-tin alloys having lower melting points have been researched in bioimaging applications because of their chemical stability and being inert toward the water at room temperature. Recently, liquid metal formulations have shown their advantage toward the photoacoustic effect. In addition, liquid metal formulation generating reactive oxygen species along with heat under near-infrared have enabled them to be used in effective cancer cell elimination. However, its elevated toxicity is a major impediment to a widespread application (Chechetka et al. 2017).

2.3 Magnetic Nanoparticles (MPNs)

MNPs are also a major category of NPs in recent theranostic research because of their intrinsic biocompatibility and cost-effectiveness (Wang et al. 2012; de Jesus et al. 2019). Moreover, MPNs have been utilized as gene targeting, tissue engineering, and biosensor agents in the biomedical field (Gao et al. 2007). Hence, general characteristics of MNPs including nanometric size, enhanced permeability retention, the high surface area for molecular therapeutic binding, and surface functionalization with cancer-homing ligands for cancer treatments give greater advantage toward theranostics (Greish 2007). At present, MNPs such as iron oxides, gadolinium, manganese, and nickel-based nanoformulations have given much more attention toward theranostics (Kim et al. 2013; Yu et al. 2015; Zhang et al. 2016a, b). Most importantly, their unique magnetic properties under external magnetic fields provide a novel direction for diagnosis and therapeutics using MNPs (Veiseh et al. 2010).

Ferromagnetism and superparamagnetism are typical magnetic properties of MNPs. Superparamagnetism, in a sense, is a property shown by particles in the presence of an external magnetic field but changes back to nonmagnetic state when it is removed. This is of prime importance when MPNs are introduced into living systems because the disappearance of magnetization in the absence of an external magnetic field can avoid the agglomeration in target drug delivery (Arruebo et al. 2007; Belyanina et al. 2017). Interestingly, ferromagnetic nanoparticles show magnetic properties even in the absence of an external magnetic field below the Curie temperature (Estelrich et al. 2015). Hence, superparamagnetic nanoparticles are widely used in theranostic applications. Apart from the above unique properties, size and surface charge of the MNPs are critical factors in biological applications. Charged nanoparticle can result in the nonspecific binding to cells while neutral MNPs can enhance the circulation time in the bloodstream (Belyanina et al. 2017). Considering this nature, most of the nanoformulations in theranostic applications are formulated in the neutralized state.

Therapeutic sensing is primarily based on the selective interaction between surface-coated MPNs and tumor-specific moieties. Therefore, the coating of MNPs is important for their applications as it favors in enhancing their properties. Primarily, organic polymers (polyethylene glycol (PEG), chitosan), organic surfactants, metallic nanoparticles, inorganic nanoparticles, and bioactive molecules are increasingly utilized (Shubayev et al. 2009).

At present, surface-functionalized MNPs are widely applied as contrast agents in magnetic resonance imaging (MRI), positron emission tomography (PET), and single-photon emission computed tomography (SPECT) (Choi et al. 2008; Srinivas et al. 2010; Xie et al. 2011). In addition, MNPs are widely used for inducing heat at the target sites. More recently, magnetofection (i.e., gene transfection using MNPs) is a broadly studied area for tumor-targeted drug delivery, which may be combined effectively with target drug delivery (Reddy et al. 2012).

2.3.1 Iron Oxide Nanoparticles (IONPs)

IONPs have remarkable attention toward personalized medicine due to their magnetic properties. Superparamagnetic IONPs have become the common and important candidate for biomedical applications. It includes therapeutic applications such as target drug delivery, hyperthermia, and target gene delivery. Other than that, it is a highly recommendable and favorable applicant for MRI enhancement as well as in photoacoustic imaging (Kang et al. 2017; Wang et al. 2017; Burgum et al. 2018). Their applications may result in different formulations for theranostics via superparamagnetic iron oxide particles (SPIO), ultra-small superparamagnetic iron oxide particles (USPIO), very small superparamagnetic iron oxide particles (VSOP), monocrystalline iron oxide particles (MION), and cross-linked iron oxide (CLIO) depending on their crystalline structure, size, coating, and higher order organization (Strijkers et al. 2007). In addition, these properties can affect their performances as a theranostic agent. Primarily magnetite (Fe_3O_4) and maghemite (γ-Fe_2O_3) are the most commonly used iron oxide nanomaterials (Wang et al. 2012).

Due to the superior behaviors exerted by IONPs, it has been extensively utilized in medical applications as a magnetic resonance imaging contrasting agent. For instance, IONPs are widely known as a T2 contrasting agent. Therefore obtaining both negative and positive contrasting images with varied brightness is possible with minimized signal deterioration (Dolci et al. 2013). Magnetic movement of IONPs creates a local magnetic field around the target site of the biological system, in which it affects the nearby water molecules to incur the variable latitudinal and longitudinal relaxation times. These variable times can be detected by MRI equipment and this forces to generate a well-defined relaxation time map with an image of the target site to be tested (Neuwelt et al. 2015). Recently, iron oxide nanocages encapsulated within methoxypolyethylene glycol-thiol (mPEG) polymer were used for cancer theranostics and they have demonstrated excellent properties as MRI contrast agent. Hence, these nanoformulations exhibit enhanced contrast properties which are important as an efficient multifunctional nanocarrier for integrated imaging and therapy (Thorat et al. 2016). More recently, iron oxide "nanobricks" (IONBs) has been patented as a potential candidate for MRI contrast agent and for hyperthermia applications (Hegmann et al. 2016). Moreover, at present, IONPs have been approved for trial studies on humans (Xi et al. 2014).

Furthermore, IONPs possess high proton relaxivity for high molar extinction coefficient potential which is a vital property in photoacoustic imaging (Sano 2017). Certain studies have shown that due to its modest absorption, ultra-small nanometric size and longer retention time with small dose make them to be more advantageous as a better contrasting agent in photoacoustic imaging (Reddy et al. 2012). Xi et al. (2014) formulated an antibody-conjugated IONPs for photoacoustic imaging (HER2-targeted imaging). The lower toxicity and enhanced biocompatibility uplift their applications in imaging modalities. In addition, imaging modalities have advanced toward multimodal imaging using IONPs (Xi et al. 2014). Wu et al. reported that IONPs can be used as a multifunctional nanoprobe for near-infrared fluorescence, photoacoustic imaging, and magnetic resonance imaging, which is activated with near-infrared multidentate polymers (Wu et al. 2016).

Magnetic hyperthermia of MNPs under alternating magnetic fields resulted in the generation of a localized heat at the target site. Usually, hysteresis loss, Néel relaxation, and Brownian relaxation are involved in the warming of NPs under alternating magnetic field. In addition, the mechanism involved in the hyperthermal stimulation depends on the particle size, geometry, coating composition, and physical configuration (Shi et al. 2015; Peng et al. 2017). Among various types of hyperthermia agents, IONPs are regarded as an excellent candidate for photothermal therapy. Nanoformulations such as gold nanorods are regarded as poor photostable agents due to "melting" that is resulted due to point and planar effects (Link et al. 2000). In addition, gold nanoshells may fail to fulfill the requirement of enhanced permeability and retention effect (Papahadjopoulos et al. 1991). Chang et al. (2015) developed an effective photothermal therapeutic nanoplatform using IONPs integrated with IR806 dye for dual imaging-guided photothermal therapy of cancer. Here, the combination of IONPs with IR806 significantly improved the thermal conversion efficiency under lower light irradiation dose highlighting the potential of this nanoplatform in "precision medicine."

Remarkably, IONPs administered with photosensitizers have been found to induce the generation of ROS which causes the destruction of tumor cells. Hence, the photodynamic therapy application of IONPs turned them into a highly demanding nanomaterial for nanotheranostics. Mostly, magnetic NPs amended with photoactive compounds provide a higher efficacy in photodynamic therapy but, on the other hand, it may induce the agglomeration under aqueous media leading to reduced delivery as well as insufficient cancer cell selectivity. However, these restrictions can be minimized with the novel advancements in nanotechnology for efficient therapeutic applications (Penon et al. 2016). For instance, Huang et al. (2011) developed one of the second-generation photoactive compounds, chlorine e6, providing expedient biocompatibility and water solubility with no cytotoxicity for biological systems. In addition, the most fascinating feature of IONPs is their ability to be used as a better agent for target drug delivery in the presence of an external magnetic field.

2.3.2 *Gadolinium Nanoparticles*

Gadolinium nanoparticles are common and known for their potential to act as a T1 contrasting agent for bioimaging. Therefore, most of their formulations are used for magnetic resonance imaging. As it was mentioned elsewhere in this chapter, gadolinium nanoparticles have multidisciplinary applications in theranostics. More recently, researchers synthesized gadolinium oxide–gold nanoclusters hybrid systems stabilized by bovine serum albumin. Here, bovine serum albumin surrounding the nanoparticles makes Gd_2O_3-AuNCs a brilliant carrier for the delivery of indocyanine green (ICG) for magnetic resonance and X-ray computed tomography imaging for tumor cell imaging. Other than that, gold nanoclusters generate red fluorescence and singlet oxygen under the treatment with NIR at 808 nm, enabling them to be used as an excellent component for therapies. Therefore, the results demonstrate that Gd_2O_3-AuNCs-ICG nanoplatform is an efficient applicant for cancer diagnosis and therapy (Han et al. 2017).

2.4 Polymeric Nanoparticles

Polymeric NPs are widely recognized as a class of integrated nanocarriers for diagnosis and therapy (Fang and Zhang 2011). These NPs can be formulated by integrating multiple functional units to soluble macromolecules via self-assemblage of copolymers. Hence, polymeric formulations can be loaded with a variety of therapeutic and imaging agents. But polymeric NPs require successful strategies to minimize the immunogenicity and antigenicity as well as to enhance the residence time and stability inside the biological system. Therefore, the most common strategy is to modify nanocarriers with PEG. PEG provides a shielding effect to the polymeric nanocarrier, preventing it from the destruction by steric hindrance as well as from renal clearance (Peer et al. 2007; Wang et al. 2012). Conventional natural polymers including chitosan, gelatin, albumin, sodium alginate, and synthetic polymers such as polylactic acid (PLA), poly(lactic-co-glycolic acid) (PLGA), poly-glutamic acid, polyglycolide, poly aspartic acid, hyaluronic acid (HA), and poly anhydride are widely used in the synthesis of polymeric NPs (Pan et al. 2010; Choi et al. 2011; Na et al. 2011; Yang et al. 2011; Sonali et al. 2018). Therefore, the polymeric nanoparticles thus formulated can be obtained as nanocapsules, nanospheres, polymeric micelles, drug-polymer conjugates, dendrimers, polymersomes, and polyplexes (Prabhu et al. 2015). Moreover, the physiochemical properties such as crystallinity, molecular weight, hydrophobicity, and polydispersity index regulate the dissolution and drug delivery kinetics of the polymeric NPs (Sonali et al. 2018).

Out of their superior properties, polymeric NPs can be amended for different theranostic applications. Levering with photoactive compounds, iron oxide NPs, gadolinium NPs, gold NPs, and iodine enables them to be used in near-infrared imaging, MRI, CT imaging, and single-photon emission computed tomography, respectively (Kojima et al. 2012; Li et al. 2014; Ray et al. 2018). The fascinating feature about polymeric NPs is that they can be manipulated to create hydrophobic environments to encapsulate hydrophobic drugs (DOX and paclitaxel, etc.) to the target site (Lu et al. 2011). For example, polymeric dendrimers can effectively administer 3,4-difluoro benzylidene curcumin (CDF) (curcumin possesses antibacterial, antioxidant, anti-inflammatory, and anticancer properties) which are poor in water solubility (Madusanka et al. 2015; Ray et al. 2018). Furthermore, conjugation with other nanomaterials such as metallic and magnetic nanoparticles confers them to be utilized in PTT, PDT, gene therapy, etc. These modifications enable polymeric NPs to be utilized in successful disease management applications.

2.4.1 Dendrimers

Of the numerous polymer-based nanomaterials, dendrimers have shown a remarkable promise as nanocarriers for tumor-specific drug delivery. Dendrimers are highly branched with a monodispersed weight distribution having a precise architecture and composition (Zhu et al. 2018). Dendrimers possess strong positive charges due to which transform them to perform as transfection agents. Moreover, the surface containing functional groups allow them to functionalize with antibodies, peptides, folate, and other targeting molecules (Palmerston et al. 2017). They have been extensively exploited as drug and gene carriers, but the presence of functional groups enables them to coordinate toward diagnosis and therapy. For instance, the administration of gold nanostar-stabilized dendrimers and complexing short interfering RNA delivering allows CT imaging, PTT, as well as gene silencing of tumors. Moreover, administration of gadolinium-bearing dendrimers enables their use as MRI contrasting agents (Wei et al. 2016; Zhang et al. 2017).

Even though we have defined it as a promising candidate for nanotheranostics, it could show some limitations for their application. Dendrimers are capable of interacting with nanosized cellular components such as cell membrane, cell organelles, and proteins. Simultaneously, dendrimers having cationic surface groups may interact with the lipid bilayer and enhance its permeability while decreasing its integrity (Rittner et al. 2002; Fischer et al. 2003; Mecke et al. 2005; Madaan et al. 2014).

2.4.2 *Micelles*

In the past decades, polymeric micelles have been paid much attention as a multifaceted nanosystem aiming for cancer treatments. Currently, these nanocarriers are successfully being claimed in preclinical and clinical studies. Simply polymeric micelles are spheroid nanoplatforms with a hydrophilic shell and a hydrophobic core. Because of their thermodynamic stability, kinetic stability, and high payload with smaller dimensions, they are attracted more toward disease treatments (Sonali et al. 2018). Polymeric micelles entrapped with superparamagnetic IONPs and DOX within its core can be used as a multifunctional agent for MRI and therapeutic delivery (Guthi et al. 2009). Moreover, PEG–polylactic acid micelles can operate as a three-in-one nanocarrier system for hydrophobic drugs such as paclitaxel,17-allyamino-17-demethoxygeldanamycin and rapamycin in cancer therapy. In addition, polyethylene glycol-block-poly-ε-caprolactone (PEG-b-PCL) micelles entrapped with carbocyanine drug enable them to be utilized as NIR optical imaging agent (Cho and Kwon 2011). Hence, these approaches show how polymeric micelles can be manipulated for different theranostic applications.

2.4.3 *Polymersomes*

Polymersomes are a novel group of thin-shelled capsules based on block copolymer chemistry. Hence, polymersomes can be defined as self-assembled nanocarriers having amphiphilic block copolymers with a tougher and thicker membrane with higher stability than liposomes. Therefore, the combination of hydrophobic drugs to thick wall and hydrophilic drugs into the vesicular lumen will develop to synergistic effects such as cocktails. In addition, the physicochemical properties of polymersomes attracted them as good candidates for targeted drug delivery (Pang et al. 2010). Furthermore, another group of researchers demonstrated that IONP stabilized by lipo-polymersomes can be used for magnetically guided gene delivery (Hu et al. 2014). In addition to that, several studies have shown their potential as a nanocarrier for efficient drug delivery.

2.5 Semiconducting Nanoparticles or Quantum Dots

Nanotechnology has acquired its recognition as a multidisciplinary entity where different forms of nanoparticles can be transformed easily into several applications. Therefore, this has given license for different nanometric forms to apply in different fields (Tan et al. 2011). In such a way, quantum dots (QD) have emerged as viable agents for theranostic applications. Simply, QDs are semiconducting nanocrystals that exhibit superior fluorescent properties (Iga et al. 2007). Hence, QDs encompass with biological imaging as well as in therapeutic applications with the merit of resistant to photobleaching capabilities over molecular dyes (Chen et al. 2008; Medintz et al. 2008; Zhu et al. 2018). Up to date, different kinds of QDs from groups II–VI Zn (S, Se) and Cd (S, Se, Te), IV–VI Pb (S,Se), I–VI Ag 2 (S,Se), II–V Cd 3 (P,As) 2, and III–V In (P, As) are being successfully used in biomedical fields. In addition, ternary I–III–VI QDs (where I = Cu or Ag, III = Ga or In, VI = S or Se) were also developed for these applications (Ji et al. 2014; Jing et al. 2014). However, the composition can vary according to the intended use and manufacturer preference. Fundamentally, QDs build up with a fluorescent core and an outer crystal shell in order to sheath the core from the ionization process within the biological system (Chen et al. 2008). For the biological applications, it is necessary to have a stable distribution in aqueous media, tolerability to different pH ranges, and ideally the water solubility. Advancement in nanotheranostics has developed effective strategies to generate QDs having enhanced properties for their applications (Burgum et al. 2018).

Semiconductor nanocrystals exhibit size-dependent optical properties compared to their bulk materials' Bohr radius. The band gap enhancement due to quantum confinement results in size-dependent QD absorption and emission spectra having band edge features both shifting higher energies with reducing particle size (Chen et al. 2010). QDs exhibit remarkable optical properties for biomedical applications such as narrow and symmetrical emission profiles advantageous for color purity and precise tenability of emission; broad excitation range combined with high molar absorption coefficients enables them to be used in high-throughput detection. Furthermore, high photoluminescence (PL) quantum yield (QY) and resistance against photobleaching provide more chances for long-term visualization and tracking of biological processes. In addition, they have relatively long PL lifetimes which enable exclusion of autofluorescence (Tan et al. 2011). Therefore, it has been suggested that QDs can be appreciated as multimodal contrast imaging agents that can employ infrared fluorescence, positron emission tomography, CT, MRI, and PA imaging applications (Tan et al. 2011; Zhang et al. 2016a, b; Guo et al. 2017).

Quantum dots are novel nanoparticles that have gathered extensive investigations as drug delivery vehicles. Hence, same as in other nanoparticles described in the previous sections in this chapter, QDs can be fabricated with different biological molecules (antibodies, peptides, and aptamers) which can enhance its biocompatibility for target drug delivery (Savla et al. 2011). Wang et al. (2014) had demonstrated that graphene QDs conjugated with folic acid can be utilized to load antitumor drugs. Due to their biocompatibility, inherent optical properties, higher surface-area-to-volume ratio, and presence of carboxylic groups for conjugation with different anchoring molecules potentiate them for target drug delivery as well as real-time monitoring of drug delivery. Additionally, certain studies have shown that QDs can be used as a carrier molecule to cargo siRNA (short, double-stranded small interfering-RNAs) for nondrugable targets which can induce the RNA interference whereby this can inhibit the protein translation. Moreover, they can serve as photostable beacons to monitor siRNA delivery (Derfus et al. 2007).

Therapeutic efficacy of QDs have given opportunities to apply them in photothermal (PTT) and photodynamic (PDT) therapy (Chen et al. 2010; Zhang et al. 2016a, b; Guo et al. 2017; Zhu et al. 2018). In order to execute PTT, it requires a near-infrared (NIR) absorbing platforms which can induce hyperthermia on the target site causing cellular damage while PDT requires photosensitizers to generate ROS to destroy cancer tissues. Although the following applications provided with promising results, it may have limitations with other NPs. For instance, efficiency PDT can deviate along with the time due to the depletion of oxygen level in tissues. Other than that, PTT efficacy can get weaker due to hyperthermia-induced heat shock, thus suppressing the apoptosis of tumor cells. Therefore nanoplatforms such as $Cu_2(OH)PO_4$ quantum dots have been identified as promising agents to overcome these drawbacks for efficient disease treatments (Guo et al. 2017). Moreover, QDs can support with the same effect for therapeutics following accompanying with carbon and graphene (Zhang et al. 2016a, b). Yong et al. (2015) demonstrated that the use of tungsten sulfide QDs for synergistic PTT and radiotherapy along with dual modal imaging has given a greater promise for theranostics application. Same as in other nanomaterials, a number of studies have shown toxicological effects of QDs, especially of formulations based on heavy metals. Therefore, the application of QDs may require deep down studies before starting its clinical applications.

2.6 Concluding Remarks and Future Implications

The drastic development in nanotechnology has granted a tremendous value on medical applications, especially in cancer. Hence, it requires well-improved platforms for efficient performance because dealing with biological systems is a serious process and requires to be sharply focused. Since conventional treatment methodologies still failed to overcome these limitations, it has resulted in a variety of side effects. Nanotheranostics has been a successful application in order to overcome these limitations. Inherent properties of nanoparticles enable them to mitigate diagnostic and therapeutic applications into a single platform. Therefore, this triggers for targeted drug delivery via enhancing its biocompatibility and pharmacokinetics potential toward the treatments. Diagnosis enables to focus on and to get precise information for clinical applications. At present nanotheranostics has advancements toward real-time monitoring of the disease treatment, which is a greater achievement for therapeutics. Cancer therapy using nanotheranostics is at present stepping toward treatment without application of drugs. These may potentiate nanotheranostics at its peak. Currently, the use of liquid metal nanoparticles, nanoparticles, and quantum dots for theranostics applications is being widely investigated. Even though at present people discuss their value and effectiveness for a wide range of fields, nanotoxicological aspects have a bigger concern over its applications. Then physiochemical properties such as size, shape, surface functionality, and stability can lead to cause toxicity on biological systems that simultaneously depend upon extrinsic factors such as organismal characteristics including age, nature of the target site, and pH of the internal environment that nanoparticles are exposed to. Therefore, the proper formulation and manipulation can advance nanotheranostics toward the next generation medicine.

References

Agarwal A, Huang SW, O'donnell M, Day KC, Day M, Kotov N, Ashkenazi S. Targeted gold nanorod contrast agent for prostate cancer detection by photoacoustic imaging. J Appl Phys. 2007;102(6):064701. https://doi.org/10.1063/1.2777127.

Arruebo M, Fernández-Pacheco R, Ibarra MR, Santamaría J. Magnetic nanoparticles for drug delivery. Nano Today. 2007;2(3):22–32.

Ashiq A, Adassooriya NM, Sarkar B, Rajapaksha AU, Ok YS, Vithanage M. Municipal solid waste biochar-bentonite composite for the removal of antibiotic ciprofloxacin from aqueous media. J Environ Manag. 2019;236:428–35.

Austin LA, Mackey MA, Dreaden EC, El-Sayed MA. The optical, photothermal, and facile surface chemical properties of gold and silver nanoparticles in diagnostics, therapy, and drug delivery. Arch Toxicol. 2014;88(7):1391–417.

Barreto JA, O'Malley W, Kubeil M, Graham B, Stephan H, Spiccia L. Nanomaterials: applications in cancer imaging and therapy. Adv Mater. 2011;23(12):18–40.

Belyanina I, Kolovskaya O, Zamay S, Gargaun A, Zamay T, Kichkailo A. Targeted magnetic nanotheranostics of cancer. Molecules. 2017;22(6):975.

Bhattacharya R, Patra CR, Earl A, Wang S, Katarya A, Lu L, Kizhakkedathu JN, Yaszemski MJ, Greipp PR, Mukhopadhyay D, Mukherjee P. Attaching folic acid on gold nanoparticles using noncovalent interaction via different polyethylene glycol backbones and targeting of cancer cells. Nanomedicine. 2007;3(3):224–38.

Bhushan B, Gopinath P. Tumor-targeted folate-decorated albumin-stabilised silver nanoparticles induce apoptosis at low concentration in human breast cancer cells. RSC Adv. 2015;5(105): 86242–53.

Burgum, M.J., Evans, S.J., Jenkins, G.J., Doak, S.H. and Clift, M.J., (2018). Considerations for the human health implications of nanotheranostics. In Handbook of nanomaterials for cancer theranostics (pp. 279–303). Elsevier, Amsterdam.

Cabral RM, Baptista PV. Anti-cancer precision theranostics: a focus on multifunctional gold nanoparticles. Expert Rev Mol Diagn. 2014;14(8):1041–52.

Chang Y, Li X, Kong X, Li Y, Liu X, Zhang Y, Tu L, Xue B, Wu F, Cao D, Zhao H. A highly effective in vivo photothermal nanoplatform with dual imaging-guided therapy of cancer based on the charge reversal complex of dye and iron oxide. J Mater Chem B. 2015;3(42):8321–7.

Chechetka SA, Yu Y, Zhen X, Pramanik M, Pu K, Miyako E. Light-driven liquid metal nanotransformers for biomedical theranostics. Nat Commun. 2017;8:15432.

Chen Y, Vela J, Htoon H, Casson JL, Werder DJ, Bussian DA, Klimov VI, Hollingsworth JA. "Giant" multishell CdSe nanocrystal quantum dots with suppressed blinking. J Am Chem Soc. 2008;130(15):5026–7.

Chen JY, Lee YM, Zhao D, Mak NK, Wong RNS, Chan WH, Cheung NH. Quantum dot-mediated photoproduction of reactive oxygen species for cancer cell annihilation. Photochem Photobiol. 2010;86(2):431–7.

Cheng X, Sun R, Yin L, Chai Z, Shi H, Gao M. Light-triggered assembly of gold nanoparticles for photothermal therapy and photoacoustic imaging of tumors in vivo. Adv Mater. 2017;29(6):1604894.

Cho H, Kwon GS. Polymeric micelles for neoadjuvant cancer therapy and tumor-primed optical imaging. ACS Nano. 2011;5(11):8721–9.

Choi JS, Park JC, Nah H, Woo S, Oh J, Kim KM, Cheon GJ, Chang Y, Yoo J, Cheon J. A hybrid nanoparticle probe for dual-modality positron emission tomography and magnetic resonance imaging. Angew Chem. 2008;47(33):6259–62.

Choi KY, Min KH, Yoon HY, Kim K, Park JH, Kwon IC, Choi K, Jeong SY. PEGylation of hyaluronic acid nanoparticles improves tumor targetability in vivo. Biomaterials. 2011;32(7):1880–9.

Cobley CM, Chen J, Cho EC, Wang LV, Xia Y. Gold nanostructures: a class of multifunctional materials for biomedical applications. Chem Soc Rev. 2011;40(1):44–56.

Cole AJ, Yang VC, David AE. Cancer theranostics: the rise of targeted magnetic nanoparticles. Trends Biotechnol. 2011;29(7):323–32.

Conde J, Bao C, Cui D, Baptista PV, Tian F. Antibody–drug gold nanoantennas with Raman spectroscopic fingerprints for in vivo tumour theranostics. J Control Release. 2014;183:87–93.

Derfus AM, Chen AA, Min DH, Ruoslahti E, Bhatia SN. Targeted quantum dot conjugates for siRNA delivery. Bioconjug Chem. 2007;18(5):1391–6.

Di Pietro P, Strano G, Zuccarello L, Satriano C. Gold and silver nanoparticles for applications in theranostics. Curr Top Med Chem. 2016;16(27):3069–102.

Dolci S, Ierardi V, Gradisek A, Jaglicic Z, Remskar M, Apih T, Cifelli M, Pampaloni G, Alberto Veracini C, Domenici V. Precursors of magnetic resonance imaging contrast agents based on cystine-coated iron-oxide nanoparticles. Curr Phys Chem. 2013;3(4):493–500.

Duncan TV. Applications of nanotechnology in food packaging and food safety: barrier materials, antimicrobials and sensors. J Colloid Interface Sci. 2011;363(1):1–24.

Estelrich J, Escribano E, Queralt J, Busquets M. Iron oxide nanoparticles for magnetically-guided and magnetically-responsive drug delivery. Int J Mol Sci. 2015;16(4):8070–101.

Fang RH, Zhang L. Dispersion-based methods for the engineering and manufacture of polymeric nanoparticles for drug delivery applications. J Nanoeng Nanomanuf. 2011;1(1):106–12.

Fernández-López C, Mateo-Mateo C, Alvarez-Puebla RA, Pérez-Juste J, Pastoriza-Santos I, Liz-Marzán LM. Highly controlled silica coating of PEG-capped metal nanoparticles and preparation of SERS-encoded particles. Langmuir. 2009;25(24):13894–9.

Fischer D, Li Y, Ahlemeyer B, Krieglstein J, Kissel T. In vitro cytotoxicity testing of polycations: influence of polymer structure on cell viability and hemolysis. Biomaterials. 2003;24(7):1121–31.

Gao L, Zhuang J, Nie L, Zhang J, Zhang Y, Gu N, Wang T, Feng J, Yang D, Perrett S, Yan X. Intrinsic peroxidase-like activity of ferromagnetic nanoparticles. Nat Nanotechnol. 2007;2(9):577–83.

Gao S, Zhang L, Wang G, Yang K, Chen M, Tian R, Ma Q, Zhu L. Hybrid graphene/Au activatable theranostic agent for multimodalities imaging guided enhanced photothermal therapy. Biomaterials. 2016;79:36–45.

García MA. Surface plasmons in metallic nanoparticles: fundamentals and applications. J Phys D Appl Phys. 2011;44(28):283001.

Gopinath P, Gogoi SK, Chattopadhyay A, Ghosh SS. Implications of silver nanoparticle induced cell apoptosis for in vitro gene therapy. Nanotechnology. 2008;19(7):075104.

Greish K. Enhanced permeability and retention of macromolecular drugs in solid tumors: a royal gate for targeted anticancer nanomedicines. J Drug Target. 2007;15(7–8):457–64.

Guo J, Rahme K, He Y, Li LL, Holmes JD, O'Driscoll CM. Gold nanoparticles enlighten the future of cancer theranostics. Int J Nanomedicine. 2017;12:6131–51.

Guthi JS, Yang SG, Huang G, Li S, Khemtong C, Kessinger CW, Peyton M, Minna JD, Brown KC, Gao J. MRI-visible micellar nanomedicine for targeted drug delivery to lung cancer cells. Mol Pharm. 2009;7(1):32–40.

Han L, Xia JM, Hai X, Shu Y, Chen XW, Wang JH. Protein-stabilized gadolinium oxide-gold nanoclusters hybrid for multimodal imaging and drug delivery. ACS Appl Mater Interfaces. 2017;9(8):6941–9.

Hegmann T, Worden M, Miller DW. Aqueous synthesis of polyhedral "brick-like" iron oxide nanoparticles for hyperthermia and T2 MRI contrast enhancement, and for targeting endothelial cells for therapeutic delivery. 2016. https://patents.google.com/patent/US20180297857A1/en

Heo DN, Yang DH, Moon HJ, Lee JB, Bae MS, Lee SC, Lee WJ, Sun IC, Kwon IK. Gold nanoparticles surface-functionalized with paclitaxel drug and biotin receptor as theranostic agents for cancer therapy. Biomaterials. 2012;33(3):856–66.

Her S, Jaffray DA, Allen C. Gold nanoparticles for applications in cancer radiotherapy: mechanisms and recent advancements. Adv Drug Deliv Rev. 2017;109:84–101.

Hoejgaard L, Hesse B. Hybrid imaging: conclusions and perspectives. Curr Med Imaging Rev. 2011;7(3):252–3.

Homan KA, Shah J, Gomez S, Gensler H, Karpiouk AB, Brannon-Peppas L, Emelianov SY. Silver nanosystems for photoacoustic imaging and image-guided therapy. J Biomed Opt. 2010;15(2):021316.

Hu SH, Hsieh TY, Chiang CS, Chen PJ, Chen YY, Chiu TL, Chen SY. Surfactant-free, lipo-polymersomes stabilized by iron oxide nanoparticles/polymer interlayer for synergistically targeted and magnetically guided gene delivery. Adv Healthc Mater. 2014;3(2):273–82.

Huang P, Li Z, Lin J, Yang D, Gao G, Xu C, Bao L, Zhang C, Wang K, Song H, Hu H. Photosensitizer-conjugated magnetic nanoparticles for in vivo simultaneous magnetofluorescent imaging and targeting therapy. Biomaterials. 2011;32(13):3447–58.

Hwang S, Nam J, Jung S, Song J, Doh H, Kim S. Gold nanoparticle-mediated photothermal therapy: current status and future perspective. Nanomedicine. 2014;9(13):2003–22.

Iga AM, Robertson JH, Winslet MC, Seifalian AM. Clinical potential of quantum dots. BioMed Research International. 2008;2007.

de Jesus PDCC, Pellosi DS, Tedesco AC. Magnetic nanoparticles: applications in biomedical processes as synergic drug-delivery systems. In: Materials for biomedical engineering. Amsterdam: Elsevier; 2019. p. 365–90.

Jain PK, Lee KS, El-Sayed IH, El-Sayed MA. Calculated absorption and scattering properties of gold nanoparticles of different size, shape, and composition: applications in biological imaging and biomedicine. J Phys Chem B. 2006;110(14):7238–48.

Ji X, Peng F, Zhong Y, Su Y, He Y. Fluorescent quantum dots: synthesis, biomedical optical imaging, and biosafety assessment. Colloids Surf B Biointerfaces. 2014;124:132–9.

Jiang Z, Le ND, Gupta A, Rotello VM. Cell surface-based sensing with metallic nanoparticles. Chem Soc Rev. 2015;44(13):4264–74.

Jing L, Ding K, Kershaw SV, Kempson IM, Rogach AL, Gao M. Magnetically engineered semiconductor quantum dots as multimodal imaging probes. Adv Mater. 2014;26(37):6367–86.

Jo SD, Ku SH, Won YY, Kim SH, Kwon IC. Targeted nanotheranostics for future personalized medicine: recent progress in cancer therapy. Theranostics. 2016;6(9):1362–77.

Joshi PN, Agawane S, Athalye MC, Jadhav V, Sarkar D, Prakash R. Multifunctional inulin tethered silver-graphene quantum dots nanotheranostic module for pancreatic cancer therapy. Mater Sci Eng C. 2017;78:1203–11.

Kang T, Li F, Baik S, Shao W, Ling D, Hyeon T. Surface design of magnetic nanoparticles for stimuli-responsive cancer imaging and therapy. Biomaterials. 2017;136:98–114.

Khan S, Alam F, Azam A, Khan AU. Gold nanoparticles enhance methylene blue-induced photodynamic therapy: a novel therapeutic approach to inhibit Candida albicans biofilm. Int J Nanomedicine. 2012;7:3245–57.

Kim D, Jeong YY, Jon S. A drug-loaded aptamer-gold nanoparticle bioconjugate for combined CT imaging and therapy of prostate cancer. ACS Nano. 2010;4(7):3689–96.

Kim TH, Lee S, Chen X. Nanotheranostics for personalized medicine. Expert Rev Mol Diagn. 2013;13(3):257–69.

Kojima C, Cho SH, Higuchi E. Gold nanoparticle-loaded PEGylated dendrimers for theragnosis. Res Chem Intermediate. 2012;38(6):1279–89.

Kottegoda N, Munaweera I, Madusanka N, Karunaratne V. A green slow-release fertilizer composition based on urea-modified hydroxyapatite nanoparticles encapsulated wood. Curr Sci. 2011;101(1):73–8.

Kottegoda N, Madusanka N, Sandaruwan C. Two new plant nutrient nanocomposites based on urea coated hydroxyapatite: efficacy and plant uptake. Indian J Agr Sci. 2016;86:494–9.

Li W, Chen X. Gold nanoparticles for photoacoustic imaging. Nanomedicine. 2015;10(2):299–320.

Li J, Cai P, Shalviri A, Henderson JT, He C, Foltz WD, Prasad P, Brodersen PM, Chen Y, DaCosta R, Rauth AM. A multifunctional polymeric nanotheranostic system delivers doxorubicin and imaging agents across the blood–brain barrier targeting brain metastases of breast cancer. ACS Nano. 2014;8(10):9925–40.

Link S, Wang ZL, El-Sayed MA. How does a gold nanorod melt? J Phys Chem B. 2000;104(33): 7867–70.

Lu PL, Chen YC, Ou TW, Chen HH, Tsai HC, Wen CJ, Lo CL, Wey SP, Lin KJ, Yen TC, Hsiue GH. Multifunctional hollow nanoparticles based on graft-diblock copolymers for doxorubicin delivery. Biomaterials. 2011;32(8):2213–21.

Ma Y, Huang J, Song S, Chen H, Zhang Z. Cancer-targeted nanotheranostics: recent advances and perspectives. Small. 2016;12(36):4936–54.

Madaan K, Kumar S, Poonia N, Lather V, Pandita D. Dendrimers in drug delivery and targeting: drug-dendrimer interactions and toxicity issues. J Pharm Bioallied Sci. 2014;6(3):139–50.

Madusanka N, Sandaruwan C, Kottegoda N, Karunaratne V. Synthesis of Ag nanoparticle/Mg-Al-layered double hydroxide nanohybrids. Eur Int J Appl Sci Technol. 2014;1(1):1–7.

Madusanka N, de Silva KN, Amaratunga G. A curcumin activated carboxymethyl cellulose–montmorillonite clay nanocomposite having enhanced curcumin release in aqueous media. Carbohydr Polym. 2015;134:695–9.

Madusanka N, Shivareddy SG, Hiralal P, Eddleston MD, Choi Y, Oliver RA, Amaratunga GA. Nanocomposites of TiO_2/cyanoethylated cellulose with ultra high dielectric constants. Nanotechnology. 2016;27(19):195402.

Madusanka N, Sandaruwan C, Kottegoda N, Sirisena D, Munaweera I, De Alwis A, Karunaratne V, Amaratunga GA. Urea–hydroxyapatite-montmorillonite nanohybrid composites as slow release nitrogen compositions. Appl Clay Sci. 2017a;150:303–8.

Madusanka N, Shivareddy SG, Eddleston MD, Hiralal P, Oliver RA, Amaratunga GA. Dielectric behaviour of montmorillonite/cyanoethylated cellulose nanocomposites. Carbohydr Polym. 2017b;172:315–21.

Mallidi S, Larson T, Tam J, Joshi PP, Karpiouk A, Sokolov K, Emelianov S. Multiwavelength photoacoustic imaging and plasmon resonance coupling of gold nanoparticles for selective detection of cancer. Nano Lett. 2009;9(8):2825–31.

Mecke A, Majoros IJ, Patri AK, Baker JR, Banaszak Holl MM, Orr BG. Lipid bilayer disruption by polycationic polymers: the roles of size and chemical functional group. Langmuir. 2005;21(23):10348–54.

Medintz IL, Mattoussi H, Clapp AR. Potential clinical applications of quantum dots. Int J Nanomedicine. 2008;3(2):151–67.

Meir R, Shamalov K, Betzer O, Motiei M, Horovitz-Fried M, Yehuda R, Popovtzer A, Popovtzer R, Cohen CJ. Nanomedicine for cancer immunotherapy: tracking cancer-specific T-cells in vivo with gold nanoparticles and CT imaging. ACS Nano. 2015;9(6):6363–72.

Melamed JR, Edelstein RS, Day ES. Elucidating the fundamental mechanisms of cell death triggered by photothermal therapy. ACS Nano. 2015;9(1):6–11.

Mura S, Couvreur P. Nanotheranostics for personalized medicine. Adv Drug Deliv Rev. 2012; 64(13):1394–1416.

Na JH, Koo H, Lee S, Min KH, Park K, Yoo H, Lee SH, Park JH, Kwon IC, Jeong SY, Kim K. Real-time and non-invasive optical imaging of tumor-targeting glycol chitosan nanoparticles in various tumor models. Biomaterials. 2011;32(22):5252–61.

Neuwelt A, Sidhu N, Hu CAA, Mlady G, Eberhardt SC, Sillerud LO. Iron-based superparamagnetic nanoparticle contrast agents for MRI of infection and inflammation. AJR Am J Roentgenol. 2015;204(3):302–13.

Padmanabhan P, Kumar A, Kumar S, Chaudhary RK, Gulyás B. Nanoparticles in practice for molecular-imaging applications: an overview. Acta Biomater. 2016;41:1–16.

Palmerston ML, Pan J, Torchilin V. Dendrimers as nanocarriers for nucleic acid and drug delivery in cancer therapy. Molecules. 2017;22(9):1401.

Pan J, Liu Y, Feng SS. Multifunctional nanoparticles of biodegradable copolymer blend for cancer diagnosis and treatment. Nanomedicine. 2010;5(3):347–60.

Pang Z, Feng L, Hua R, Chen J, Gao H, Pan S, Jiang X, Zhang P. Lactoferrin-conjugated biodegradable polymersome holding doxorubicin and tetrandrine for chemotherapy of glioma rats. Mol Pharm. 2010;7(6):1995–2005.

Papahadjopoulos D, Allen TM, Gabizon A, Mayhew E, Matthay K, Huang SK, Lee KD, Woodle MC, Lasic DD, Redemann C. Sterically stabilized liposomes: improvements in pharmacokinetics and antitumor therapeutic efficacy. Proc Natl Acad Sci U S A. 1991;88(24):11460–4.

Peer D, Karp JM, Hong S, Farokhzad OC, Margalit R, Langer R. Nanocarriers as an emerging platform for cancer therapy. Nat Nanotechnol. 2007;2(12):751–60.

Peng C, Zheng L, Chen Q, Shen M, Guo R, Wang H, Cao X, Zhang G, Shi X. PEGylated dendrimer-entrapped gold nanoparticles for in vivo blood pool and tumor imaging by computed tomography. Biomaterials. 2012;33(4):1107–19.

Peng H, Tang J, Zheng R, Guo G, Dong A, Wang Y, Yang W. Nuclear-targeted multifunctional magnetic nanoparticles for photothermal therapy. Adv Healthc Mater. 2017;6(7):1601289.

Penon O, Marín MJ, Amabilino DB, Russell DA, Pérez-García L. Iron oxide nanoparticles functionalized with novel hydrophobic and hydrophilic porphyrins as potential agents for photodynamic therapy. J Colloid Interface Sci. 2016;462:154–65.

Popovtzer R, Agrawal A, Kotov NA, Popovtzer A, Balter J, Carey TE, Kopelman R. Targeted gold nanoparticles enable molecular CT imaging of cancer. Nano Lett. 2008;8(12):4593–6.

Prabhu RH, Patravale VB, Joshi MD. Polymeric nanoparticles for targeted treatment in oncology: current insights. Int J Nanomedicine. 2015;10:1001–18.

Ray S, Li Z, Hsu CH, Hwang LP, Lin YC, Chou PT, Lin YY. Dendrimer-and copolymer-based nanoparticles for magnetic resonance cancer theranostics. Theranostics. 2018;8(22):6322–49.

Reddy LH, Arias JL, Nicolas J, Couvreur P. Magnetic nanoparticles: design and characterization, toxicity and biocompatibility, pharmaceutical and biomedical applications. Chem Rev. 2012;112(11):5818–78.

Rienzie R, Adassooriya NM. Toxicity of nanomaterials in agriculture and food. In: Nanomaterials: ecotoxicity, safety, and public perception. Cham: Springer; 2018. p. 207–34.

Rittner K, Benavente A, Bompard-Sorlet A, Heitz F, Divita G, Brasseur R, Jacobs E. New basic membrane-destabilizing peptides for plasmid-based gene delivery in-vitro and in-vivo. Mol Ther. 2002;5(2):104–14.

Sano K. Development of molecular probes based on iron oxide nanoparticles for in vivo magnetic resonance/photoacoustic dual imaging of target molecules in tumors. Yakugaku zasshi: Journal of the Pharmaceutical Society of Japan. 2017;137(1):55–60.

Savla R, Taratula O, Garbuzenko O, Minko T. Tumor targeted quantum dot-mucin 1 aptamer-doxorubicin conjugate for imaging and treatment of cancer. J Control Release. 2011;153(1): 16–22.

Scholl JA, Koh AL, Dionne JA. Quantum plasmon resonances of individual metallic nanoparticles. Nature. 2012;483(7390):421–7.

Schweitzer VG, Somers ML. PHOTOFRIN-mediated photodynamic therapy for treatment of early stage (Tis-T2N0M0) SqCCa of oral cavity and oropharynx. Lasers Surg Med. 2010;42(1):1–8.

Selvan ST, Narayanan K. Introduction to nanotheranostics. In: Introduction to nanotheranostics. Singapore: Springer; 2016. p. 1–6.

Sharma H, Mishra PK, Talegaonkar S, Vaidya B. Metal nanoparticles: a theranostic nanotool against cancer. Drug Discov Today. 2015;20(9):1143–51.

Shi J, Votruba AR, Farokhzad OC, Langer R. Nanotechnology in drug delivery and tissue engineering: from discovery to applications. Nano Lett. 2010;10(9):3223–30.

Shi D, Sadat ME, Dunn AW, Mast DB. Photo-fluorescent and magnetic properties of iron oxide nanoparticles for biomedical applications. Nanoscale. 2015;7(18):8209–32.

Shubayev VI, Pisanic TR II, Jin S. Magnetic nanoparticles for theragnostics. Adv Drug Deliv Rev. 2009;61(6):467–77.

Sonali MKV, Singh RP, Agrawal P, Mehata AK, Datta Maroti Pawde N, Sonkar R, Muthu MS. Nanotheranostics: emerging strategies for early diagnosis and therapy of brain cancer. Nanotheranostics. 2018;2(1):70–86.

Srinivas M, Aarntzen EHJG, Bulte JWM, Oyen WJ, Heerschap A, De Vries IJM, Figdor CG. Imaging of cellular therapies. Adv Drug Deliv Rev. 2010;62(11):1080–93.

Stevenson MJ, Heffern MC. Sounding out dysfunctional oxygen metabolism: a small-molecule probe for photoacoustic imaging of hypoxia. Biochemistry. 2018;57(6):893–4.

Strijkers GJ, Mulder M, Willem J, Van Tilborg F, Geralda A, Nicolay K. MRI contrast agents: current status and future perspectives. Anti Cancer Agents Med Chem. 2007;7(3):291–305.

Tan A, Yildirimer L, Rajadas J, De La Peña H, Pastorin G, Seifalian A. Quantum dots and carbon nanotubes in oncology: a review on emerging theranostic applications in nanomedicine. Nanomedicine. 2011;6(6):1101–14.

Templeton AC, Pietron JJ, Murray RW, Mulvaney P. Solvent refractive index and core charge influences on the surface plasmon absorbance of alkanethiolate monolayer-protected gold clusters. J Phys Chem B. 2000;104(3):564–70.

Thorat ND, Lemine OM, Bohara RA, Omri K, El Mir L, Tofail SA. Superparamagnetic iron oxide nanocargoes for combined cancer thermotherapy and MRI applications. Phys Chem Chem Phys. 2016;18(31):21331–9.

Usov OA, Sidorov AI, Nashchekin AV, Podsvirov OA, Kurbatova NV, Tsekhomsky VA, Vostokov AV. SPR of Ag nanoparticles in photothermochromic glasses. In Plasmonics: metallic nanostructures and their optical properties VII 7394: 73942J, International Society for Optics and Photonics. 2009

Veiseh O, Gunn JW, Zhang M. Design and fabrication of magnetic nanoparticles for targeted drug delivery and imaging. Adv Drug Deliv Rev. 2010;62(3):284–304.

Wang LS, Chuang MC, Ho JAA. Nanotheranostics—a review of recent publications. Int J Nanomedicine. 2012;7:4679–95.

Wang X, Sun X, Lao J, He H, Cheng T, Wang M, Wang S, Huang F. Multifunctional graphene quantum dots for simultaneous targeted cellular imaging and drug delivery. Colloids Surf A Physicochem Eng Asp. 2014;122:638–44.

Wang J, Tao W, Chen X, Farokhzad OC, Liu G. Emerging advances in nanotheranostics with intelligent bioresponsive systems. Theranostics. 2017;7(16):3915–9.

Wei P, Chen J, Hu Y, Li X, Wang H, Shen M, Shi X. Dendrimer-stabilized gold nanostars as a multifunctional theranostic nanoplatform for CT imaging, photothermal therapy, and gene silencing of tumors. Adv Healthc Mater. 2016;5(24):3203–13.

Wu W, Zhou T, Berliner A, Banerjee P, Zhou S. Smart core–shell hybrid nanogels with Ag nanoparticle core for cancer cell imaging and gel shell for pH-regulated drug delivery. Chem Mater. 2010;22(6):1966–76.

Wu Y, Gao D, Zhang P, Li C, Wan Q, Chen C, Gong P, Gao G, Sheng Z, Cai L. Iron oxide nanoparticles protected by NIR-active multidentate-polymers as multifunctional nanoprobes for NIRF/PA/MR trimodal imaging. Nanoscale. 2016;8(2):775–9.

Xi L, Grobmyer SR, Zhou G, Qian W, Yang L, Jiang H. Molecular photoacoustic tomography of breast cancer using receptor targeted magnetic iron oxide nanoparticles as contrast agents. J Biophotonics. 2014;7(6):401–9.

Xie J, Liu G, Eden HS, Ai H, Chen X. Surface-engineered magnetic nanoparticle platforms for cancer imaging and therapy. Acc Chem Res. 2011;44(10):883–92.

Xu X, Chong Y, Liu X, Fu H, Yu C, Huang J, Zhang Z. Multifunctional nanotheranostic gold nanocages for photoacoustic imaging guided radio/photodynamic/photothermal synergistic therapy. Acta Biomater. 2019;84:328–38.

Yang HM, Oh BC, Kim JH, Ahn T, Nam HS, Park CW, Kim JD. Multifunctional poly (aspartic acid) nanoparticles containing iron oxide nanocrystals and doxorubicin for simultaneous cancer diagnosis and therapy. Colloids Surf A Physicochem Eng Asp. 2011;391(1-3):208–15.

Yang Z, Song J, Tang W, Fan W, Dai Y, Shen Z, Lin L, Cheng S, Liu Y, Niu G, Rong P. Stimuli-responsive nanotheranostics for real-time monitoring drug release by photoacoustic imaging. Theranostics. 2019;9(2):526.

Yong Y, Cheng X, Bao T, Zu M, Yan L, Yin W, Ge C, Wang D, Gu Z, Zhao Y. Tungsten sulfide quantum dots as multifunctional nanotheranostics for in vivo dual-modal image-guided photothermal/radiotherapy synergistic therapy. ACS Nano. 2015;9(12):12451–63.

Yu J, Yin W, Zheng X, Tian G, Zhang X, Bao T, Dong X, Wang Z, Gu Z, Ma X, Zhao Y. Smart MoS_2/Fe_3O_4 nanotheranostic for magnetically targeted photothermal therapy guided by magnetic resonance/photoacoustic imaging. Theranostics. 2015;5(9):931–45.

Zhang Z, Jia J, Lai Y, Ma Y, Weng J, Sun L. Conjugating folic acid to gold nanoparticles through glutathione for targeting and detecting cancer cells. Bioorganic Med Chem. 2010;18(15):5528–34.

Zhang Z, Wang S, Xu H, Wang B and Yao C. Role of 5-aminolevulinic acid-conjugated gold nanoparticles for photodynamic therapy of cancer. J Biomed Opt. 2015;20(5):051043.

Zhang P, Hu C, Ran W, Meng J, Yin Q, Li Y. Recent progress in light-triggered nanotheranostics for cancer treatment. Theranostics. 2016a;6(7):948–68.

Zhang XF, Liu ZG, Shen W, Gurunathan S. Silver nanoparticles: synthesis, characterization, properties, applications, and therapeutic approaches. Int J Mol Sci. 2016b;17(9):E1534.

Zhang S, Zheng Y, Fu DY, Li W, Wu Y, Li B, Wu L. Biocompatible supramolecular dendrimers bearing a gadolinium-substituted polyanionic core for MRI contrast agents. J Mater Chem B. 2017;5(22):4035–43.

Zhu J, Wang G, Alves CS, Tomás H, Xiong Z, Shen M, Rodrigues J, Shi X. Multifunctional dendrimer-entrapped gold nanoparticles conjugated with doxorubicin for pH-responsive drug delivery and targeted computed tomography imaging. Langmuir. 2018;34(41):12428–35.

Chapter 3
Nanotheranostics Approaches in Antimicrobial Drug Resistance

Juan Bueno

Abstract The theranostics as a combination of diagnosis and therapy is a trend of personalized medicine that seeks to develop a precision medical care, also this approach has found in nanotechnology a possibility for the conjugation of several molecules with different functionalities that will allow diagnosis, treatment, monitoring, and prediction of the patient condition in the same nanosystem. Equally, fields such as biosensors and novel therapies and thermotherapy have been integrated into the multiple variants that this platform presents. It is thus that as before the technological developments should be the problems where the designed solutions are practiced, and nanotheranostics can be used in a multidisciplinary way to address one of the major public health problems such as antimicrobial resistance. This implies not only developing tools to detect the infectious disease, it also requires introducing medications and treatment alternatives; in the same way the devices implemented must act to prevent the appearance and spread of pathogens. Similarly, by means of nanotheranostic it will be possible to have more sophisticated and effective antimicrobial and vaccination protocols that will allow an adequate control of the microorganisms causing disease. Finally, in this chapter, the different approaches with translational possibility that this exciting field of nanobiotechnology has allowed to face the great health threats of our time will be integrated in a multi-trans-interdisciplinary approach.

Keywords Theranostics · Nanotechnology · Biosensors · Pharmacogenomics · Antimicrobial drug resistance

---

J. Bueno (✉)
Research Center of Bioprospecting and Biotechnology for Biodiversity Foundation (BIOLABB), Armenia, Quindío, Colombia

M. Rai, B. Jamil (eds.), *Nanotheranostics*,
https://doi.org/10.1007/978-3-030-29768-8_3

3.1 Introduction

Theranosis is defined as the ability to affect the treatment of the disease through its diagnosis (Jagtap et al. 2017). In this way, the nanotheranostic approach seeks to use the biomedical applications of nanomaterials to develop new diagnostic methods capable of changing the natural history of the disease due to the potential to direct the treatment (Pedrosa et al. 2015). This is how a new field in personalized medicine is developed to improve the bioavailability of active agents and in which the stages of healing can be monitored in real time as well as avoiding the appearance of complications (Kevadiya et al. 2018). In addition, as information about the current patient condition can be obtained, complex biological systems can be observed, which opens an unprecedented opportunity in public health prevention (Sahlgren et al. 2017). This complies with the postulates of precision medicine outlined in the four Ps (predictive, preventive, personalized, and participatory), which opens the possibility of managing patients in their functionality (Vicini et al. 2016). Likewise, the theranostic promises to increase the adherence of patients to treatments by their control of adverse reactions to medications (Zazo et al. 2016). This is because it measures cellular changes without chemically or genetically altering the physiology of the biological system, which makes it a more relevant diagnostic platform (Hillger et al. 2017). For this reason, prospecting for new biomarkers able to determine the course of treatment and healing process is necessary, this in order to achieve precision medicine homeostasis (Zhang et al. 2017). These biomarkers should also have the capacity to be coupled with nanobiomaterials that will allow them to determine the routes of change and cell adaptation during treatment (Singh et al. 2018).

This novel approach can be applied to solve the major public health problems of our time, such as antimicrobial resistance (AMR) developed by microbes causing common infections such as bacterial pneumonia, postoperative infections, and higher mortality infections including malaria, HIV, and tuberculosis that are becoming increasingly difficult to treat due to the increase in resistance (Inoue and Minghui 2017). In addition AMR represents a risk for the possibility of transmission of drug-resistant microorganisms from hospitals and health care centers (Col and Brig 2010; Tournier et al. 2019). Therefore, it is necessary to implement new diagnostic strategies to identify AMR, determine antimicrobial susceptibility, discover new drugs, and monitor treatment (Burnham et al. 2017). In this way, theranostic platforms have been implemented that seek to detect multiresistant microorganisms in tissues, as well as monitor their treatment and eliminate by both pharmacological and physical means (Xie et al. 2017; Vangara et al. 2018). Likewise, the use of nanomaterials assembled in devices for early detection and treatment due to the possibility of being coated with antimicrobial molecules are a great possibility in avoiding and controlling nosocomial infections (Li et al. 2017). Also models of screening for natural pathogenic stages of infection such as biofilms have been tested, because these formations are a source of AMR and clinical study for the development of new drugs is limited (Gomes et al. 2018). These microbial associations can now be recognized and evidenced by the application of diagnostic and treatment platforms based on theranostics approaches (Ribeiro et al. 2016; Magana et al. 2018). Additionally, in this effort, innovative treatment alternatives such as photo-antimicrobial technologies, which have a wide field of application, have been evaluated (Wainwright et al. 2017). In the same way, photodynamic antimicrobial therapy can be included in nanoformulations that guarantee adequate treatment without adverse effects (Gonzalez-Delgado et al. 2015). On the other hand, the photodynamic inactivation in its theranostic variant has a promising application in a great public health problem such as foodborne infectious diseases, which kill 420,000 people per year all over the world (Silva et al. 2018).

Because of the aforementioned facts, the aim of this chapter is to give a multidisciplinary and integral perspective of the theranostics in conjugation with nanotechnology as a translational tool for the prevention and cure of major threats to public health; that is why it is necessary to evaluate both its efficacy and its safety in the time to apply it in clinical practice.

3.2 Antimicrobial Nanotheranostics

In this order of ideas, the primary objectives of the antimicrobial nanotheranostics include (Larrañeta et al. 2016; Gao et al. 2018):

- Diagnostics of the presence of infectious disease.
- Locates the site of infection.
- Deliver medications at the site of infection.
- Performs antimicrobial therapy at the site of infection.
- Prevents the spread of infectious microbial agents.

In this way, this anti-infectious technology has focused on three approaches: nanomaterials (NMs) fused with imaging agents, NMs fused with antimicrobial drugs, and NMs fused with targeting ligands (Tonga et al. 2014). Additionally, the construction of nanosystems has made it possible to increase the efficacy of low doses of drugs and improve their bioavailability, so the development of these nanostructures become an important tool in theranostics (Grumezescu et al. 2014). Likewise, the design and synthesis of fluorescent organic nanoparticles (FONs) and fluorescent inorganic nanoparticles (FINs) (Ag, Au, Zn, Cu, Ti, Mg, Ni, Ce, Se, Al, Cd, Y, Pd, and super-paramagnetic Fe) (Lee et al. 2017; Hemeg 2017; Gao et al. 2018) used in protective coatings for the prevention of intrahospital transmission of microorganisms (Ge et al. 2017) has been considered a promising field to integrate diagnosis, treatment, and prevention in the same antimicrobial platform; for that reason, it can be projected as a theranostics approach (Sahlgren et al. 2017; Primiceri et al. 2018). Likewise, the development of novel formulations for the release of antimicrobial drugs is an urgent investigation to improve the activity of medicines and reduce their toxicity (Kavanagh et al. 2018), such as those with a polymeric configuration that allow activity without inducing resistance and with high biosafety (Cao et al. 2018).

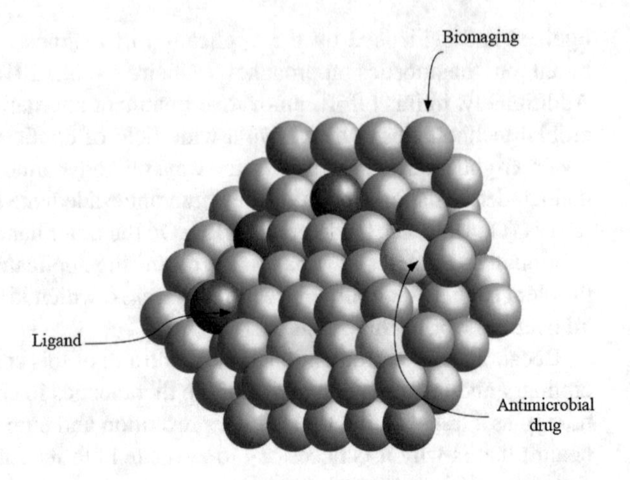

Fig. 3.1 Configuration of a nanocluster can be composed of a single or multiple elements

Other interesting approach is the implementation of nanocluster technology as nanopyramids (DPs), which combine red-emissive glutathione-protected gold nanoclusters (GSH-Au NCs) for bacterial detection and actinomycin D (AMD) as antimicrobial agent in order to have larger coupling sites and increase the functionality of the theranostic platform (Setyawati et al. 2014). This nanocluster technology (Fig. 3.1.) can combine several molecules by increasing the synergistic effect of the formulations, as well as include fluorescent compounds that allow the location of the pathogen at the site of infection, because they are composed of several elements (Huma et al. 2018; Mirahmadi-Zare et al. 2019). For this reason, the new era of antimicrobial treatment is determined by the potentiation of activity through the use of nanotechnological platforms, as well as the early diagnosis of infectious disease (Muzammil et al. 2018).

3.2.1 Microbial Detection

In the microbiological diagnosis there are two approaches that influence the therapeutic behavior, which are: the detection and isolation of the pathogen. This is how detection can be carried out through the development of nanostructures such as silver, gold, and zinc nanoparticles coupled to colorimetric, fluorescent, or bioluminescent markers; in the same way, the isolation of the agent, crucial step for drug sensitivity tests, can be carried out in a sensitive and specific way with the use of magnetic nanoparticles to be separated by an electromagnetic field (Mocan et al. 2017). Likewise, this field of action lends itself to the design of all types of devices with the use of biosensors with applications at the point of care for obtaining an early diagnosis as is the case of optical and electrochemical biosensors (Arduini et al. 2017; Craciun et al. 2017). Other interesting approach is the implementation of lab-on-chip (LoC) technology that can integrate several analyses (biochemical, chemical synthesis, and DNA) into a single chip, which can provide a large amount

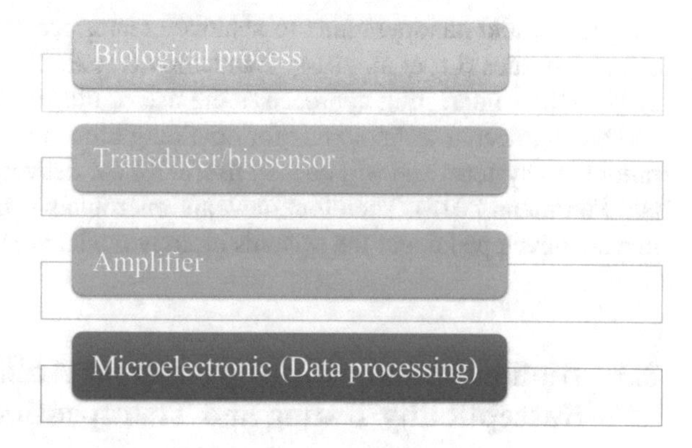

Fig. 3.2 Composition of the operation process of lab-on-chip (LoC) technology

of information in the clinical diagnosis and treatment of infectious diseases (Fig. 3.2.), also LoC allows the miniaturization to be coupled with microfluidic platforms that can be used in the diagnosis and treatment of diseases such as tuberculosis and HIV (Gupta et al. 2016). Just as exciting is the possibility of integrating these devices to cell phones to track pathogens in clinical samples, as well as those from food and environment (Nasseri et al. 2018). This indicates that the diagnosis at the point of care will be fundamental for the prevention and treatment of infectious diseases in the coming years (Drain et al. 2014).

3.2.2 *From Drug Combinations to Antimicrobial Cluster Bombs*

The combination of two microbicides is possible by their integration in novel cluster-type formulations, which allows the union of several functional pharmacological groups without altering their mechanism of action and stability (Zheng et al. 2016). Also a great advantage of the nanoclusters is having a higher ratio of the area of action per volume of distribution than the conventional NMs (Xu et al. 2016). Likewise, its microbicidal activity is due to the increase of reactive oxygen species that destroy the cell membranes of infectious agents (Yuan et al. 2014). But it is important to keep in mind that absorption, distribution, metabolism, and excretion (ADME) and pharmacokinetic (PK) of these particles is dependent on the contact surface and its three-dimensional structure that can cause the formulation to be trapped by proteins of the reticuloendothelial system, so it is necessary to evaluate more biocompatible models or other administration routes (Wong et al. 2013). In this way, innovative technologies as nanocluster (NC) aerosol dry powder can be an option when avoiding inappropriate PK using other routes such as inhaled for the treatment and diagnosis of lung infections (Ruiz et al. 2016). Additionally, these nanoclusters can be designed in the form of nanorods, nanohorns, nanospheres, nanocubes, and nanopyramids to achieve greater penetration and interaction with affected tissues (Lu et al. 2014; Trandafilović et al. 2014; Nowlin and LaJeunesse 2017). Other interesting approach is the use of biogenic, biomimetic, bioinspired, and bioengineered technologies for obtaining biocompatible drug carriers that can mimic cell systems and will be very useful for the delivery of nanoantibiotics (Goes and Fuhrmann 2018). Therefore, new antimicrobials with greater biocompatibility that are developed under the systems biology model are necessary (Arnold 2018).

3.3 Biofilm Theranostic Approach for Antimicrobial Susceptibility Testing and Therapeutics

Biofilm is a model of microbial and survival persistence. This persistence is defined inside of matrix barrier conformed in this structure in which genetic and metabolic exchange occurs (Flemming et al. 2016). As solidarity expression, biofilm shows the ability of bacterial and fungal cells to cooperate for space colonization. For that reason, it has become a public health problem because it can produce chronic and resistant infections (He et al. 2014). In this way it becomes necessary to implement new strategies for diagnostics and treatment of biofilms that avoid the emergence of resistance. Theranostic applications can be an interesting approach looking for solutions for biofilm control (Şen Karaman et al. 2018). Theranostic combine diagnostics and selection of adequate treatment in the same platform; in this order of ideas, in a biofilm screening platform it would be important to integrate microbiology, biosensors, optoelectronics, nanotechnology, and metabolomics, which can provide rapid results with this end to give information for selecting the best therapy (Ramasamy and Lee 2016). Thus, a new antibiofilm susceptibility testing (ABST) is required, which is rapid, portable, with the ability to be miniaturized, and can offer results in detection and treatment in a simple device (Trivedi et al. 2017).

In this way, the use of beta-emitting radioactive molecule as ^{131}I to load antibiotics has been useful in monitoring the behavior of antimicrobials within the biofilm and also presenting conjugated activity against mature biofilms and microorganisms inside the biofilm (Avcıbaşı et al. 2018). On the other hand, photothermal therapy (PTT), which uses the ability of materials (gold nanoparticles, carbon nanotubes, and graphene) to convert light into thermal energy, is a promising method for the destruction of biofilms using light at a wavelength of 700–1000 nm (Wang et al. 2019). It is important to note that these photoactivatable formulations can be designed in nanostructures to obtain a synergistic effect in the detection as in the microbicidal activity in combination with nanoparticles, antibodies, and antibiotics, which makes them a platform of wide use in the elimination of biofilms in different locations (Meeker et al. 2018). So in biofilms the great challenges are early detection, effective treatment as well as later eradication and prevention of adherence for their formation (Wolfmeier et al. 2017).

3.3.1 Antimicrobial Physics: The Solution to Control Biofilm Problem?

Antimicrobial physics is an interesting approach in which anti-infective agents are combined with physical sterilizing procedures such as phototherapy and photo-acoustic that have shown great potential to destroy biofilms (Barroso et al. 2016; Shi et al. 2019). In this way new nanostructures are designed and applied, containing an antibiotic plus photoactivatable compounds that are activated under light or heat (Smeltzer et al. 2015). Likewise, photoactivatable nanostructures using lasers can be coupled to antibodies to increase their specificity and activity, as well as decrease toxicity (Meeker et al. 2018). In this order of ideas, photodynamic agents have been used for the design and development of these nanostructures as tetrapyrroles, which contain chlorophyll and cytochromes which makes them photoactivatable; also photothermic agents have been used successfully in these nanocomposites such as metal nanoparticles and carbon-based nanoparticles, which produce heat by light stimulation (Chitgupi et al. 2017). Thus, in the search for new biocompatible nanostructures, the use of biomimetic strategies that are inspired by nature have opened the possibility of the use of natural products as a valuable source of functional molecules with antimicrobial activity, which can penetrate microbial communities such as biofilms (Vitiello et al. 2018). Likewise, PTT will be of radical importance in the control of healthcare-associated infections (HAI) that are one of the major causes of mortality and morbidity in the hospital environment (Percival et al. 2015).

3.3.2 Functional Coatings, the Prevention of Biofilm Formation

In this order of ideas, the development of new materials capable of interacting with the biological environment to detect and avoid threats from pathogens is an attractive innovation, as the implementation of zwitterionic surface coatings become an alternative due to its stability and repellent capacity (Huang et al. 2015). Likewise, graphene-based materials have aroused interest as a protective coating due to their chemical resistance, adsorptive capacity, and antimicrobial properties (Nine et al. 2015; Madni et al. 2018). In addition to the above, nano-assemblies, such as nanodiamonds, can also be used. When they are covered with polymers, they are converted into particles with theranostics applications (Neburkova et al. 2017). Other interesting approach is the use of superparamagnetic iron oxide nanoparticles (SPIONs), particles with great penetrability in biofilms and ability to be conjugated on the surfaces of cationic polyacrylamide (Taresco et al. 2015). On the other hand, molecular biomimicry approach can also be used to obtain functional coatings with higher microbicidal capacity; in this case, peptidopolysaccharides that mimic bacterial peptidoglycan, a fundamental component of its cell walls, can be used in coatings designed to destroy microbial cells that adhere to these surfaces (Pranantyo et al. 2018). This is how functional coatings are a strategy for reducing the microbial load that will allow the development of new materials capable of reducing the spread of pathogens within communities (Kratochvil et al. 2017).

3.4 Nanotheranostic Approach for Antimicrobial Susceptibility Testing

Antimicrobial susceptibility testing (AST) as diagnostic method is used both to determine the resistance profiles of the infecting microorganisms, as well as to guide the antibiotic therapy and to predict the clinical response to the antimicrobial treatment (Syal et al. 2017). In this order of ideas, the new AST methods must be oriented toward a precision medicine capable of giving rapid results at the medical care site and with high clinical predictability (Rello et al. 2018). Thus, phenotypic methods for the early determination of antibiotic activity have been implemented, as is the case of the optical methods developed to measure the deformation of the microbial cell membrane at the nanometer scale, which can obtain results after 20 min of exposure to the antibiotic (Iriya et al. 2017). Also, an important strategy in AST is the application of optical biosensors that using both surface plasmon resonance (SPR) and surface enhanced Raman spectroscopy (SERS) have the ability of measure the aggregation of gold nanoparticles coupled to receptors which capture pathogenic microorganisms in different samples, with the advantage of miniaturization as it can be coupled with microfluidic technology (Yoo and Lee 2016). Likewise, the use of synthetic biology tools will allow the development of better diagnostic techniques due to the possibility of synthesizing biomolecular machines such as living cell biosensors that detect resistant microorganisms (Bueno 2014; Courbet et al. 2016). In this way, the biosensors will play a key role in the new AST techniques for their ability to determine susceptibility more quickly and their possibility to be miniaturized in microfluidic platforms, as well as being designed in multiple format with the cantilever system (Sabhachandani et al. 2017).

3.5 Multidrug-Resistant Microbes (MDR)

Multidrug-resistant (MDR) microorganisms are one of the greatest threats to the public health of populations, and hence theranostic approaches have been developed for diagnosis and treatment of MDR as the case of Verigene® which is a gold nanoparticles platform (AuNPs) and can identify Gram-negative bacteria from blood samples (Baptista et al. 2018). Another interesting multifunctional nanoplatform combines fluorescent detection with photodestruction, for the diagnosis and treatment of *Salmonella* in infected blood; this method also has the ability to perform magnetic separation of pathogens to perform fluorescent detection (Dai et al. 2013). In addition, magnetic separation can be coupled with other nanosystems as carbon dots (CD) for obtaining fluorescent magneto-CD nanoparticles as hybrid NM with capacity to detect, isolate, and destroy methicillin-resistant *Staphylococcus aureus* (MRSA) (Pramanik et al. 2017). Fullerenes nanostructures can also be photoactivated to release oxygen radicals capable of destroying MDR microorganisms. But its biggest problem continues to be hydrophobicity, which requires new formulations that increase the solubility of this NM (Hamblin 2018). For this reason, the nanotheranostic approach is fundamental for the design and development of applicable tools of high impact in clinical medicine that decrease the appearance of MDR strains and also treat the different outbreaks.

3.6 Nanoplatforms in Antimicrobial Drug Delivery

An important approach in the search for new strategies to reduce mortality due to infectious diseases is the development of nanoformulations that allow the antibiotic to be brought to the affected site; in this way, new nanoplatforms using zinc oxide (ZnO) in conjunction with semiconductor quantum dots (QDs) have shown a promising potential to improve the antimicrobial activity of the nanosystems where they are applied; that is how ZnO QDs structures possess bactericidal, antibiofilm, and antifungal activity (Martínez-Carmona et al. 2018). Likewise, the use of natural products has also been of added value to these nanoplatforms, as is the case of encapsulated curcumin nanoformulations with high penetration both through the skin barrier and intracellularly (Krausz et al. 2015). Also nanostructures have been implemented with this end, in this order of ideas nanostructured lipid carriers (NLCs) which is a solid lipid matrix which represent a novel formulation to be coupled to drugs and fluorescent molecules to be administered both dermally and orally because it increases penetration through mucous barriers (Beloqui et al. 2016). Equally this DNA-based assemblies as DNA origami nanostructures which belong to the emerging DNA nanotechnology it allows a greater functionality of the molecules for its use as a biosensor, delivery of drugs and nanomachines (Linko et al. 2015). Another important approach in drug delivery is the use of peptide–drug conjugates (PDCs) that integrates proteins with drugs in order to increase their effectiveness. In addition, conjugates with aggregation-induced emission fluorogens are also possible (AIEgens), which are more photostable and decrease background interference (Wang et al. 2017).

Finally, the use of a supramolecular architecture will create a new field of synergism in which the biological activities of all the components within the nanostructure can be potentiated and allowed to have in the same scaffold carbohydrates, proteins, DNA, polymers, and synthetic molecules (Ma et al. 2016).

3.7 Anti-Infectious Therapy and Pharmacogenomics

Pharmacogenomics studies the interindividual variability in pharmacological treatment, with which a more effective medication can be administered (Siest and Schallmeiner 2014). This is how the new precision medicine requires predicting the effectiveness and adverse effects of the treatments, with the use of biomarkers that applied in clinical practice favor the medical outcome (Pirmohamed 2014). In this way, the theranostic approach can be the bridge that brings pharmacogenomic tests to primary healthcare (Bartlett et al. 2012). Because it has been difficult to translate the pharmacogenomic data into clinical practice, an important example has been the biomarker HLA-B∗5701 allele as a pharmacogenetic marker for antiviral drug abacavir hypersensitivity, but for the development of a personalized medicine useful in pharmacogenomics, it requires fast, safe, and reproducible molecular biology tests (Ventola 2011). An advance in this regard is the development of biosensors that employ aptamers, which are single-stranded nucleic acid molecules (ssDNA or RNA) with the ability to recognize specific targets including pathogenic microorganisms and pharmacogenomic biomarkers (Molefe et al. 2018). Also, these methods can be adapted to biosensing platforms such as microfluidics that allow the development of multiple character devices with which to obtain the most information in the point-of-care (Mauk et al. 2018). The prediction of clinical outcome in the therapy using pharmacogenomic tools will allow to increase the cure rates and decrease the incidence of antimicrobial resistance.

3.8 Nanomedicine and Vaccine Development

One of the major and more successful strategies in infectious disease control is the development of new vaccines (Kennedy and Read 2017). Currently, one of the great challenges of nanotechnological applications is to increase the immunogenicity of current vaccines in an effective and safe way; also the use of nanoparticle formulations such as polymeric, inorganic, liposomes, and emulsions has shown advantages as immunopotentiators and delivery of antigens (Zhao et al. 2014; Yan et al. 2019). The vaccine platforms capable of presenting the antigen in lymphatic tissues have been developed as is the case of SPIONs included in porous silicon (pSi) which has demonstrated its ability to stimulate and regulate the response of T cells to a given antigen (Lundquist et al. 2014). Equally, the use of microneedle (MN) patches promises an effective and safe immunogenic system that together with thermostable formulations such as polyplex can be an alternative to conventional intramuscular injections (Liao et al. 2017). On the other hand, a new approach is photodynamic vaccination (PDV), which consists of a type of immunotherapy that uses photosensitizers as vaccine carriers, which have shown activity against *Leishmania* (Viana et al. 2018). It is also possible to couple vaccine adjuvants in nanoformulations to increase the immunogenicity of antigens when administered in other more cost-effective routes such as intranasal (Qu et al. 2018). Thus, vaccination is the most cost-effective intervention in public health, so this is an area where applications of theranostics are of great impact and benefits (Pang 2012).

3.9 Antimicrobial Drug Discovery

It is a fact that drug discovery requires drug screening tools with high reproducibility (Moffat et al. 2017). So, QDs present great advantages for the design and development of trials using bioluminescence and fluorescence probes for which they are considered promising for the implementation of new platforms for drug screening; for that reason QDs has been widely used in high-throughput screening (HTS), specially for the implementation of multi-parameter HTS (MPHTS) experiments that can measure multiple targets in the same assay (Yao et al. 2018; Lombardo et al. 2019). Another important approach for the implementation of new drug screening tools is the use of aptamers, which can bind specifically to therapeutic targets and replace antibodies in this function; equally this method in conjugation with SELEX technique that can determinate the right interaction between RNA/DNA library and targets is very promising for the development of future methodologies in the discovery of new antimicrobial drugs (Ashrafuzzaman 2014). Equally biosensing technologies is another approach useful in drug discovery, where LoC devices with 3D structures linked to microfluidics and integrated with living cells will be primordial for the increase of efficiency in the process due to advantages as it is a faster analysis with lower sample volume (Khalid et al. 2017).

3.10 Monitoring Therapy Response to Antimicrobial Drugs

For an adequate monitoring of antimicrobial therapy, it requires the identification of biomarkers to predict the clinical response to treatment in different trials (El Bairi et al. 2019). Likewise, the use of integrated nanoplatforms to biomarkers will be an indispensable tool for precision medicine; in this way, the coupling of fluorescent molecules can be useful to evaluate bioavailability in tissues, as well as healing; in this order of ideas, nanosystems that use fluorescence resonance energy transfer (FRET)-based drug delivery system have been tested (Chen et al. 2018). Similarly, nanostructure-based mass spectrometry imaging (MSI) is an interesting nanoplatform for the determination of metabolomic spectra of different clinical samples useful for the monitoring of antimicrobial therapy; also this platform can detect various compounds in tissues unlike other traditional techniques, which makes it a promising technology to perform metabolomic diagnosis in clinical practice (Gustafsson et al. 2017). Equally, drug monitoring in real time of bioactive molecules can be performed using nanostructures conformed by mesoporous silica nanoparticles (MSNs) that will allow the development of new devices (Lai et al. 2015). Similarly, MSNs can be monitored by other techniques such as optical imaging, magnetic resonance imaging (MRI), and positron emission tomography (PET), what makes them an alternative with versatility (Narayan et al. 2018). In this manner, the development of new nanoplatforms capable of monitoring antimicrobial therapy during the medical care process will go through the possibility of implementing metabolomics-on-a-chip (MoC) devices that can determine the different biomarkers that predict the clinical response (Bueno 2017).

3.11 Risk Assessment Process

Therefore, an important aspect to take into account when implementing a new diagnostic approach is to demonstrate its safety at the moment of applying it. Thus, the greatest risk with the use of NMs is the alteration of biological systems, despite the evidence is still uncertain safety for patients and the environment of the nanostructures, so it is necessary for a real assessment of the possible presence of the nanorisk (Fadeel et al. 2018; Hauser et al. 2019). In this context, it is the need of hour to study the risk assessment of nanoparticles, as well as the management thereof (Roco et al. 2011). In this way, this process must contemplate the following aspects (Wiesner and Bottero 2011):

- Generate the correct forecasts for the different levels of uncertainty, in order to develop models of immediate response.
- Take into consideration all possible and relevant sources of NMs.
- A comprehensive assessment of the impacts of the production of the NM should be carried out in order to determine its effects on life cycles.
- The ability to update risk information as new scientific evidence becomes available.
- Implement information collection and analysis methods to improve access to it.
- Implement feedback capabilities to improve the design and production of NMs.

Additionally, any nanotechnological innovation effort should have an approach based on the 3S (small, smart, and safe) and it should also include protocols both in vitro and in vivo to evaluate immunotoxicity, genotoxicity, and epigenetic toxicity (Dusinska et al. 2017).

3.11.1 Genotoxicity

NMs used in theranostic approach as SPIONs induce ROS generation and genotoxicity, although it is possible to decrease the intensity of toxicity by exercising and natural antioxidants (Ansari et al. 2018). Equally silver nanoparticles can induce

DNA damage and alteration of cytokine secretion after exposure (De Matteis et al. 2018). This is how DNA damage caused by oxidative stress should always be considered when designing and evaluating a theranostic impact platform for patient care (Gonzalez-Hunt et al. 2018).

3.11.2 Immunotoxicity

The interaction between the nanoparticles and the immune system depends on the physical properties of the NMs, as well as the payload and the administration route. In this order of ideas, nanoparticles with polymeric coating have presented immunotoxicity. In addition, it has been shown that products that have more time in the blood circulation and have low clearance in lymphoid organs have less immunotoxicity, which can be a predictive factor when designing new nanotechnological approaches (Elsabahy and Wooley 2015). Also, it is very important to assess the immunotoxicity of graphene family nanomaterials (GFNs) due to toxic effect that produce mediated by oxidative stress (Yan et al. 2017).

3.11.3 Dermal Toxicity

The skin is usually one of the first places where the toxicity of NMs can be observed, where they can cause irreversible pigmentation known as argyria; in this way the use of photodynamic agents should also be evaluated for the possible dermal toxicity that these can cause (Schulte et al. 2018). It is equally necessary to evaluate the carcinogenic potential in skin of NMs, due to the proinflammatory effects that present this kind of compounds (Saifi et al. 2018).

3.11.4 Epigenetic Toxicity

One aspect of concern is the ability of nanomaterials to affect the epigenetic regulation of gene expression; also the heavy metals present in the inorganic nanoparticles are considered as epimutagen, i.e., they alter DNA methylation and histone modification (Smolkova et al. 2017). Finally, the stressors that alter the epigenetics can lead to mutations that lead to intergenerational effects with impact on the following generations; therefore, the use of long-term NMs should be guided to diminish the epigenetic alterations derived from their exposure (Barouki et al. 2018).

3.12 Conclusions and Future Perspectives

Nanotheranostic is one of the most promising tools to alter the deadly course of infectious disease and decrease the onset of resistance. This aspect of nanobiotechnology should not be approached as a single discipline but as the transdisciplinary integration of many areas that will eventually lead to the translation of knowledge from basic research to clinical application.

This is how new approaches in this field should be evaluated for the design and implementation of new devices, as is the case of nanovesicles capable of transporting different classes of molecules and performing functions of cell signaling and antigenic presentation; also this approach can be promising for the development of new nanostructures that will serve for the design of theranostics platforms (He et al. 2018; Yang et al. 2019). Equally, the use of cell membrane-based drug delivery systems as carrier RBCs (red blood cells) will help the design of biocompatible and non-immunogenic platforms with which to implement diagnostic techniques and treatment protocols safer and more effective (Tan et al. 2015). In the case of the elimination of biofilms, the application of platforms formed by milli/microrobots that can reach areas where conventional medicines cannot access to perform a specific therapy is a future tool of promising and fascinating scope (Vikram Singh and Sitti 2016). Other fascinating approach is the use of biomimicry, in which nature is used as a guide and as a mentor in order to develop bioinspired products with which to have more biocompatible platforms (Evangelopoulos et al. 2018; Chen et al. 2019). This aspect is important to be able to make a correct translation of nanomedicine to clinical practice (Lagarce 2015).

In addition, the toxicity and safety assessment platforms for the use of these tools in clinical practice should be considered. The evaluation of the effects of theranostics applications on host autoimmunity/allergy as well as microbiome should be implemented at the same time for the development of these approaches for the diagnosis and treatment of infectious diseases (Wypych and Marsland 2018).

Acknowledgment The author thanks to CF Honeypot for her collaboration and invaluable support during the writing of this chapter.

References

Ansari MO, Ahmad MF, Shadab GGHA, Siddique HR. Superparamagnetic iron oxide nanoparticles based cancer theranostics: a double edge sword to fight against cancer. J Drug Deliv Sci Technol. 2018;45:177–83.

Arduini F, Cinti S, Scognamiglio V, Moscone D. Based electrochemical devices in biomedical field: recent advances and perspectives. Compr Anal Chem. 2017;77:385–413.

Arnold FH. Directed evolution: bringing new chemistry to life. Angew Chem Int Ed. 2018;57:4143–8.

Ashrafuzzaman M. Aptamers as both drugs and drug-carriers. Biomed Res Int. 2014;2014:697923.

Avcıbaşı U, Demiroğlu H, Sakarya S, Tekin V, Ateş B. The effect of radiolabeled antibiotics on biofilm and microorganism within biofilm. J Radioanal Nucl Chem. 2018;316:275–87.

Baptista PV, McCusker MP, Carvalho A, Ferreira DA, Mohan NM, Martins M, Fernandes AR. Nano-strategies to fight multidrug resistant bacteria—"A Battle of the Titans". Front Microbiol. 2018;9:1441.

Barouki R, Melén E, Herceg Z, Beckers J, Chen J, Karagas M, Puga A, Xia Y, Chadwick L, Yan W, Audouze K, Slama R, Heindel J, Grandjean P, Kawamoto T, Nohara K. Epigenetics as a mechanism linking developmental exposures to long-term toxicity. Environ Int. 2018;114:77–86.

Barroso Á, Grüner M, Forbes T, Denz C, Strassert CA. Spatiotemporally resolved tracking of bacterial responses to ROS-mediated damage at the single-cell level with quantitative functional microscopy. ACS Appl Mater Interfaces. 2016;8:15046–57.

Bartlett G, Antoun J, Zgheib NK. Theranostics in primary care: pharmacogenomics tests and beyond. Expert Rev Mol Diagn. 2012;12:841–55.

Beloqui A, Solinís MÁ, Rodríguez-Gascón A, Almeida AJ, Préat V. Nanostructured lipid carriers: promising drug delivery systems for future clinics. Nanomedicine. 2016;12:143–61.

Bueno J. Biosensors in antimicrobial drug discovery: since biology until screening platforms. J Microb Biochem Technol. 2014;S10:2.

Bueno J. The future of metabolomics and individual monitoring in antimicrobial therapy. J Microb Biochem Technol. 2017;9:e132.

Burnham CAD, Leeds J, Nordmann P, O'Grady J, Patel J. Diagnosing antimicrobial resistance. Nat Rev Microbiol. 2017;15:697–703.

Cao B, Xiao F, Xing D, Hu X. Polyprodrug antimicrobials: remarkable membrane damage and concurrent drug release to combat antibiotic resistance of methicillin-resistant *Staphylococcus aureus*. Small. 2018;14:1802008.

Chen F, Hableel G, Zhao ER, Jokerst JV. Multifunctional nanomedicine with silica: role of silica in nanoparticles for theranostic, imaging, and drug monitoring. J Colloid Interface Sci. 2018;521:261–79.

Chen Y, Feng Y, Deveaux JG, Masoud MA, Chandra FS, Chen H, Zhang D, Feng L. Biomineralization forming process and bio-inspired nanomaterials for biomedical application: a review. Fortschr Mineral. 2019;9:68.

Chitgupi U, Qin Y, Lovell JF. Targeted nanomaterials for phototherapy. Nano. 2017;1:38–58.

Col SDL, Brig VKR. Bioterrorism: a public health perspective. Med J Armed Forces India. 2010;66:255–60.

Courbet A, Renard E, Molina F. Bringing next-generation diagnostics to the clinic through synthetic biology. EMBO Mol Med. 2016;8:987–91.

Craciun AM, Focsan M, Magyari K, Vulpoi A, Pap Z. Surface plasmon resonance or biocompatibility—key properties for determining the applicability of noble metal nanoparticles. Materials. 2017;10:836.

Dai X, Fan Z, Lu Y, Ray PC. Multifunctional nanoplatforms for targeted multidrug-resistant-bacteria theranostic applications. ACS Appl Mater Interfaces. 2013;5:11348–54.

De Matteis V, Cascione M, Toma C, Leporatti S. Silver nanoparticles: synthetic routes, in vitro toxicity and theranostic applications for cancer disease. Nano. 2018;8:319.

Drain P, Hyle E, Noubary F, Freedberg K, Wilson D, Bishai W, Rodriguez W, Bassett I. Diagnostic point-of-care tests in resource-limited settings. Lancet Infect Dis. 2014;14:239–49.

Dusinska M, Tulinska J, El Yamani N, Kuricova M, Liskova A, Rollerova E, Rundén-Pran E, Smolkova B. Immunotoxicity, genotoxicity and epigenetic toxicity of nanomaterials: new strategies for toxicity testing? Food Chem Toxicol. 2017;109:797–811.

El Bairi K, Atanasov AG, Amrani M, Afqir S. The arrival of predictive biomarkers for monitoring therapy response to natural compounds in cancer drug discovery. Biomed Pharmacother. 2019;109:2492–8.

Elsabahy M, Wooley KL. Data mining as a guide for the construction of cross-linked nanoparticles with low immunotoxicity via control of polymer chemistry and supramolecular assembly. Acc Chem Res. 2015;48:1620–30.

Evangelopoulos M, Parodi A, Martinez J, Tasciotti E. Trends towards biomimicry in theranostics. Nano. 2018;8:637.

Fadeel B, Farcal L, Hardy B, Vázquez-Campos S, Hristozov D, Marcomini A, Lynch I, Valsami-Jones E, Alenius H, Savolainen K. Advanced tools for the safety assessment of nanomaterials. Nat Nanotechnol. 2018;13:537–43.

Flemming HC, Wingender J, Szewzyk U, Steinberg P, Rice SA, Kjelleberg S. Biofilms: an emergent form of bacterial life. Nat Rev Microbiol. 2016;14:563–75.

Gao T, Zeng H, Xu H, Gao F, Li W, Zhang S, Liu Y, Luo G, Li M, Jiang D, Chen Z, Wu Y, Wang W, Zeng W. Novel self-assembled organic nanoprobe for molecular imaging and treatment of gram-positive bacterial infection. Theranostics. 2018;8:1911–22.

Ge H, Zhang J, Yuan Y, Liu J, Liu R, Liu X. Preparation of organic–inorganic hybrid silica nanoparticles with contact antibacterial properties and their application in UV-curable coatings. Prog Org Coat. 2017;106:20–6.

Goes A, Fuhrmann G. Biogenic and biomimetic carriers as versatile transporters to treat infections. ACS infectious diseases. 2018;4:881–92.

Gomes IB, Meireles A, Gonçalves AL, Goeres DM, Sjollema J, Simões LC, Simões M. Standardized reactors for the study of medical biofilms: a review of the principles and latest modifications. Crit Rev Biotechnol. 2018;38:657–70.

Gonzalez-Delgado JA, Kennedy PJ, Ferreira M, Tome JP, Sarmento B. Use of photosensitizers in semisolid formulations for microbial photodynamic inactivation: miniperspective. J Med Chem. 2015;59:4428–42.

Gonzalez-Hunt C, Wadhwa M, Sanders LH. DNA damage by oxidative stress: measurement strategies for two genomes. Curr Opin Toxicol. 2018;7:87–94.

Grumezescu A, Gesta M, Holban A, Grumezescu V, Vasile B, Mogoanta L, Iordache F, Bleotu C, Dan Mogosanu G. Biocompatible Fe_3O_4 increases the efficacy of amoxicillin delivery against gram-positive and gram-negative bacteria. Molecules. 2014;19:5013–27.

Gupta S, Ramesh K, Ahmed S, Kakkar V. Lab-on-Chip Technology: a review on design trends and future scope in biomedical applications. Int J Bio Sci Bio Technol. 2016;8:311–22.

Gustafsson OJR, Guinan TM, Rudd D, Kobus H, Benkendorff K, Voelcker NH. Metabolite mapping by consecutive nanostructure and silver-assisted mass spectrometry imaging on tissue sections. Rapid Commun Mass Spectrom. 2017;31:991–1000.

Hamblin MR. Fullerenes as photosensitizers in photodynamic therapy: pros and cons. Photochem Photobiol Sci. 2018;17:1515–33.

Hauser M, Li G, Nowack B. Environmental hazard assessment for polymeric and inorganic nanobiomaterials used in drug delivery. J Nanobiotechnol. 2019;17:56.

He C, Zheng S, Luo Y, Wang B. Exosome theranostics: biology and translational medicine. Theranostics. 2018;8:237–55.

He X, McLean J, Guo L, Lux R, Shi W. The social structure of microbial community involved in colonization resistance. ISME J. 2014;8:564–74.

Hemeg HA. Nanomaterials for alternative antibacterial therapy. Int J Nanomedicine. 2017;12:8211–25.

Hillger JM, Lieuw WL, Heitman LH, IJzerman AP. Label-free technology and patient cells: from early drug development to precision medicine. Drug Discov Today. 2017;22:1808–15.

Huang CJ, Chu SH, Wang LC, Li CH, Lee TR. Bioinspired zwitterionic surface coatings with robust photostability and fouling resistance. ACS Appl Mater Interfaces. 2015;7:23776–86.

Huma ZE, Gupta A, Javed I, Das R, Hussain SZ, Mumtaz S, Hussain I, Rotello VM. Cationic silver nanoclusters as potent antimicrobials against multidrug-resistant bacteria. ACS Omega. 2018;3:16721–7.

Inoue H, Minghui R. Antimicrobial resistance: translating political commitment into national action. Bull World Health Organ. 2017;95:242–242A.

Iriya R, Syal K, Jing W, Mo M, Yu H, Haydel SE, Wang S, Tao N. Real-time detection of antibiotic activity by measuring nanometer-scale bacterial deformation. J Biomed Opt. 2017;22:126002.

Jagtap P, Sritharan V, Gupta S. Nanotheranostic approaches for management of bloodstream bacterial infections. Nanomedicine. 2017;13:329–41.

Kavanagh ON, Albadarin AB, Croker DM, Healy AM, Walker GM. Maximising success in multidrug formulation development: a review. J Control Release. 2018;283:1–19.

Kennedy DA, Read AF. Why does drug resistance readily evolve but vaccine resistance does not? Proc R Soc B Biol Sci. 2017;284:20162562.

Kevadiya BD, Ottemann BM, Thomas MB, Mukadam I, Nigam S, McMillan J, Goranthia S, Bronich T, Gendelman HE. Neurotheranostics as personalized medicines. Adv Drug Deliv Rev. 2018; S0169-409X(18): 30261–8

Khalid N, Kobayashi I, Nakajima M. Recent lab-on-chip developments for novel drug discovery. Wiley Interdiscip Rev Syst Biol Med. 2017;9:e1381.

Kratochvil MJ, Yang T, Blackwell HE, Lynn DM. Nonwoven polymer nanofiber coatings that inhibit quorum sensing in Staphylococcus aureus: toward new nonbactericidal approaches to infection control. ACS Infect Dis. 2017;3:271–80.

Krausz A, Adler B, Cabral V, Navati M, Doerner J, Charafeddine R, Chandra D, Liang H, Gunther L, Clendaniel A, Harper S, Friedman J, Nosanchuk J, Friedman A. Curcumin-encapsulated nanoparticles as innovative antimicrobial and wound healing agent. Nanomedicine. 2015;11:195–206.

Lagarce F. Nanomedicines: are we lost in translation? Eur J Nanomed. 2015;7:77–8.

Lai J, Shah BP, Zhang Y, Yang L, Lee KB. Real-time monitoring of ATP-responsive drug release using mesoporous-silica-coated multicolor upconversion nanoparticles. ACS Nano. 2015;9:5234–45.

Larrañeta E, McCrudden MT, Courtenay AJ, Donnelly RF. Microneedles: a new frontier in nanomedicine delivery. Pharm Res. 2016;33:1055–73.

Lee S, Lin M, Lee A, Park Y. Lanthanide-doped nanoparticles for diagnostic sensing. Nano. 2017;7:411.

Li Q, Wu Y, Lu H, Wu X, Chen S, Song N, Yang Y, Gao H. Construction of supramolecular nanoassembly for responsive bacterial elimination and effective bacterial detection. ACS Appl Mater Interfaces. 2017;9:10180–9.

Liao JF, Lee JC, Lin CK, Wei KC, Chen PY, Yang HW. Self-assembly DNA polyplex vaccine inside dissolving microneedles for high-potency intradermal vaccination. Theranostics. 2017;7(10):2593–605.

Linko V, Ora A, Kostiainen MA. DNA nanostructures as smart drug-delivery vehicles and molecular devices. Trends Biotechnol. 2015;33:586–94.

Lombardo D, Kiselev MA, Caccamo MT. Smart nanoparticles for drug delivery application: development of versatile nanocarrier platforms in biotechnology and nanomedicine. J Nanomater. 2019;2019:3702518.

Lu R, Zou W, Du H, Wang J, Zhang S. Antimicrobial activity of Ag nanoclusters encapsulated in porous silica nanospheres. Ceram Int. 2014;40:3693–8.

Lundquist CM, Loo C, Meraz IM, Cerda JDL, Liu X, Serda RE. Characterization of free and porous silicon-encapsulated superparamagnetic iron oxide nanoparticles as platforms for the development of theranostic vaccines. Med Sci (Basel). 2014;2:51–69.

Ma W, Cheetham AG, Cui H. Building nanostructures with drugs. Nano Today. 2016;11:13–30.

Madni A, Noreen S, Maqbool I, Rehman F, Batool A, Kashif PM, Rehman M, Tahir N, Khan MI. Graphene-based nanocomposites: synthesis and their theranostic applications. J Drug Target. 2018;26:858–83.

Magana M, Sereti C, Ioannidis A, Mitchell CA, Ball AR, Magiorkinis E, Chatzipanagiotou S, Hamblin MR, Hadjifrangiskou M, Tegos GP. Options and limitations in clinical investigation of bacterial biofilms. Clin Microbiol Rev. 2018;31:e00084-16.

Martínez-Carmona M, Gun'ko Y, Vallet-Regí M. ZnO nanostructures for drug delivery and theranostic applications. Nano. 2018;8:268.

Mauk M, Song J, Liu C, Bau H. Simple approaches to minimally-instrumented, microfluidic-based point-of-care nucleic acid amplification tests. Biosensors. 2018;8:17.

Meeker DG, Wang T, Harrington WN, Zharov VP, Johnson SA, Jenkins SV, Oyibo SE, Walker CM, Mills WB, Shirtliff ME, Beenken KE, Chen J, Smeltzer MS. Versatility of targeted antibiotic-loaded gold nanoconstructs for the treatment of biofilm-associated bacterial infections. Int J Hyperth. 2018;34:209–19.

Mirahmadi-Zare SZ, Allafchian AR, Jalali SAH. Core–shell fabrication of an extra-antimicrobial magnetic agent with synergistic effect of substrate ligand to increase the antimicrobial activity of Ag nanoclusters. Environ Prog Sustain Energy. 2019;38:237–45.

Mocan T, Matea CT, Pop T, Mosteanu O, Buzoianu AD, Puia C, Iancu C, Mocan L. Development of nanoparticle-based optical sensors for pathogenic bacterial detection. J Nanobiotechnol. 2017;15:25.

Moffat JG, Vincent F, Lee JA, Eder J, Prunotto M. Opportunities and challenges in phenotypic drug discovery: an industry perspective. Nat Rev Drug Discov. 2017;16:531.

Molefe P, Masamba P, Oyinloye B, Mbatha L, Meyer M, Kappo A. Molecular application of aptamers in the diagnosis and treatment of cancer and communicable diseases. Pharmaceuticals. 2018;11:93.

Muzammil S, Hayat S, Fakhar-E-Alam M, Aslam B, Siddique MH, Nisar MA, Saqalein M, Atif M, Sarwar A, Khurshid A, Amin N, Wang Z. Nanoantibiotics: future nanotechnologies to combat antibiotic resistance. Front Biosci (Elite Ed). 2018;10:352–74.

Narayan R, Nayak U, Raichur A, Garg S. Mesoporous silica nanoparticles: a comprehensive review on synthesis and recent advances. Pharmaceutics. 2018;10:118.

Nasseri B, Soleimani N, Rabiee N, Kalbasi A, Karimi M, Hamblin MR. Point-of-care microfluidic devices for pathogen detection. Biosens Bioelectron. 2018;117:112–28.

Neburkova J, Vavra J, Cigler P. Coating nanodiamonds with biocompatible shells for applications in biology and medicine. Curr Opinion Solid State Mater Sci. 2017;21:43–53.

Nine MJ, Cole MA, Tran DN, Losic D. Graphene: a multipurpose material for protective coatings. J Mater Chem A. 2015;3:12580–602.

Nowlin K, LaJeunesse DR. Fabrication of hierarchical biomimetic polymeric nanostructured surfaces. Mol Syst Design Eng. 2017;2:201–13.

Pang T. Theranostics, the 21st century bioeconomy and 'one health'. Expert Rev Mol Diagn. 2012;12:807–9.

Pedrosa P, Vinhas R, Fernandes A, Baptista PV. Gold nanotheranostics: proof-of-concept or clinical tool? Nanomaterials. 2015;5:1853–79.

Percival SL, Suleman L, Vuotto C, Donelli G. Healthcare-associated infections, medical devices and biofilms: risk, tolerance and control. J Med Microbiol. 2015;64:323–34.

Pirmohamed M. Personalized pharmacogenomics: predicting efficacy and adverse drug reactions. Annu Rev Genomics Hum Genet. 2014;15:349–70.

Pramanik A, Jones S, Pedraza F, Vangara A, Sweet C, Williams M, Ruppa-Kasani V, Risher S, Sardar D, Ray P. Fluorescent, magnetic multifunctional carbon dots for selective separation, identification, and eradication of drug-resistant superbugs. ACS Omega. 2017;2:554–62.

Pranantyo D, Xu LQ, Kang ET, Chan-Park MB. Chitosan-based peptidopolysaccharides as cationic antimicrobial agents and antibacterial coatings. Biomacromolecules. 2018;19:2156–65.

Primiceri E, Chiriacò MS, Notarangelo FM, Crocamo A, Ardissino D, Cereda M, Bramanti AP, Bianchessi MA, Giannelli G, Maruccio G. Key enabling technologies for point-of-care diagnostics. Sensors. 2018;18:3607.

Qu W, Li N, Yu R, Zuo W, Fu T, Fei W, Hou Y, Liu Y, Yang J. Cationic DDA/TDB liposome as a mucosal vaccine adjuvant for uptake by dendritic cells in vitro induces potent humoural immunity. Artif Cells Nanomed Biotechnol. 2018;46:852–60.

Ramasamy M, Lee J. Recent nanotechnology approaches for prevention and treatment of biofilm-associated infections on medical devices. Biomed Res Int. 2016;2016:1851242.

Rello J, van Engelen TSR, Alp E, Calandra T, Cattoir V, Kern WV, Netea MG, Nseir S, Opal SM, van de Veerdonk FL, Wilcox MH, Wiersinga WJ. Towards precision medicine in sepsis: a position paper from the European Society of Clinical Microbiology and Infectious Diseases. Clin Microbiol Infect. 2018;24:1264–72.

Ribeiro SM, Felício MR, Boas EV, Gonçalves S, Costa FF, Samy RP, Santos NC, Franco OL. New frontiers for anti-biofilm drug development. Pharmacol Ther. 2016;160:133–44.

Roco M, Mirkin C, Hersam M. Nanotechnology research directions for societal needs in 2020: summary of international study. J Nanopart Res. 2011;13:897–919.

Ruiz SI, El-Gendy N, Bowen LE, Berkland C, Bailey MM. Formulation and characterization of nanocluster ceftazidime for the treatment of acute pulmonary melioidosis. J Pharm Sci. 2016;105:3399–408.

Sabhachandani P, Sarkar S, Zucchi PC, Whitfield BA, Kirby JE, Hirsch EB, Konry T. Integrated microfluidic platform for rapid antimicrobial susceptibility testing and bacterial growth analysis using bead-based biosensor via fluorescence imaging. Microchim Acta. 2017;184:4619–28.

Sahlgren C, Meinander A, Zhang H, Cheng F, Preis M, Xu C, Salminen TA, Toivola D, Abankwa D, Rosling A, Karaman DŞ, Salo-Ahen OMH, Österbacka R, Eriksson JE, Willför S, Petre I, Peltonen J, Leino R, Johnson M, Rosenholm J, Sandler N. Tailored approaches in drug development and diagnostics: from molecular design to biological model systems. Adv Healthc Mater. 2017;6:1–34.

Saifi MA, Khan W, Godugu C. Cytotoxicity of nanomaterials: using nanotoxicology to address the safety concerns of nanoparticles. Pharmaceut Nanotechnol. 2018;6:3–16.

Schulte PA, Kuempel ED, Drew NM. Characterizing risk assessments for the development of occupational exposure limits for engineered nanomaterials. Regul Toxicol Pharmacol. 2018;95:207–19.

Şen Karaman D, Manner S, Rosenholm JM. Mesoporous silica nanoparticles as diagnostic and therapeutic tools: how can they combat bacterial infection? Ther Deliv. 2018;9:241–4.

Setyawati MI, Kutty RV, Tay CY, Yuan X, Xie J, Leong DT. Novel theranostic DNA nanoscaffolds for the simultaneous detection and killing of Escherichia coli and Staphylococcus aureus. ACS Appl Mater Interfaces. 2014;6:21822–31.

Shi X, Zhang C Y, Gao J, Wang Z. Recent advances in photodynamic therapy for cancer and infectious diseases. Wiley Interdiscip Rev Nanomed Nanobiotechnol. 2019:11(5):e1560.

Siest G, Schallmeiner E. Pharmacogenomics and theranostics in practice: a summary of the Euromedlab-ESPT (The European Society of Pharmacogenomics and Theranostics) satellite symposium, May 2013. EJIFCC. 2014;24:85.

Silva AF, Borges A, Giaouris E, Graton Mikcha JM, Simões M. Photodynamic inactivation as an emergent strategy against foodborne pathogenic bacteria in planktonic and sessile states. Crit Rev Microbiol. 2018;44:667–84.

Singh AV, Gemmate D, Kanase A, Pandey I, Misra V, Kishore V, Jahnke T, Bill J. Nanobiomaterials for vascular biology and wound management: a review. Veins Lymphat. 2018;7:7196.

Smeltzer MS, Zharov V, Galanzha E, Chen J, Meeker D, Beenken K. U.S. Patent Application No. 14/728,849. 2015.

Smolkova B, Dusinska M, Gabelova A. Nanomedicine and epigenome. Possible health risks. Food Chem Toxicol. 2017;109:780–96.

Syal K, Mo M, Yu H, Iriya R, Jing W, Guodong S, Wang S, Grys TE, Haydel SE, Tao N. Current and emerging techniques for antibiotic susceptibility tests. Theranostics. 2017;7:1795–805.

Tan S, Wu T, Zhang D, Zhang Z. Cell or cell membrane-based drug delivery systems. Theranostics. 2015;5:863–81.

Taresco V, Francolini I, Padella F, Bellusci M, Boni A, Innocenti C, Martinelli A, D'Ilario L, Piozzi A. Design and characterization of antimicrobial usnic acid loaded-core/shell magnetic nanoparticles. Mater Sci Eng C. 2015;52:72–81.

Tonga GY, Moyano DF, Kim CS, Rotello VM. Inorganic nanoparticles for therapeutic delivery: trials, tribulations and promise. Curr Opin Colloid Interface Sci. 2014;19:49–55.

Tournier JN, Peyrefitte CN, Biot F, Merens A, Simon F. The threat of bioterrorism. Lancet Infect Dis. 2019;19:18–9.

Trandafilović LV, Whiffen RK, Dimitrijević-Branković S, Stoiljković M, Luyt AS, Djoković V. ZnO/Ag hybrid nanocubes in alginate biopolymer: synthesis and properties. Chem Eng J. 2014;253:341–9.

Trivedi U, Madsen JS, Rumbaugh KP, Wolcott RD, Burmølle M, Sørensen SJ. A post-planktonic era of *in vitro* infectious models: issues and changes addressed by a clinically relevant wound like media. Crit Rev Microbiol. 2017;43:453–65.

Vangara A, Pramanik A, Gao Y, Gates K, Begum S, Chandra Ray P. Fluorescence resonance energy transfer based highly efficient theranostic nanoplatform for two-photon bioimaging and two-photon excited photodynamic therapy of multiple drug resistance bacteria. ACS Appl Bio Mater. 2018;1:298–309.

Ventola CL. Pharmacogenomics in clinical practice: reality and expectations. Pharm Therapeut. 2011;36:412–50.

Viana SM, Celes FS, Ramirez L, Kolli B, Ng DK, Chang KP, De Oliveira CI. Photodynamic vaccination of BALB/c mice for prophylaxis of cutaneous leishmaniasis caused by *Leishmania amazonensis*. Front Microbiol. 2018;9:165.

Vicini P, Fields O, Lai E, Litwack ED, Martin AM, Morgan TM, Pacanowski MA, Papaluca M, Perez OD, Ringel MS, Robson M, Sakul H, Vockley J, Zaks T, Dolsten M, Søgaard M. Precision medicine in the age of big data: the present and future role of large-scale unbiased sequencing in drug discovery and development. Clin Pharmacol Therapeut. 2016;99:198–207.

Vikram Singh A, Sitti M. Targeted drug delivery and imaging using mobile milli/microrobots: a promising future towards theranostic pharmaceutical design. Curr Pharm Des. 2016;22:1418–28.

Vitiello G, Silvestri B, Luciani G. Learning from nature: bioinspired strategies towards antimicrobial nanostructured systems. Curr Top Med Chem. 2018;18:22–41.

Wainwright M, Maisch T, Nonell S, Plaetzer K, Almeida A, Tegos GP, Hamblin MR. Photoantimicrobials—are we afraid of the light? Lancet Infect Dis. 2017;17:e49–55.

Wang Y, Cheetham AG, Angacian G, Su H, Xie L, Cui H. Peptide–drug conjugates as effective prodrug strategies for targeted delivery. Adv Drug Deliv Rev. 2017;110:112–26.

Wang Y, Jin Y, Chen W, Wang J, Chen H, Sun L, Li X, Ji J, Yu Q, Shen L, Wang B. Construction of nanomaterials with targeting phototherapy properties to inhibit resistant bacteria and biofilm infections. Chem Eng J. 2019;358:74–90.

Wiesner MR, Bottero JY. A risk forecasting process for nanostructured materials, and nanomanufacturing. Comptes Rendus Physique. 2011;12:659–68.

Wolfmeier H, Pletzer D, Mansour SC, Hancock RE. New perspectives in biofilm eradication. ACS Infect Dis. 2017;4:93–106.

Wong OA, Hansen RJ, Ni TW, Heinecke CL, Compel WS, Gustafson DL, Ackerson CJ. Structure–activity relationships for biodistribution, pharmacokinetics, and excretion of atomically precise nanoclusters in a murine model. Nanoscale. 2013;5:10525–33.

Wypych TP, Marsland BJ. Antibiotics as instigators of microbial dysbiosis: implications for asthma and allergy. Trends Immunol. 2018;39:697–711.

Xie S, Manuguri S, Proietti G, Romson J, Fu Y, Inge AK, Wu B, Zhang Y, Häll D, Ramström O, Yan M. Design and synthesis of theranostic antibiotic nanodrugs that display enhanced antibacterial activity and luminescence. Proc Natl Acad Sci. 2017;114:8464–9.

Xu D, Wang Q, Yang T, Cao J, Lin Q, Yuan Z, Li L. Polyethyleneimine capped silver nanoclusters as efficient antibacterial agents. Int J Environ Res Public Health. 2016;13:334.

Yan J, Chen L, Huang CC, Lung SC, Yang L, Wang WC, Lin PH, Suo G, Lin CH. Consecutive evaluation of graphene oxide and reduced graphene oxide nanoplatelets immunotoxicity on monocytes. Colloids Surf B: Biointerfaces. 2017;153:300–9.

Yan Y, Wang X, Lou P, Hu Z, Qu P, Li D, Li Q, Xu Y, Niu J, He Y, Zhong J, Huang Z. A nanoparticle-based HCV vaccine with enhanced potency. J Infect Dis. 2019;pii:jiz228

Yang B, Chen Y, Shi J. Exosome biochemistry and advanced nanotechnology for next-generation theranostic platforms. Adv Mater. 2019;31:1802896.

Yao J, Li P, Li L, Yang M. Biochemistry and biomedicine of quantum dots: from biodetection to bioimaging, drug discovery, diagnosis, and therapy. Acta Biomater. 2018;74:36–55.

Yoo SM, Lee SY. Optical biosensors for the detection of pathogenic microorganisms. Trends Biotechnol. 2016;34:7–25.

Yuan X, Setyawati MI, Leong DT, Xie J. Ultrasmall Ag+-rich nanoclusters as highly efficient nanoreservoirs for bacterial killing. Nano Res. 2014;7:301–7.

Zazo H, Colino CI, Lanao JM. Current applications of nanoparticles in infectious diseases. J Control Release. 2016;224:86–102.

Zhang L, Wan S, Jiang Y, Wang Y, Fu T, Liu Q, Cao Z, Qiu L, Tan W. Molecular elucidation of disease biomarkers at the interface of chemistry and biology. J Am Chem Soc. 2017;139:2532–40.
Zhao L, Seth A, Wibowo N, Zhao CX, Mitter N, Yu C, Middelberg AP. Nanoparticle vaccines. Vaccine. 2014;32:327–37.
Zheng K, Setyawati MI, Lim TP, Leong DT, Xie J. Antimicrobial cluster bombs: silver nanoclusters packed with daptomycin. ACS Nano. 2016;10:7934–42.

Chapter 4
Theranostic Nanoplatforms as a Promising Diagnostic and Therapeutic Tool for *Staphylococcus aureus*

Bushra Uzair, Anum Shaukat, and Safa Mariyam

Abstract The Gram-positive microorganism *Staphylococcus aureus* is the major infective agent responsible for life-threatening infections. *Staphylococcus aureus* have developed resistance against last resort therapeutics. Nonetheless, there is a great urge to synthesize novel antimicrobial agents with proficient activity and great biocompatibility for clinical administrations. Advanced technologies emphasize more on detecting and identifying diseases for successful therapy combined with the diagnostic agents itself. Theranostics is a novel approach in which the exact treatment is conjugated with specific diagnostic tests. With a key spotlight on patient-focused consideration, theranostics alter traditional prescription to a contemporary personalized and accurate medicine approach. The research has been supported toward the development of enhanced imaging agents as carbon quantum dots, nanoparticles, and other new theranostic drugs. Better diagnosis helps in prescreening the profile of target particles to create biomarkers dependent on disease-specific therapy, thus making it less expensive with less off-target toxicity and high efficiency and specificity with continuous analysis for detailed observation and guidance and examining impacts and side effects to develop further alternatives. In this chapter, numerous theranostics methods are discussed to cure ailments caused by *Staphylococcus aureus*.

Keywords *Staphylococcus aureus* · Theranostics · Bacterial imaging · Antibacterial activity · Nanotechnology · Bacterial infections

B. Uzair (✉) · A. Shaukat · S. Mariyam
Department of Biological Sciences, International Islamic University, Islamabad, Pakistan
e-mail: bushra.uzair@iiu.edu.pk

M. Rai, B. Jamil (eds.), *Nanotheranostics*,
https://doi.org/10.1007/978-3-030-29768-8_4

4.1 Introduction

Staphylococcus aureus (*S. aureus*) is a very diverse and harmful infectious agent for humans that is responsible for a variety of clinical manifestations (Corey 2009). It *is* a circular, nonsporulating, immotile bacterium. When it is observed beneath microscope, it appears in sets, grape-like clusters, and in short chains. These facultative aerobic and anaerobic microscopic organisms grow at the temperature range between 18 °C and 40 °C. On growth medium, the colonies of *S. aureus* are often yellowish or golden (Taylor and Unakal 2019). Various biochemical tests are performed to identify these organisms. All pathogenic species of *Staphylococcus* are catalase positive. Coagulase-positive test is used to discriminate between *Staphylococcus aureus* and other *Staphylococcus* strains. Whereas, novobiocin-sensitive test and mannitol fermentation-positive test are used to differentiate it from *Staphylococcus saprophyticus* and *Staphylococcus epidermidis* (Tong et al. 2019). Staphylococci are omnipresent and can be found in dust, sewage, water, biological surfaces, air, and different animals. It is the normal skin flora and mostly present at the nasal area of healthy persons.

S. aureus including drug-resistant strains are skin colonizer and humans are the main reservoirs for these microorganisms. Human population is colonized around 30% with *S. aureus* (Hennekinne et al. 2012). Almost 15% of human population carry these organisms in the anterior nares. However, immunocompromised individuals, diabetic patients, and health care workers have high risk of *S. aureus* colonization that is up to 80%. Chronic carriage is related with an increased possibility of disease. Nasal carriage has contributed to the spread of methicillin-resistant *S. aureus* strains. It can be passed on from individual to individual by fomites or through direct contact (Rasigade and Vandenesch 2014).

S. aureus causes broad range of skin diseases. It may also cause pneumonia, bloodstream infections, and bone and joints diseases (Bilung et al. 2018). Based on the strains involved and site of the infection, these organisms can cause toxin-mediated diseases and invasive infections. *Staphylococcus aureus* possesses an extensive array of virulence factors including capsule, biofilm, techoic and lipotechoic acids, hemolysins, exfoliative toxins, pathogenicity islands, biofilm, and superantigens (Turner et al. 2019).

The rates of infections brought by staphylococci, both hospital-acquired and community strains, are expanding consistently. Treatment of these diseases is becoming progressively troublesome in view of the expanding predominance of multidrug-resistant strains (Boswihi and Udo 2018). The need for new antimicrobial therapeutics is becoming increasingly urgent. Theranostics have emerged as a novel platform in drug improvement and development.

Based on the above considerations, in this chapter we discuss some multifunctional bacterial theranostic systems with proficient activity and good biocompatibility for clinical managements.

4.2 Infections Caused by *Staphylococcus aureus*

S. aureus generates a variety of life-threatening diseases as follows:

4.2.1 Bacteremia

Bacteremia caused by *S. aureus* is called as S. aureus bacteremia (SAB). It can also lead to infective endocarditis. Age is a dominant element of SAB, with the most hoisted rates of infections occurring at either extreme of life (Tong et al. 2019). The population which is infected with HIV and hemodialysis patients have a significantly high rate of SAB. The transcendent hazard factors responsible for these infections are intravascular devices and explicitly the usage of tunneled and cuffed catheter for the purpose of dialysis (Fitzgerald et al. 2011). Notwithstanding among other host factors are neutrophil impairment, iron over-burden, diabetes, and extended rates of colonization that may enhance the probability of intrusive *S. aureus* infections (Boelaert et al. 1990; Vanholder et al. 1991; Zimakoff et al. 1996).

4.2.2 Skin and Soft Tissue Infections

S. aureus is the eminent infectious agent that causes a huge range of life-threatening skin diseases like cutaneous abscesses, cellulitis, impetigo and also isolated from infections at surgical sites (Bangert et al. 2012; Tong et al. 2019).

The neutrophil response is major protection against the infections caused by *S. aureus* pathogen. When *S. aureus* infuses into the skin, both the macrophages and neutrophils move to the disease site. *S. aureus* escapes from the neutrophil reaction in a substantial number of ways; it blocks the chemotaxis of leukocytes, sequestering host antibodies and avoids the recognition through polysaccharide capsules and apposes demolition after absorption by phagocytosis. Numerous harmful elements seem to contribute, like Panton-Valentine leukocidin (PVL) that causes breakdown of white blood cells (WBCs) of humans. Secondly, alpha-hemolysin, also known as alpha-poison, that produces pores in several cells of humans. Thirdly, it prompts cell lysis, and the presence of phenol-dissolvable modulins (PSMs) in *S. aureus* also aid in the breakdown of human neutrophils and erythrocytes (Peschel and Otto 2013).

Staphylococci scalded skin syndrome (SSS) is also called Ritter von Ritterschein infection. It is common in youngsters, however uncommon in grownups (Spaulding et al. 2013).

4.2.3 *Osteoarticular Infections*

Osteoarticular infections are the infections of bones. There are three noteworthy classes of osteoarticular disease, osteomyelitis (OM), local septic joint pain, and prosthetic joint disease. In each of these classes *S. aureus* is the most well-known pathogen. Staphylococcal osteoarticular diseases are more common in kids and have critical clinical and board issues as compared to elders. Osteomyelitis is a bone disease that prompts incendiary annihilation, bone putrefaction, and the development of new bones (Westberg et al. 2012). Waldvogel portrays three kinds of OM: Hematogenous OM, adjoining center OM, and OM with vascular inadequacy. *S. aureus* is the transcendent reason for OM in these classifications and is identified in 30 to 60% of cases (Waldvogel et al. 1970).

4.2.4 *Pulmonary Infections*

Pneumonia is a typical disease that may be caused by *Staphylococcus aureus* and it is conceivably hazardous. The presence of methicillin-resistant *S. aureus* (MRSA) in almost 50% of the samples has made the situation gruesome and the management of staphylococcal pneumonia has become very difficult (Ragle et al. 2010).

4.2.5 *Food Poisoning*

Food that is contaminated with viruses, parasite, bacteria, or their toxins may cause food poisoning or foodborne illness (Argudín et al. 2010). Staph food poisoning is caused by eating food that is contaminated with enterotoxins generated by *S. aureus.* Staph is considered to be a normal skin flora, and food contamination is brought about by direct contact of the colonized individual with food-processing units. Contaminated food provides the nutrition for bacterial multiplication and toxins are produced. These toxins actually cause the disease. Bacteria can be destroyed during cooking procedure but toxins are usually heat resistant and persist there. Food contaminated with these toxins does not have any smell or appear spoiled (Boswihi and Udo 2018).

4.3 Methicillin-Resistant *S. aureus* (MRSA)

The first β-lactam drug, penicillin was used successfully for decades to cure ailments caused by *S. aureus* (Lowy 1998). However, *S. aureus* penicillin-resistant strains emerged and invalidated it (Shanson 1981). The enzyme penicillinase that is generated by the bacteria is the main cause of acquired resistance against penicillin because the enzyme inactivates the drug.

The quest for antimicrobial agents which are active against *S. aureus* penicillin-resistant isolates has led to the production of methicillin. Methicillin is a penicillin derivative formed in the late 1950s by an adjustment in the structure of penicillin. Similar to penicillin, methicillin prevents the bacterial cell wall synthesis and kills bacteria.

The rise in methicillin-resistant *S. aureus* (MRSA) strains in the UK during the 1960s made the drug clinically ineffective (Jevons 1961). The acquisition of mecA or mecC gene by previously sensitive isolates results in methicillin resistance (García-Álvarez et al. 2011). mecA gene is present on Staphylococcal cassette chromosome which is the mobile genetic element. mecA encodes for penicillin binding protein 2A. It has very low affinity for methicillin and penicillin.

mecC gene is mecA's homolog. MRSA isolates harboring mecC have been isolated from both human and animals. The discovery of mecC in the research lab may be tricky. mecA-MRSA shows resistance to both cefoxitin and oxacillin, while mecC-MRSA only expresses resistance to cefoxitin (Paterson et al. 2014).

4.4 Resistance Mechanism in *S. aureus*

It is evident that *Staph aureus* like other microorganisms develops resistance after exposure to antibiotics. Various mechanisms have been reported for acquisition of resistance, for instance, activation of antimicrobial efflux, bacterial enzymes that breakdown the drug, reduced drug penetration inside the bacteria, and development of biofilm as shown in Fig. 4.1 (Khameneh et al. 2016).

4.5 Recent Methods Used for Bacterial Identification and Treatment

The in vivo identification and cure of ailments caused by drug-resistant pathogens is not facile. Many diverse kind of antimicrobials such as silver nanoparticles, cationic polymers, and antimicrobial peptides are used for this purpose (Marr et al. 2006; Bazzaz et al. 2014; Wang et al. 2015). Their advantages and disadvantages are described in Table 4.1 (Mukherjee et al. 2014; Dai et al. 2016; Cao et al. 2017; Chen et al. 2017).

Emerging antimicrobial resistance is a serious issue to deal with, but it is also very vital to develop novel techniques for high-pitched bacterial identification (Allegranzi et al. 2011). The recent methods used for microbial identification along with their drawbacks are described in Table 4.2 (Reller et al. 2007; Oethinger et al. 2011; Cheng et al. 2016; Gao et al. 2018).

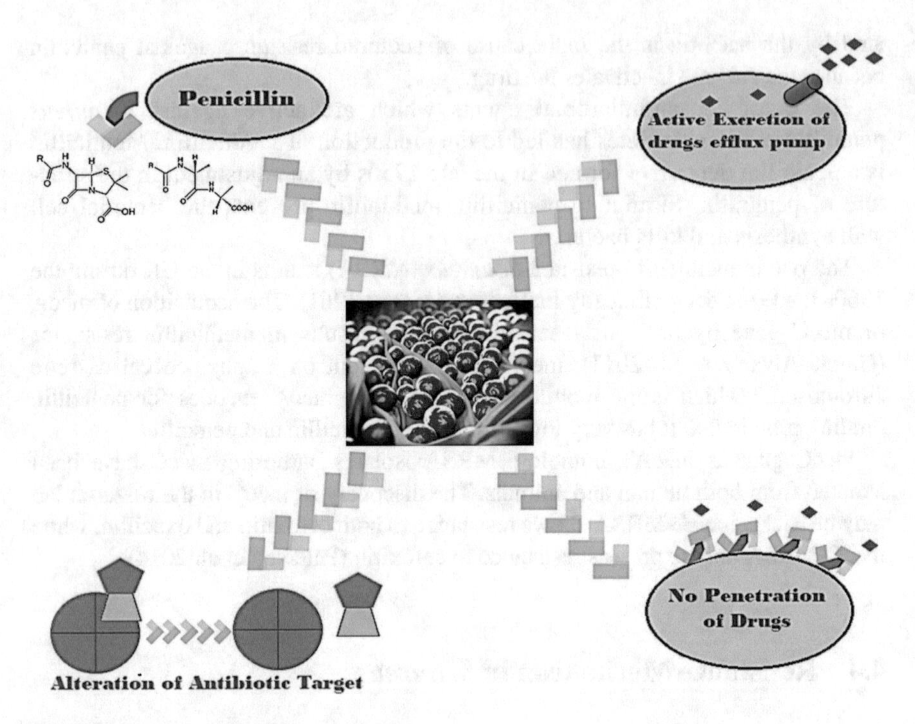

Fig. 4.1 Antimicrobial-resistant mechanism

Table 4.1 Advantages and disadvantages of antimicrobial materials

Antimicrobial agents	Merits	Demerits
Cationic polymer	• Broad-spectrum antibacterial activity • Easy surface functionalization	• Cytotoxic to human cells and cause hemolysis
Antimicrobial peptides	• Very efficient	• High cost • Limited photolytic stability • Poorly studied toxicology • Poor pharmacokinetics
Silver nanoparticles	• Increased antibacterial activity against multidrug-resistant (MDR) microbes	• Toxic to humans, causes spasms and gastrointestinal diseases • Death • Potential immunotoxicity

S. aureus and MRSA are among the most diverse infective agents to cause infections. The hazard of these life-threatening ailments can be significantly diminished if they can be delicately analyzed in vivo and treated at the beginning (Zhao et al. 2017). Therefore, there is a great urge to design new antimicrobial materials with proficient activity and good biocompatibility for patient managements.

Table 4.2 Methods used for bacterial identification

Methods	Merits	Demerits
Gram staining	Distinguish microbes into • Gram-positive • Gram-negative	• Susceptible to create false positive results
• Enzyme-linked immunosorbent assay (ELISA) • Polymerase chain reaction (PCR)	• Increased sensitivity • Increased reproducibility to detect microbes	• Time consuming • Very lengthy and difficult procedures
Fluorescence assay Antibiotics (vancomycin and daptomycin)-modified fluorophores nanoparticles	• Detection of specific microbes	• MDR cases are not detected so generating false negative results • High level doses of antibiotics may indorse antimicrobial resistance

4.6 Theranostics with Dual Function of Detection and Treatment

Theranostics is a novel approach in which the treatment is conjugated with specific diagnostic tests. With a key spotlight on patient-focused consideration, theranostics alter traditional prescription to a contemporary personalized and accurate medicine approach. The theranostics utilize nanotechnology to combine both identification and therapeutic methods to form a single agent, thus permitting diagnosis, delivery of drug, and monitoring of treatment response simultaneously. Various theranostic techniques to combat infections caused by *Staphylococcus aureus* and MRSA are mentioned below.

4.6.1 Theranostic Nanoprobes

NPs have incredible antimicrobial potential and can also identify pathogens if specific probes are attached because of an extensive surface-to-volume proportion and versatile surface chemistry (Li et al. 2012; Cavalieri et al. 2014; Natan and Banin 2017). As compared to conventional drugs, nano-antimicrobial compounds are less inclined to develop antimicrobial resistance.

NPs have the potential of self-assembly. Molecular self-assembly is basically the combination of unconstrained mix of individual segments into very much arranged structures that are maintained by non-covalent cooperations, including electrostatic connection, hydrogen bonding, association between β and β, collaborations between charge and exchange, and hydrophobic impacts (Bhattacharya and Samanta 2016). Self-assembled small molecules have gotten extensive consideration because of their biocompatibility, decent variety and adaptability in molecular structure and photostability among the announced self-assembled nanomaterials (Zhao et al. 2017; Gao et al. 2018).

4.6.1.1 Self-Assembled TPIP-FONs Nanoprobes

Fluorescent organic nanoparticles have phenomenal antimicrobial ability and diminished bacterial burden in diseased sites. TPIP is made by combining the core of tetraphenyl imidazole with the quaternary ammonium group for rapid bacterial recognition and cure of infections. TPIP contains three parts: (1) the substituted imidazole is utilized as AIEgen with antimicrobial action; (2) the alkyl chain can change the spatial arrangement of the positively charged particles and lessen the harmful effects of cationic atoms; (3) the pyridinium salt group which acts as a hydrophilic terminal has an antibacterial effect. The self-gathering of TPIP was explored by utilizing transmission electron microscopy and dynamic light scattering in aqueous phase. The laser confocal microscope was used to take image of bacteria. Scanning electron microscopy was utilized to examine the antimicrobial mechanism. TPIP could be self-assembled in an aqueous solution into a rectangular structure of nanoparticles (TPIP-FONs) with a specific aggregation-induced emission (AIE).

TPIP-FONs picture the bacteria without the need of the washing procedure because of predominant optical properties. By observing the SEM results it can be seen that NPs cause damage to the cytoplasmic layer and cytoplasm leakage. In vitro antibacterial actions confirmed that TPIP-FONs had phenomenal antimicrobial activity against *S. aureus*. With mammalian red blood cells TPIP-FONs displayed characteristic biocompatibility. In light of the low toxic effects and minor hemolysis, TPIP-FONs can be utilized as an antimicrobial molecule in vivo. "Self-assembled small molecules" novel design of microbial imaging and antimicrobial activity provide a unique methodology toward the development of "theranostics" framework to cure microbial ailments (Chen et al. 2018; Gao et al. 2018).

4.6.1.2 SiO_2-Cy-Van Nanoprobes

Bacteria-activated SiO_2-Cy-Van nanoprobes against the infections caused by MRSA were designed in another study by vancomycin-modified polyelectrolyte-cypate complexes. Silica nanoparticles were initially used in the probe designing and SiO_2-Cy-Van was coated over it. These synthesized nanoprobes were activated by microbial responsive separation of the polyelectrolyte from silica nanoparticles.

The nanoprobes were nonfluorescent in aqueous phase because of the conglomeration of hydrophobic cypate fluorophores on silica nanoparticles to prompt ground-state quenching. MRSA potentially haul out the vancomycin-modified polyelectrolyte-cypate components from silica nanoparticles and induce them onto their own surface; because of this, the cypate condition was altered from off to on and prompted in vitro MRSA activation near-infrared fluorescence (NIRF) and photothermal removal.

In vivo trials demonstrated that this novel designed nanoprobe allowed rapid NIRF imaging with great affectability and efficient photothermal treatment of MRSA diseases. These nanoprobes provided a long-term follow-up of the advancement of MRSA diseases and permitted a source of constant estimation of microbial burden in infected tissues.

This mechanism of the dissociation of polyelectrolyte activated by bacteria from nanoparticles could also be used as a common practice for the synthesis and manufacture of functional nanomaterials responsive to drug-resistant bacterial infections (Zhao et al. 2017).

4.6.2 Theranostic Antibiotic Nanodrug

Theranostic nanodrugs displayed enhanced antibacterial activities. In an experiment conducted by Xie and colleagues, perfluoroaryl ring, ciprofloxacin, and a phenyl ring were connected by an amidine bond. These compounds were propeller shaped and readily assembled into stable nanoaggregates for the transformation of ciprofloxacin derivatives into AIE—active luminogens upon precipitation into water. The nanoaggregates showed prolonged luminescence and used to take bacterial image. These nanodrugs confirmed improved antimicrobial activity and also reduce the MIC of drug against Gram-positive bacteria (Xie et al. 2017).

4.6.3 Aptamers

Macromolecules such as DNA and RNA are responsible of encoding and transmitting hereditary information. In addition to that, they can also be utilized as recognition elements for different targets such as peptides, lipids, molecules, and even entire cells (Pan et al. 2018). Aptamers are the nucleic acid-based frameworks including RNA or single-stranded DNA (ssDNA). The three-dimensions oligonucleotide aptamers noncovalently bind to different targets with increased affinity. Aptamers can be practically used as agonists, antagonists, or as chemicals (Sundaram et al. 2013; Bruno 2015).

Furthermore, several aptamers of ssDNA were generated for the identification of diseases caused by bacterial toxins such as staphylococcal enterotoxins (SEs):

- **aptamer APTSEB1** is synthesized for staphylococci enterotoxin B (DeGrasse 2012).
- **aptamer C10** is synthesized for enterotoxin C1 (Huang et al. 2015).
- **aptamer R12.06** is synthesized for alpha-toxin (Huang et al. 2015).
- **aptamer Antibac1 and Antibac2** are for peptidoglycan (Ferreira et al. 2014).
- **aptamer PA#2/8** is synthesized for protein A (Stoltenburg et al. 2015).

Cao et al. designed the first single-stranded DNA aptamers to target entire cells of *S. aureus* (Cao et al. 2009; Vivekananda et al. 2014). Likewise, Borsa developed a SA20 silica nanoparticle-aptamer conjugate that was also able to detect *S. aureus* cells at low concentrations (Borsa et al. 2016). SA17 was utilized to determine sepsis. Aptamer S3 has a capacity to detect infections caused by *S. aureus* and enterotoxin A (Wang et al. 2017).

Anti S. *aureus* aptamers are also used as the recognition components in a drug delivery method. Aptamer-gated nanocapsules were designed to target *S. aureus* with a sustainable discharge of drug. The advancement of ssDNA aptamer-functionalized gold nanorods (Apt@Au NRs) to deactivate methicillin-resistant *S. aureus* isolates by photothermal therapy was designed by Oscay. They restrained the aptamers onto gold nanorods and showed that Apt@Au NRs act as a targeting and photothermal agent to recognize and effectively inactivate MRSA (Kavruk et al. 2015; Ocsoy et al. 2017).

4.6.4 Small Organic Molecules

The discovery of these molecules has brought unlimited benefits. Here we will discuss only those small organic molecules that have biological applications.

4.6.4.1 Carbon Quantum Dots (CQDs)

These innovative natural carbonaceous nanoparticles have gained significant reputation owing to their vital attributes, such as astonishingly minute size, excellent brightness, tranquil manufacturing, photostability, and low cell toxicity (Lim et al. 2015). They have a lot of applications in biomedical and sensor technology, for instance, in substantial metal particles identification, fluorescent marking, cells imaging, and targeted supply of drug to cancerous cells (Gao et al. 2015). Due to unique attributes of CQDs, they have also been utilized for bacterial marking and identification of live/dead bacteria (Hua et al. 2017). Carbon dots activated by visible light effectively kill bacteria (Meziani et al. 2016). In numerous antimicrobial applications they act as a unique class of quantum nanoparticles with novel optical properties and generally lower toxic reactions.

4.6.4.2 Aggregation-Induced Emission (AIE) Molecules

This emission (fluorescence) is produced by non-emissive molecules when they aggregate strongly in solution. Restriction of intramolecular motion is generally considered as their main mechanism. They are emerging as promising alternative to conventional fluorophores. These molecules have a variety of optical, electrical, and biomedical applications. Consequently, AIEgen are luminogens having AIE. A wide variety of AIEgen have been discovered since the initiation of concept of AIE in 2001. Tetraphenylethene (TPE), triphenylamine (TPA), and tetraphenylpyrazine (TPP) are few examples of AIEgen. They kill pathogens by photodynamic inactivation (PDI) or ultimately by yielding reactive oxygen species (ROS).

For the recognition and photodynamic inactivation of bacteria zinc (II)-dipicolylamine (AIE–ZnDPA) based probe was designed with AIEgen that displayed potent activity without inducing any cytotoxicity. The positive charge of the probe

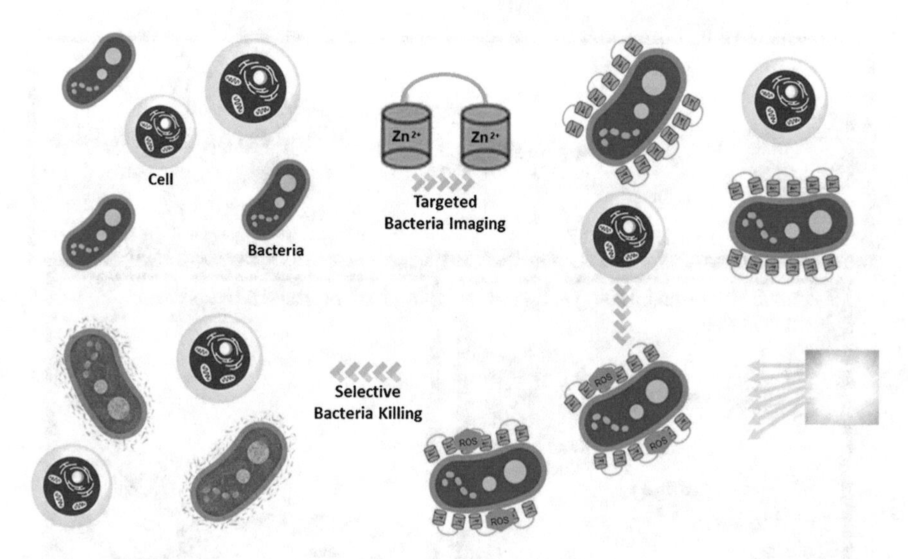

Fig. 4.2 AIE-ZnDPA-specific targeting, imaging, and killing of bacteria over mammalian cells

interacted more strongly with bacterial membranes as compared to mammalian cells. Electrostatic interaction between bacteria and AIE–ZnDPA triggered aggregation of AIE–ZnDPA on the membranes of bacteria (Gao et al. 2015) (Fig. 4.2). The increased antibacterial action was due to the toxicity of positive groups of ZnDPA and the phototoxicity of ROS.

4.6.5 Inorganic Metal-Free Nanoparticles

Metal-free nanoparticles such as silicon nanoparticles (SiNPs) have been used broadly in numerous fields. Because of tremendous fluorescence and photostability, SiNPs have been utilized generally as fluorescent probe to image the bacteria. In addition, they built up a synergistic catching procedure for microbial identification and removal by forming silicon-based 3D nanowire substrate. In an experiment, quaternized SiNPs were prepared by the covalent connection between amine-containing fluorescent SiNPs and carboxyl-containing *N*-alkyl betaines (Fig. 4.3). The formed nanoparticles were able to effectively eliminate and image Gram-positive microorganisms (Zhang et al. 2016).

4.6.6 Natural Antimicrobial Compounds

These compounds have indisputable ability to destroy microorganisms. Because of their reproducibility, biocompatibility, biodegradability, and low immunogenicity, they have already been utilized in different fields such as medical, farming, food, and pharmaceuticals.

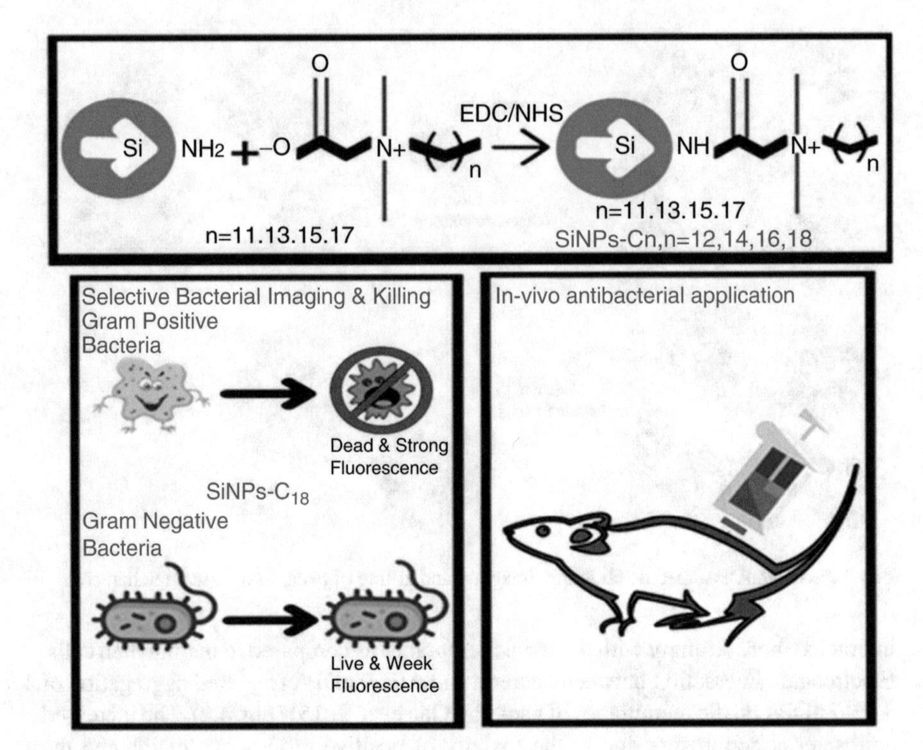

Fig. 4.3 SiNPs manufacturing path, selective bacterial killing, and antimicrobial activity in vivo

4.6.6.1 Chitosan

In 1859, Rouget explored chitosan, a natural cationic polymer. Chitosan is highly toxic to variety of bacteria, yeasts, and fungus. Corrole grafted-chitosan films were manufactured by covalent linkage among chitosan and pentafluorophenyl corrole, followed by solvent casting. Corrole is a sweet-smelling natural compound having a few alluring photophysical attributes like visible absorption, high luminescent quantum yields, and large stokes shifts. These films demonstrated improved antimicrobial impact and retained their fluorescence (Barata et al. 2016).

4.6.6.2 Polyethylenimine

It is a cationic synthetic polymer with powerful antimicrobial activity, including various amino functions. It can be combined with various particles for desired physicochemical attributes because of numerous active amino groups. There is a great need to develop biodegradable PEI because increased molecular weight and lot of amines make them toxic to human cells. Due to tremendous solubility, great antimicrobial activity, and minor toxicity, polyguanidines and polybiguanides are considered an important type of antibacterial polymers. A phototoxic nanocomplex based on PEI and polyguanidines was designed which can produce reactive oxygen species to enhance antimicrobial activity under daylight regulation. In the system, PEI was covalently linked to arginine (Arg) and cyclodextrin (CD) in order to obtain PEI-CD-Arg, then modified by 4,4′-(1,2-diphenylethene-1,2-diyl) bis(1,4-phenylene) diboronic acid (TPEDB).

Generally, Arg significantly reduces PEI toxic effects. However, the copolymer's light-induced reactive oxygen species aggravated cell membrane dissociation and DNA destruction and deactivated proteins in Gram-positive bacteria. TPEDB, an AIE molecule, can selectively image killed bacteria by absorbing into the membrane and joining with DNA inside the cytoplasm. Once bacteria are added, the gradual adhesion and assembly of PEI-CD-Arg-TPEDB with membranes of bacteria resulted in a steadily increased fluorescence emission. When bacterial damage occurred, the fluorescence emissions gradually increased but relatively higher level than the initial level. This complex demonstrated the capacity of contemporary bacteria to be controlled and monitored (Chen et al. 2018).

4.7 Conclusion

Staphylococcus aureus causes variety of life-threatening diseases. Not only there is rise of antibiotics resistance, but the range of clinical ailments also continues to change. The advancement of nanotechnology has proposed a great breakthrough to endorse the improvement of novel theranostics. Theranostic approaches of "customized drugs," which conjugate identification and treatment, have been established to improve the drugs efficiency while reducing its harmful effects. Giving treatment on right time depends very much on right-time identification with high specificity and accuracy. Theranostics holds a bright future with constant improvement in technology, research, and advancement.

Apparently, huge advancement in microbial identification and prevention has been accomplished so far. However, there is still a need for further investigation and proper optimization and utilization for biomedical application. Prosperous application of antimicrobial materials and high-quality studies can help to achieve better healthy life.

References

Allegranzi B, Nejad SB, Combescure C, Graafmans W, Attar H, Donaldson L. Burden of endemic health-care-associated infection in developing countries: systematic review and meta-analysis. Lancet. 2011;377:228–41.

Argudín MA, Mendoza MC, Rodicio MR. Food poisoning and *Staphylococcus aureus* enterotoxins. Toxins. 2010;2:1751–73.

Bangert S, Levy M, Hebert AA. Bacterial resistance and impetigo treatment trends: a review. Pediatr Dermatol. 2012;29:243–8.

Barata JF, Pinto RJ, Vaz Serra VI, Silvestre AJ, Trindade T, Neves MG, Cavaleiro JA, Daina S, Sadocco P, Freire CS. Fluorescent bioactive corrole grafted-chitosan films. Biomacromolecules. 2016;17:1395–403.

Bazzaz BSF, Khameneh B, Jalili-Behabadi MM, Malaekeh-Nikouei B, Mohajeri SA. Preparation, characterization and antimicrobial study of a hydrogel (soft contact lens) material impregnated with silver nanoparticles. Cont Lens Anterior Eye. 2014;37:149–52.

Bhattacharya S, Samanta SK. Soft-nanocomposites of nanoparticles and nanocarbons with supramolecular and polymer gels and their applications. Chem Rev. 2016;116:11967–2028.

Bilung LM, Tahar AS, Kira R, Rozali AM, Apun K. High occurrence of *Staphylococcus aureus* isolated from fitness equipment from selected gymnasiums. J Environ Public Health. 2018;2018:1–5.

Boelaert JR, Daneels RF, Schurgers ML, Matthys EG, Gordts BZ, Van Landuyt HW. Iron overload in haemodialysis patients increases the risk of bacteraemia: a prospective study. Nephrol Dial Transplant. 1990;5:130–4.

Borsa BA, Tuna BG, Hernandez FJ, Hernandez LI, Bayramoglu G, Arica MY. *Staphylococcus aureus* detection in blood samples by silica nanoparticle-oligonucleotides conjugates. Biosens Bioelectron. 2016;86:27–32.

Boswihi SS, Udo EE. Methicillin-resistant *Staphylococcus aureus*: an update on the epidemiology, treatment options and infection control. Curr Med Res Pract. 2018;8:18–24.

Bruno JG. Predicting the uncertain future of aptamer-based diagnostics and therapeutics. Molecules. 2015;20:6866–87.

Cao F, Ju E, Zhang Y, Wang Z, Liu C, Li W. An efficient and benign antimicrobial depot based on silver-infused MoS2. ACS Nano. 2017;11:4651–9.

Cao X, Li S, Chen L, Ding H, Xu H, Huang Y. Combining use of a panel of ssDNA aptamers in the detection of *Staphylococcus aureus*. Nucleic Acids Res. 2009;37:4621–8.

Cavalieri F, Tortora M, Stringaro A, Colone M, Baldassarri L. Nanomedicines for antimicrobial interventions. J Hosp Infect. 2014;88:183–90.

Chen S, Li Q, Wang X, Yang Y, Gao H. Multifunctional bacterial imaging and therapy systems. J Mater Chem B. 2018;6:5198–214.

Chen Z, Yuan H, Liang H. Synthesis of multifunctional cationic poly (p-phenylenevinylene) for selectively killing bacteria and lysosome-specific imaging. ACS Appl Mater Interfaces. 2017;9:9260–4.

Cheng D, Yu M, Fu F, Han W, Li G, Xie J. Dual recognition strategy for specific and sensitive detection of bacteria using aptamer-coated magnetic beads and antibiotic-capped gold nanoclusters. Anal Chem. 2016;88:820–5.

Corey GR. Staphylococcus aureus bloodstream infections: definitions and treatment. Clin Infect Dis. 2009;48:254–9.

Dai X, Guo Q, Zhao Y, Zhang P, Zhang T, Zhang X. Functional silver nanoparticle as a benign antimicrobial agent that eradicates antibiotic-resistant bacteria and promotes wound healing. ACS Appl Mater Interfaces. 2016;8:25798–807.

DeGrasse JA. A single-stranded DNA aptamer that selectively binds to Staphylococcus aureus enterotoxin B. PLoS One. 2012;7:e33410.

Ferreira IM, de Souza Lacerda CM, de Faria LS, Correa CR, de Andrade AS. Selection of peptidoglycan-specific aptamers for bacterial cells identification. Appl Biochem Biotechnol. 2014;174:2548–56.

Fitzgerald SF, O'Gorman J, Morris-Downes MM, Crowley RK, Donlon S, Bajwa R, Smyth EG, Fitzpatrick F, Conlon PJ, Humphreys H. A 12-year review of *Staphylococcus aureus* bloodstream infections in haemodialysis patients: more work to be done. J Hosp Infect. 2011;79:218–21.

Gao M, Hu Q, Feng G, Tomczak N, Liu R, Xing B, Tang BZ, Liu B. A multifunctional probe with aggregation-induced emission characteristics for selective fluorescence imaging and photodynamic killing of bacteria over mammalian cells. Adv Healthc Mater. 2015;4:659–63.

Gao T, Zeng H, Xu H, Gao F, Li W, Zhang S, Liu Y, Luo G, Li M, Jiang D, Chen Z, Wu Y, Wang W, Zeng W. Novel self-assembled organic nanoprobe for molecular imaging and treatment of gram-positive bacterial infection. Theranostics. 2018;8(7):1911–22.

García-Álvarez L, Holden MT, Lindsay H. Methicillin-resistant *Staphylococcus aureus* with a novel mecA homologue in human and bovine populations in the UK and Denmark: a descriptive study. Lancet Infect Dis. 2011;11:595–603.

Hennekinne JA, Buyser ML, Dragacci S. *Staphylococcus aureus* and its food poisoning toxins: characterization and outbreak investigation. FEMS Microbiol Rev. 2012;36:815–36.

Hua XW, Bao YW, Wang HY, Chen Z, Wu FG. Bacteria-derived fluorescent carbon dots for microbial live/dead differentiation. Nanoscale. 2017;9:2150–61.

Huang Y, Chen X, Duan N, Wu S, Wang Z, Wei X, Wang Y. Selection and characterization of DNA aptamers against Staphylococcus aureus enterotoxin C1. Food Chem. 2015;166:623–9.

Jevons MP. Celbenin-resistant staphylococci. BMJ. 1961;1:124–5.

Kavruk M, Celikbicak O, Ozalp VC, Borsa BA, Hernandez FJ, Bayramoglu G. Antibiotic loaded nanocapsules functionalized with aptamer gates for targeted destruction of pathogens. Chem Commun (Camb). 2015;51:8492–5.

Khameneh B, Diab R, Ghazvini K, Bazzaz BSF. Breakthroughs in bacterial resistance mechanisms and the potential ways to combat them. Microb Pathogen. 2016;95:32–42.

Li L, Sun J, Li X, Zhang Y, Wang Z, Wang C. Controllable synthesis of monodispersed silver nanoparticles as standards for quantitative assessment of their cytotoxicity. Biomaterials. 2012; 33:1714–21.

Lim SY, Shen W, Gao Z. Carbon quantum dots and their applications. Chem Soc Rev. 2015;44: 362–81.

Lowy FD. Staphylococcus aureus infections. N Engl J Med. 1998;339:520–32.

Marr AK, Gooderham WJ, Hancock RE. Antibacterial peptides for therapeutic use: obstacles and realistic outlook. Curr Opin Pharmacol. 2006;6:468–72.

Meziani MJ, Dong XL, Zhu L, Jones LP, LeCroy GE, Yang F, Wang SY, Wang P, Zhao YP, Yang LJ, Tripp RA, Sun YP. ACS Appl Mater Interfaces. 2016;8:10761–6.

Mukherjee S, Chowdhury D, Kotcherlakota R, Patra S. Potential theranostics application of biosynthesized silver nanoparticles (4-in-1 system). Theranostics. 2014;4:316–35.

Natan M, Banin E. From nano to micro: using nanotechnology to combat microorganisms and their multidrug resistance. FEMS Microbiol Rev. 2017;41:302–22.

Ocsoy I, Yusufbeyoglu S, Yilmaz V, McLamore ES, Ildiz N, Ulgen A. DNA aptamer functionalized gold nanostructures for molecular recognition and photothermal inactivation of methicillin-resistant *Staphylococcus aureus*. Colloids Surf B Biointerfaces. 2017;159:16–22.

Oethinger M, Warner DK, Schindler SA, Kobayashi H, Bauer TW. Diagnosing periprosthetic infection: false-positive intraoperative Gram stains. Clin Orthop Relat Res. 2011;469:954–60.

Pan Q, Luo F, Liu M, Zhang XL. Oligonucleotide aptamers: promising and powerful diagnostic and therapeutic tools for infectious diseases. J Infect. 2018;77:83–98.

Paterson GK, Morgan FJE, Harrison EM. Prevalence and characterization of human mecC methicillin-resistant Staphylococcus aureus isolates in England. J Antimicrob Chemother. 2014;69:907–10.

Peschel A, Otto M. Phenol-soluble modulins and staphylococcal infection. Nat Rev Microbiol. 2013;11:667.

Ragle BE, Karginov VA, Wardenburg JB. Prevention and treatment of *Staphylococcus aureus* pneumonia with a β-cyclodextrin derivative. Antimicrob Agents Chemother. 2010;54(1):298–304.

Rasigade JP, Vandenesch F. Staphylococcus aureus: a pathogen with still unresolved issues. Infect Genet Evol. 2014;21:510–4.

Reller LB, Weinstein MP, Petti CA. Detection and identification of microorganisms by gene amplification and sequencing. Clin Infect Dis. 2007;44:1108–14.

Shanson DC. Antibiotic-resistant *Staphylococcus aureus*. J Hosp Infect. 1981;2:11–36.

Spaulding AR, Salgado-Pabón W, Kohler PL, Horswill AR, Leung DY, Schlievert PM. Staphylococcal and streptococcal superantigen exotoxins. Clin Microbiol Rev. 2013;26:422–47.

Stoltenburg R, Schubert T, Strehlitz B. In vitro selection and interaction studies of a DNA aptamer targeting protein a. PLoS One. 2015;10:e0134403.

Sundaram P, Kurniawan H, Byrne ME, Wower J. Therapeutic RNA aptamers in clinical trials. Eur J Pharm Sci. 2013;48:259–71.

Taylor TA, Unakal CG. Staphylococcus aureus. StatPearls. Treasure Island (FL): StatPearls Publishing; 2019.

Tong SYC, Davis JS, Eichenberger E, Holland TL, Fowler VG. *Staphylococcus aureus* infections: epidemiology, pathophysiology, clinical manifestations, and management. Clin Microbiol Rev. 2019;28:603–61.

Turner NA, Sharma-Kuinkel BK, Maskarinec SA, Shah PP, CarugatI M, Holland TL. Methicillin-resistant Staphylococcus aureus: an overview of basic and clinical research. Nat Rev Microbiol. 2019;17:203–18.

Vanholder R, Ringoir S, Dhondt A, Hakim R. Phagocytosis in uremic and hemodialysis patients: a prospective and cross sectional study. Kidney Int. 1991;39:320–7.

Vivekananda J, Salgado C, Millenbaugh NJ. DNA aptamers as a novel approach to neutralize Staphylococcus aureus alpha-toxin. Biochem Biophys Res Commun. 2014;444:433–8.

Waldvogel FA, Medoff G, Swartz MN. Osteomyelitis: a review of clinical features, therapeutic considerations and unusual aspects. N Engl J Med. 1970;282:198–206.

Wang J, Wu H, Yang Y, Yan R, Zhao Y, Wang Y. Bacterial species-identifiable magnetic nanosystems for early sepsis diagnosis and extracorporeal photodynamic blood disinfection. Nanoscale. 2017;10:132–41.

Wang M, Zhou C, Chen J, Xiao Y, Du J. Multifunctional biocompatible and biodegradable folic acid conjugated poly (ε-caprolactone)–polypeptide copolymer vesicles with excellent antibacterial activities. Bioconjug Chem. 2015;26:725–34.

Westberg M, Grogaard B, Snorrason F. Early prosthetic joint infections treated with debridement and implant retention: 38 primary hip arthroplasties prospectively recorded and followed for median 4 years. Acta Orthop. 2012;83:227–32.

Xie S, Manuguria S, Proiettia G, Romsona J, Fub Y, Ingec AK, Wud B, Zhanga Y, Hälla D, Ramströma O, Yana M. Design and synthesis of theranostic antibiotic nanodrugs that display enhanced antibacterial activity and luminescence. Proc Natl Acad Sci USA. 2017;114:8464–9.

Zhang X, Chen X, Yang J, Jia HR, Li YH, Chen Z, Wu FG. Quaternized silicon nanoparticles with polarity-sensitive fluorescence for selectively imaging and killing gram-positive bacteria. Adv Fun Mat. 2016;26:5958–70.

Zhao Z, Yan R, Yi X, Li J, Rao J, Guo Z, Yang Y, Li W, Li YQ, Chen C. Bacteria-activated theranostic nanoprobes against methicillin-resistant *Staphylococcus aureus* infection. ACS Nano. 2017;11:4428–38.

Zimakoff J, Bangsgaard PF, Bergen L, Baago-Nielsen J, Daldorph B, Espersen F, Gahrn Hansen B, Hoiby N, Jepsen OB, Joffe P, Kolmos HJ, Klausen M, Kristoffersen K, Ladefoged J, Olesen-Larsen S, Rosdahl VT, Scheibel J, Storm B, Tofte-Jensen P. Staphylococcus aureus carriage and infections among patients in four haemo- and peritoneal-dialysis centres in Denmark. The Danish Study Group of Peritonitis in Dialysis (DASPID). J Hosp Infect. 1996;33:289–300.

Chapter 5
Current Status and Prospects of Chitosan: Metal Nanoparticles and Their Applications as Nanotheranostic Agents

Dilipkumar Pal and Supriyo Saha

Abstract Nanotheranostic is an approach to merge nanotechnology, therapeutics, and diagnosis for the betterment of mankind. Chitosan is a natural polysaccharide procured from chitin, which has a great impact as carrier of drug materials. Chitosan-gold nanoparticles have been applied in carcinoma, detection of solid tumors, prostate cancer, glucose, amplified nucleic acid, cancer cell imaging, etc. Chitosan-silver nanoparticles have also been used for inhibition of bacterial growth, tissue regeneration process, wound healing process, etc. Technetium-radiolabeled chitosan nanoparticles have directly been used to study the biodistribution of molecule, specialized delivery system as nose to brain, etc. This chapter mainly emphasizes on the nanotheranostic applications of chitosan-gold/silver or technetium-radiolabeled chitosan nanoparticles.

Keywords Nanotheranostic agent · Chitosan-gold nanoparticles · Chitosan-silver nanoparticles · Technetium-radiolabeled chitosan nanoparticles · Biosensor · Tumor detector · Biodistribution

5.1 Introduction

The word "Nanotheranostic" comprises nanotechnology, therapeutics, and diagnosis. This field mainly focuses on the applicability of nanotechnology for efficient diagnosis of disease, site-specific delivery of therapeutic agents, imaging of solid

D. Pal (✉)
Department of Pharmaceutical Sciences, Guru Ghasidas Vishwavidyalaya
(Central University), Bilaspur, Chhattisgarh, India

S. Saha
School of Pharmaceutical Sciences & Technology, Sardar Bhagwan Singh University,
Dehradun, Uttarakhand, India

M. Rai, B. Jamil (eds.), *Nanotheranostics*,
https://doi.org/10.1007/978-3-030-29768-8_5

tumor, proper distribution of drugs, diagnosis of heavy metals, specific delivery of DNA molecule to achieve greater efficacy (Hamburg and Collins 2010), and minimizing the pharmacokinetic load on the therapeutic agent (Swierczewska et al. 2011). It is a very well-known fact that maximum part of the drug molecule excrete out of the body without showing any therapeutic response (Swierczewska et al. 2012). This nanotheranostic approach delivers the molecule into its site of absorption for better therapeutic index and lowest adverse effects. Particle size in the nano range as delivered by nanotechnology promotes greater absorption (Akhter et al. 2012), biodistribution, bioavailability, and greater surface-area-to-volume ratio, which is correlated with therapeutic window of the molecule (Yang et al. 2013). Initially the term "theranostic" deals with the magnetic resonance imaging, computed tomography, and positron emission tomography using fluorescent dyes application in the imaging technology. The amalgamation of nanotechnology and theranostic approach is used in the imaging studies, biomolecular sensing, drug delivery, etc. (Lim et al. 2015). We know that chitosan is a heterogeneous polymer of amino sugar glucosamine and a monosaccharide *N*-acetyl glucosamine (Nayak and Pal 2012; Nayak et al. 2012). Source of chitin and isolation of chitosan from chitin regulate the quality of chitosan. Chitosan is largely found in crabs, shrimps, and prawns and after alkali treatment or after reaction with papain, it is extracted out; after the hydrolytic cleavage of acetamide group and heat treatment, chitin is converted into chitosan with greater degree of deacetylation. After the molecular level modification, solubility problem of chitosan is modified to achieve its applicability spectrum as diversified polymer. In case of metal, we select three metals, namely gold, silver, and technetium. Gold is a transition metal (Au) of group 11 and gold nanoparticle has diversified applications in the photodynamic therapy, delivery of therapeutic agent, colorimetric sensation of heavy metal, detection of solid tumor, and cancer progression (Nayak and Pal 2011; Nayak and Pal 2015a). Silver (Ag) is a soft, transition metal with greater electrical and thermal conductivity, and silver nanoparticle has applications in diagnosis, antibacterial application, conductive application, and optical application (Nayak and Pal 2016). Technetium (Tc) is a silvery-gray metal with oxidation states of +7,+5,+4 (Pal and Nayak 2010). It has four different radioisotopes, namely ^{99m}Tc, ^{98m}Tc, ^{97m}Tc, and ^{95m}Tc, where m stands for metastate (Nayak and Pal 2014a). But here technetium is not directly applied in the chitosan nanoparticle formation, and is mainly applied as image developer using its ability to emit gamma radiation (Nayak and Pal 2015b; Pal and Nayak 2015). The importance of chitosan in this approach is to crosslink the nanoparticle with polymer for better applicability (Nayak and Pal 2014b, c). Chitosan itself is applied in wound healing process, tissue engineering, stem cell technology, carrier for vaccines, etc. (Nayak et al. 2013; Nayak and Pal 2013). In this chapter, we emphasize on the various applications of chitosan-gold/silver/technetium (radiolabeled) nanoparticles as nanotheranostic agent.

5.2 Application of Chitosan-Gold Nanoparticles as Nanotheranostic Agents

5.2.1 Chitosan/Gold Nanoparticles as the Detector of Carcinoma Antigen (CA-125)

Pakchin et al. (2018) developed a chitosan-gold nanoparticle as carcinoma antigen 125 (CA125) detector (Scheme 5.1). The nanoparticles were developed by the reaction of chloroauric acid and chitosan. The nanocomposite was performed as substrate for immobilization of protein. Electrochemical behavior of the electrode was observed at a low electrode potential (0.034 V) due to peroxidation. Lactate oxidase enzyme was used for the first time in sandwich-type immunosensor. In the optimum condition, the designed immunosensor exhibited two linear ranges. Chronoamperometric data showed that immunosensor was observed with two linear ranges, i.e., 0.01–0.5 U/mL and 0.5–100 U/mL. Immunosensor was observed with good accuracy, precision, reproducible data characteristic, and easily detectable CA125 level in human serum (Pakchin et al. 2018).

5.2.2 Chitosan-Gold Nanoparticles as Detector of Amplified Nucleic Acid

Tammam et al. (2017) developed chitosan-gold nanoparticles by the reaction of chitosan and gold (III) chloride trihydrate in acetic acid solution. Atomic force microscopy, scanning electron microscopy, and transmission electron microcopy

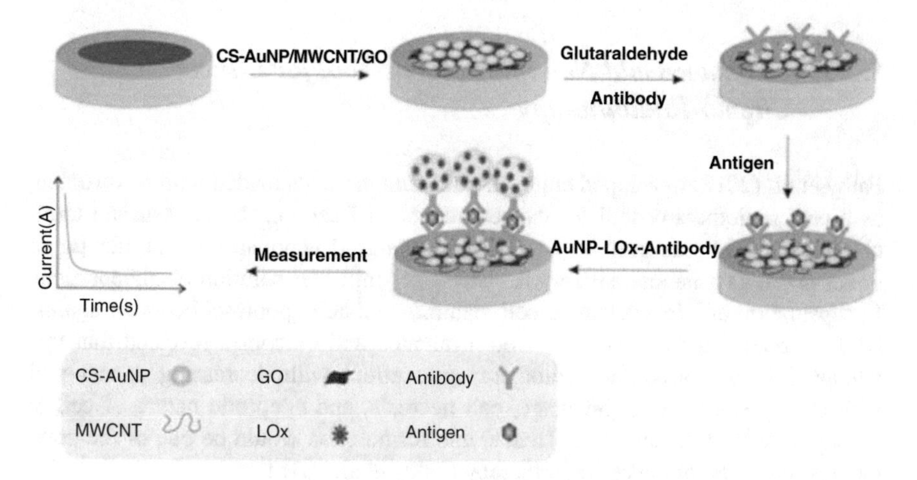

Scheme 5.1 Development of chitosan-gold nanoparticle (Source: Pakchin et al., copyright © 2018 with permission from Elsevier B.V.)

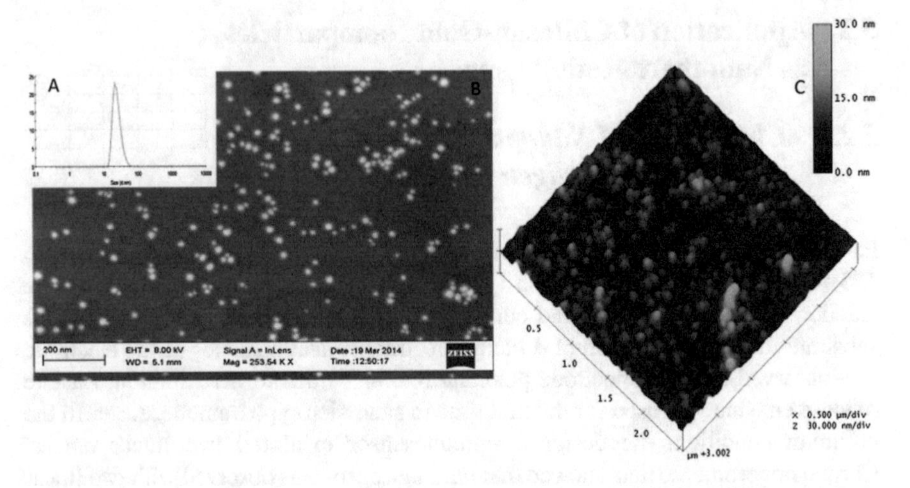

Fig. 5.1 SEM and AFM tomographic image of chitosan-gold nanoparticle (Source: Tammam et al., copyright © 2017 with permission from Elsevier B.V.)

were applied to characterize the nanoparticles (Fig. 5.1). *Mycobacterium tuberculosis* was considered as the target. TB sputum samples were collected from Abassia Chest Hospital in Cairo and amplified. Chitosan-free nonconjugated DNA molecule with interaction was considered as negative control and showed red color. Amplified DNA molecule possessed with gold nanoparticles was blue in color. Gel electrophoresis data suggested that most of the samples were observed with positive response, so it would behave as cheap polymer to increase the sensitivity of gold nanoparticle detection of nucleic acids (Tammam et al. 2017).

5.2.3 Chitosan-Gold Nanoparticles as Tool for Cancer Chemo-Radiotherapy

Fathy et al. (2018) developed chitosan-gold nanoparticles loaded with doxorubicin as a new radiotherapy tool for cancer treatment. Reaction of chitosan and tetrachloroauric acid was used to develop chitosan-gold nanoparticle and 125 μg of doxorubicin as a nanosensitizer was added per milliliter solution of nanoparticle. Encapsulation efficiency, cancer cell viability, and cell apoptosis behavior against breast cancer cell line (MCF-7) were evaluated and outcomes revealed that this nanoparticle increased the radiotherapeutic effect with decreasing cancer cell viability by DNA strand breakage, cell necrosis, and apoptotic nature of cell at 0.5 Gy, 6 MV of X-rays dose. Hence, this formulation would be one of the good medication tools for cancer radiotherapy (Fathy et al. 2018).

5.2.4 *Chitosan-Gold Nanoparticles for Hepatocellular Carcinoma*

Salem et al. (2018) developed chitosan-based gold nanoparticles using hydrogen tetrachloroaurate trihydrate ($HAuCl_4{\cdot}3H_2O$, 98%, MW: 393.83) and chitosan (75% of deacetylation) at 100 °C, and it was found that after complete reduction, the color of the nanoparticles was changed from colorless to ruby-red (Scheme 5.2). 5-Fluorouracil was loaded into the nanoparticles and final nanocomposite was evaluated by particle size, zeta potential, TEM, FTIR, HPLC, and in vitro cytotoxicity assay against human hepatocellular carcinoma (HepG2) cell. In the higher amount of chitosan as 0.1%, 0.5%, and 1% w/v, zeta potential was greater than 30 mV, hydrodynamic diameter was greater than 100 nm, and polydispersity index was greater than 0.5. These outcomes revealed a clear correlation between chitosan concentrations and positively charged amino groups on nanoparticles. The maximum values of zeta potential, hydrodynamic diameter, and polydispersity index PDI were showed at 0.2% chitosan concentration. The stability of

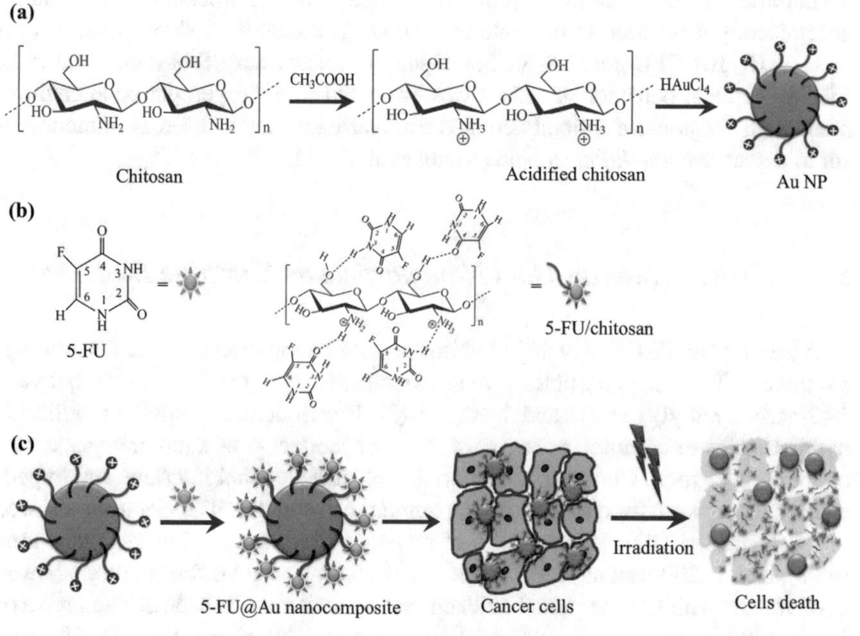

Scheme 5.2 (**a**) Chemical structure of chitosan and acidified chitosan with illustration of Au NPs formation, (**b**) chemical structure of anticancer drug, 5-FU, and conjugation of 5-FU with acidified chitosan-coated Au NPs, and (**c**) schematic illustration of the treatment of cancer cells with 5-FU@ Au NPs for chemophotothermal therapy (Source: Salem et al., copyright © 2018 with permission from Elsevier B.V.)

nanoparticles was observed at acidic and neutral media with zeta potential greater than 30 mV, and exposure to laser radiation potentiated the inhibition of HepG2 cell with respect to laser radiation-free 5-fluorouracil nanocomposite. These results clearly stated that chitosan-gold-5-fluorouracil nanoparticle was effective against hepatocellular carcinoma but laser radiation application would create a great impact (Salem et al. 2018).

5.2.5 Chitosan-Gold Nanoparticles for Electrochemical Detection of *Salmonella*

Xiang et al. (2015) developed an ultrasensitive immunosensor as detector of *Salmonella*. Gold nanoparticles were obtained from the reaction of hydrochloroauric acid and chitosan (1% w/v). A suspension of chitosan-gold nanoparticles was poured into the clean glass electrode and the composite was immobilized with capture antibody (Ab1). Further incubation was done using *Salmonella* and horseradish peroxidase conjugated secondary antibody (Ab2), so that a sandwich immunosensor was obtained. Cyclic voltammetry and electrochemical impedance spectroscopy were used to evaluate the sensor. The outcomes showed that sensor creates 10–10^5 CFU mL^{-1} with 3:1 signal-to-noise ratio (Tables 5.1 and 5.2). Electrochemical behavior of sensor was evaluated in hydrogen peroxide environment. Anti-*Salmonella* antibody crusted nanoparticles may be used as immunosensor to detect *Salmonella* in sample (Xiang et al. 2015).

5.2.6 Novel Chitosan-Gold Nanoparticle as Xanthine Biosensor

Dervisevic et al. (2017*)* developed chitosan-gold nanoparticle for the detection of xanthine. The nanoparticles were obtained by the reaction between 1-(2-carboxyethyl) pyrrole and 1-ethyl-3-(3- dimethylaminopropyl)carbodiimide hydrochloride at equimolar concentration. Concentration of xanthine was source of spoilage of food. Compact, soft, portable electrochemical interface and impedance analyzer were used to identify the amount of xanthene. The electrode showed low 0.25 mM (LOD), less than 8 s of response time, 1.4 nA low response time (~8 s), and 1–200 mm as linear range (Table 5.3). The final formulation showed good results with fish, beef, and chicken real-sample measurements. The prepared formulation was the good formulation for xanthine oxidase identification (Dervisevic et al. 2017).

Table 5.1 Comparison between the proposed method and the reported immunosensor for the detection of *Salmonella*

Immunosensor platform	Bio-receptor of immobilization	LOD (CFU/mL)	Fabrication and analysis time (h)
SPR	Antibody	5×10^6	36
Fluorescence	Oligonucleotide	30	27
Electrochemical	Antibody	500	14.5
Electrochemical	Antibody	42	5
Electrochemical	Antibody	143	2
Electrochemical	Antibody	10	26
Electrochemical	Antibody	5	4

Source: Xiang et al. (2015), copyright © 2015 with permission from Elsevier B.V.

Table 5.2 Comparison of the proposed detection immunosensor assay and plate counting methods

Samples	Number	Immunosensor (CFU/mL)	Plate counting method (CFU/mL)
Tapped water	1	1.2×10^3	1.1×10^3
	2	1.5×10^3	1.4×10^3
	3	2.0×10^3	2.1×10^3
	4	3.3×10^3	3.2×10^3
	5	4.1×10^3	4.1×10^3
Milk	1	1.4×10^3	1.3×10^3
	2	1.9×10^3	1.8×10^3
	3	2.3×10^3	2.2×10^3
	4	3.7×10^3	3.5×10^3
	5	4.8×10^3	4.9×10^3

Source: Xiang et al. (2015), copyright © 2015 with permission from Elsevier B.V.

Table 5.3 Analytical data of chitosan-gold xanthine biosensor

Xanthine added (mM)	Xanthine found (mM)	RSD (%)	Recovery %
1	0.98	2.45	98.0
10	9.75	3.1	97.5
100	103.67	3.73	103.67
200	194.8	2.88	97.4

Source: Dervisevic et al. (2017)

5.2.7 *Chitosan-Gold Nanoparticles as Nanocarrier for Erlotinib Delivery*

Fathi et al. (2018) suggested a chitosan-gold nanoparticle for the delivery of erlotinib (mainly used to treat non-small lung cancer and pancreatic cancer) against A549 cell line. In this case, chitosan copolymer was developed by reacting chitosan and sodium dodecyl sulfate, and by maleoylation reaction the polymeric behavior of chitosan

Table 5.4 The kinetics models used to fit the ETB release data from CGH NPs at different temperatures

Kinetics model	Equation	Coefficient of determination (R^2)	
		$T = 25$ °C	$T = 37$ °C
Zero order	$F = k_0 t$	0.9083	0.9394
First order	$\ln(1 - F) = -k_f t$	0.9784	0.9549
Higuchi	$F = k_H \sqrt{t}$	0.9911	0.9801
Power law	$\ln F = \ln K_P + PI\, nt$	0.9772	0.9451
Square root of mass	$1 - \sqrt{(1 - F)} = k_{1/2}\, t$	0.9815	0.9861
Hixson-Crowel	$1 - \sqrt[3]{(1 - F)} = k_{1/3}\, t$	0.9919	0.9862
Three seconds root mass	$1 - \sqrt[3]{(1 - F)^2} = k_{2/3}\, t$	0.9626	0.9769
Weibull	$\ln(-\ln(1 - F)) = -\beta \ln t_d + \beta \ln t$	0. 897	0.8253
Reciprocal powered time	$\left(\frac{1}{F} - 1\right) = \frac{m}{t^b}$	0.7826	0.6794

Source: Fathi et al. (2018)

copolymer was initiated by maleic anhydride. In the next step, the grafting was done by the reaction of oleic acid and maleoyl-sodium dodecyl chitosan and the final product was developed by the reaction of chitosan copolymer and chloroauric acid. Formulations were characterized by dynamic light scattering, atomic absorption spectroscopy, transmission electron microscopy, in vitro drug release and in vitro cytotoxicity against A549 cell line, and also cellular uptake analysis by flow cytometry (Table 5.4). The outcomes revealed that 80–100 nm was the particle size of the formulation with greater stability in different pH value without any sign of agglomeration. Erlotinib was released from the nanoparticles in a thermoresponsive manner with 85.81% of cellular uptake as observed by flow cytometry. In vitro cytotoxicity data confirmed formulation with greater antiproliferative efficacy against A549 cell line. These data surely indicate the larger prospect of the nanoparticle as anticancer agent (Fathi et al. 2018).

5.2.8 *Chitosan-Gold Nanoparticles with Paclitaxel for Cancer Cell Imaging*

Manivasagan et al. (2016) developed a chitosan-gold nanoparticle embedded with paclitaxel which was used as image developer of cancer cell (Scheme 5.3). The polymeric nanoparticles were developed by the reaction of chitosan and gold (III) chloride trihydrate ($HAuCl_4$, $3H_2O$) and the prepared formulation was further reacted with paclitaxel in dimethyl sulfoxide. The final formulation was lyophilized and characterized by X-ray diffraction, scanning electron microscopy, and FTIR, which showed formulation with spherical shape and 61.86 ± 3.01 nm average

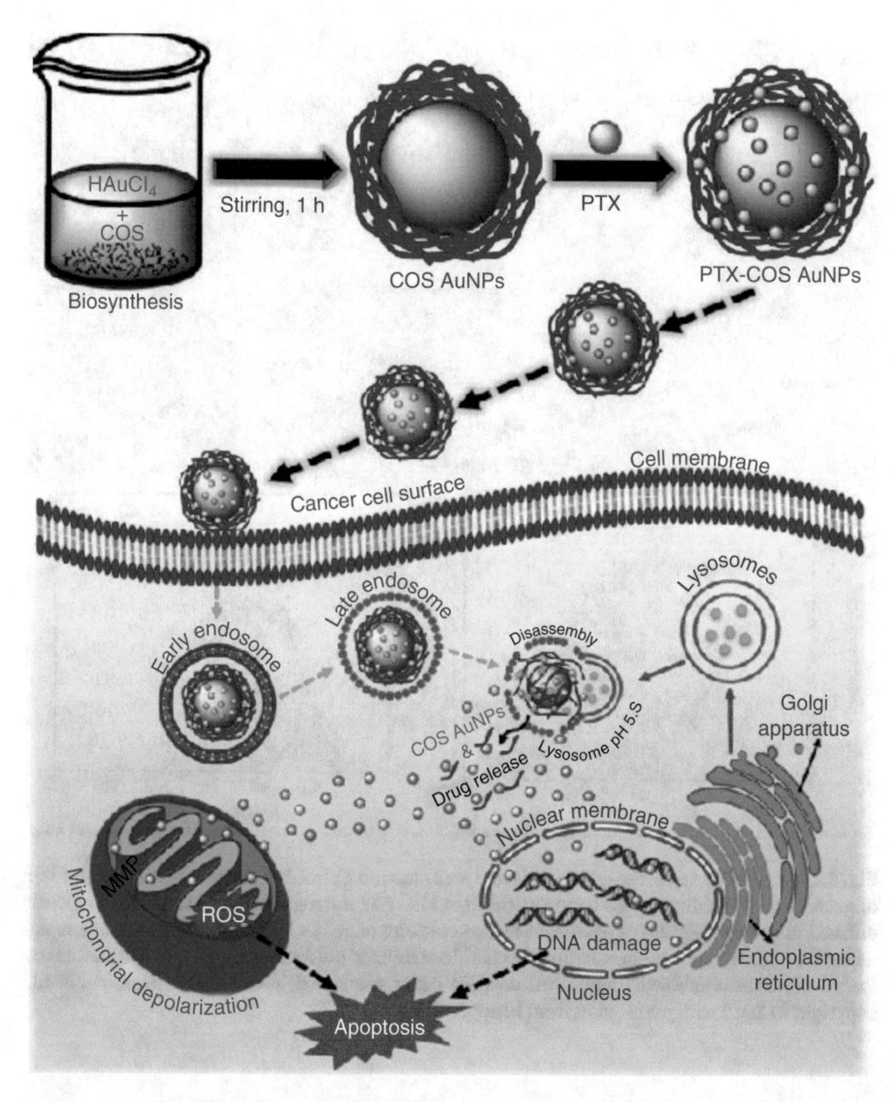

Scheme 5.3 Overall scheme for the biosynthesis of gold nanoparticles using chitosan oligosaccharide, subsequent loading of paclitaxel (PTX) on chitosan oligosaccharide-stabilized gold nanoparticles (COS AuNPs) and the possible mechanism for cellular uptake of paclitaxel-loaded chitosan oligosaccharide-stabilized gold nanoparticles (PTX-COS AuNPs) in MDA-MB-231 cancer cells (Source: Manivasagan et al., copyright © 2016 with permission from Elsevier B.V.)

particle size (Fig. 5.2) and pH-dependent drug release. The formulation was also observed with greater antiproliferation activity against MDA-MB-231 cell line. Flow cytometry data showed the cellular internalization inside the formulation. These outcomes clearly indicated the development of a new optical contrasting agent for photoacoustic imaging.

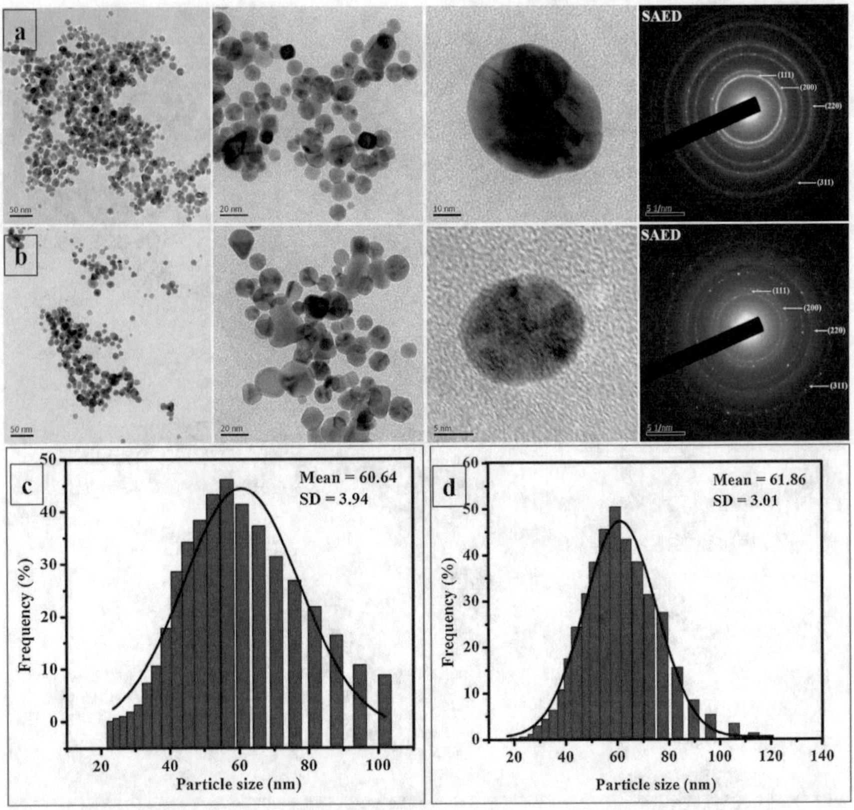

Fig. 5.2 (a) HRTEM micrograph and selected area electron diffraction pattern (SAED) of chitosan oligosaccharide-stabilized gold nanoparticles. (b) HRTEM micrograph and selected area electron diffraction pattern (SAED) of paclitaxel-loaded chitosan oligosaccharide-stabilized gold nanoparticles. DLS histogram of chitosan oligosaccharide-stabilized gold nanoparticles (c) and paclitaxel-loaded chitosan oligosaccharide-stabilized gold nanoparticles (d) (Source: Manivasagan et al., copyright © 2016 with permission from Elsevier B.V.)

5.2.9 Chitosan-Gold Nanoparticles as Detector of Prostate Cancer

A new detector molecule of prostate cancer was developed by Suresh et al. (2018). The nanoparticles were fabricated on electrodes as a reaction with chitosan and gold (III) chloride trihydrate ($HAuCl_4$, $3H_2O$). Then the electrodes containing nanoparticles were immersed in monoclonal prostrate primary antibody, which was further immersed in horseradish peroxidase (HRP)-tagged secondary antibody. The final nanoformulation was characterized by scanning electron microscopy (Fig. 5.3), energy dispersive

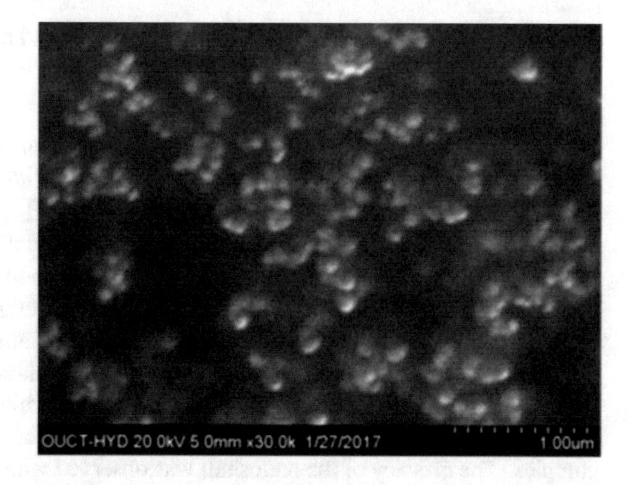

Fig. 5.3 SEM image of AuNPs/CHI film on SPE (Source: Suresh et al., copyright © 2017 with permission from Elsevier B.V.)

X-ray spectroscopy, cyclic voltammetry, and electrochemical impedance spectroscopy. Methylene blue was used as redox mediator to identify the electron movement from electrode to horseradish peroxidase enzyme. The amount of hydrogen peroxide and pH of the medium greatly impacted the immunosensor ability. The hydrogen peroxidase concentration of 2.5 mM showed remarkable effect on prostate-specific antigen concentration and the developed immunosensor was observed with 1–18 ng/mL of linear response with LOD value of 0.001 ng/mL. These results confirmed that this developed nanoformulation was used as a potential biomarker for prostate cancer (Suresh et al. 2018).

5.2.10 Chitosan-Gold Nanoparticles for Diagnosis of Invasive Aspergillosis

Bhatnagar et al. (2018) formulated chitosan-gold nanoparticles for the diagnosis of Aspergillosis. Invasive Aspergillosis was developed by spreading of *Aspergillus fumigatus* into the blood vessel and the formulation of electrochemical nanobiosensor with the detection ability of virulent glip-T gene. The nanosensor was developed using 1,6-hexanedithiol and chitosan-gold nanoparticle with glip-P gene probe. Ultraviolet-visible spectroscopy, cyclic voltammetry, and electrochemical impedance spectroscopy were used to characterize the nanoformulation. Toluidine blue indicator was used to analyze the signal and detect the glip-T gene by hybridization reaction. The nanosensor was observed with dynamic range between 1×10^{-14} and 1×10^{-2} M with LOD of $0.32 \pm 0.01 \times 10^{-14}$. This ability perfectly showed the development of a good nanobiosensor of glip-T gene (Bhatnagar et al. 2018).

5.2.11 Glycol Chitosan-Gold Template for Anti-Cancer Theranostics

Shanavas et al. (2018) suggested the development of glycol chitosan-gold nanoparticles as anticancer theranostic agent. The synthesis of nanoparticles was done by the reaction with poly lactic-co-glycolic acid, glycol chitosan, and gold hydrochloride with ascorbic acid as reducing agent. Photothermal therapy, X-ray-CT imaging, and in vitro biocompatibility test [against mouse fibroblast cell (L929)] were the way of nanoformulation evaluation. The average size and shape of the nanoparticle were 100 nm and spherical. Presence of amino groups regulated the formation of continuous and branched gold nanoformulation with 0.025% of glycol chitosan concentration. If the concentration of glycol chitosan was 0.25%, then the nanoshell became merged due to abnormal distribution of tetrachloroaurate complex. The efficacy of the nanoshell was observed with greater cell death against breast cancer cell line. These data confirmed the development of new gold nanoshell as anticancer theranostic agent (Shanavas et al. 2018).

5.2.12 Chitosan-Gold Nanoparticles for STAT3 siRNA Delivery

Labala et al. (2016) developed chitosan-gold nanoparticles for STAT3siRNA delivery through transcutaneous iontophoretic application. The nanoparticles were synthesized using tetrachloroaurate trihydrate and chitosan with polyvinyl pyrrolidine as stabilizing agent and finally adsorbed with siRNA through HEPES buffer. Gel retardation assay was performed to determine the N/P (nitrogen to phosphorous) ratio in nanoparticle and siRNA, which showed that 10:1 ratio is needed for optimal complexation. STAT3siRNA was replaced by siRNA as control. Encapsulation efficiency was determined by biospectrometer at 260 nm. STAT3siRNA was evaluated against murine melanocyte cell viability, western blot analysis was used to identify the in vitro gene silencing, FITC-Annexin V/PI apoptosis assay was used to perform the cell proliferation, and Cy3-labeled siRNA was used to evaluate skin penetration value using excised porcine ear skin method. The formulation was observed with 150 ± 10 nm and 35 ± 6 mV of average particle size and zeta potential, respectively. At 0.25 nm concentration 49.0% and 0.50 nm concentration 66.0% of cell inhibition against murine melanocyte cell were observed. Confocal microscope and skin cryosections showed that application of anodal iontophoresis of 0.47 mA/cm^2 was observed with greater skin penetration. So, this nanoformulation was observed as theranostic agent to treat melanoma (Labala et al. 2016).

5.2.13 Chitosan-Gold Nanoparticles Probes as Melamine Sensor

In 2013, a new probe for the sensation of melamine was developed by Guan et al. The chitosan-gold nanoparticles were synthesized by the reaction of chloroauric acid tetrahydrate, chitosan and tripolyphosphate as reducing agent, where chitosan was used as stabilizer. Finally, the color of the solution was changed from pale yellow to dark red. X-ray diffraction, transmission electron microscopy, and UV-visible spectroscopy were used to characterize the nanoformulation. The nanoparticles were spherical in shape and particle size was less than 20 nm. At room temperature, 50 μL of processed milk was reacted with 450 μL of chitosan-gold nanoparticle for 10 min to detect the presence of melamine in liquid milk. Outcomes revealed that A650/A520 of absorbance ratio was linearly correlated with melamine concentration with minimal concentration of melamine (6×10^{-6} g/L as detected by UV-visible spectroscopy) (Fig. 5.4). This nanoprobe was able to detect minimal to large concentration of melamine in liquid milk. This was the most simple and rapid visual colorimetric method. These data surely indicated that this nanoprobe is best for detection of melamine in nano level (Guan et al. 2013).

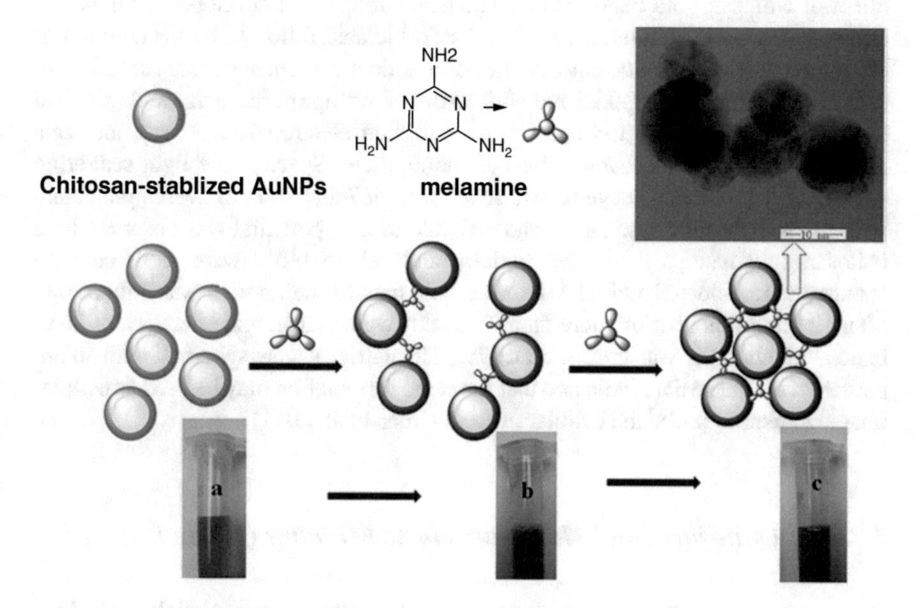

Fig. 5.4 Schematic representation of the colorimetric mechanism for melamine detection. The insert is photographs of solution of tube (**a**) chitosan-stabilized AuNPs, tube (**b**) chitosan-stabilized AuNPs with melamine, tube (**c**) chitosan-stabilized AuNPs with melamine and TEM image of chitosan-stabilized AuNPs with melamine (Source: Guan et al., copyright © 2013 with permission from Elsevier B.V.)

5.2.14 Prussian Blue–Chitosan-Gold Nanoparticles as β-Glucan Biosensor

Wang et al. (2014) developed a new chitosan-gold-embedded Prussian blue as beta-glucan biosensor. The biosensor development was initiated by PB–CS/Au, which was formed by electrodeposition of Prussian blue-chitosan on gold disk electrode and then gold-chitosan nanoparticle (50 nm of average particle size) was chronoamperometrically deposited on PB-CS/Au with the reaction of chloroauric acid tetrahydrate-chitosan. Linear range of 6.25–93.75 μM was maintained by biosensor and detection limit was 1.56 μM with 1.0 mM Michaelis-Menten constant (Km). The developed biosensor exhibited greater response toward beta-glucan with greater sensitivity, selectivity, and stability (Wang et al. 2014).

5.2.15 Chitosan-Gold Nanoparticles as Stabilizer in Cytotoxic Effects

Boca et al. (2011) used Chinese hamster ovary cell as target to produce cytotoxic effect of chitosan-gold nanoparticle. The nanoparticle was developed by the reaction of chloroauric acid tetrahydrate and ascorbic acid followed by the addition of chitosan in acetic acid. The color of the nanoparticle was changed from colorless to dark blue and finally as pinkish red with 50 nm of average particle size and spherical in shape. Optical extinction spectra, transmission electron microscopy, and zeta potential were used to characterize the nanoparticle. Spectrum of light scattering was observed with large asymmetric shape with 587 and 700 nm overlapping band with surface plasmon resonance, and +37 mV of zeta potential was observed with chitosan-gold nanoparticle. Cell toxicity and cell viability were assessed with standard citrate-coated and CTAB-coated gold nanoformulation. It was noticed that after 10 min of incubation more than 95% cells were viable, and after 19 h of incubation, cell viability was greater than 87%. The particles were spherical with 50 nm particle size. These data indicated that the same formulation may be used as cellular imaging agent to probe the cellular process (Boca et al. 2011).

5.2.16 Chitosan-Gold Nanoparticle as Mercury (II) Sensor

Tian et al. (2017) developed chitosan-gold nanoparticle-embedded 2,6-pyridinedicarboxylic acid as the biosensor of mercury (II). The nanoparticle was developed using chitosan, hydrochloroauric acid tetrahydrate with 2,6-pyridinedicarboxylic acid for the sensitization of mercury (II) with colorimetric process. Pyridine of 2,6-pyridinedicarboxylic acid was responsible for strong affinity toward mercury (II) and the final 2,6-pyridinedicarboxylic acid-embedded chitosan-gold nanoparticle was able to form complex Hg^{2+} ion. This agglomerate was identified by change of color from red to blue. The agglomerate was characterized by transmission electron microscopy and ultraviolet spectroscopy with 300 nM to 5 μM and greater selectivity toward Hg^{2+} in the river water sample. These data confirmed the development of new biosensor of mercury (II) (Tian et al. 2017).

5.2.17 Chitosan-Gold Nanoparticles Biocompatible Film as Detector of Aflatoxin B1

Ma et al. (2016) developed a new generation chitosan-reduced gold nanoparticle (10 nm of particle size) biofilm as aflatoxin B1 detector. The detector was developed using chronoamperometric method with the reaction of chitosan and chloroauric acid tetrahydrate fixed within a microelectrode chip at −1.1 V for 300 s. Anti-aflatoxin B1 antibody was activated upon reaction of *N*-hydroxysuccinimide and *N*-(3-dimethylaminopropyl)-*N′*-ethylcarbodiimide and it was immobilized into nanoparticle. Cyclic voltammetry and electrochemical impedance spectroscopy were used to evaluate the electrochemical measurement of the formulation. Also FTIR and SEM were used to characterize the morphology of the nanoparticle (Ma et al. 2016).

5.2.18 Chitosan-Gold Nanoparticle-Modified Electrode as Nonenzymatic Glucose Sensor

Kangkamano et al. (2017) developed a chitosan-reduced gold nanoparticle-modified electrode for nonenzymatic glucose sensation. The first multiwall carbon nanotube was developed using electrostatic interaction between gold nanotube with citric acid with the subsequent transformation of Au^{3+} to Au^0 and final gold nanoparticle (60–100 nm) was obtained. The obtained gold nanorod was electrochemically cleaned with sulfuric acid, which was further reacted with chitosan solution with glutaraldehyde as crosslinking agent and cryogelation was developed on gold electrode, which was characterized by transmission electron microscope (Fig. 5.5). The electrochemical characterization was performed by μAutolab type III controlled by GPES 4.9 software with the platinum reference electrode. The presence of glucose was detected by amperometric process with the Autolab. The biosensor was observed with linear range between 0.001 and 1.0 mM with LOD of 0.5 μM. The glucose sensation created a good correlation employed by hexokinase method. These data suggested that this nanoparticle was feasible for glucose detection (Kangkamano et al. 2017).

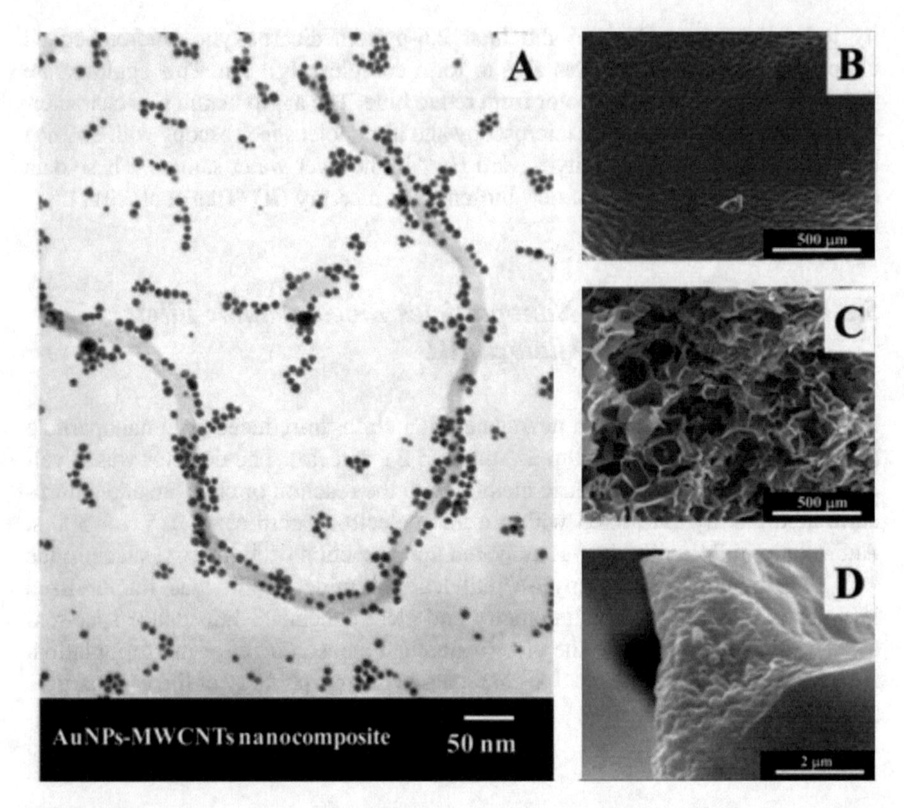

Fig. 5.5 (**a**) TEM image of AuNPs-MWCNTs nanocomposite; SEM images of (**b**) AuNPs-MWCNTs-CS noncryogel and (**c**) the top view of AuNPs-MWCNTs-CS cryogel modified layer at ×80 magnification showing the porous structure of cryogel and (**d**) the inside pore view at ×20,000 magnification showing the surface of the chitosan cryogel with coated protrusions of the embedded MWCNTs (Source: Kangkamano et al., copyright © 2013 with permission from Elsevier B.V.)

5.2.19 Chitosan-Gold Nanoparticle as Heavy Metal Ion Sensor

Sugunan et al. (2005) developed a chitosan-reduced gold nanoparticle for heavy metal sensation. The nanoparticle was developed upon reaction with hydrochloroauric acid, monosodium glutamate, and chitosan (3.3%), whereas copper sulfate and zinc sulfate were used to detect heavy metal by ultraviolet spectroscopic analysis within 350–1000 nm wavelength. At 519 nm wavelength, particle size was 10–20 nm in diameter. TEM analysis data was observed with approximately 20 nm in diameter without any sign of agglomeration, and greater detection of Cu^{2+} and Zn^{2+} ions were observed at 650 nm wavelengths. These outcomes clearly indicate the usefulness of this nanoparticle as heavy metal sensor (Sugunan et al. 2005).

5.2.20 Chitosan-Gold Nanoparticle Film as Glucose Biosensor

Du et al. (2007) developed a new chitosan-fabricated gold nanoparticle film for the detection of glucose. The first phase of the development included chitosan solution preparation, done by reacting chitosan with hydrochloric acid with pH adjustment to 5.0 by sodium hydroxide solution, the average particle size of the formulation was 50 nm with 10 nm of roughness layer. CHI electrochemical instrument was used for the measurement of overnight mutarotated glucose, and amperometric studies were performed under phosphate buffer solution with 0.70 V. The deposition of gold nanoparticle on glass carbon electrode was done by potentiostatic procedure within −0.5 to −3.0 V, and glucose oxidase enzyme was immobilized over formulation. Atomic force microscopy and cyclic voltammetry were used to characterize the developed electrode. The outcomes were recorded at 5.0×10^{-5} to 1.30×10^{-3} M linear range, 3.5 mM (Michaelis-Menten constant) and 13 μM (LOD) of glucose sensation (Fig. 5.6). These data suggested that this newly developed formulation would be one of the best detectors of glucose (Du et al. 2007).

5.2.21 Chitosan-Gold Nanomaterial as Efficient Antifilarial Agent

Saha et al. (2017) utilized a black pepper-embedded gold nanoparticle as an efficient antifilarial agent. The formulation was developed upon reaction of chitosan, *Piper nigrum* extract, and chloroauric acid. Absorption maxima with 531–557 nm confirmed the formation of gold nanoparticle. Size distribution was observed by dynamic light scattering method with 50–400 nm particle size; TEM images delivered a sphere of nanoparticle with 10 nm particle size range. Antifilarial activity was assessed against *Setaria cervi* roundworm and by MTT assay. At minimum effective concentration (5 μg/mL), cell viability was reduced up to 5.56% in adult parasite. There was a sign of apoptosis which was evidenced by DNA ladder assay, TUNEL assay, Hoechst staining, and Western blot analysis. Reactive oxygen species generation was confirmed by H2DCFDA assay. These data confirmed the development of nanomaterials with greater antifilaric activity (Saha et al. 2017).

5.2.22 Chitosan-Gold Nanoparticles as Cholesterol Sensor

Gomathi et al. (2011) developed a cholesterol biosensor based on chitosan-gold nanoparticle. The first step of the development was chitosan nanofiber formation by the reaction of hydrochloric acid, ammonium persulfate, petroleum ether, and chitosan, then the nanofiber (50–100 nm of spherical particle) was immobilized on

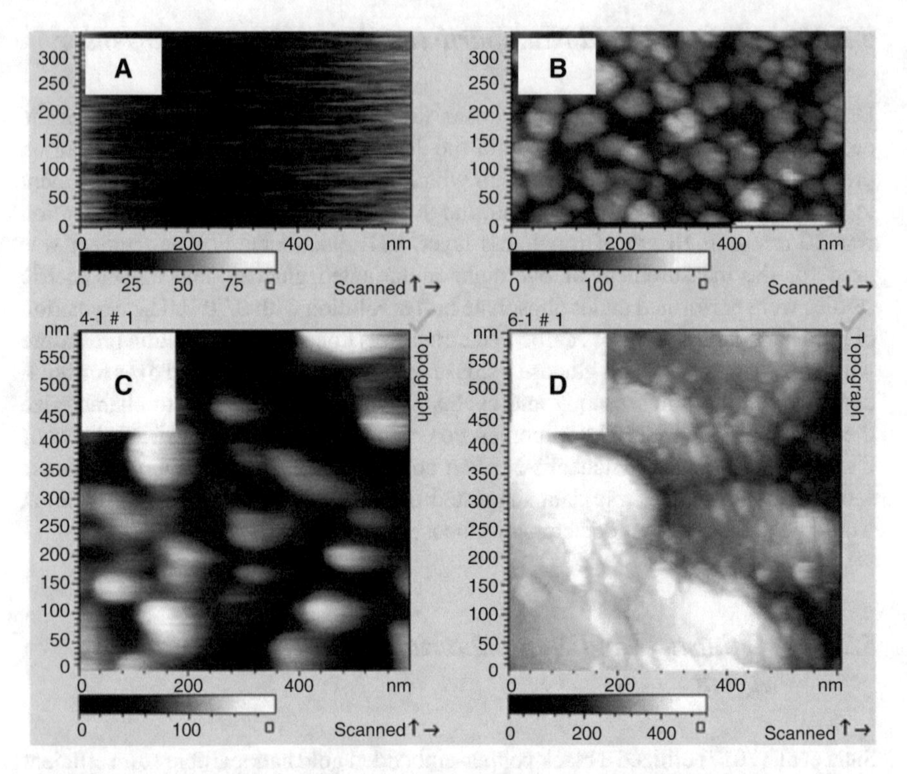

Fig. 5.6 AFM images of bare gold substrate (**a**), electrochemically deposited gold nanoparticles (**b**), chitosan (**c**), chitosan-gold nanoparticles (**d**) on gold surfaces (Source: Du et al., copyright © 2007 with permission from Elsevier B.V.)

indium-doped tin oxide-coated glass electrode followed by amalgamation of chloroauric acid. Finally, cholesterol oxidase was immobilized on the nanoparticle, which was characterized by field emission SEM, and the electrochemical behavior of the biosensor was examined by hydrodynamic voltameter and amperometry, which resulted in linear response of 1–45 M concentration within less than 5 s response time followed by extended stability. These data clearly suggest the greater electrochemical detection of cholesterol by the biosensor (Gomathi et al. 2011).

5.2.23 Chitosan-Gold Nanoparticle as Bacillus cereus *Electrochemical Immunosensor*

Kang et al. (2013) developed a new generation electrochemical biosensor for the detection of *Bacillus cereus*. The biosensor was developed by the reaction between gold nanoparticle, horseradish peroxidase, and thionine chitosan film. *Bacillus cereus*-immunized SP2/0 myeloma and spleen cells were used to prepare hybridoma cell.

The fabrication of *Bacillus cereus* on glass carbon electrode was done by the reaction of chitosan on acetate solution followed by the reaction of sodium hydroxide solution. Transmission electron microscopy and UV–visible spectroscopy were used to characterize the formulation and the outcomes showed that particles were spherical, epigranular with 15 nm particle size. Cyclic voltagram revealed that redoc current was linear with scan rate. The sensation of the nanoparticle was 5.0×10^1 to 5.0×10^4 cfu/mL as evaluated by chronoamperometric method. The results revealed that this immunosensor would be the best tool for *B. cereus* identification in the food sample (Kang et al. 2013).

5.2.24 Chitosan–Gold Nanoparticles as Carcinoembryonic Antigen Immunosensor Application

Gao et al. (2011) developed a chitosan-crosslinked gold nanoparticle immunosensor for the identification of carcinogenic antigen. The formulation was created by the reaction of carboxylated multiwalled carbon nanotube and chitosan solution followed by the addition of gold colloid. Application of constant voltage −1.5 V for 3 min and fabrication of chitosan-gold nanoparticle (with 16 nm of average diameter) over glass carbon nanotube was done. Immunosensation toward carcinogenic antigen of the nanoparticle was measured by electrochemical impedance spectroscopy and cyclic voltameter using $K_3[Fe(CN)_6]/K_4[Fe(CN)_6]$ mixture in phosphate buffer solution; which was observed with 0.1–2.0 and 2.0–200.0 ng mL^{-1} of two linear expression with LOD value of 0.04 ng mL^{-1}. These outcomes revealed the development of new carcinogenic antigen immunosensor (Gao et al. 2011).

5.2.25 Chitosan-Gold Nanoparticle as Mucosal Vaccine Delivery

Barhate et al. (2013) developed a new targeted delivery agent made of chitosan-gold nanoparticle for antigen tetanus toxoid along with *Quillaja saponaria* extract with mucosal vaccine delivery. The nanoparticle was developed upon reaction of chitosan and chloroauric acid. Stability of the nanoparticle was characterized by zeta potential and pH value, which was observed with 40 nm spherical size and +35 mV of positively charged portions along with 65% of tetanus toxoid and 0.01% of *Q. saponaria* extract without any alteration of tetanus toxoid secondary structure. ELISA data was observed with antigen specificity. Tetanus toxoid amalgamated with *Q. saponaria* extract increased the activity by 28 times. These data indicated that co-administration of tetanus toxoid and *Q. saponaria* extract embedded with nanoparticle would work as a remarkable approach for oral delivery of vaccine (Barhate et al. 2013).

5.2.26 Chitosan-Gold Nanoparticle as In Vitro Cytotoxic Agent Against Human Fibroblasts Cells

Yang and Li (2015) developed a chitosan-crosslinked gold nanoparticle as in vitro cytotoxic agent against human fibroblasts cells. The nanoparticle was obtained by chitosan oligosaccharide and chloroauric acid tetrahydrate. High-resolution transmission electron microscopy and FTIR were used to characterize the morphology and particle size of the nanoparticle. MTT assay on human fibroblast cell was used to evaluate in vitro toxicity assay and computed tomography was used to evaluate the in vivo biodistribution of chitosan oligosaccharide. The shifting of absorbance was observed with 542–566 nm at pH value 2.90 and it showed the presence of anionic form of chloroauric acid $(AuCl_4)^-$. Average particle size of the nanoparticle was 115.21 ± 16.87 nm with spherical shape. At 15.6 μg/mL, 31.3 μg/mL, 62.5 μg/mL, 125.0 μg/mL, 250.0 μg/mL, and 500.0 μg/mL concentration of nanoformulation, it was treated with human fibroblast cell for 24 h and observed with dose-dependent toxicity above 62.5 μg/mL concentration, whereas at a concentration above 500.0 μg/mL, cell morphology was destroyed with the sign of apoptosis (Yang and Li 2015).

5.2.27 Carboxymethyl Chitosan-Gold Nanoparticle as Targeted Anticancer Therapy

Madhusudhan et al. (2014) developed doxorubicin immobilized chitosan-gold nanoparticle for the targeted anticancer therapy. The gold nanoparticle was developed upon reaction of carboxymethylated chitosan and chloroauric acid tetrahydrate and the mixture was autoclaved; it was indicated by red color formation and the particle size of the formulation was 9 nm with spherical shape. The immobilization of doxorubicin into the nanoparticle was done by incubation for 24 h followed by centrifugation at 20,000 rpm for 20 min and at 480 nm absorbance of doxorubicin was recorded by UV-visible spectroscopy. HeLa cell was used to evaluate the anticancer efficacy. The results showed that 83.3% drug was loaded with doxorubicin and nanoparticle was observed with 95% cell viability. Doxorubicin-embedded gold nanoparticle was observed with cervical cancer cell inhibition, which was higher at acidic condition as induced by nigericin. These observations clearly suggested the efficiency of doxorubicin-loaded gold nanoparticle as anticancer theranostic agent (Madhusudhan et al. 2014).

5.2.28 Chitosan-Gold Nanoparticle as Targeted DNA Delivery

Bhattarai et al. (2008) developed a chitosan-crosslinked gold nanoparticle as targeted DNA delivery. The formation of nanoparticle was done by the reaction of chloroauric acid and *N*-acetylated chitosan with sodium borohydride as reducing agent. Plasmid DNA with bacterial-beta-galactosidase within cytomegalovirus promoter along with *E. coli* as host cell were considered for the amplification of plasmids. Mouse embryo cell, colon cancer cell, and breast cancer cell were considered for cell toxicity effect using MTT assay. Outcomes revealed that well-dispersed spherical particles were formed with 10–12 nm particle size with no sign of cytotoxicity at low concentration. Surface potential was reduced by 50–66.6% upon addition of plasmid DNA with increase in hydrodynamic diameter by 13.33%. Higher cell toxicity effect was observed by gold nanoparticle against breast cancer cell line and maximum activity was observed at 4.5 mg/mL concentration as similar with in vivo transfection kit. These outcomes showed that these nanoparticles would be the possible source of DNA delivery (Bhattarai et al. 2008).

5.3 Application of Chitosan-Silver Nanoparticle as Nanotheranostic Agents

5.3.1 Chitosan-Silver Sulfadiazine Nanoparticles as Wound Healer

El-Feky et al. (2017) developed a chitosan-crosslinked silver sulfadiazine nanoparticle for the fast healing of burn. The nanoparticle was created upon reaction between positively charged amino groups of chitosan and negatively charged carboxymethyl-beta-cyclodextrin and silver sulfadiazine was loaded into the formulation with 1:1 molar ratio. Transmission electron microscopy and association efficiency were used to characterize the nanoparticle. Wound healing efficiency was assessed by using Egyptian cotton gauze padded with chitosan-silver sulfadiazine nanoparticle using pad-dry-cure process followed by thermal treatment at 85 °C for 5 min and 150 °C for 3 min. Water absorbency, nitrogen percentage, tensile strength, weight, thickness, FTIR, XRD, in vitro drug release, and assessment of fungicidal activity were used to characterize the chitosan-silver sulfadiazine nanoparticle. Results expressed that chitosan concentration (0.2% w/v) was responsible for nanoparticle formation whereas TEM images come with spherical molecule and 40.56–87.67 nm of particle size diameter. Highest association efficiency was observed with 7 mg/mL of carboxymethyl-beta-cyclodextrin concentration with −0.2% w/v of chitosan. All silver sulfadiazine nanoparticle-embedded dressing material showed 4.9–17 mm and 4–14 mm of inhibition zone for gram-positive and gram-negative bacteria, respectively, and also observed with fungal strain inhibition against *C. albicans*. These data confirmed the development of chitosan-silver sulfadiazine nanoparticle for wound healing (El-Feky et al. 2017).

5.3.2 Chitosan-Silver Nanoparticles as Antitoxoplasmic Agent

Gaafar et al. (2014) developed new chitosan and silver nanoparticle for the inhibition of intracellular protozoan infection caused by *Toxoplasma gondii* by single or in combination. Chitosan nanoparticle was developed by the reaction of chitosan and tripolyphosphate using ionotropic gelation method and silver nanoparticle was developed by the reaction of silver nitrate and trisodium citrate, after that nanoparticles were characterized by transmission electron microscopy and ultraviolet spectroscopy. Mice were immunized with nanoparticles and infected with *T. gondii* tachyzoites and immunological assay was performed by interferon gamma and toxicity was determined by atomic absorption spectrometer. TEM analysis was observed with 90.3–97.00 nm spherical particles. In case of infected control group, mean parasite count was 9.1 and 11.3 found in liver and spleen, respectively, and it was observed that parasite count was correlated with combination therapy of silver and chitosan nanoparticle. The degree of deformity and immobilization of organism were profound for combination therapy with increased amount of interferon gamma. Highest quantity of silver was obtained with liver without any traces of that in brain in case of silver nanoparticle-immunized animals. These data clearly stated that the combination of chitosan and silver nanoparticle will create a positive response toward *T. gondii* (Gaafar et al. 2014).

5.3.3 Chitosan-Silver Nanoparticles as Small Molecules Enhanced Delivery

Levi-Polyachenko et al. (2016) developed chitosan-silver nanoparticle for the delivery of small molecules. The chitosan-silver nanoparticle films were developed by the reaction of chitosan and hexagonal silver nanoparticle and were analyzed by transmission electron microscopy, dynamic light scattering, zeta potential, and differential scanning calorimetry (Fig. 5.7). The outcomes showed 15 nm of average diameter with spherical shape. HepG2 cells were used to evaluate the cell toxicity of the formulation and the intracellular heat generation capability was also recorded by Rhod-conjugated dextran for fast delivery of small molecules (Levi-Polyachenko et al. 2016).

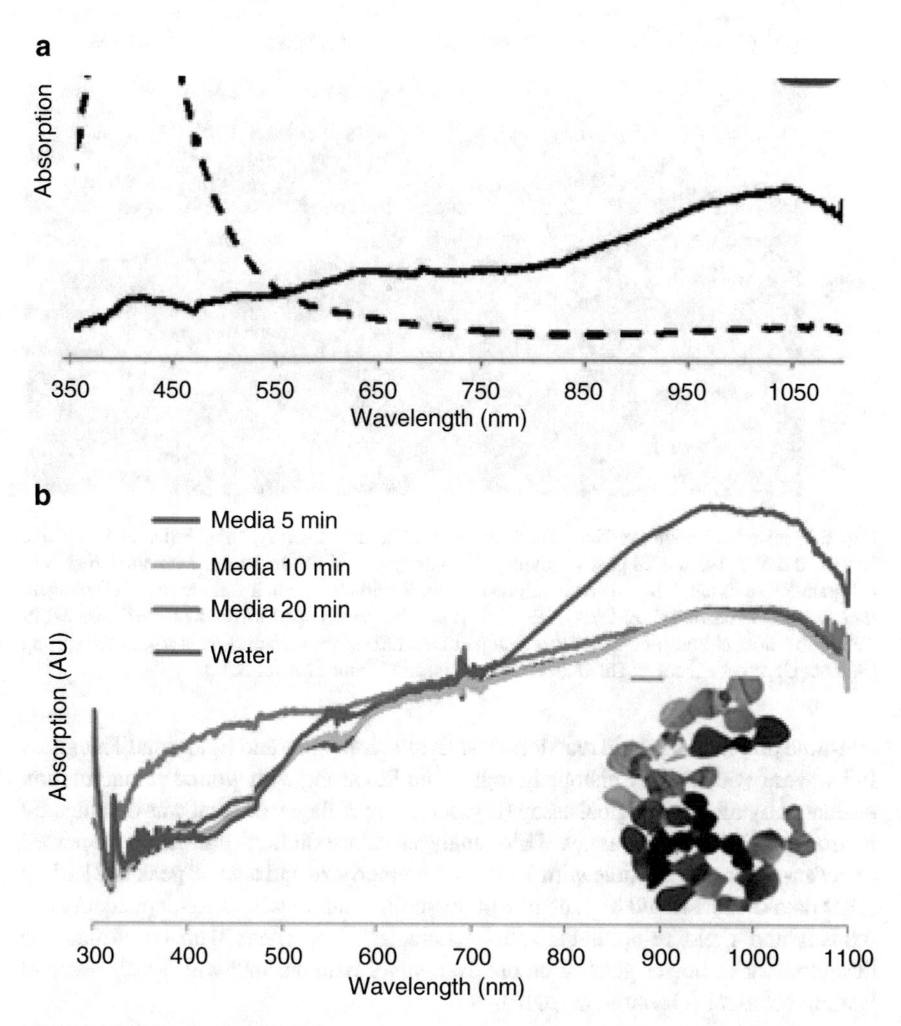

Fig. 5.7 (**a**) TEM image of the hexagonal Ag NP with an approximate edge length of about 100 nm. (**b**) Stability of the hexagonal AgNP in tissue culture media was evaluated over time by UV–vis (Source: Levi-Polyachenko et al., copyright © 2016 with permission from Elsevier B.V.)

5.3.4 Chitosan-Silver Nanoparticles as Wound Healer

Oryan et al. (2018) developed chitosan-silver nanoparticle as burn wound healer. The nanoparticle was created upon reaction between sodium borohydride-reduced silver nitrate and chitosan. FTIR, XRD, and TEM were used to analyze the nanoparticle. Free radical scavenging assay was performed using DPPH method and 7–9-week-old male Sprague-Dawley rats with weight between 180 and 210 g were considered for the evaluation of burn healing process followed by calculation of percent wound closure. Also the mRNA level in each target was calculated by

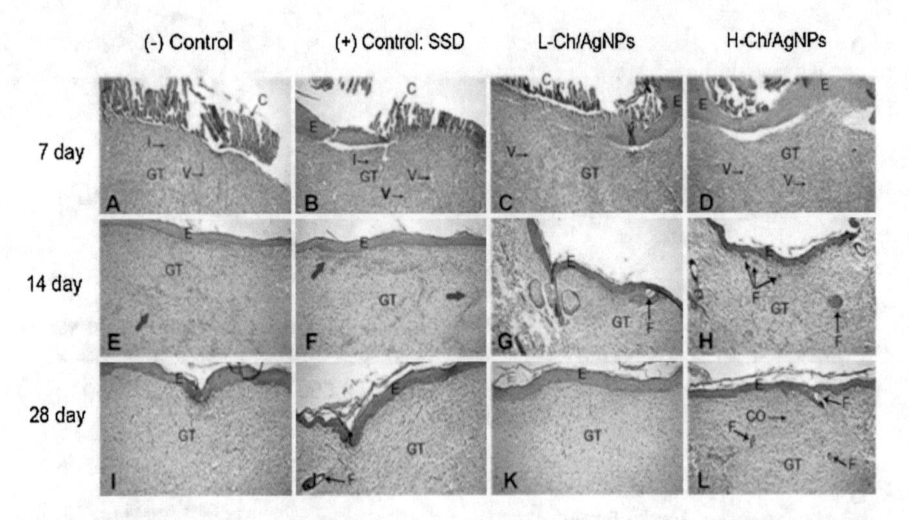

Fig. 5.8 Histopathologic sections from the H-Ch/AgNPs, L-Ch/AgNPs, SSD, and negative control at day 7, 14, and 28 post wounding (Hematoxylin and Eosin, 100∈). *I* inflammatory cell, *GT* granulation tissue, *V* blood vessel, *C* crusty scab, *E* epidermis, *F* hair follicle cell, *CO* collagen, *Red arrows* hemorrhage, *H-Ch/AgNPs* high dose chitosan-capped silver nanoparticles, *L-Ch/AgNPs* low dose chitosan-capped silver nanoparticles, *SSD* silver sulfadiazine. $n = 4$, in each group (Source: Oryan et al., copyright © 2018 with permission from Elsevier B.V.)

real-time polymerase chain reaction. Scar Evaluation Index and Epidermal Thickness Index were evaluated by histopathological studies along with wound contamination evaluated by microbiological assay (Fig. 5.8). The collagen content was obtained by hydroxyproline content assay. TEM analysis data exhibited uniformly dispersed chitosan-silver nanoparticle with 15 mm of average size and a weak peak at 21.8° by XRD data demonstrated high degree of crystalline nature with dose-dependent anti-oxidant and rapid re-epithelialization characteristics. These data confirmed the development of newer generation chitosan-silver nanoparticle with greater wound healing behavior (Oryan et al. 2018).

5.3.5 Chitosan-Silver Nanoparticles as Antibacterial Nanocarrier

Sharma (2017) developed a chitosan-crosslinked silver nanoparticle as nanocarrier for antibacterial drugs. The nanoparticle was obtained upon reaction of silver nitrate and chitosan by ionic gelation method followed by addition of sodium hydroxide solution. TEM, dynamic light scattering, and laser Doppler electrophoresis were used to assess the nanoparticles. Gram-positive bacterial strains as *Bacillus cereus* (MTCC1305), *Enterococcus faecalis* (MTCC439), and gram-negative bacterial strains *E. coli*, *Enterobacter aerogenes* (MTCC2822), *Pseudomonas aeruginosa*

(MTCC2488) were used to assess the antibacterial efficiency of the nanoparticle. AAS studies indicated the role of Ag NPs rather than leached out Ag+ ions for the killing activity against bacteria. The data obtained from TEM analysis confirmed the average particle size with 162.3 ± 35 nm and the flow cytometry using GFP-propidium iodide data revealed that nanoparticles were affected by the integrity of cell wall proceeding to cell lysis and greater activity against both gram-positive and gram-negative bacterial strain. These findings suggested the development of chitosan-crossslinked silver nanoparticle as an effective nanocarrier for antibacterial agents (Sharma 2017).

5.3.6 Chitosan-Silver Nanoparticle as Anti-Tubercular Drug Delivery System

Praphakar et al. (2018) developed rifampicin-pyrazinamide-embedded chitosan-crosslinked silver nanoparticle as antitubercular drug delivery system. At first, silver nanoparticle was formulated using silver nitrate and polyvinylpyrrolidone. Cetyl alcohol-maleic anhydride was developed using cetyl alcohol, maleic anhydride, and dimethylaminopyridine, then pyrazinamide was added into the solution. Then, ethyl-dimethylaminopropyl carbodiimide and *N*-hydroxysuccinimide were used as coupling agents to react between cetyl alcohol-maleic anhydride-pyrazinamide and chitosan solution; finally, the mixture was amalgamated with silver nanoparticles. SEM, TEM, XRD, NMR, dynamic light scattering, encapsulation efficiency, and atomic force microscopy were used to characterize the nanocomposite (Fig. 5.9). Swelling studies, in vitro drug release, antibacterial activity against *E. coli*, *S. aureus*, *K. pneumoniae*, and *B. streptococci*, cytotoxicity assessment against human monocytic cell line, and cell apoptosis assessment using Hoechst 33342 and propidium iodide were also checked as the bioactivities of nanoformulation. Outcomes revealed that 84.23 and 86.70% of rifampicin and pyrazinamide were encapsulated within the 10–20 nm spherical nanocomposite, and at pH 7.4, a maximum of 20% of drugs were released from the system. Greater antibacterial and cytotoxic effects were observed for the nanocomposites. These data clearly expressed the credits of the nanocomposite as delivery system for antitubercular agents (Praphakar et al. 2018).

5.3.7 Chitosan-Silver Nanoparticles as SERS Substrate for Biomedical Applications

Jung et al. (2013) developed surface-enhanced Raman scattering-fabricated chitosan-crosslinked silver nanoparticle of 20–100 nm size with three-dimensional porous structure as atopic dermatitis genetic marker for different biomedical application. The nanoparticle was created using chitosan and silver nitrate, reduced by sodium boro-

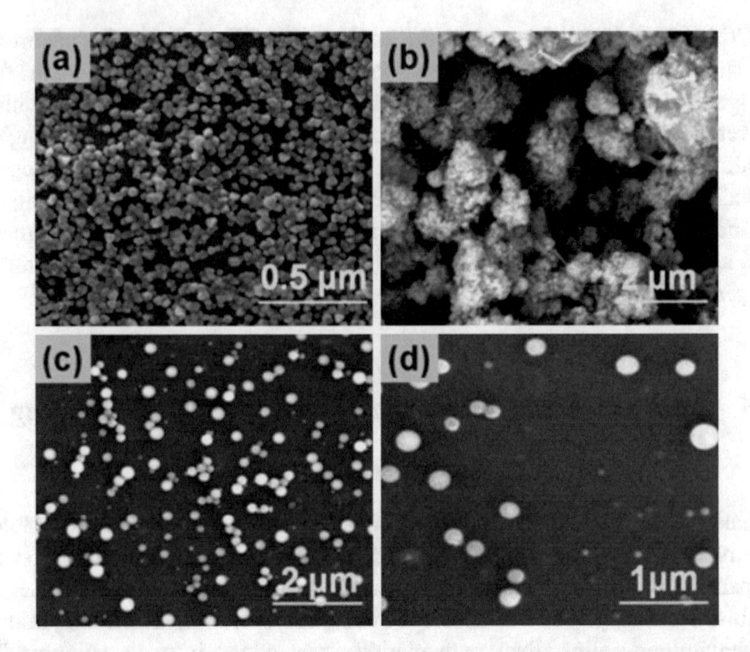

Fig. 5.9 SEM images of (**a**) Ag NPs, (**b**) CS-g-(CA-MA-PZA), and Ag NPs and RF loaded CS-g-(CA-MA-PZA) at (**c**) 2 μm, (**d**) 1 μm (Source: Praphakar et al., copyright © 2018 with permission from Elsevier B.V.)

hydride. SEM, TEM, and XRD were used to study the properties of nanoparticle. Rhodamine B and chemokine CL17 were used as fluorescence dye for Raman spectra analysis as analyzed by SENTERRA confocal Raman system. SERS activity was evaluated using Rhodamine B and thiolated single-stranded DNA as marker. The pore size was 40–100 μm. Four peaks at 38°, 44°, 64°, and 77° confirmed the presence of face-centered cubic silver crystal and peak at 21.9° suggested low crystalline nature of the structure. SERS enhancement factor was 6.7×10^4. These data suggested the development of chitosan-silver nanoparticle as SERS substrate for responsible biomedical application (Jung et al. 2013).

5.3.8 Chitosan-Silver Nanoparticles as Anti-Cytotoxic Effect on MCF-7

Nayak et al. (2016) developed a combination of vitamin A, vitamin C, and chitosan-silver nanoparticle as effective inhibitor of breast cancer cell line (MCF-7). The nanoparticle was developed upon reaction of chitosan, sodium tripolyphosphate, glucose, and ascorbic acid with the further addition of vitamin E and catechol which was amalgamated with *H. rosa-sinensis* extracted silver nanoparticle, 155.9 and 182.8 nm of particle size. Polydispersity index, surface charge, field emission SEM, attenuated total reflection-FTIR, and XRD were used to chemically study the nanoparticle. In vitro antioxidant assay by DPPH radical scavenging, hydrogen peroxide scavenging, and nitric oxide scavenging; hemocompatibility test by hemolysis, hemagglutination, and erythrocyte aggregation studies followed by MTT assay on MCF-7 cell line and in vitro reactive oxygen species by dichloro-dihydro-fluorescein diacetate assay were used to biologically characterize the nanoparticle. The outcomes from MTT assay revealed that IC_{50} was observed as 53.36 ± 0.36 μg/mL by chitosan-ascorbic acid-glucose nanocomposite, 55.28 ± 0.85 μg/mL by chitosan-vitamin E nanocomposite, 63.72 ± 0.27 μg/mL by chitosan-catechol nanocomposite, and 58.53 ± 0.55 μg/mL by chitosan-silver nanoparticles. These findings confirmed the synergistic effect of antioxidant, silver and chitosan nanoparticle as targeted delivery for breast cancer cell line (Nayak et al. 2016).

5.3.9 Chitosan-Silver Nanocomposites as Potential Wound Healer

Hernandez-Rangel et al. (2019) developed chitosan-silver nanoparticle 7–50 nm of average size and spherical shaped films as potential dressing material for wound. The nanoparticle was prepared upon reaction of chitosan and silver nitrate and nanoparticle was poured into a petri dish and dried at 60 °C to obtain the film. TEM, ICP-OES, FTIR, swelling property, vapor transmission rate, and biodegradability data were used to characterize the film by means of morphology and physical characteristic. In vitro drug release from the film, antibacterial activity against gram-positive *S. aureus* (ATCC 25923) and *S. epidermidis* (ATCC 14990) and gram-negative *E. coli* (ATCC 33780) and *P. aeruginosa* (ATCC 43636), and immunofluorescence using human fibroblast cell were used to evaluate the release properties, antibacterial efficacy, and biocompatibility behavior, respectively. TEM analysis data showed that spherical, 7–50 nm silver nanoparticles were dispersed into the film and 260% of swelling was observed by the film. Antibacterial efficacies of the film were increased in a dose-dependent manner and cell viability was greater than 90% with positive sign of tropoelastin, procollagen type I, and Ki-67. These data stated that silver concentration of 0.04–0.20% weight would be profound for wound dressing (Hernandez-Rangel et al. 2019).

5.3.10 Chitosan-Silver Nanocomposites as Antibacterial and Tissue Regeneration Agent

Luna-Hernandez et al. (2017*)* developed chitosan-silver nanocomposite as potential burn healer with the inhibition of bacterial growth and tissue regeneration for fast recovery. The nanocomposite was prepared by the reaction of chitosan and silver nitrate, poured in petri dish and dried to develop film. UV-visible spectroscopy, FTIR, TEM, and atomic force microscopy were used to analyze the formulation. Antibacterial efficacy was evaluated against *S. aureus* and *P. aeruginosa* followed by thermal burn and histological assay. Silver content of 0.018% in chitosan-silver nanocomposite was observed with 7.01–33.09 nm of average particle size. A complete inhibition of bacterial growth was observed after 1.5 h of exposure. After 7 days of treatment with chitosan-silver nanocomposite on the thermal burn induced in rats, tissue regeneration was evidenced by increased amount of myofibroblast formation, collagen remodeling, and blood vessel formation indicating faster healing. These data clearly proved the formation of chitosan-silver nanoparticle for tissue regeneration (Luna-Hernandez et al. 2017).

5.3.11 Chitosan-Silver Nanoparticles for Effective Delivery of Gene

Sarkar et al. (2015) developed chitosan-polyacrylamide copolymerized silver nanoparticle for the effective delivery of gene to treat diseases. The formulation was developed upon reaction of chitosan-polyacrylamide copolymer and silver nitrate in a one-pot synthesis using polyethylene glycol as stabilizer. The dynamic light scattering and transmission electron microscope were used to evaluate the morphology of the formulation. pDNA was immobilized on the nanoparticle and binding of pDNA on nanoparticle was evaluated by agarose gel electrophoresis. In vitro cytotoxicity against HeLa and A549 cell line using MTT assay was evaluated. Average size of the nanoparticle was 38 ± 4 nm with greater distribution of particle into the formulation. Peptide sequence of Arg–Gly–Asp–Ser (RGDS) was used to enhance the gene transfection with 42 ± 4% and 30 ± 3% of transfection in HeLa and A549 cells, respectively. These data clearly indicated that copolymerized silver nanoparticle was with minimal induced toxicity and effectively delivered gene for targeted therapy (Sarkar et al. 2015).

5.4 Application of Technetium-Radiolabeled Chitosan Nanoparticle as Nanotheranostic Agents

5.4.1 Ultrafine Chitosan-Technetium Nanoparticle for Biodistribution into Bone Marrow

Banerjee et al. developed a radiolabeled technetium (^{99m}Tc) chitosan nanoparticle for the biodistribution into bone marrow for bone imaging. The chitosan nanoparticle was developed using sodium bis(ethylhexyl) sulfosuccinate and chitosan in

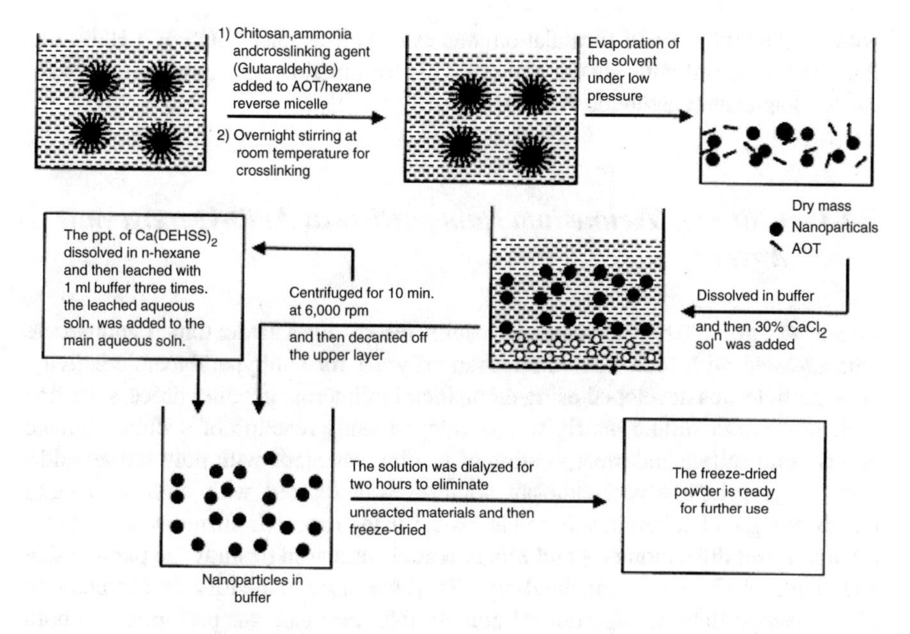

Fig. 5.10 Flowchart of preparation of crosslinked chitosan nanoparticles (Source: Banerjee et al., copyright © 2002 with permission from Elsevier B.V.)

acetic acid and Tris–HCl. Quasi-elastic light scattering (QELS), scanning electron microscope, and FTIR were used to characterize the size and morphology of the nanoparticle, and the radiolabeling efficiency of ^{99m}Tm was checked by thin layer chromatography, whereas gamma imaging study was performed through the dorsal ear vein of animal (Fig. 5.10). The particle size of the nanoparticle was 30–110 nm in diameter upon all amines when crosslinked. TEM images observed with spherical particle without aggregation and 85% of technetium was complexed with nanoparticle. Gamma studies observed that after 2 h of exposure, sample was visible in liver, kidney, bladder, and vertebral column. These biodistribution data confirmed the development of nanoparticle to identify biodistribution in bone marrow (Banerjee et al. 2002).

5.4.2 Chitosan-Technetium Nanoparticle for Radiolabeling

Gundogdu et al. (2015) developed technetium-radiolabeled chitosan nanoparticle for radiolabeling. The chitosan nanoparticle was developed by ionotropic gelation method with reaction with sodium tripolyphosphate and nanoformulation was developed. Photon correlation spectroscopy and scanning electron microscopy were used to characterize the formulation. Particle size diameter of the formulation was within 100–800 nm and formulation was evaluated against U_2OS and H209 cell lines and observed with greater activity. This formulation was highly effective for cancer diagnosis (Gundogdu et al. 2015).

5.4.3 Chitosan-Technetium Nanoparticle as Antihyperglycemic Agent

Lopes et al. (2017) developed insulin-fused alginate/dextran nanoparticle amalgamated with radiolabeled chitosan polymer for antihyperglycemic activity. Nanoparticle was developed using emulsification/internal gelation process. At first alginate-dextran sulfate matrix was developed using reaction of sodium alginate and dextran sulfate and incorporation of insulin was made with polyoxamer addition. Polyethylene glycol-chitosan solution was reacted with alginate-dextran matrix along with albumin solution at pH 5.1 using magnetic stirrer at 800 rpm for 60 min. Laser diffractometry and zeta potential were used to study the particle size and diameter. Field environmental-cryo-SEM was used to identify the morphology of the nanoparticle. Intraperitoneal glucose tolerance test was performed on both type 1 and type 2 diabetic rat by using 50 IU/kg insulin-entrapped nanoparticle and outcomes were observed with prolonged antihyperglycemic effect within 8–12 h period as compared to that of noninsulin-entrapped nanoparticle. The outer layer of the formulation was bovine serum albumin-type structure, which was radiolabeled by ^{99m}Technetium for biodistribution studies in mice balb-c identified by gamma camera and monitored for 1440 min throughout the GI tract. As intestine was the site of absorption for insulin, greater absorption was observed at 120–180 min. These data suggested that 160–170 nm sized nanoparticle was observed with greater antihyperglycemic activity and the 99mTc-ALB-NP formulation created a positive connection between intestinal wall and mucoadhesiveness of chitosan. So this insulin-entrapped nanoformulation would be greatly influential on the antihyperglycemic disease (Lopes et al. 2017).

5.4.4 Chitosan-Technetium Nanoparticle for Nose-to-Brain Delivery of Thymoquinone

Alam et al. (2012) developed a thymoquinone-loaded chitosan-targeted nose-to-brain delivery system for Alzheimer's disease. The nanoparticle was developed by the reaction of amino group of chitosan and tripolyphosphate with the amalgamation of thymoquinone using ionic gelation method. Dynamic light scattering and X-ray diffraction were used to characterize the nanoparticle, which showed particle diameter between 150 and 200 nm range, and it also comes with a good correlation between particle size and drug:chitosan ratio whereas tripolyphosphate:chitosan ratio was less correlated with particle size. In vitro drug release from the nanoparticle was performed by dialysis sac, ex vivo permeation through nasal passage was observed by HPLC method and the biodistribution of the nanoparticle was monitored through ^{99m}Technetium. Percent brain targeting efficiency (% DTE) and brain targeting potential (% DTP) were evaluated as the parameter for nose-to-brain delivery, which showed 3318.24 ± 65.79 and 96.99 ± 3.64 data, respectively. Drug entrapment efficiency and loading were observed with 63.3% and 31.23%, respectively, and were negatively correlated with drug:chitosan ratio. Amorphous thymoquinone was dispersed into the nanoparticle. This study comes with a conclusion that this formulation delivers thymoquinone more rapidly with proper cerebrospinal fluid penetration with greater extent of applicability against Alzheimer's disease (Alam et al. 2012).

5.4.5 Chitosan-Technetium Nanoparticle as Oral Delivery System

Chauhan and Bhatt (2016) developed amphotericin B-loaded chitosan nanoparticle as an oral delivery system. The nanoparticle was developed by the reaction of chitosan, bovine serum albumin, and tripolyphosphate in amphotericin B solution by ionic gelation method and the biodistribution of the nanoparticle was done by ^{99m}Technetium with stannous chloride as reducing agent. Nanoparticles were observed with 440 nm of mean size, +22 mV of zeta potential, 9.02% of loading capacity, and 58% of entrapment efficiency as well as the gamma scintigraphic images correlated with drug localization effect. Biodistribution data revealed that amphotericin B was largely accumulated in stomach and intestine, and after 10 h, maximum molecule had remained in the GIT. These data confirmed the nanoformulation as better oral drug delivery system (Chauhan and Bhatt 2016).

5.4.6 Chitosan-Technetium Nanoparticle for Breast Cancer Diagnosis

Ekinci et al. (2015) developed methotrexate-loaded chitosan nanoparticle for the diagnosis of breast cancer progression. The nanoparticle was developed using chitosan and tripolyphosphate and the resultant mixture was centrifuged with methotrexate to obtain final molecule. The formulation was characterized by polydispersity index and zeta potential and the formulations were kept at 25 ± 2 °C temperature/60 ± 5% humidity and 40 ± 2 °C temperature/75 ± 5% humidity for 6 months to assess the accelerated stability profiles. FTIR, SEM, and encapsulation efficiency were used to characterize the formulation. The mean diameter and zeta potential of the formulations were 169 nm, 427.63 nm and 20.133 mV, 29.067 mV, respectively, with 35% and 64% of encapsulation efficiency (Figs. 5.11, 5.12). The nanoformulation was radiolabeled with ^{99m}Technetium with suitable emission of gamma rays and greater than 90% labeling efficiency and stability up to 6 h were observed. Antiproliferation

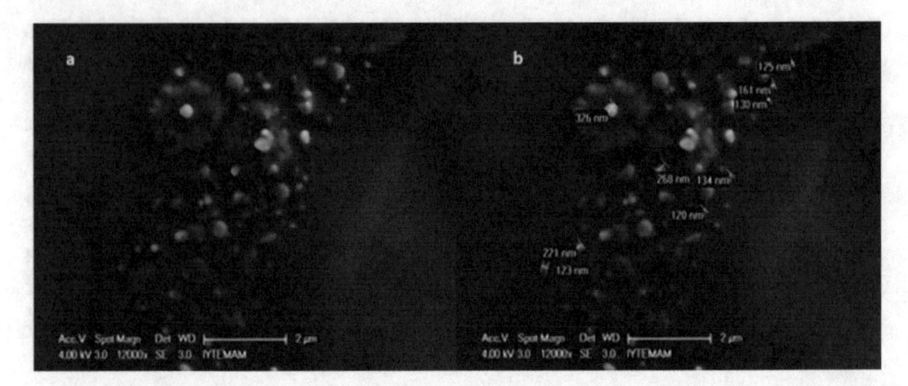

Fig. 5.11 SEM imaging of F1 (**a**: F1 images, **b**: F1 images with particle size) (Source: Ekinci et al., copyright © 2015 with permission from Elsevier B.V.)

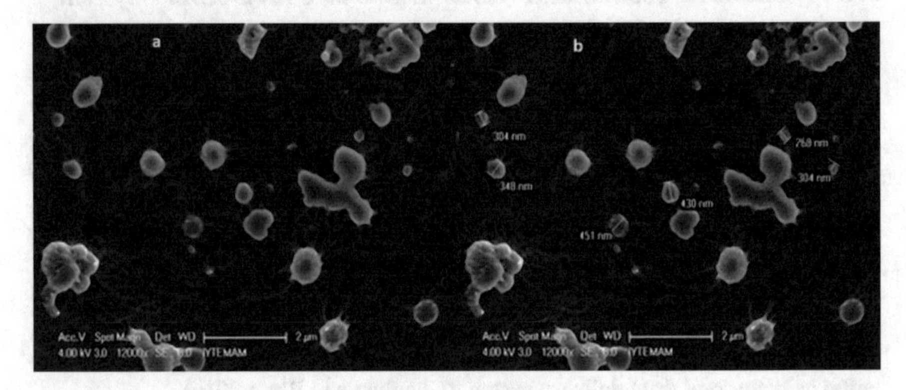

Fig. 5.12 SEM imaging of F2 (**a**: F2 images, **b**: F2 images with particle size) (Source: Ekinci et al., copyright © 2015 with permission from Elsevier B.V.)

efficiency was evaluated against human breast cancer cell line (MCF-7) and human keratinocyte cell line (HaCaT). The radiolabeled chitosan nanoparticle was found two times more active in breast cancer cell than that of normal cell. These data clearly indicated the efficiency of methotrexate-loaded chitosan nanoparticle in relevance to breast cancer (Ekinci et al. 2015).

5.5 Conclusion

This chapter provides a brief and compressive data about the nanotheranostic applicability of chitosan-gold/silver nanoparticle or technetium-radiolabeled chitosan nanoparticle. Chitosan-gold nanoparticle has been applied in carcinoma, detection of solid tumors, prostate cancer, glucose, amplified nucleic acid, and cancer cell imaging and also used as immunosensor, cholesterol sensor, electrochemical sensor, etc. Chitosan-silver nanoparticle has also been applied in inhibition of bacterial growth, tissue regeneration process, wound healing process, etc. Technetium-radiolabeled chitosan nanoparticle has directly been used to study the biodistribution of molecule, specialized delivery system of thymoquinone, antiparkinson drugs for nose-to-brain delivery, etc. The main limitation of nanoparticle includes toxicity on prolong exposure of it which was considered as nuisance dust. Technetium may be considered as highly appreciable for biodistribution study, but it can cause serious damage, it may even cause cancer or create genetic mutation, if proper protection is not taken. Despite these limitations, chitosan-gold/silver or technetium radiolabeled nanoparticles are extensively used in the field of controlled drug delivery system, cell imaging, anticancer therapy, and radiopharmaceutical applications.

5.6 Future Scope

Nowadays, natural polymers are the primary selection for scientists to develop new pharmaceutical agent(s) for human due to their biocompatibility, biodegradability along with diversified spectrum of activity. So, if we focus on the natural polymer as plant derived, microorganism derived or animal derived and make them amalgamated with metal nanoparticles such as gold, silver, copper, zinc, and titanium, they will appear as novel nanotheranostic agents with greater activity, lesser adverse effect, and higher diagnostic properties, which will act as boon for the mankind. In future various other polymers derived from microbial source as alginate or plant-derived polymers as albumin or chondroitin sulfate used as crosslinking agent for nanotheranostic application of drug molecules, the spectrum of activity will increase.

References

Akhter S, Ahmad MZ, Ahmad FJ, Storm G, Kok RJ. Gold nanoparticles in theranostic oncology: current state-of-the-art. Expert Opin Drug Deliv. 2012;9(10):1225–43.

Alam S, Khan ZI, Mustafa G, Kumar M, Islam F, Bhatnagar A, Ahmad FJ. Development and evaluation of thymoquinone encapsulated chitosan nanoparticles for nose-to brain targeting: a pharmacoscintigraphic study. Int J Nanomedicine. 2012;7:5705–18.

Banerjee T, Mitra S, Singh AK, Sharma RK, Maitra A. Preparation, characterization and biodistribution of ultrafine chitosan nanoparticles. Int J Pharm. 2002;243:93–105.

Barhate G, Gautam M, Gairola S, Jadhav S, Pokharkar V. Quillaja saponaria extract as mucosal adjuvant with chitosan functionalized gold nanoparticles for mucosal vaccine delivery: stability and immunoefficiency studies. Int J Pharm. 2013;441:636–42.

Bhatnagar I, Mahato K, Ella KKR, Asthana A, Chandra P. Chitosan stabilized gold nanoparticle mediated self-assembled gliP nanobiosensor for diagnosis of invasive Aspergillosis. Int J Biol Macromol. 2018;110:449–56.

Bhattarai SR, Bahadur KCR, Aryal S, Bhattarai N, Kim SY, Yi HK, Hwang PH, Kim HY. Hydrophobically modified chitosan/gold nanoparticles for DNA delivery. J Nanopart Res. 2008;10:151–62.

Boca SC, Potara M, Toderas F, Stephan O, Baldeck PL, Astilean S. Uptake and biological effects of chitosan-capped gold nanoparticles on Chinese hamster ovary cells. Mater Sci Eng C. 2011;31:184–9.

Chauhan M, Bhatt N. Amphotericin- B loaded chitosan nanoparticles intended for oral delivery: biodistribution, pharmacokinetic and scintigraphy study in rabbit model. Int J Eng Tech Sci Res. 2016;3(11):1–6.

Dervisevic M, Dervisevic E, Cevik E, Senel M. Novel electrochemical xanthine biosensor based on chitosan polypyrrole gold nanoparticles hybrid bio-nanocomposite platform. J Food Drug Anal. 2017;25(3):510–9.

Du Y, Luo XL, Xu JJ, Chen HY. A simple method to fabricate a chitosan-gold nanoparticles film and its application in glucose biosensor. Bioelectrochemistry. 2007;70:342–7.

Ekinci M, Ilem-Ozdemir D, Gundogdu E, Asikoglu M. Methotrexate loaded chitosan nanoparticles: preparation, radiolabeling and in vitro evaluation for breast cancer diagnosis. J Drug Deliv Sci Technol. 2015;30:107–13.

El-Feky GS, Sharaf SS, Shafei AE, Hegazy AA. Using chitosan nanoparticles as drug carriers for the development of a silver sulfadiazine wound dressing. Carbohydr Polym. 2017;158:11–9.

Fathi M, Zangabad PS, Barar J, Aghanejad A, Erfan-Niya H, Omidi Y. Thermo-sensitive chitosan copolymer-gold hybrid nanoparticles as a nanocarrier for delivery of erlotinib. Int J Biol Macromol. 2018;106:266–76.

Fathy MM, Mohamed FS, Elbialy N, Elshemey WM. Multifunctional chitosan-capped gold nanoparticles for enhanced cancer chemo-radiotherapy: an invitro study. Phys Med. 2018;48:76–83.

Gaafar MR, Mady RF, Diab RG, Shalaby TI. Chitosan and silver nanoparticles: promising anti-toxoplasma agents. Exp Parasitol. 2014;143:30–8.

Gao X, Zhang Y, Wu Q, Chen H, Chen Z, Lin X. One step electrochemically deposited nanocomposite film of chitosan–carbon nanotubes–gold nanoparticles for carcinoembryonic antigen immunosensor application. Talanta. 2011;85:1980–5.

Gomathi P, Ragupathy D, Choi JH, Yeum JH, Lee SC, Kim JC, Lee SH, Ghim HD. Fabrication of novel chitosan nanofiber/gold nanoparticles composite towards improved performance for a cholesterol sensor. Sensor Actuator B Chem. 2011;153:44–9.

Guan H, Yu J, Chi D. Label-free colorimetric sensing of melamine based on chitosan-stabilized gold nanoparticles probes. Food Control. 2013;32:35–41.

Gundogdu E, Ilem-Ozdemir D, Ekinci M, Ozgenc E, Asikoglu M. Radiolabeling efficiency and cell incorporation of chitosan nanoparticles. J Drug Deliv Sci Technol. 2015;29:84–9.

Hamburg MA, Collins FS. The path to personalized medicine. N Engl J Med. 2010;363(4):301–4.

Hernandez-Rangel A, Silva-Bermudez P, Espana-Sanchez BL, Luna-Hernandez E, Almaguer-Flores A, Ibarra C, Garcia-Perez VI, Velasquillo C, Luna-Barcenas G. Fabrication and in vitro behavior of dual-function chitosan/silver nanocomposites for potential wound dressing applications. Mater Sci Eng C. 2019;94:750–65.

Jung GB, Kim JH, Burm JS, Park HK. Fabrication of chitosan-silver nanoparticle hybrid 3D porous structure as a SERS substrate for biomedical applications. Appl Surf Sci. 2013;273:179–83.

Kang X, Pang G, Chen Q, Liang X. Fabrication of Bacillus cereus electrochemical immunosensor based on double-layer gold nanoparticles and chitosan. Sensor Actuator B Chem. 2013;177:1010–6.

Kangkamano T, Numnuam A, Limbut W, Kanatharana P, Thavarungkul P. Chitosan cryogel with embedded gold nanoparticles decorated multiwalled carbon nanotubes modified electrode for highly sensitive flow based non-enzymatic glucose sensor. Sensor Actuator B Chem. 2017;246:854–63.

Labala S, Jose A, Vamsi V, Venuganti K. Transcutaneous iontophoretic delivery of STAT3 siRNA using layer-by-layer chitosan coated gold nanoparticles to treat melanoma. Colloid Surf B Biointerfaces. 2016;146:188–97.

Levi-Polyachenko N, Jacob R, Day C, Kuthirummal N. Chitosan wound dressing with hexagonal silver nanoparticles for hyperthermia and enhanced delivery of small molecules. Colloid Surf B Biointerfaces. 2016;142:315–24.

Lim EK, Kim T, Paik S, Haam S, Huh YM, Lee K. Nanomaterials for theranostics: recent advances and future challenges. Chem Rev. 2015;115(1):327–94.

Lopes M, Aniceto D, Abrantes M, Simoes S, Branco F, Vitoria I, Botelho MF, Seiça R, Veiga F, Ribeiro A. In vivo biodistribution of antihyperglycemic biopolymer-based nanoparticles for the treatment of type 1 and type 2 diabetes. Euro J Pharm Biopharm. 2017;113:88–96.

Luna-Hernandez E, Cruz-Soto ME, Padilla-Vaca F, Mauricio-Sanchez RA, Ramirez-Wong D, Munoz R, Granados-Lopez L, Ovalle-Flores LR, Menchaca-Arredondo JL, Hernandez-Rangel A, Prokhorov E, Garcıa-Rivas JL, Espana-Sanchez BL, Luna-Barcenas G. Combined antibacterial/tissue regeneration response in thermal burns promoted by functional chitosan/silver nanocomposites. Int J Biol Macromol. 2017;105(1):1241–9.

Ma H, Sun J, Zhang Y, Bian C, Xia S, Zhen T. Label-free immunosensor based on one-step electrodeposition of chitosan-gold nanoparticles biocompatible film on Au microelectrode for determination of aflatoxin B1 in maize. Biosens Bioelectron. 2016;80:222–9.

Madhusudhan A, Reddy GB, Venkatesham M, Veerabhadram G, Dudde AK, Natarajan S, Yang MY, Hu A, Singh SS. Efficient pH dependent drug delivery to target cancer cells by gold nanoparticles capped with carboxymethyl chitosan. Int J Mol Sci. 2014;15:8216–34.

Manivasagan P, Bharathiraja S, Bui NQ, Lim IG, Oh J. Paclitaxel-loaded chitosan oligosaccharide-stabilized gold nanoparticles as novel agents for drug delivery and photoacoustic imaging of cancer cells. Int J Pharm. 2016;511:367–9.

Nayak A, Pal D. Development of pH sensitive tamarind seed polysaccharide alginate composite beads for controlled diclofenac sodium delivery using response surface methodology. Int J Biol Macromol. 2011;49:784–93.

Nayak A, Pal D. Ionotropically-gelled mucoadhesive beads for oral metformin HCl. Int J Pharm Sci Rev Res. 2012;1:1–7.

Nayak AK, Pal D, Malakar J. Development, optimization and evaluation of floating beads using natural polysaccharides blend for controlled drug release. Polym Eng Sci. 2012;53:238–50.

Nayak A, Pal D, Das S. Calcium pectinate-fenugreek seed mucilage mucoadhesive beads for controlled delivery of metformin HCl. Carbohydr Polym. 2013;96:349–57.

Nayak A, Pal D. Fenugreek seed gum-alginate mucoadhesive beads of metformin –HCl: design, optimization and evaluation. Int J Biol Macromol. 2013;54:144–54.

Nayak A, Pal D. Ispaghula mucilage-gellan mucoadhes polysaccharide-gellan mucoadhesive beads for controlled release of metformin HCl. Carbohydr Polym. 2014a;103:41–50.

Nayak A, Pal D. Tamarind seed polysaccharide-gellan mucoadhesive beads for controlled release of metformin HCl. Carbohydr Polym. 2014b;103:154–63.

Nayak A, Pal D. Development of calcium pectinate-tamarind seed polysaccharide mucoadhesive beads containing metformin HCl. Carbohydr Polym. 2014c;101:220–30.

Nayak AK, Pal D. Polymeric hydrogels as smart biomaterials part of the Springer Series on Polymer and composite materials book series (SSPCM). 2015a. 105–51.

Nayak AK and Pal D. Plant-derived polymers: ionically-gelled sustained drug release systems. In Encyclopedia of biomedical polymers and polymer biomaterials. Taylor & Francis. ISBN 9781439898796. 2015b;8: 6002–17.

Nayak AK and Pal D (2016) Sterculia gum-based hydrogels for drug delivery applications. In Springer Series on Polymer and composite materials—polymeric hydrogels as smart biomaterials. ISBN 978-3-319-25320-6.

Nayak D, Minz AP, Ashe S, Rauta PR, Kumari M, Chopra P, Nayak B. Synergistic combination of antioxidants, silver nanoparticles and chitosan in a nanoparticle based formulation: characterization and cytotoxic effect on MCF-7 breast cancer cell lines. J Colloid Interface Sci. 2016;470:142–52.

Oryan A, Alemzadeh E, Tashkhourian J, Ana SFN. Topical delivery of chitosan-capped silver nanoparticles speeds up healing in burn wounds: a preclinical study. Carbohydr Polym. 2018;200:82–92.

Pakchin PS, Ghanbari H, Saber R, Omidi Y. Electrochemical immunosensor based on chitosan/gold nanoparticle-carbon nanotube as platform and lactate oxidase as the label for the detection of CA125. Biosens Bioelectron. 2018;122:68–74.

Pal D, Nayak AK. Nanotechnology for targeted delivery in cancer therapeutics. Int J Pharm Sci Rev Res. 2010;1(1):1–7.

Pal D, Nayak AK. Alginate, blends and microspheres: controlled drug delivery. In Encyclopedia of biomedical polymers and polymer biomaterials, 11 volume set. 1st edn. Taylor & Francis. ISBN 9781439898796. 2015; 89–98.

Praphakar RA, Jeyaraj M, Ahmed M, Kumar SS, Rajan M. Silver nanoparticle functionalized CS-g-(CA-MA-PZA) carrier for sustainable anti-tuberculosis drug delivery. Int J Biol Macromol. 2018;118(B):1627–38.

Saha SK, Roy P, Mondal MK, Roy D, Gayen P, Chowdhury P, Babu SPS. Development of chitosan based gold nanomaterial as an efficient antifilarial agent: a mechanistic approach. Carbohydr Polym. 2017;157:1666–76.

Salem DS, Sliem MA, El-Sesy M, Shouman SA, Badr Y. Improved chemo-photothermal therapy of hepatocellular carcinoma using chitosan-coated gold nanoparticles. J Photochem Photobiol B Biol. 2018;182:92–9.

Sarkar K, Banerjee SL, Kundu PP, Madrasa G, Chatterjee K. Biofunctionalized surface-modified silver nanoparticles for gene delivery. J Mater Chem B. 2015;3:5266–76.

Shanavas A, Rengan AK, Chauhan D, George L, Vats M, Kaur N, Yadav P, Mathur P, Chakraborty S, Tejaswini A, De A, Srivastava R. Glycol chitosan assisted in situ reduction of gold on polymerictemplate for anti-cancer theranostics. Int J Biol Macromol. 2018;110:392–8.

Sharma S. Enhanced antibacterial efficacy of silver nanoparticles immobilized in a chitosan nanocarrier. Int J Biol Macromol. 2017;104(B):1740–5.

Sugunan A, Thanachayanont C, Dutta J, Hilborn JG. Heavy-metal ion sensors using chitosan-capped gold nanoparticles. Sci Tech Adv Mater. 2005;6:335–40.

Suresh L, Brahman PK, Reddy KR, Bondili JS. Development of an electrochemical immunosensor based on gold nanoparticles incorporated chitosan biopolymer nanocomposite film for the detection of prostate cancer using PSA as biomarker. Enz Microbiol Tech. 2018;112:43–51.

Swierczewska M, Lee S, Chen X. Moving theranostics from bench to bedside in an interdisciplinary research team. Ther Deliv. 2011;2(2):165–70.

Swierczewska M, Liu G, Lee S, Chen X. High-sensitivity nanosensors for biomarker detection. Chem Soc Rev. 2012;41(7):2641–55.

Tammam SN, Khalil MAF, Gawad EA, Althani A, Zaghloul H, Azzazy HME. Chitosan gold nanoparticles for detection of amplified nucleic acid isolated from sputum. Carbohydr Polym. 2017;164:57–63.

Tian K, Siegel G, Tiwari A. A simple and selective colorimetric mercury (II) sensing system based on chitosan stabilized gold nanoparticles and 2,6-pyridinedicarboxylic acid. Mater Sci Eng C. 2017;71:195–9.

Wang B, Ji X, Zhao H, Wang N, Li X, Ni R, Liu Y. An amperometric β-glucan biosensor based on the immobilization of bi-enzyme on Prussian blue–chitosan and gold nanoparticles–chitosan nanocomposite films. Biosens Bioelectron. 2014;55:113–9.

Xiang C, Li R, Adhikari B, She Z, Li Y, Kraatz HB. Sensitive electrochemical detection of Salmonella with chitosan-gold nanoparticles composite film. Talanta. 2015;140:122–7.

Yang J, Chen J, Pan D, Wan Y, Wang Z. pH-sensitive interpenetrating network hydrogels based on chitosan derivatives and alginate for oral drug delivery. Carbohydr Polym. 2013;92:719–25.

Yang N, Li WH. Preparation of gold nanoparticles using chitosan oligosaccharide as a reducing and capping reagent and their in vitro cytotoxic effect on human fibroblasts cells. Mater Lett. 2015;138:154–7.

Chapter 6
Nanomaterials for Selective Targeting of Intracellular Pathogens

Muhammad Ali Syed and Nayab Ali

Abstract Many infectious diseases are caused by intracellular pathogens such as *Mycobacterium tuberculosis*, *Salmonella enterica* serovar Typhi, *Listeria monocytogenes*, *Plasmodium* species, *Toxoplasma gondii*, *Brucella* species, and *Cryptococcus neoformans*. Infections caused by such pathogens are treated with antimicrobial agents. Nevertheless, selective targeting of intracellular pathogens is difficult due to the reason that the drug has to enter the infected host cells in order to target or kill the infectious agent. Further, nonspecific interaction of the antimicrobial agent also affects noninfected body cells. On one side there is substantial loss of drug in the body due to nonspecific interaction while, on the other hand, many drugs find it difficult to enter the host cells. Targeted drug delivery using nanomaterials offers unique and efficient opportunity to deliver the drug loaded in the nanocarriers such as liposomes, polymeric nanoparticles, or micelles into the host cells infected with intracellular pathogens. Furthermore, sustained drug release inside the infected cells may solve the issues of bioavailability and patient compliance. Research studies conducted by different groups have shown promising results of drug-loaded nanocarriers against intracellular pathogens in a number of studies. This chapter discusses various types of intracellular pathogens, nanocarriers, and their role in targeted drug delivery of intracellular pathogens.

Keywords Intracellular pathogens · Nanomaterials · Nanocarriers · Targeted drug delivery · Liposomes

Nomenclature

AuNPs	Gold nanoparticles
CAMP	Cationic antimicrobial peptides
CPP	Cell-penetrating peptides
LPG	Lipophosphoglycan
MPS	Mononuclear phagocytic system

M. A. Syed (✉) · N. Ali
Department of Microbiology, The University of Haripur, Haripur, Pakistan

M. Rai, B. Jamil (eds.), *Nanotheranostics*,
https://doi.org/10.1007/978-3-030-29768-8_6

MSNPs	Mesoporous silica nanoparticles
PEG	Polyethylene glycol
PLA	Poly lactic acid
PLGA	Poly lactide-co-glycolide
PMA	Polymethacrylic acid
PMs	Polymeric micelles
PV	Parasitophorous vacuole
SCV	Salmonella containing vacuole
SLNPs	Solid lipid nanoparticles
T3SS	Type III secretion system

6.1 Introduction

Nanomaterials have attracted great attention in the last few decades due to their unique optical, physiochemical, electronic, thermal, and magnetic properties. Among many other applications in medicine, they are widely being investigated for their utilization as therapeutic drug delivery vehicles including targeted drug delivery systems in a number of diseases and disorders (Hubbell and Chilkoti 2012; Chen et al. 2013; Kumari et al. 2016). The most important features of nanocarriers are their improved pharmacokinetics, stability, controlled release, and usage in site-specific toxicity of the drug they carry (Din et al. 2017; Atbiaw et al. 1965).

One of the main problems with the antibiotics is nonspecific toxicity to human cells. In human body, drugs have to travel to the site of their action whereby they encounter a number of barriers including rapid filtration by kidneys and reticuloendothelial system or transport from blood stream to the target cells. Further, drugs are also supposed to be stable in harsh acidic environment inside the host cells (Gagliardi 2016). Nanomaterial-based carrier systems have emerged as a promising alternative to existing drug delivery systems to cope with such problems (Hubbell and Chilkoti 2012; Xie et al. 2014). Drugs to be administered in the target cells is either adsorbed onto or impregnated, or encapsulated into the nanomaterials such as liposomes, nanoparticles, and micelles (Singh and Lilliard 2009). Of particular importance are the biodegradable, polymeric nanocarriers that are nontoxic, easy to functionalize, and have controllable features such as charge, size, and hydrophobicity (Singh and Lilliard 2009; Moritz and Gezske-Moritz 2015).

Intracellular pathogens such as *Mycobacterium tuberculosis*, *Listeria monocytogenes*, *Brucella* spp., *Salmonella typhi, Plasmodium* spp., and *Toxoplasma gonidii* have the ability to grow and reproduce inside the host cells and hence are particularly more difficult to eradicate from the body as compared to other pathogenic microorganisms. Intracellular pathogens account for millions of illnesses such as listeriosis, Legionnaires' disease, toxoplasmosis, leishmaniasis, typhoid, and tuberculosis with higher morbidity rate (Casadevall 2008). They are special focus of the scientists not only due to the health burden, but also due to hurdles in treating such infections, since many antibiotics are unable to enter the infected cells (Thi et al. 2012). In spite of the discovery of new antibiotics, eradication of intracellular parasites still remains problematic (Ladaviere and Gref 2015; Atbiaw et al. 1965).

Over the last few degrades, biodegradable nanomaterials, such as liposomes, and polymeric nanosystems, e.g., polymeric micelles, niosomes, solid lipid nanoparticles (SLNPs), and dendrimers, have been explored for intracellular delivery of antimicrobial agents that have shown promising bactericidal effects in the target cells (Lamprecht et al. 2001; Xie et al. 2014). These nanocarriers have the ability to reach and accumulate in the infected target cells including macrophages and release drug content to kill the intracellular pathogens efficiently (Xie et al. 2017).

The concept of targeted drug delivery dates back to 1906, when Paul Ehrlich introduced magic bullets to target infectious agents selectively. Targeted drug delivery to intracellular pathogens requires the materials carrying the drug to the infected cells be stable and not cleared immediately from the body (Ranjan et al. 2012). Nanomaterials have been found to be active agents to carry the drug to the target cells when coupled with antibodies or aptamers against the infected cells (Fahmy et al. 2005). This chapter describes basic concepts and some developments made in the field of targeted drug delivery for intracellular pathogens using nanomaterials.

6.2 Intracellular Pathogens

Pathogen adaptation or resistance to antibiotic therapy, increasing migration, and increase in the number of immunosuppressed patients are some of the factors leading to an increase in infectious diseases. Infectious agents can be categorized as extracellular, obligate, or facultative intracellular pathogens (Silva 2012). Proliferation of extracellular pathogens takes place in extracellular body fluids that is highly conducive to the growth of these pathogens and is characterized by the lack of intracellular invasion. In contrast to these, intracellular pathogens such as *Mycobacterium tuberculosis* have the ability to invade host cells. These pathogens replicate within endosomal vesicles or in the cytoplasm of various body cells like macrophages, dendritic cells, erythrocytes, fibroblasts, neutrophils, epithelial, or endothelial cells. Just like viruses, certain bacteria, fungi, and protozoan pathogens exist with the potential of intracellular replication (Dehio et al. 2012). The intracellular invasion gives the pathogens a selective advantage of being protected from host cell defense. Some intracellular pathogens can replicate both inside and outside the host cell and are considered as facultative intracellular, while others are obligate intracellular and can only replicate inside the host cells (Niller et al. 2017).

Intracellular pathogens have evolved several mechanisms that facilitate their intracellular survival and protection from host defense system. Bacteria coordinately express the genes that are responsible for their intracellular survival and this gives the bacteria a selective advantage for their survival (Ernst et al. 1999). A brief discussion of some common intracellular pathogens is given below.

6.2.1 Bacterial Pathogens

A number of bacterial diseases such as tuberculosis, typhoid, and meningitis are caused by intracellular bacterial pathogens. These bacteria have evolved mechanisms to survive inside the host cells and escape from immune system. Examples of some intracellular bacterial agents are discussed in the next few paragraphs.

6.2.1.1 *Salmonella enterica* serovar Typhi

Salmonella enterica serovar Typhi is a facultative intracellular bacterium with the ability to replicate inside phagocytic and nonphagocytic cells. *Salmonella* utilize Type III secretion system (T3SS) for successful intracellular invasion. Membrane ruffle formation and intracellular internalization are mediated by effector proteins like SipA, SipC, SopD, and SopE2. After internalization the bacterium survives in salmonella containing vacuole (SCV). Avoidance of intracellular antibacterial activity is mediated by a number of virulence factors located on Salmonella pathogenicity island, most notably are T3SS and T1SS that facilitate its survival in host cells in a protected manner (Ibarra and Mortimer 2009).

6.2.1.2 *Mycobacterium tuberculosis*

Mycobacterium tuberculosis survives in host macrophages by various mechanisms. One such mechanism that aid in its intracellular survival, is that after internalization by macrophages and phagosomes containing *Mycobacteria* resist fusion with lysosomes, an important step in their intracellular survival (Walburger et al. 2004). After internalization by host macrophages, pathogenic *Mycobacteria* secrete eukaryotic-like threonine/serine protein kinase G, thus inhibiting phagosome fusion with lysosomes and survive intracellularly (Walburger et al. 2004).

6.2.1.3 *Neisseria meningitidis*

For intracellular survival, capsule of *Neisseria meningitidis* plays an important role. The pathogens remain protected from the components of host immune system like cationic antimicrobial peptides (CAMPs), cathelicidins, defensins, and protegrins by means of their capsule (Spinosa et al. 2007).

6.2.1.4 *Listeria monocytogenes*

L. monocytogenes is a foodborne pathogen with the intracellular mode of replication. Of the total genome, approximately 17% is responsible for intracellular survival in the host cell. Gene expression by the pathogen in the intracellular environment encodes for glucose limitation, reduction in mRNA expression having role in central metabolism, and expression of alternate carbon utilization pathway genes and its regulation (Chatterjee et al. 2006).

6.2.2 *Protozoan Parasites*

In order to survive, protozoan parasites induce various changes in host cell signaling pathways and trafficking mechanism. Some protozoan parasites inhibit phagosomes fusion with lysosomes in order to get protected from hydrolytic action of lysosomal enzymes and others produce certain inhibitory factors to resist hydrolytic enzymes (Mauel 1984). One mechanism involves the formation of lipid-rich organelles called lipid bodies within the host cell cytoplasm. When parasite interacts with an immune or nonimmune host cell, lipid bodies also accumulate around the parasitophorous vacuole (PV). These lipid bodies may serve as a source for the growth of parasites and also induce inflammatory mediators that cause deactivation of host immune response (Toledo et al. 2016). Some medically important species of intracellular protozoans are discussed in following.

6.2.2.1 *Leishmania* species

Leishmania infantum chagasi is an obligate intracellular pathogen that is responsible for visceral leishmaniasis in South America. One of the intracellular survival strategies of these parasites involves transient delay in maturation of phagolysosome such that enough time is available for promastigotes to be converted into amastigotes, and amastigotes are highly resistant to degradation (Ueno 2011).

Leishmania donovani is an important species of *Leishmania* genus. Its promastigote forms when its phagosome poorly fuses with lysosomes and endosomes. Also *Leishmania* species display lipophosphoglycan (LPG) molecules on their cell surface that has a role in inhibition process of phagosome maturation (Desjardins and Descoteaux 1998).

6.2.2.2 Malarial Parasites

Plasmodium, the causative agent of malaria, can cause the infection via successful replication within erythrocytes. This parasite has evolved molecular and cellular pathways for successful multiplication within host cells as well as to protect itself from host immune responses (Miller et al. 1994). Their survival is mostly based on changes in number of genetic factors like allelic variation, intracellular replication, and bimolecular exposure of proteins (Gomes et al. 2016).

6.2.2.3 *Toxoplasma gondii*

Toxoplasma gondii is a highly potent intracellular parasite that resides within host cell vacuole. Like other pathogens, *Toxoplasma gondii* survive intracellularly via inhibition of phagosome fusion with lysosomes and also via modification of acidity of phagosomes that contain them (Sibley et al. 1985).

6.2.3 Fungal Intracellular Pathogens

Pathogenic fungi have enormous impact on human health as many fungal species may cause serious and life-threatening infections. Many pathogenic fungi may also grow intracellularly. Some medically important fungal species that are intracellular parasites are discussed here.

6.2.3.1 *Candida* species

Like other microorganisms that behave as intracellular pathogens, some fungi can also exist with intracellular mode of replication. Many *Candida* species can evade engulfment by host cell macrophages and reduce formation of inflammatory cytokines that can otherwise inhibit their growth. This helps them to survive in host cells (Brothers et al. 2013).

6.2.3.2 *Cryptococcus neoformans*

One of the intracellular fungal pathogens is *Cryptococcus neoformans*. Similar to other intracellular pathogens, *Cryptococcus neoformans* also have evolved mechanisms for their intracellular survival. One of the strategies for their intracellular survival in host cell involves accumulation of polysaccharide vesicles in the cytoplasm followed by intracellular replication, leading to the formation of special phagosomes. In these phagosomes, multiple replicated *Cryptococcal neoformans* cells are present in a protected manner (Alvarez and Casadevall 2006).

6.2.3.3 *Aspergillus* species

In the case of *Aspergillus* species, after their internalization into phagosomes, maturation and acidification is inhibited through an unknown mechanism that helps them to survive intracellularly (Morton et al. 2012). Secondary metabolite production by *A. fumigatus* plays a role in its residence in phagosome. Various toxins are produced by *A. fumigatus* with immunosuppressive properties, preventing macrophages from their functioning (Gilbert et al. 2015).

6.3 Treatment Options and Challenges

Introduction of antibiotics into medical practice has resulted in the control of various diseases and saved millions of human lives. However, with the passage of time many pathogens have evolved ways of protection against antibiotics. Further, some pathogens have the ability to survive intracellularly within immune and nonimmune cells and thus remain protected within these cells. This results in disease relapses and treatment failure (Maurin and Raoult 2001). Macrophages are the main target site for many intracellular pathogens, despite the fact that macrophages produce an antimicrobial response against the pathogens. There is cooperation between neutrophils and macrophages in generating defense, as neutrophils transfer their preformed antimicrobial granules to macrophages, thus producing an innate immune response against intracellular pathogens. It also complements various pathways involving antimicrobial peptide delivery to macrophages (Tan et al. 2006).

Treatment of intracellular pathogens such as *M. tuberculosis* involves use of antibiotics or other antimicrobials that are capable of entering the infected host cells to kill bacterial agents (Carryn et al. 2003). Some antibiotics such as quinolones, rifampicin, and streptomycin have shown better effect against intracellular bacteria; it is a very limited option available to treat such infections (Kamaruzzaman and Kendall 2017).

As stated previously, one of the challenges of treating intracellular pathogens is reduced drug efficacy inside the host cells. For example, amino glycosides are effective against extracellular infections, but ineffective or poorly effective against intracellular infections. The reason behind this is their exclusive concentration within lysosomes and inactivation due to local acidic environment, as these antibiotics are highly sensitive to pH (Maurin and Raoult 2001). Another problem encountered in treatment practices is delivery of therapeutic agents to their target site. Conventional methods of drug utilization are associated with poor distribution, lack of sensitivity, limited effectiveness, and side effects. An alternative approach to this is the targeted delivery of therapeutic agents. This approach employs loading of antibiotics into carrier molecules like nanoparticles, micelles, and liposomes to efficiently and selectively deliver therapeutic agents to their targets (Atbiaw et al. 1965). In the recent years, significant progress has been made by nanocarrier-based technology in targeted drug delivery for the treatment of intracellular infections (Lin et al. 2015).

Conventional methods for the treatment of these diseases are associated with combination of drugs and long-term therapy practices that result in antibiotic resistance and various side effects (Armstead and Li 2011). Salmonellosis, one of the most common intracellular infections, is difficult to treat because the pathogen remains protected in intracellular compartments and it is difficult for various antibiotics to cross the selective permeable plasma membrane and target these pathogens intracellularly (Klemm et al. 2018).

Nanotechnology offers great promise to overcome these limitations as in this technology nanomaterials can be fabricated with antibiotics and thus an efficient drug delivery to their targets (Ranjan et al. 2012). Recently developed sol-gel technique showed great efficacy for the treatment of intracellular infections. Biologically active agents are incorporated into silica matrix that delivers these therapeutic agents to their target (Saleem et al. 2009a, b). Another effective strategy for treatment of intracellular pathogens involves the use of cell-penetrating peptides (CPP). Cell-penetrating peptides enhance the delivery of large macromolecules intracellular. These arginine- or lysine-rich cationic peptides have successful applications in the field of medicine for targeted drug delivery. These peptides have the potential to cross various barriers like blood brain barrier, neurons, and retina, thus having advantage over other therapeutic agents that cannot cross these barriers (Ye et al. 2016).

6.4 Nanomaterials as an Alternative Treatment Option

Targeted drug delivery in the body using nanomaterials is one of the key focus areas of nanomedicine (Abrhaley and Miku 2018; Bei et al. 2010; Hillaireau and Couvreur 2009; Giri et al. 2011). According to an estimate, over 90% of the drugs suffer from poor pharmacokinetics and biopharmaceutical properties that require therapeutically active drug to be delivered into the target cells and tissues (Ruggiero et al. 2010). Targeted drug delivery into the cells infected with infectious agents is an efficient way of delivering drug without affecting the normal body cells. Furthermore, nanocarriers also increase the drug solubility in blood and protect from degradation in the body (Li et al. 2017). By doing so, the therapeutics index will be reduced and safety profile improved (Baruah et al. 2017; Oeztuerk-Atar et al. 2018).

Nanocarriers to be used for selective targeting of pathogens inside the infected cells are usually bioconjugated with either antibodies or aptamers having high affinity for the receptors on the target host cells. More importantly, they should have higher ability to be internalized through endocytosis and biodegradability (Yu et al. 2016; Senapati et al. 2018; Sinha et al. 2006).

The treatment options for infectious diseases caused by intracellular pathogens may be improved, as we enter the new era of nanotechnology. Biodegradable nanoparticles have been a widely investigated material for targeted drug delivery among others (Hans and Lowman 2002). Some drugs like fluoroquinolones and macrolides have the ability to penetrate the host cells, but they are poorly retained and efficacious. There is a need of alternative drug options with both improved penetration and retention inside the hosts. Here, nanocarriers such as nanoparticles or liposomes seem to be a promising strategy to deliver the drug selectively to the infected host cells, such as macrophages, dendritic cells, and red blood cells, infected with intracellular pathogens. The drugs may be encapsulated, attached, or incorporated in these nanocarriers and released in the infected cells in the site-specific manner. Furthermore, the properties of the nanocarriers such as size, shape, hydrophobicity, and other chemical features may also be tuned according to the requirement (Armstead and Li 2011; Urban et al. 2011; Li et al. 2017).

Treatment of intracellular pathogens (e.g., *M. tuberculosis*) requires long-term drug administration that may cause side effects on patients. Patient compliance is a major issue in treating such infections and controlling the disease burden. This may be compensated with reduced drug doses per day using solutions offered by nanotechnology. Sustained drug release from then nanocarriers causes better therapeutic level for prolonged period of time, so the shorter half-life of drug elimination may be prolonged (Ranjan et al. 2012; Ladaviere and Gref 2015).

Biocompatibility is an important feature of a drug delivery system that is aimed at minimizing nonspecific effects on the host cells other than the target. Nanoparticles have been made of a number of biocompatible materials such as poly(lactide-co-glycolide) (PLGA), poly lactic acid (PLA), polyethylene glycol (PEG), and polymethacrylic acid (PMA) and natural polymers such as gelatin, chitosan, and alginate, as well as gold and silica (Armstead and Li 2011).

One of the prerequisites of the smart nanocarriers is their prolonged retention in the body and resistance to clearing action of reticuloendothelial cells. This is achieved by attaching PEG molecules (PEGylation) on the nanocarrier surface. Nevertheless, PEGylation reduces uptake of the nanocarriers by the target cells, a twist known as PEGylation dilemma (Hossen et al. 2018). The life of nanocarriers (e.g., liposomes) may be increased by attaching PEG groups with them, so that they remain protected engulfment by phagocytes (Peer et al. 2007).

In addition, nanocarriers are bioconjugated with some ligands such as monoclonal antibodies, aptamers, sugars, peptides, or any other biomolecule having affinity for the surface receptors of the infected host cells. Therefore, ease of bioconjugation with the desired ligands may be considered a feature of the smart nanocarrier, as it prevents accumulation of nanocarriers in liver, kidneys, or other sites (Lin et al. 2015; Hossen et al. 2018).

An important issue to be considered is the release of encapsulated drug from endosome into the cytoplasm. Rapid release of nanocarriers from endosome into cytoplasm may be achieved by incorporating cell-penetrating peptides such as fusogenic lipids or listeriolysin-O onto the nanocarriers (Paulo et al. 2011; Ranjan et al. 2012).

In the studies carried out on the use of drug-loaded nanocarriers for the clearance of intracellular pathogens from mice, it has been found that nanocarriers required lesser amount of drug to clear the pathogen as compared to drug alone (De Steenwinkel et al. 2007). Furthermore, some groups have also reported better drug uptake by the infected cells carried by the nanocarriers as compared to drugs alone (Toti et al. 2011; Xie et al. 2017).

6.4.1 *Types of Nanocarriers*

A number of nanomaterials may be used as nanocarriers in targeted drug delivery systems (Uddin et al. 2017) (Fig. 6.1). Detail of some of them is given in the proceeding sections.

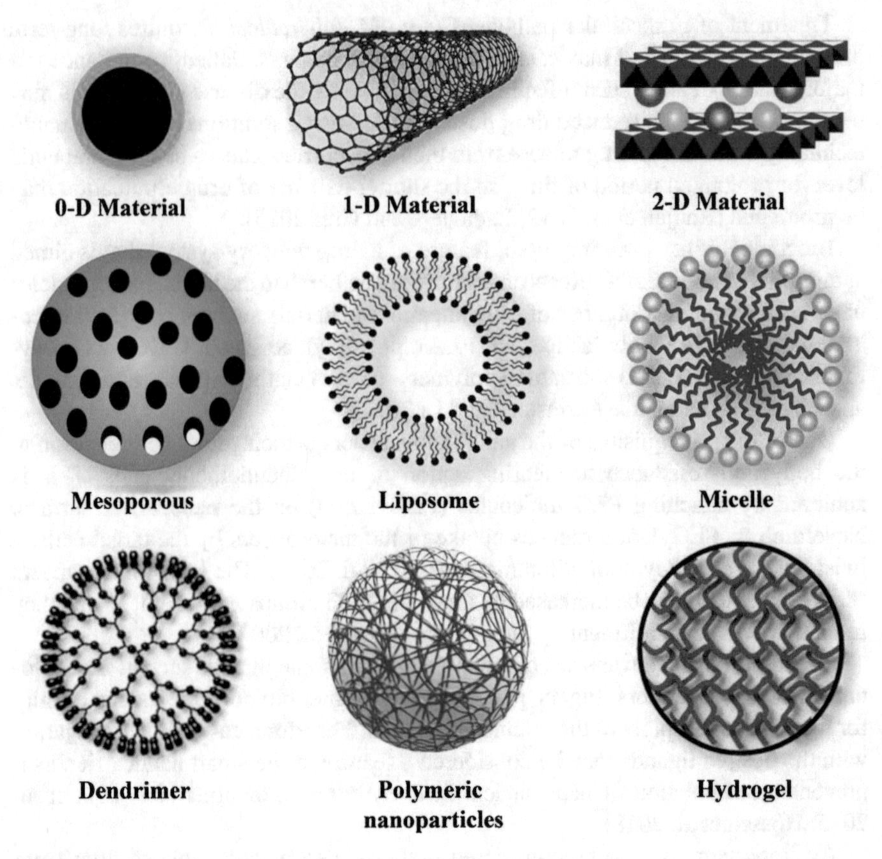

Fig. 6.1 Different types of nanomaterials that may be used as nanocarriers for targeted drug delivery (Source: Senapati et al. 2018)

6.4.1.1 Polymeric Nanoparticles

Polymeric nanoparticles are among the most widely investigated types of nanomaterial that find a range of applications in targeted drug delivery systems (Karlsson et al. 2018). They are prepared from biodegradable and biocompatible polymers (Nagavarma et al. 2012; Gopalasatheeskuma et al. 2017). Initial use of polymeric nanoparticles was limited to nonbiodegradable nanoparticles (e.g., polystyrene, polyacrylamide, poly(methyl methacrylate)). These nonbiodegradable nanoparticles were required to be removed from body after use through urine and feces and they were not supposed to accumulate in the tissues at toxic level. Chronic toxicity and inflammation reactions were observed in the case of nonbiodegradable nanoparticles that lead to consider the biodegradable nanoparticles owing to their reduced toxicity and interference with drugs as well as biocompatibility (Banik et al. 2016).

There are two major categories of polymeric nanoparticles, namely nanocapsules and nanospheres. Nanocapsules have a drug reservoir in their liquid or nonliquid core, while polymeric shell covers it. In contrast, in case of nanospheres, the entire particles have a polymeric matrix-like structure retaining the drug inside it or on its surface (El-Say and El-Sawy 2017).

There are several advantages of polymeric nanoparticles. For examples, they are easy and economical to synthesize. The choice of polymer and controlled drug release at the site of infection have made them an ideal candidate of nanocarriers in drug delivery systems. Drug may easily be impregnated into them and effectively released into the target cells. Further, they are easy to conjugate with biomolecules such as antibodies (Chan et al. 2010; Dalbhanjan and Bomble 2013; Nagavarma et al. 2012).

6.4.1.2 Liposomes

Since their first introduction by British hematologist Alec Bangham in 1961 (Bangham et al. 1965), liposomes have been the most widely investigated type of nanocarriers for the targeted drug delivery due to their greater physiochemical and biophysical properties. They have been used to stabilize the drug as well as improve cellular uptake and biodistribution in the body (Sercombe et al. 2015). Liposomes are phospholipid vesicles consisting of one or two bilayer membranes with discrete aqueous space. Liposomal particles possess unique property of entrapping diverse types of drugs including both hydrophilic and hydrophobic molecules. The biocompatible lipid exterior and aqueous interior of the liposomes permit delivery of a variety of biomolecules such as antibiotics, anticancer drugs, proteins, enzymes, and DNA. They are usually considered as pharmacologically inactive with minimum side effects or toxicity. They offer the advantage of large payloads, protection of drug from degradation, and inactivation as well as least immunogenicity inside the human body (Sercombe et al. 2015; Alavi et al. 2017).

Chemically, liposomes are phospholipid bilayers with aqueous center. Lipids are amphipathic molecules with hydrophobic and hydrophilic ends; thus, their self-assembly results in separate orientation of both hydrophobic and hydrophilic ends in aqueous phage (Alavi et al. 2017). Several methods of liposome synthesis have been described including lipid hydration, ethanol injection, freeze thawing, and reverse phase evaporation. Initially, liposome had some disadvantages such as difficulties with mass production, short shelf life, and poor stability. However, these all issues have been resolved with continuous research and optimization of synthesis procedures. To date, about ten liposome-based drugs have reached the market and many more are in the phase of clinical trials (Kim 2016).

6.4.1.3 Gold Nanoparticles

Gold nanoparticles (AuNPs) are among the most widely investigated type of nanomaterials. They possess a range of applications in medicine including drug delivery systems, imaging, diagnosis, as well as treatment of the diseases

(Dreaden et al. 2012; Kong et al. 2017; Syed and Bokhari 2011). AuNPs are non-toxic and inert in body fluids. They may be synthesized in different sizes and shapes and bioconjugated with a single or multiple drugs (Kong et al. 2017).

AuNPs may also be decorated with fluorescent dyes, antibodies, and drugs. Cell internalization peptide TAT may also be attached with the AuNPs. The intracellular location of the AuNPs may also be traced using fluorescent microscopy (Singh et al. 2016).

6.4.1.4 Micelles

Polymeric micelles (PMs) were introduced in drug delivery system by Kataoka's group as doxorubicin-encapsulated nanocarriers of copolymers in 1990s. PMs consist of a core and a shell made of block copolymers. The hydrophobic core encapsulates poorly water-soluble drug, whereas the hydrophilic shell protects it from body fluids and reticuloendothelial system (Kedar et al. 2010).

Majority of the drugs are poorly soluble in water and their application in drug delivery is often challenging. Micelles are a group of nanocarriers that may carry drugs that are poorly water soluble (Zhang et al. 2014). The amphipathic MPs are formed by copolymers in aqueous phase. The drug to be incorporated is inside the hydrophobic core (Xu et al. 2013). They may be used for both active and passive delivery of the drugs (Kedar et al. 2010). Variation in the chemistry of the core-forming copolymer of the PMs may be performed to improve drug loading, stability, and controlled release (Aliabadi and Lavasanifar 2006).

6.4.1.5 Solid Lipid Nanoparticles

Solid lipid nanoparticles (SLNPS) are at the forefront of nanoparticles research for drug delivery system (Calderon-Colon et al. 2015; Mukherjee et al. 2009). They are lipid nanoparticles of the size range 50–1000 nm. These nanoparticles are prepared by dispersing the solid lipid in water, whereas emulsifiers are used to stabilize the dispersion (Uddin et al. 2017). These nanoparticles may be used to overcome the limitations of the polymeric nanoparticles and liposomes. These nanoparticles are more useful for lipophilic drugs (Mukherjee et al. 2009). Nevertheless, these nanoparticles may be used for the delivery of both hydrophobic and hydrophilic drugs as well as proteins and DNA (Dolatabadi et al. 2015). SLNs have already been used for the controlled and targeted drug delivery for the treatment of cancer and intracellular pathogens. The core of SLNs is hydrophobic with single layer of phospholipids and the drug is usually impregnated in the core (Dolatabadi et al. 2015; Mishra et al. 2018). SLNs carry more drug than the liposome and protect it from degradation.

Monocytes and macrophages are target of a number of intracellular pathogens. As natural carriers, liposomes may target the cells of mononuclear phagocytic system (MPS) including macrophages for targeted delivery of antimicrobial agents.

The macrophages express a number of surface receptors on their surface such as integrins, mannose receptors, and Fc portion receptors. Addition of ligands of these receptors to the liposome facilitates selective attachment of drug-loaded liposomes on the surface of macrophages (Kelly et al. 2011).

6.5 Current Research Trends

A number of dedicated research groups have published excellent articles with promising results on intracellular delivery of antibiotics into the infected cells (Delsol et al. 2004; Lemmer et al. 2015; Choi 2017). The goal of intracellular drug delivery is sustained drug release at subcellular level that is not going to be affected by low endosomal pH (Ranjan et al. 2012).

As mentioned previously, tuberculosis is one of the most notorious intracellular pathogens that invade macrophages. A number of studies have been carried out that reported efficient delivery of antimycobacterial drugs such as first-line antibiotics using nanocarriers. For example, a recent study conducted by Vieira et al. (2017) used lipid nanoparticles decorated with mannose sugar for the delivery of rifampin antibiotics into the macrophages. These mannosylated lipid nanoparticles were more efficient in uptake by the macrophages for inhibiting the microbial survival inside the cells (Vieira et al. 2017). A previous study conducted by Clemens et al. (2012) used mesoporous silica nanoparticles (MSNPs) for the efficient delivery of isoniazid into the human macrophages. These MSNPs used acid gates that open in the acidic environment and release its antibiotic content for bacterial killing.

Another example of targeting highly pathogenic microorganisms using nanoparticles is the study conducted by Saleem et al. (2009a, b). They applied polymeric nanoparticles loaded with doxycycline and streptomycin that exhibited promising effects while clearing microbes inside the host macrophage cells. Application of two doses of antibiotics-loaded nanoparticles cleared *Brucella melitensis* significantly from the balb/c mice.

Salmonella species are notorious for the range of infections they cause in humans. *Salmonella enterica* serovar *Typhimurium*, *Salmonella enteric serovar Typhi*, and *Salmonella enterica Paratyphi* are the bacterial species that have gained resistance against a number of commonly used antibiotics. Bactericidal drug concentration for these intracellular pathogens is higher than the extracellular bacteria. Salmonella invade macrophages and reside in a vacuole called Salmonella-containing vacuole. Salmonella in the SCVs have increased arginine requirement causing scavenge the host cells of arginine starvation. In case of presence of *Salmonella* in SCV, there is an upregulation of CAT 1 transporters responsible for enhanced arginine uptake. A study conducted by Mudakavi et al. (2017) synthesized arginine decorated mesoporous nanoparticles loaded with ciprofloxacin used to target Salmonella infected macrophages. The results of their study revealed that these nanoparticles exhibited better antibacterial effect of the drug against the target bacteria as compared to free ciprofloxacin. In another study conducted by Saleem et al. (2009a, b) gentamycin-loaded silica xerogel was synthesized to target *Salmonella enterica* sero*var Typhimurium* inside mice. The results of their study revealed that the drug was active inside the nanoparticles and there was significant reduction in bacterial number in spleen and liver of the infected AJ 646 mice at lower drug dose as in nanocarriers as compared to free drug.

One of the most remarkable areas of applications of targeted drug delivery is the cure of malaria, treatment of which has been hampered due to the problem of emerging drug-resistant strains (Fernandez-Busquets 2016). Nanocarriers may offer opportunity to encapsulate various drugs in a single cargo to selectively deliver to the infected RBCs (Urban et al. 2015; Baruah et al. 2017). A combined formulation of more than two antimalarial drugs will also increase the rate of patient compliance, as patient may avoid taking different tablets separately. Sustained drug release may be achieved by increasing or decreasing the drug discharge from the carrier in the cells (Thakkar and Brijesh 2016). Moles et al. (2016) developed immunoliposomes as drug carriers to selectively target *Plasmodium falciparum* infected RBCs. A 70% reduction in all parasitic forms in culture was observed after 30 min of incubation with drug-loaded immunoliposomes. In another such study, Urban et al. (2014) used poly(amidoamine) biodegradable nanocarriers for selectively targeting plasmodium infected red blood cells for drug delivery. Intraperitoneal administration of drug-loaded nanocarriers cured the infected mice, while mice with free drug did not survive.

The intracellular parasite of the genus *Leishmania* cause cutaneous, subcutaneous, and visceral disease and pose a serious public health threat globally. *Leishmania donovani*, causative agent of visceral leishmaniasis, is an intracellular pathogen that infects macrophages of reticuloendothelial system. In a study conducted by Mukherjee et al. (2004), a parasite-specific 51 kDa protein was isolated from *L. donovani*-infected macrophages. Active delivery of doxorubicin to the infected macrophages was carried out using immunoliposomes prepared by grafting anti 51 kDa antibody onto the liposome surface. A 45-day-old mouse model of visceral leishmaniasis was used to study the effects of targeted drug delivery in vivo. Complete elimination of spleen parasite burden was achieved by delivering doxorubicin-loaded immunoliposomes on four consecutive days at the drug concentration of 250 μg/kg/day.

AuNPs have also been used as nanocarriers for selective targeting intracellular pathogens. For example, a study conducted by Chowdhury et al. (2017) conjugated AuNPs with antimicrobial peptide VG16KRKP. The peptide not only facilitated the intracellular uptake by the macrophages and epithelial cells but also exhibited antimicrobial activity against intracellular pathogen *Salmonella*. The results of their study revealed that there was a reduced level of bacterial recovery from the cells.

6.6 Theranostic Approaches for Intracellular Pathogens

In 2002, the term theranostic was for the first time coined by Funkhouser (Kelkar and Reineke 2011). In theranostic approach, the diagnostic and therapeutic properties of nanoparticles are combined for personalized and more specific disease

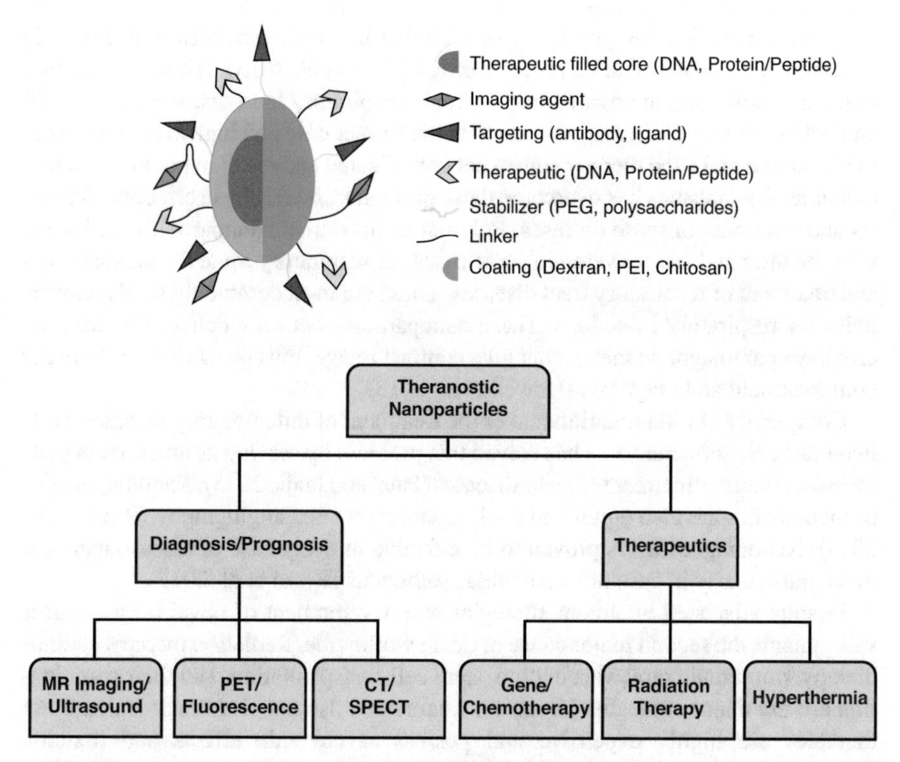

Fig. 6.2 Schematic illustration of nanoparticle-based theranostics and their potential applications (Source: Howell et al. 2013)

management. The nanoparticles can serve as multifunctional system in theranostics (Yang et al. 2013; Fig. 6.2). In this approach, therapeutic drugs and the diagnostic agents are delivered in an equal dose at the same time (Kelkar and Reineke 2011). For making nanoparticles-based theranostics, carbon nanotubes, quantum dots, gold nanoparticles, and silica nanoparticles are well investigated (Chen et al. 2014; Liu et al. 2018; Xie et al. 2010).

Bacterial infections of a blood stream possess a greater threat to economy and public health. Emergence of antibiotic resistance makes the targeting of these bacterial infections a difficult task. Furthermore, for early intervention serology and blood culture techniques are not much sensitive. To overcome these problems, nanotheranostic is a novel strategy for the diagnosis and provide an effective treatment platform (Jagtap et al. 2017).

Multidrug-resistant bacteria like *Salmonella typhi* possess a great threat to healthcare. To save lives, early detection of the bacteria in the blood stream and development of novel antibiotics are of prime importance. Dai et al. reported multifunctional nanocarrier platform consisting of a magnetic core-plasmonic shell nanocarrier, multidrug-resistant bacteria (MDRB) *Salmonella* DT104-specific antibody, and a methylene blue bound aptamer. The reported nanocarrier platform is capable of targeted separation of the pathogen from the blood and in the detection and multimode killing of multidrug-resistant bacteria (Dai et al. 2013).

Nanotheranostics has proven to be effective in the therapeutics with less side effects in comparison with therapeutic drugs (Yang et al. 2013). It has the potential in the diagnosis and the treatment of chronic respiratory tract diseases such as TB and lung cancer. A multidrug resistance in the former case and high dose of drugs in the latter one make the therapy somewhat complicated and result in poor therapeutic outcome. Theranostics has overcome these problems and results in effective diagnosis and treatment of these diseases. Polymer or liposomal nanoparticles are loaded with the drug and are delivered in some cases via respiratory route for the diagnosis and treatment of respiratory tract diseases. These are most commonly used nanoparticles for respiratory infections. These nanoparticles not only deliver the drug but also have paramagnetic metals that give contrast image, thus providing platform for both treatment and diagnosis (Howell et al. 2013).

Complexity of inflammation makes the treatment of inflammatory diseases a difficult task. Nanotheranostics has solved this problem by serving as an efficient platform for effective treatment of these diseases (Patel and Janjic 2015). Nanodiagnostics or theranostics are also employed for lymphography and angiography (Rizzo et al. 2013). Nanodiagnostics is proven to be effective in every case of disease management, particularly in case of intracellular pathogens (Rizzo et al. 2013).

Despite advanced treatment strategies and development of novel drugs, cancer still remains the second major cause of death worldwide. Radiation therapy, chemotherapy, immunotherapy, vaccination, stem cell transplantation, and their combination are the major strategies for cancer treatment. Also one drawback is that these therapies are highly expensive and possess severe side effects and toxicity. Phytochemical-based green synthesized gold nanoparticles have proven to be effective in the diagnosis and treatment of cancer. These nanoparticles have the potential to be used in diagnostics, therapy, imaging, and as a delivery system for drugs in case of cancer treatment (Ovais et al. 2017). Gold nanoparticles have gained the attention of researchers for cancer treatment because of its unique physiochemical and optical properties. Additionally, these nanocarriers have been effectively used in cancer therapies and diagnosis because of their multimodality imaging, as a carrier for drug delivery, and tumor targeting (Akhter et al. 2012). Same strategy may be employed in the cases of chronic infections of Hepatitis B and C as well as latent infections of tuberculosis (Fig. 6.2).

6.7 Conclusion and Future Perspectives

There has been substantial development in the field of nanomedicine in the last decade. Explosive research in nanomaterials and their applications to solve the problems of biomedical sciences has resulted in many breakthroughs. Dozens of nanodrugs have acquired FDA approval and the list of drugs in clinical trials is ever increasing. Despite being widely investigated interdisciplinary science, nanomedicine is still in infancy and most of its products are in pipeline and did not yet reach the commercialization. Despite potential benefits, nanodrugs also face some challenges such as need for better characterization, toxicity of nanomaterials, lack of regulatory bodies, cost-benefit considerations, as well as lack of enthusiasms by healthcare professionals in some cases. Nevertheless, the future of nanomaterials in targeted drug delivery seems to be bright. The number of nanodrugs approved by the FDA is disproportionately smaller as compared to the massive investment in the last two decades. It may be justified by the fact that in the past many of the discoveries in medicine have taken many long years to reach the market and patients (Ventola 2017).

Successful targeting of intracellular pathogens demands smart strategies and deeper understanding of selective binding of the nanocarriers on the infected cells, intracellular release of drugs from nanocarriers, and microbial killing. Drug released in the endosome may or may not kill microbe, as microbe may be present in cytoplasm or nucleus and it may interact endoplasmic reticulum or Golgi apparatus. Therefore, bacterial interaction with the drug inside the host cell must be deeply investigated.

The unique features and advantages offered by different types of nanomaterials in drug delivery to intracellular parasites have been effective and promising in a number of studies. Encapsulation of concentrated antibiotics in nanocarriers will not only reduce the risk of toxicity to normal cells but it is also a very effective mode of sustained drug delivery to intracellular pathogens. Further, the issue of patient compliance in the case of long-term treatment in the diseases such as tuberculosis may be resolved if nanocarrier-based drug delivery systems are used to carry antibiotics to the pathogen.

References

Abrhaley A, Mitku F. Review on targeted drug delivery against intracellular pathogen. Pharm Pharmacol Int J. 2018;6(3):183–9.

Akhter S, Ahmad MZ, Ahmad FJ, Storm G, Kok RJ. Gold nanoparticles in theranostic oncology: current state-of-the-art. Expert Opin Drug Deliv. 2012;9(10):1225–43.

Alavi M, Karimi N, Safae M. Application of various types of liposomes in drug delivery systems. Adv Pharm Bull. 2017;7(1):3–9.

Aliabadi HM, Lavasanifar A. Polymeric micelles for drug delivery. Expert Opin Drug Deliv. 2006;3(1):139–62.

Alvarez M, Casadevall A. Phagosome extrusion and host cell survival after *Cryptococcus neoformans* phagocytosis by macrophages. Curr Biol. 2006;16:2161–5.

Armstead AL, Li B. Nanomedicine as an emerging approach against intracellular pathogens. Int J Nanomedicine. 2011;6:3281–93.

Atbiaw N, Aman E, Dessalegn B, Masrie O, Debalke B, Enbiyale G, Yirga A, Tekilu G, Bangham AD, Standish MM, Watkins JC. Diffusion of univalent ions across the lamellae of swollen phospholipids. J Mol Biol. 1965;13:238–52.

Banik BL, Fattahi P, Brown JL. Polymeric nanoparticles: the future of nanomedicine. Wiley Interdiscip Rev Nanomed Nanobiotechnol. 2016;8(2):271–99.

Baruah UK, Gowthamrajan K, Vanka R, Kari VVSK, Selvaraj K, Jo Jo GM. Malaria treatment using novel nano-based drug delivery systems. J Drug Target. 2017;25:567–81.

Bei D, Meng J, Youan BC. Engineering nanomedicines for improved melanoma therapy: progress and promises. Nanomedicine (Lond). 2010;5(9):1385–99.

Brothers KM, Gratacap RL, Barker SE, Newman ZR, Norum A, Wheeler RT. NADPH oxidase-driven phagocyte recruitment controls *Candida albicans* filamentous growth and prevents mortality. PLoS Pathog. 2013;9:1–17.

Calderon-Colon X, Raimondi G, Benkoski JJ, Patrone JB. Effective use of nanocarriers as drug delivery systems for the treatment of selected tumors. J Vis Exp. 2015;2015(105):1–8.

Carryn S, Chanteux H, Seral C, Mingeot-Leclercq MP, Van Bambecke F, Tulkens PM. Intracellular pharmacodynamics of antibiotics. Infect Dis Clin. 2003;17(3):615–34.

Casadevall A. Evolution of intracellular pathogens. Ann Rev Microbiol. 2008;62:19–33.

Chan JM, Valencia PM, Zhang L, Langer R, Farokhzad OC. Polymeric nanoparticles for drug delivery. Methods Mol Biol. 2010;624:163–75.

Chatterjee SS, Hossain H, Otten S, Kuenne C, Kuchmina K, Machata S, Domann E, Chakraborty T, Hain T. Intracellular gene expression profile of *Listeria monocytogenes*. Infect Immun. 2006;74(2):1323–38.

Chen F, Ehlerdin EB, Cai W. Theranostic nanoparticles. J Nucl Med. 2014;55(12):1919–22.

Chen S, Zhang Q, Hou Y, Zhang J, Liang XJ. Nanomaterials in medicine and pharmaceuticals: nanoscale materials developed with less toxicity and more efficacy. Eur J Nanomed. 2013;5(2):61–79.

Choi SR, Britigan BE, Morgan DM, Narayanasamy P. Gallium nanoparticles facilitate phagosome maturation and inhibit growth of virulent *Mycobacterium tuberculosis* in macrophages. PLoS One. 2017;12(5):1–20.

Chowdhury R, Ilyas H, Ghosh A, Ali H, Ghorai A, Midya A, Jana NR, Das S, Bhunia A. Multivalent gold nanoparticle–peptide conjugates for targeting intracellular bacterial infections. Nanoscale. 2017;9:14073–93.

Clemens DL, Lee BY, Xue M, Thomas CR, Meng H, Ferris D, Nel AE, Zink JI, Horwitz MA. Targeted intracellular delivery of antituberculosis drugs to Mycobacterium tuberculosis-infected macrophages via functionalized mesoporous silica nanoparticles. Antimicrob Agents Chemother. 2012;56(5):2535–45.

Dai X, Fan Z, Lu Y, Ray PC. Multifunctional nanoplatforms for targeted multidrug-resistant-bacteria theranostic applications. ACS Appl Mater Interfaces. 2013;5(21):11348–54.

Dalbhanjan RR, Bomble SD. Biomedical approach of nanomaterials for drug delivery. Int J Chem Chem Eng. 2013;3(2):95–100.

De Steenwinkel JE, Van Vianen W, Ten Kate MT, Verbruch HA, Van Agtmael MA, Schifellers RA, Bakker-Woudenberg IAJM. Targeted drug delivery to enhance efficacy and shorten treatment duration in disseminated *Mycobacterium avium* infection in mice. J Antimicrob Chemother. 2007;60:1064–73.

Dehio C, Berry C, Bartenschlager R. Persistent intracellular pathogens. FEMS Microbiol Rev. 2012;36(12):513.

Delsol AA, Woodward MJ, Roe J. Effect of a 5 day enrofloxacin treatment on *Salmonella enterica* serotype Typhimurium DT104 in the pig. Antimicrob Chemother. 2004;54:692–3.

Desjardins M, Descoteaux A. Survival strategies of *Leishmania donovani* in mammalian host macrophages. J Immunol Res. 1998;149(7–8):689–92.

Din F, Aman W, Ullah I, Qureshi OS, Mustapha O, Shafique S, Zeb A. Effective use of nanocarriers as drug delivery systems for the treatment of selected tumors. Int J Nanomedicine. 2017;12:7291–309.

Dolatabadi JEN, Valizadeh H, Hamishehkar H. Solid lipid nanoparticles as efficient drug and gene delivery systems: recent breakthroughs. Adv Pharm Bull. 2015;5(2):151–9.

Dreaden EC, Austin LA, Mackey MA, El-Syed MA. Size matters: gold nanoparticles in targeted cancer delivery. Ther Deliv. 2012;3(4):457–78.

El-Say KM, El-Sawy HS. Polymeric nanoparticles: promising platform for drug delivery. Int J Pharm. 2017;528(1–2):675–91.

Ernst RK, Guina T, Miller SI. How intracellular bacteria survive: surface modifications that promote resistance to host innate immune responses. J Infect Dis. 1999;179(2):S326–30.

Fahmy TM, Fong PM, Goyal A, Saltzman WM. Targeted for drug delivery. Mater Today. 2005;8(8):18–26.

Fernandez-Busquets X. Novel strategies for Plasmodium-targeted drug delivery. Expert Opin Drug Deliv. 2016;13(7):919–22.

Gagliardi M. Novel biodegradable nanocarriers for enhanced drug delivery. Ther Deliv. 2016;7(12):809–26.

Gilbert AS, Wheeler RT, May RC. Fungal pathogens: survival and replication within macrophages. Cold Spring Harb Perspect. 2015;5:1–13.

Giri N, Tomar P, Karwasara VS, Pandey RS, Dixit VK. Targeted novel surface-modified nanoparticles for interferon delivery for the treatment of hepatitis B. Acta Biochim Biophys Sin. 2011;43:877–83.

Gomes PS, Bhardwaj J, Correa JR, Freire-De-Lima CG. Immune Escape Strategies of malaria parasites. Front Micbiol. 2016;7:1–7.

Gopalasatheeskuma K, Komala S, Mahalakshmi M. An overview on polymeric nanoparticles used in the treatment of diabetes mellitus. Pharma Tutor. 2017;5(12):40–6.

Hans ML, Lowman AM. Biodegradable nanoparticles for drug delivery and targeting. Curr Opin Solid State Mater Sci. 2002;6(4):319–27.

Hillaireau H, Couvreur P. Nanocarriers' entry into the cell: relevance to drug delivery. Cell Mol Life Sci. 2009;66:2873–96.

Hossen S, Hossain MK, Basher MK, Mia MNH, Rehman MT, Uddin MJ. Smart nanocarrier-based drug delivery systems for cancer therapy and toxicity studies: a review. J Adv Res. 2018;15:1–18.

Howell M, Wang C, Mahmoud A, Hellermann G, Mohapatra SS, Mohapatra S. Dual-function theranostic nanoparticles for drug delivery and medical imaging contrast: perspectives and challenges for use in lung diseases. Drug Deliv Transl Res. 2013;3:352–63.

Hubbell JA, Chilkoti A. Nanomaterials for drug delivery. Science. 2012;339:303–5.

Ibarra JA, Mortimer OS. *Salmonella*—the ultimate insider *Salmonella* virulence factors that modulate intracellular survival. Cell Microbiol. 2009;11(11):1579–86.

Jagtap P, Sritharan V, Gupta S. Nanotheranostic approaches for management of bloodstream bacterial infections. Nanomedicine. 2017;13(1):329–41.

Kamaruzzaman F, Kendall S. Targeting the hard to reach: challenges and novel strategies in the treatment of intracellular bacterial infections. Br J Pharmacol. 2017;174:2225–36.

Karlsson J, Vaughan HJ, Green JJ. Biodegradable polymeric nanoparticles for therapeutic cancer treatments. Annu Rev Chem Biomol Eng. 2018;9:105–27.

Kedar U, Phutane P, Shidhaye S, Kadam V. Advances in polymeric micelles for drug delivery and tumor targeting. Nanomedicine. 2010;6:714–29.

Kelkar SS, Reineke TM. Theranostics: combining imaging and therapy. Bioconjug Chem. 2011;22(10):1879–903.

Kelly C, Jefferies C, Cryan SA. Targeted liposomal drug delivery to monocytes and macrophages. J Drug Deliv. 2011;2011:1–11.

Kim JS. Liposomal drug delivery system. J Pharm Investig. 2016;46(4):387–92.

Klemm EJ, Shakoor S, Page AJ, Qamar FN, Judge K, Saeed DK, Wong VK, Dallman TJ, Nair S, Baker S, Shaheen G, Qureshi S, Yousafzai MT, Saleem MK, Hasan Z, Dougan G, Hasan R. Emergence of an extensively drug resistant *Salmonella* enterica serovar Typhi clone harboring a promiscuous plasmid encoding resistance to fluoroquinolones and third-generation cephalosporins. MBio. 2018;9(1):1–10.

Kong FY, Zhang JW, Li RF, Wang ZX, Wang WJ, Wang W. Unique roles of gold nanoparticles in drug delivery, targeting and imaging applications. Molecules. 2017;22:1–13.

Kumari P, Ghosh B, Biswas S. Nanocarriers for cancer-targeted drug delivery. J Drug Target. 2016;24(3):179–91.

Ladaviere C, Gref R. Toward an optimized treatment of intracellular bacterial infections: input of nanoparticulate drug delivery systems. Fut Med. 2015;10(9):3033–55.

Lamprecht A, Urich N, Yamamoto H, Schaefer U, Takeuchi H, Maincent P, Kawashima Y, Lehr CM. Biodegradable nanoparticles for targeted drug delivery in treatment of inflammatory bowel disease. J Pharmacol Exp Ther. 2001;299(2):775–81.

Lemmer Y, Kalombo L, Pietersen DY, Jones AT, Semete-Makokotlela B, Wyngaardt SV, Ramalapa B, Stoltz AC, Baker B, Verschoor JA, Swai HS, Chastellier C. Mycolic acids, a promising

mycobacterial ligand for targeting of nanoencapsulated drugs in tuberculosis. J Control Release. 2015;211:94–104.
Li Z, Tan S, Li S, Shen Q, Wang K. Cancer drug delivery in the nano era: an overview and perspectives (review). Oncol Rep. 2017;38:611–24.
Lin YS, Lee MY, Yang CH, Huang KS. Active targeted drug delivery for microbes using nanocarriers. Curr Top Med Chem. 2015;15(15):1525–31.
Liu P, Xu LQ, Xu G, Pranantyo D, Neoh KG, Kang ET. pH-sensitive theranostic nanoparticles for targeting bacteria with fluorescence imaging and dual-modal antimicrobial therapy. ACS Appl Nano Mater. 2018;1:6187–96.
Mauel J. Mechanisms of survival of protozoan parasites in mononuclear phagocytes. Parasitology. 1984;88(4):579–92.
Maurin M, Raoult D. Use of aminoglycosides in treatment of infections due to intracellular bacteria. Antimicrob Agents Chemother. 2001;45(11):2977–86.
Miller LH, Good MF, Milon G. Malaria pathogenesis. Science. 1994;264(5167):1878–83.
Mishra V, Bansal K, Verma A, Yadav N, Thakur S, Sudhakar K, Rosenholm J. Solid lipid nanoparticles: emerging colloidal nano drug delivery systems. Pharmaceutics. 2018;10(4):1–21.
Moles E, Moll K, Ching JH, Parini P, Wahgren M, Busquest X. Development of drug-loaded immunoliposomes for the selective targeting and elimination of rosetting Plasmodium falciparum-infected red blood cells. J Control Release. 2016;241:57–67.
Moritz M, Gezske-Moritz M. Recent developments in the application of polymeric nanoparticles as drug carriers. Adv Clin Exp Med. 2015;24(5):749–58.
Morton CO, Bouzani M, Loeffler J, Rogers TR. Direct interaction studies between *Aspergillus fumigatus* and human immune cells; what have we learned about pathogenicity and host immunity? Front Microbiol. 2012;3(413):1–7.
Mudakavi RJ, Vanamali S, Chakarvortty D, Raichur AM. Development of arginine based nanocarriers for targeting and treatment of intracellular *Salmonella*. RSC Adv. 2017;7:7022–32.
Mukherjee S, Das L, Kole L, Karmakar S, Datta N, Das K. Targeting of parasite-specific immunoliposome encapsulated doxorubicin in the treatment of experimental visceral leishmaniasis. J Infect Dis. 2004;189:124–34.
Mukherjee S, Ray S, Thakur RS. Solid lipid nanoparticles: a modern formulation approach in drug delivery system. Indian J Pharm Sci. 2009;71(4):349–58.
Nagavarma BVN, Yadav HKS, Ayaz A, Vasudha LS, Shivakumar SG. Different techniques for preparation of polymeric nanoparticles –a review. Asian J Pharm Clin Res. 2012;5(3):16–23.
Niller HH, Masa R, Venkei A, Meszaros S, Minarovits J. Pathogenic mechanisms of intracellular bacteria. Curr Opin Infect Dis. 2017;30(3):309–15.
Oeztuerk-Atar K, Eroglu H, Calis S. Novel advances in targeted drug delivery. J Drug Target. 2018;26:633–42.
Ovais M, Raza A, Naz S, Islam NU, Khalil AT, Ali S, Khan MA, Shinwari ZK. Current state and prospects of the phytosynthesized colloidal gold nanoparticles and their applications in cancer theranostics. Appl Microbiol Biotechnol. 2017;101(9):3551–65.
Patel SK, Janjic JM. Macrophage targeted theranostics as personalized nanomedicine strategies for inflammatory diseases. Theranostics. 2015;5:150.
Paulo CSO, Neves RP, Ferreira LS. Nanoparticles for intracellular-targeted drug delivery. Nanotechnology. 2011;22:1–12.
Peer D, Karp JM, Hong S, Farokhzad OC, Margalit R, Langer R. Nanocarriers as an emerging platform for cancer therapy. Nat Nanotechnol. 2007;2:751–60.
Ranjan A, Pothayee N, Saleem MN, Boyle SM, Kasimnickam R, Riffle JS, Sriranganathan N. Nanomedicine for intracellular therapy. FEMS Microbiol Lett. 2012;332:1–9.
Rizzo LY, Theek B, Storm G, Kiessling F, Lammers T. Recent progress in nanomedicine: therapeutic, diagnostic and theranostic applications. Curr Opin Biotechnol. 2013;24:1159–66.
Ruggiero C, Pastorino L, Herrera OL. Nanotechnology based targeted drug delivery, 32nd Annual International Conference of the IEEE EMBS Buenos Aires, Argentina, August 31–September 4, 2010.

Saleem MN, Jain N, Pothayee N, Ranjan A, Riffle JS, Sriranganathan N. Targeting *Brucella melitensis* with polymeric nanoparticles containing streptomycin and doxycycline. FEMS Microbiol Lett. 2009a;294(1):24–31.

Saleem MN, Munosamy P, Ranjan A, Alqublan H, Pickrell G, Sriranganathan N. Silica-antibiotic hybrid nanoparticles for targeting intracellular pathogens. Antimicrob Agents Chemother. 2009b;53(10):4270–4.

Senapati S, Mahanta AK, Kumar S, Maiti P. Controlled drug delivery vehicles for cancer treatment and their performance. Signal Transduct Target Ther. 2018;3(7):1–19.

Sercombe L, Veerati T, Moheimani F, Wu SY, Sood AK, Hua S. Advances and challenges of liposome assisted drug delivery. Front Pharmacol. 2015;6:1–13.

Sibley LD, Weidner E, Krahenbuhl JL. Phagosome acidification blocked by intracellular *Toxoplasma gondii*. Nature. 1985;315:416–9.

Silva M. Classical labelling of bacterial pathogens according to their lifestyle in the host: inconsistencies and alternatives. Front Microbiol. 2012;3:1–7.

Singh L, Parboosing R, Kruger HG, Maguire GEM, Govender T. Intracellular localization of gold nanoparticles with targeted delivery in MT-4 lymphocytes. Adv Nat Sci Nanosci Nanotechnol. 2016;7:1–8.

Singh R, Lilliard JW. Toward an optimized treatment of intracellular bacterial infections: input of nanoparticulate drug delivery systems. Exp Mol Pathol. 2009;86(3):215–23.

Sinha R, Kim GJ, Nie S, Shin DM. Nanotechnology in cancer therapeutics: bioconjugated nanoparticles for drug delivery. Mol Cancer Ther. 2006;5(8):1909–17.

Spinosa MR, Progida C, Tala A, Cogli L, Alifano P, Bucci C. The *Neisseria meningitidis* capsule is important for intracellular survival in human cells. Infect Immun. 2007;75(7):3594–603.

Syed MA, Bokhari SH. Gold nanoparticles based microbial detection and identification. J Biomed Nanotechnol. 2011;7(2):229–37.

Tan BH, Meinken C, Bastian M, Bruns H, Legaspi A, Ochoa MT, Krutzik SR, Bloom BR, Ganz T, Modlin RL, Stenger S. Macrophages acquire neutrophil granules for antimicrobial activity against intracellular pathogens. J Immunol. 2006;177(3):1864–71.

Thakkar M, Brijesh S. Combating malaria with nanotechnology-based targeted and combinatorial drug delivery strategies. Drug Deliv Transl Res. 2016;6(4):414–25.

Thi EP, Lambertz U, Reiner NE. Sleeping with the enemy: how intracellular pathogens cope with a macrophage lifestyle. PLoS Pathog. 2012;2012(8):1–4.

Toledo DAM, Avila HD, Melo RCN. Host lipid bodies as platforms for intracellular survival of protozoan parasites. Front Immunol. 2016;7(174):1–6.

Toti US, Guru BR, Hali M, McPharlin C, Wykes SM, Panyam J, Whittum-Hudson JA. Targeted delivery of antibiotics to intracellular chlamydial infections using PLGA nanoparticles. Biomaterials. 2011;32(27):6606–13.

Uddin F, Aman W, Ullah I, Qureshi US, Mustapha U, Shafique S, Zeb A. Effective use of nanocarriers as drug delivery systems for the treatment of selected tumors. Int J Nanomedicine. 2017;12:7291–309.

Ueno N. Host and parasite determinants of Leishmania survival following phagocytosis by macrophages. PhD (Doctor of Philosophy) thesis, University of Iowa; 2011

Urban P, Estelrich J, Adeva A, Cortes A, Fernandez-Busquets X. Study of the efficacy of antimalarial drugs delivered inside targeted immunoliposomalanovectors. Nanoscale Res Lett. 2011;6:1–9.

Urban P, Ranucci E, Fernandez-Busquets X. Polyamidoamine nanoparticles as nanocarriers for the drug delivery to malaria parasite stages in the mosquito vector. Nanomedicine. 2015;10(22):3401–14.

Urban P, Valle Delgado JJ, Mauro N, Marques J, Manfredi A, Rottmann M, Ranucci E, Ferruti P, Fernandez-Busquets X. Use of poly(amidoamine) drug conjugates for the delivery of antimalarials to Plasmodium. J Control Release. 2014;177:84–95.

Ventola CL. Progress in nanomedicine: approved and investigational nanodrugs. Pharm Ther. 2017;42(12):742–55.

Vieira ACC, Magalhaes J, Rocha S, Cardoso MS, Santos SG, Borges M, Pinheiro M, Reis S. Targeted macrophages delivery of rifampicin-loaded lipid nanoparticles to improve tuberculosis treatment. Nanomedicine. 2017;12(24):2721–36.

Walburger A, Koul A, Ferrari G, Nguyen L, Baschong CP, Huygen K, Klebal B, Thomson C, Bacher G, Pieters J. Protein kinase G from pathogenic *Mycobacteria* promotes survival within macrophages. Science. 2004;304:1800–4.

Xie J, Lee S, Chen X. Nanoparticle-based theranostic agents. Adv Drug Deliv Rev. 2010;62:1064–79.

Xie S, Tao Y, Pan Y, Qu W, Cheng G, Huang L, Chen D, Wang X, Liu Z, Yuan Z. Biodegradable nanoparticles for intracellular delivery of antimicrobial agents. J Control Release. 2014;187:101–17.

Xie S, Yang F, Tao Y, Chen D, Qu W, Huang L, Liu Z, Pan Y, Yuan Z. Enhanced intracellular delivery and antibacterial efficacy of enrofloxacin-loaded docosanoic acid solid lipid nanoparticles against intracellular *Salmonella*. Sci Rep. 2017;7:1–9.

Xu W, Ling P, Zhang T. Polymeric micelles, a promising drug delivery system to enhance bioavailability of poorly water-soluble drugs. J Drug Deliv. 2013;2013:1–13.

Yang K, Feng L, Shi X, Liu Z. Nano-graphene in biomedicine: theranostic applications. Chem Soc Rev. 2013;42(2):530–47.

Ye J, Liu E, Yu Z, Pei X, Chen S, Zhang P, Shin MC, Gong J, He H, Yang VC. CPP-assisted intracellular drug delivery, what is next? Int J Mol Sci. 2016;17(1892):1–16.

Yu X, Trase I, Ren M, Duval K, Guo X, Chen Z. Design of nanoparticle-based carriers for targeted drug delivery. J Nanomater. 2016;2016:1–16.

Zhang Y, Huang Y, Li S. Polymeric micelles: nanocarriers for cancer-targeted drug delivery. AAPS Pharm Sci Tech. 2014;15(4):862–71.

Chapter 7
Nanoformulations: A Valuable Tool in the Therapy of Viral Diseases Attacking Humans and Animals

Josef Jampílek and Katarína Kráľová

Abstract Various viruses can be considered as one of the most frequent causes of human diseases, from mild illnesses to really serious sicknesses that end fatally. Numerous viruses are also pathogenic to animals and plants, and many of them, mutating, become pathogenic also to humans. Several cases of affecting humans by originally animal viruses have been confirmed. Viral infections cause significant morbidity and mortality in humans, the increase of which is caused by general immunosuppression of the world population, changes in climate, and overall globalization. In spite of the fact that the pharmaceutical industry pays great attention to human viral infections, many of clinically used antivirals demonstrate also increased toxicity against human cells, limited bioavailability, and thus, not entirely suitable therapeutic profile. In addition, due to resistance, a combination of antivirals is needed for life-threatening infections. Thus, the development of new antiviral agents is of great importance for the control of virus spread. On the other hand, the discovery and development of structurally new antivirals represent risks. Therefore, another strategy is being developed, namely the reformulation of existing antivirals into nanoformulations and investigation of various metal and metalloid nanoparticles with respect to their diagnostic, prophylactic, and therapeutic antiviral applications. This chapter is focused on nanoscale materials/formulations with the potential to be used for the treatment or inhibition of the spread of viral diseases caused by human immunodeficiency virus, influenza A viruses (subtypes H3N2 and H1N1), avian influenza and swine influenza viruses, respiratory syncytial virus, herpes simplex virus, hepatitis B and C viruses, Ebola and Marburg viruses, Newcastle disease virus, dengue and Zika viruses, and pseudorabies virus. Effective antiviral long-lasting and target-selective nanoformulations developed for oral, intravenous, intramuscular, intranasal, intrarectal, intravaginal, and intradermal applications are discussed.

J. Jampílek (✉)
Faculty of Natural Sciences, Department of Analytical Chemistry, Comenius University, Bratislava, Slovakia

K. Kráľová
Faculty of Natural Sciences, Institute of Chemistry, Comenius University, Bratislava, Slovakia

M. Rai, B. Jamil (eds.), *Nanotheranostics*,
https://doi.org/10.1007/978-3-030-29768-8_7

Benefits of nanoparticle-based vaccination formulations with the potential to secure cross protection against divergent viruses are outlined as well.

Keywords Antivirals · Metals · Metal oxides · Nanoformulations · Nanoparticles · Nanoparticle-based vaccines · Viruses

7.1 Introduction

Nanotechnology is a fast-growing field that provides the development of materials that have new dimensions, novel properties, and a broader array of applications. U.S. National Nanotechnology Initiative defines nanoparticles (NPs) in the range 1–100 nm (National Nanotechnology Initiative 2008). Microbial, fungal, and viral infections represent an increasing worldwide threat that is caused by both general immunosuppression of the world population and the growth of resistance of pathogens to clinically used drugs as well as development of cross-resistant or multidrug-resistant strains (Jampílek 2018). In particular, viruses caused a number of diseases worldwide, many of them with fatal termination. Although smallpox has been eradicated and for some of them vaccines were developed, viruses continuously indicate to us that conventional antiviral drugs/strategies directly targeting viral or cellular proteins have been limited, due to frequent virus variation resulting in drug resistance, because many of drugs show high specificity (e.g., De Clercq and Li 2016; Takizawa and Yamasaki 2018; Wang et al. 2018a; Deng and Wang 2018). It can be stated that design and discovery of structurally new anti-infectives with a new/innovative mode of action is time consuming and relatively risky (Jampílek 2016a, b, 2018). Moreover, many of the existing antivirals demonstrate also human toxicity and rapid clearance from the body. Thus, design and development of new safe and potent antivirals with activity against viral infection at multiple points in the viral life cycle remains a major challenge (Al-Ghananeem et al. 2013; Li et al. 2016; Kos et al. 2019).

Strategy based on the reformulation of existing anti-infectives into nanoformulations as well as investigation of various metal and metalloid nanoparticles with respect to their anti-infectious activity and potency is applied more and more frequently (Jampílek and Kráľová 2017; Pisárčik et al. 2017, 2018). In fact, this strategy is not only used for anti-infectious drugs but nanoformulations are also widely used for different classes of drugs, such as antineoplastics (Pentak et al. 2016; Jampílek and Kráľová 2019a), antipsychotics, antidepressants (Jampílek and Kráľová 2019b, c), and nootropics (Jampílek et al. 2015). Nanoscale dimension significantly modifies properties and behavior of all materials (e.g., Dolez 2015), whereby to benefits of pharmaceutical and medical nanoformulations belong primarily sustainable release of drugs, modifications of bioavailability, reduction of the required drug amount, reduction of toxicity of drugs, as well as increasing drug stability (e.g., Jampílek and Kráľová 2018; Patra et al. 2018). Thus in the light of the above mentioned facts, nanomaterials can constitute a useful tool in combat with viruses as well (Szunerits et al. 2015; Milovanovic et al. 2017; Siddiq et al. 2017; Singh et al. 2017; Zazo et al. 2017).

Recent findings related to development of virus-based nanoparticle (NP) systems as vaccines estimated for the prevention or treatment of infectious diseases, chronic diseases, cancer, and addiction were summarized by Lee et al. (2016), while Jackman et al. (2016) focused their attention on the progress in application of nanomedicine strategies to control critical stages in the virus life cycle through either direct or indirect approaches involving membrane interfaces. For example, diagnostic, prophylactic, and therapeutic applications of metal and metal oxide NPs in human immunodeficiency virus (HIV), hepatitis virus, influenza virus, and herpes simplex virus (HSV) infections were overviewed by Yadavalli and Shukla (2017) and Scherliess (2019). Antiviral activity of the metal NPs, especially the mechanism of action of AgNPs against different viruses (HSV, HIV, hepatitis B virus (HBV), metapneumovirus, respiratory syncytial virus (RSV)), was discussed by Rai et al. (2016).

In the nonviral gene therapy, the use of multifunctional NPs (e.g., lipid-based NPs, quantum dots, carbon nanotubes, magnetic NPs, silica NPs, and polymer-based NPs) is advantageous because these nonviral vectors ensure enhanced gene stability at gene delivery, shielding of cargo from nuclease degradation, and improve passive/active targeting (Lin et al. 2018a; Ariza-Saenz et al. 2018).

At the entry process of viruses into host cells, multivalent interactions with different cell surface receptors occur. Many human viruses attach to the cells through heparan sulfate (HS) proteoglycans and the attachment of the virus to HS on the cell surface start up a cascade of events ending with virus entry. Cagno et al. (2018) designed antiviral NPs having long and flexible linkers mimicking HS proteoglycans, allowing for effective viral association with a binding that was simulated to be strong and multivalent to the viral attachment ligand repeating units, generating forces that eventually lead to irreversible viral deformation and showing nanomolar irreversible activity against HSV, human papilloma virus, RSV, dengue virus, and lentivirus in vitro and were active ex vivo in human cervicovaginal histocultures infected by HSV-2 and in vivo in mice infected with RSV. Nanogels with different degrees of flexibility based on dendritic polyglycerol sulfate to mimic cellular HS designed by Dey et al. (2018) were able to multivalently interact with viral glycoproteins, shield virus surfaces, and efficiently block infection and were found to act as robust inhibitors for human viruses. Dendritic cells are crucial during development of T cell-specific responses against bacterial and viral pathogens.

Surfactant proteins A and D form an important part of the innate immune response in the lung, which can interact with NPs to modulate the cellular uptake of these particles. Unmodified polystyrene NPs of 100 nm were found to modulate surfactant proteins A and D mediated protection against influenza A infection in vitro (McKenzie et al. 2015). Substitution of sulfate groups in cellulose nanocrystals by tyrosine sulfate mimetic groups (i.e., phenyl sulfonates) led to improved viral inhibition indicating that the conjugation of target-specific functionalities to

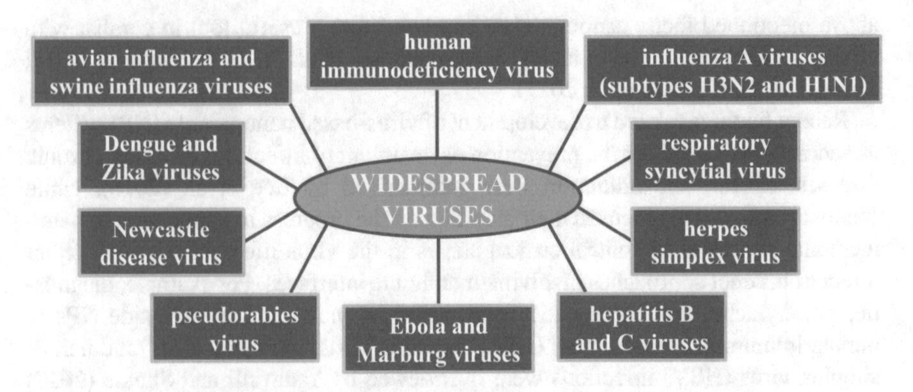

Fig. 7.1 The most common cause of viral diseases

cellulose nanocrystal surfaces provides a mean to control their antiviral activity (Zoppe et al. 2014).

A well-defined structure of virus due to its multifunctional proteinaceous shell (capsid) surrounding genomic material is a promising approach to obtain nanostructured materials, and viruses exhibit an ideal template for the formation of nanoconjugates with noble metal NPs. The interaction of the multifunctional viruses with NPs and other functional additives results in generation of bioconjugates with different properties, including possible antiviral and antibacterial activities (Parboosing et al. 2012; Capek 2015).

This contribution is focused on nanoscale materials/formulations with potential to be used to treat or inhibit the spread of viral diseases caused by human immunodeficiency virus, influenza A viruses (subtypes H3N2 and H1N1), avian influenza and swine influenza viruses, respiratory syncytial virus, herpes simplex virus, hepatitis B and C viruses, Ebola and Marburg viruses, Newcastle disease virus, dengue and Zika viruses, and pseudorabies virus (see Fig. 7.1). Effective antiviral long-lasting and target-selective nanoformulations developed for oral, intravenous, intramuscular, intranasal, intrarectal, intravaginal, and intradermal applications are discussed. Benefits of nanoparticle-based vaccination and nanoformulations with potential to secure cross protection against divergent viruses are outlined as well.

7.2 Human Immunodeficiency Virus

Human immunodeficiency virus (HIV) is a lentivirus (a subgroup of retrovirus) that causes HIV infection and over time acquired immunodeficiency syndrome (AIDS). HIV is a virus spread through certain body fluids that kills or impairs cells of the immune system, specifically the CD4 T cells, and progressively destroys the body's ability to fight infections and certain cancers (German Advisory Committee Blood 2016).

Based on the investigations of the antiretroviral (ARV) activity of carbon nanotubes (CNTs) using computational molecular approach, the strong molecular interactions suggested the efficacy of CNTs for targeting the HIV-mediated retroviral infections (Krishnaraj et al. 2014). Also, highly hydrophilic and dispersible carboxylated multiwalled carbon nanotubes (MWCNTs) bearing ARV drugs and hydrophilic functionalities were found to exhibit anti-HIV activity (Iannazzo et al. 2015). SiNPs (5–50 nm) prepared by grinding of porous silicon were found to act as efficient scavengers of HIV and RSV (Osminkina et al. 2014).

The production of AgNPs and their use as antiviral therapeutics against pathogenic viruses was reviewed by Galdiero et al. (2011). Ag nanocomplexes with anionic linear globular dendrimers that were assessed against HIV replication pathway in vitro showed also good ARV activity with nonsevere toxic effects in comparison with nevirapine as the standard drug in positive control group (Ardestani et al. 2015). Polyvinylpyrrolidone (PVP)-coated AgNPs were found to be a promising microbicidal candidate for use in topical vaginal/cervical agents to prevent HIV-1 transmission (Lara et al. 2010). Inactivation of microbial infectiousness by AgNPs-coated condom as a new approach to inhibit HIV- and HSV-transmitted infection was proposed by Fayaz et al. (2012). At treatment of HIV-1 infected cells with curcumin-stabilized AgNPs of 45 nm reduced replication of HIV by inhibition of NF-B nuclear translocation and the downstream expression of the pro-inflammatory cytokines interleukin-1β, tumor necrosis factor-α, and interleukin-6 was estimated and no similar biological effects were observed with curcumin alone and conventional AgNPs capped with citric acid (Sharma et al. 2017a). Stable and crystalline AgNPs fabricated using the aqueous leaf extract of mangrove (*Rhizophora lamarckii*) with particle sizes ranging from 12 to 28 nm inhibited HIV type 1 reverse transcriptase activity (IC_{50} of 0.4 μg/mL on the HIV-1) showing the promising potential to be used in the fight against HIV and other viruses of public health importance (Kumar et al. 2017a). AuNPs capped with sulfate-ended ligands that are able to bind HIV envelope glycoprotein gp120 and inhibit in vitro the HIV infection of T cells at nanomolar concentrations were reported by Di Gianvincenzo et al. (2010). AuNPs inhibiting HIV entry by binding with gp120 and preventing CD4 attachment were described by Vijayakumar and Ganesan (2012). HIV-1 peptides loaded onto AuNPs bearing high-mannoside-type oligosaccharides increased HIV-specific $CD4^+$ and $CD8^+$ T cell proliferation and induced highly functional cytokine secretion compared with HIV peptides alone and elicited a highly efficient secretion of pro-Th1 cytokines and chemokines, a moderate production of pro-Th2, and considerably higher secretion of pro-inflammatory cytokines such as tumor necrosis factor-α and interleukin-1β, suggesting that AuNPs that could simultaneously deliver HIV-1 antigens and high-mannoside-type oligosaccharides could be utilized as a superb vaccine delivery system (Climent et al. 2018).

Nowacek et al. (2011) performed analyses of nanoformulated ARV drugs and found that physical characteristics such as particle size, surfactant coating, surface charge, and most importantly shape are predictors of cell uptake and ARV efficacy. The progress in the development of NP-based drug delivery systems for HIV therapy which focused mainly on injectable nanocarriers enabling delivery of drug combinations that are long-lasting and target-selective in physiological contexts (in vivo) to provide safe and effective use was summarized by Gao et al. (2018). For example, tenofovir (TFV) alafenamide and elvitegravir (EVG) loaded NPs subcutaneously administered to female humanized CD34$^+$-NSG mice showed long residence time and exposure for both drugs. The $AUC_{(0–14\ day)}$ estimated for TFV and EVG using NPs was 23.1 ± 4.4 and 39.7 ± 6.7 μg h/mL, while with application of drug solutions the observed $AUC_{(0–72\ h)}$ reached 14.1 ± 2.0 and 7.2 ± 1.8 μg h/mL, respectively. Similarly, application of NPs resulted in strong increase of elimination half-life ($t_{1/2}$) that was 5.1 and 3.3 days for TFV and EVG, respectively, while the corresponding $t_{1/2}$ values for free drugs were 14.2 and 10.8 h (Prathipati et al. 2017).

Jiang et al. (2015) incorporated ARV drugs maraviroc, etravirine, and raltegravir into poly(lactic-co-glycolic) acid (PLGA) NPs and found that ARV NPs maintained potent HIV inhibition and were more effective, when used in combinations. Significantly higher antiviral potency and dose-dependent reduction against both cell-free and cell-associated HIV-1 BaL infection in vitro was estimated mainly for ARV NPs combinations involving etravirine NPs, whereby the combinations that showed large dose-reduction were identified to be synergistic. Moreover, ARV NPs combinations inhibited propagation of reverse transcriptase in simian-human immunodeficiency viruses in macaque cervicovaginal tissue and blocked virus transmission by migratory cells emigrating from the tissue. PGLA NPs were also used as a biodegradable carrier for loading with TFV in combination with efavirenz or saquinavir, which resulted in pronounced combination drug effects, and emphasized the potential of NPs for the realization of unique drug-drug activities (Chaowanachan et al. 2013). TFV and EVG-loaded PLGA NPs tested using humanized BLT mouse model were reported to be suitable for long-acting prevention of HIV-1 vaginal transmission (Mandal et al. 2017a). In an in vitro HIV-1 inhibition study, the IC_{50} value observed with emtricitabine-loaded PLGA NPs (size of <200 nm and surface charge −23 mV) was found to be 43-fold lower in TZM-bl cells (0.00043 μg/mL) and approx. fourfold lower (0.009 μg/mL) in peripheral blood mononuclear cells compared with drug solution (0.01861 and 0.033 μg/mL, respectively) and showed comparable activity with emtricitabine solution. Based on prolonged intracellular drug concentration and inhibition of HIV infection, the researchers noted that this long-acting, stable formulation could ensure once-biweekly dosing to prevent or treat HIV infection (Mandal et al. 2017b). PLGA-EVG NPs (~47 nm; zeta potential of approx. 6.74 mV) showed time- and concentration-dependent uptakes in monocytes with an approx. twofold higher drug intracellular internalization of EVG compared to free drug and also exhibited superior viral suppression over control for a prolonged period of time (Gong et al. 2017). Due to the presence of mannose receptors on the surface of macrophages, the mannosylated NPs of anti-HIV drug can target the macrophages, resulting in improvement of the therapeutic outcome and reduced toxicity of ARV bioactives. Surface-functionalized mannosylated-PLGA NPs of lamivudine (LVD) administered by intravenous route in a dose of 10 mg/kg to rats showed pronouncedly higher calculated brain/plasma ratio than PLGA NPs and continuously increased drug concentration up to 12 h (Patel et al. 2018). EFV-loaded PLGA NPs prepared using microfluidic method with particle size 73 nm, zeta potential −14.1 mV, and 10.8% drug loading showed a sustained in vitro EFV release (50% released within the first 24 h), and NPs functionalization with a transferrin receptor-binding peptide was found to be safe to blood brain barrier (BBB) endothelial and neuron cells (metabolic activity above 70%) and nonhemolytic, whereby functionalized nanosystems exhibited 1.3-fold higher drug permeability through a BBB in vitro model compared to free drug (Martins et al. 2019). EFV-loaded PLGA NPs and PLGA NPs with polyethylene glycol (PEG) coating with particle sizes 200–225 nm also retained native ARV activity of drug in vitro, while showed lower cytotoxicity against different epithelial cell lines and HIV target cells. Both types of NPs were readily taken up by colorectal cell lines and mildly reduced EFV permeation and increased membrane retention in Caco-2 and Caco-2/HT29-MTX cell models monolayer in vitro. At intrarectal administration to CD-1 mice in phosphate-buffered saline (pH 7.4), the coated PLGA NPs encapsulating EFV reached higher drug levels in colorectal tissues and lavages compared to free EFV or EFV-loaded PLGA NPs and they provided enhanced local pharmacokinetics that could be beneficial in preventing rectal HIV transmission (Nunes et al. 2018). TFV and EFV-loaded PLGA NPs incorporated alongside free TFV into fast dissolving films during film manufacturing showed higher retention in vivo in vaginal lavages and tissue when associated to film, and NPs-in-film were still able to enhance drug concentrations of EFV. Film alone also contributed to higher and more prolonged local drug levels as compared to the administration of TFV and EFV in aqueous vehicle, and once daily vaginal administration to mice of this formulation did not cause notable histological changes and major alterations in cytokine/chemokine profiles (Cunha-Reis et al. 2016).

The comparison of EVG-loaded nonadhesive and surface-modified bioadhesive poly(lactic acid)-hyperbranched polyglycerols NP formulations after intravaginal administration showed that bioadhesive NPs markedly improved and prolonged intravaginal delivery of drug and could provide sustained protection over longer durations (Mohideen et al. 2017).

Atripla-trimethyl chitosan (CS) NPs nanoconjugated with particle sizes of 177.2 ± 7.8 nm and zeta potential of −1.35 ± 0.04 mV showed a higher inhibitory effect on HIV replication than atripla alone, and at low doses it could be used for antiviral treatment resulting in reduction of drug resistance and other side effects (Shohani et al. 2017). Spherical intranasal EFV NPs prepared using CS-g-hydroxypropyl-β-cyclodextrin (198 ± 4.4 nm) with 23.28 ± 1.5% drug loading and 38 ± 1.43% entrapment efficiency (EE) exhibited sustained drug release (99.03 ± 0.30% in 8 h), 4.76-fold greater permeability through porcine nasal mucosa than plain drug solution, and 12.40-fold higher central nervous system bioavailability than drug solution administered by i.v. route (Belgamwar et al. 2018). Experiments with NPs coated with glycol chitosan and loaded with a HIV-1 inhibitor peptide tested in vitro and ex vivo using the porcine vaginal mucosa showed that such formulation reached the vaginal tissue and released the peptide within intercellular space without any side effects (Ariza-Saenz et al. 2017).

A novel NP-in-microparticle (MP) delivery system (NiMDS) comprised of pure NPs of the darunavir (DRV) and its boosting agent ritonavir (RTV) encapsulated within film-coated MPs was reported, in which pure NPs were encapsulated within calcium alginate/CS MPs that were film-coated with a series of poly(methacrylate) copolymers showing differential solubility in the gastrointestinal tract and enabling stability under gastric-like pH and sustained drug release under intestinal one. Using this formulation a 2.3-fold higher oral bioavailability of DRV with respect to both the unprocessed and the nanonized DRV/RTV combinations was estimated in albino Sprague-Dawley rats (Augustine et al. 2018). Enhanced bioavailability of darunavir-loaded lipid nanoemulsion in the brain after oral administration was described by Desai and Thakkar (2019).

Among TFV-loaded HIV-1 g120 targeting mannose responsive particles prepared through the layer-by-layer coating of $CaCO_3$ with concanavalin A (Con A) and glycogen, the one Con A layer containing system showed about twofold increase in drug release vs. control at a concentration of 25 μg/mL HIV gp120 and percent mucoadhesion estimated ex vivo on porcine vaginal tissue, ranged from 10% to 21%, depending on the number of Con A layers in the formulation (Coulibaly et al. 2017).

Hillaireau et al. (2013) referred about the potential of nanoassemblied nucleoside reverse transcriptase inhibitors (NRTIs) delivered as squalenoylated prodrugs to enhance their absorption and improve their biodistribution and also to enhance their intracellular delivery and antiviral efficacy toward HIV-infected cells. Biodegradable cationic cholesterol-ε-polylysine nanogel carriers for delivery of triphosphorylated NRTIs demonstrating high anti-HIV activity along with low neurotoxicity, warranting minimal side effects following systemic administration, were prepared by Warren et al. (2015). Nanogel modification with brain-specific peptide vectors resulted in efficient central nervous system targeting. Senanayake et al. (2015) conjugated succinate derivatives of NRTIs with cholesteryl-ε-polylysine nanogels. Nanogel conjugates of zidovudine (ZDV), LVD, and abacavir demonstrated tenfold suppression of reverse transcriptase activity in HIV-infected macrophages with EC_{90} drug levels of 2–10, 2–4, and 1–2 μM, respectively, for single nanodrugs and dual and triple nanodrug cocktails, conjugate of LVD being the most effective single nanodrugs (EC_{90} 2 μM). NPs of 50–60 nm prepared by encapsulation of ZDV in lactoferrin NPs were found to be stable in simulated gastric and intestinal fluids and the anti-HIV-1 activity of drug remained unaltered in nanoformulation in acute infection, the drug release from NPs was constant up to 96 h and bone marrow micronucleus assay showed that nanoformulation exhibited ca. twofold lower toxicity than soluble form (Kumar et al. 2015). Lactoferrin NPs loaded with a triple drug combination of ZDV, EFV, and LVD showing a diameter of 67 nm and EE >58% for each drug delivered the maximum of its payload at pH 5 with a minimum burst release in vitro, exhibited improved anti-HIV activity, were practically not toxic to the erythrocytes, and reached in in vivo experiment an approx. >4-fold increase in AUC, 30% increase in the C_{max} compared to individual drugs, and >2-fold enhancement in the half-life of each drug. They also showed improved bioavailability of all drugs with less tissue-related inflammation (Kumar et al. 2017b). Improved HIV-microbicide activity through the co-encapsulation of drugs acting as NRTIs in biocompatible metal organic framework nanocarriers containing iron(III) polycarboxylate NPs was reported by Marcos-Almaraz et al. (2017).

TFV disoproxil fumarate encapsulated in multifunctional magneto-plasmonic liposomes, a hybrid system combining liposomes and magneto-plasmonic NPs, enabled the treatment in the brain microenvironment that is inaccessible to most of the drugs, showed enhanced transmigration across an in vitro BBB model by magnetic targeting, and exhibited desired therapeutic effects against HIV-infected microglia cells, while the gold shell of liposomes showed bright positive contrast in X-ray computed tomography (Tomitaka et al. 2018).

EFV-loaded solid lipid NPs (SLNPs) with mean particle size of 108.5 nm, polydispersity index of 0.172, and 64.9% EE administered intranasally in vivo revealed increased concentration of the drug in brain as desired suggesting their potential to be used for eradication of HIV and cure of HIV-infected patients (Gupta et al. 2017). Endsley and Ho (2012) constructed CD4-targeted SLNPs providing selective binding and efficient delivery of indinavir to $CD4^+$-HIV host cells (whereby inclusion of PEG in SLNPs minimized immune recognition of peptides) that showed enhancement of anti-HIV effects even under limited time exposure. Nanoscaled NRTIs decorated with the peptide-binding brain-specific apolipoprotein E receptor demonstrated low neurotoxicity and high antiviral activity against HIV infection in the brain (Gerson et al. 2014).

A cell surface chemokine receptor CXCR4 targeting peptide 4DV3 acting as an HIV entry inhibitor and a ligand for targeted drug delivery was described by Lee et al. (2019). The use of 4DV3 as the targeting ligand resulted in enhanced endocytosis due to the uptake of 4DV3 functionalized nanocarriers combined with the allosteric interaction with CXCR4, suggesting that 4DV3 peptide could be considered as a dual function ligand.

Bayon et al. (2018) developed vaccine formulations comprising nanostructured lipid carriers (NLC) grafted with p24 antigen, together with cationic NLC optimized for the delivery of immunostimulant CpG, which was able notably enhance immune responses against p24 manifested in specific antibody production and T cell activation in mice as well as in nonhuman primates.

A tissue- and cell-targeted long-acting four-in-one nanosuspension composed of lopinavir, RTV, TFV, and LVD administered as a single injection subcutaneously to four macaques was able to exhibit persistent drug levels in lymph node mononuclear cells and peripheral blood mononuclear cells for 5 weeks and could be proposed for a long-acting treatment with the potential to target residual virus in tissues and improve patient adherence (McConnachie et al. 2018).

Biocompatible ZDV-loaded hybrid core-shell NPs of carboxymethyl cellulose (core) and Compritol®-PEG (shell) for ARV drug delivery with mean size of 161 ± 44.06 nm and 82% drug EE showed controlled drug release and were able effectively enter the brain cells (Joshy et al. 2017). The biocompatible PF-68-coated NPs of amide functionalized alginate prepared by coupling reaction with D,L-glutamic acid encapsulating ZDV with mean size of 432 ± 11.9 nm and a loading efficacy of 29.5 ± 3.2% showed slow and sustained release of drug in phosphate-buffered saline (PBS, pH 7.4) and exhibited significantly higher cellular internalization efficiency in vitro (Joshy et al. 2018a). Considerable improvement in cellular internalization (murine neuro-2a and HeLa cells) in vitro was observed also with ZDV-loaded PVP/stearic acid (SA)-PEG NPs due to the core-shell NPs prepared from lipid and polymer, suggesting that such NPs have potential to be used for antiviral drug delivery for use in HIV/AIDS therapy (Joshy et al. 2018b). ZDV-loaded core-shell dextran hybrid nanosystem expressed enhanced cellular internalization of drug-loaded hybrid NPs in comparison with free drug (Joshy et al. 2018c).

Retained intracytoplasmic nanoformulations consisting of a hydrophobic and lipophilic modified dolutegravir prodrug encapsulated into poloxamer released the drug from macrophages and suppressed viral replication and spread of virus to $CD4^+$ T cells, while the drug blood and tissue levels in BALB/cJ mice were above 64 ng/mL corresponding to IC_{90} value for 56 and 28 days, respectively (Sillman et al. 2018).

Crystals of myristoylated cabotegravir prodrug that were formulated into NPs facilitated avid monocyte-macrophage entry, retention, and reticuloendothelial system depot formulation, showed sustained protection against HIV-1 challenge, and a single 45 mg/kg intramuscular injection of these NPs at a dose of 45 mg/kg to BALB/cJ mice resulted in fourfold higher pharmacokinetic profiles compared to long-acting parenteral cabotegravir, and similar results were obtained also with rhesus macaques (*Macaca mulatta*); improved viral restriction in human adult lymphocyte-reconstituted NOD/SCID/IL2R$\gamma c^{-/-}$ mice by this nanoformulations was observed as well (Zhou et al. 2018).

7.3 Influenza A Viruses, Subtypes H3N2 and H1N1

Influenza is a respiratory illness caused by a virus. All influenza viruses are negative-sense, single-stranded, segmented RNA viruses. Viruses of influenza A type occur in humans and animals, while types B and C can be found only in humans. Influenza A viruses are classified according to the type of hemagglutinin (H) and the type of neuroaminidase (N) into subtypes H1N1, H1N2, H2N2, H3N2, H5N1, and H8N4. Each virus subtype has mutated into a variety of strains with differing pathogenic profiles using humanized BLT mouse model (Taubenberger and Morens 2010; Takizawa and Yamasaki 2018). Influenza A virus causes influenza in birds and some mammals, and is the only species of the *Alphainfluenzavirus* genus of the *Orthomyxoviridae* family of viruses. Type A influenza is a contagious viral infection that can have life-threatening complications if left untreated. An influenza could be pandemic, when an epidemic of an influenza virus spreads on a worldwide scale, e.g., "Spanish flu" of 1918 (Kuchipudi and Niessly 2018).

Inhibition of A/Human/Hubei/3/2005 (H3N2) influenza virus infection by AgNPs in vitro and in vivo was reported by Xiang et al. (2013). Feng et al. (2013) designed glycosylated metal (Ag and Au) NPs as potent influenza A virus hemagglutinin (HA) blockers. Inhibitory effects of AgNPs with particle size of 10 nm on H1N1 influenza A virus in vitro were also reported by Xiang et al. (2011).

The effect of size dependence of the AgNPs (3.5, 6.5, and 12.9 nm average diameters) that were embedded into the CS matrix on antiviral activity against H1N1 influenza A virus was also observed showing generally stronger antiviral activity with smaller AgNPs in the composites (Mori et al. 2013). Evaluation of the efficacy of AgNP-decorated silica hybrid composite (Ag30-SiO_2) with particle size of 400 nm in diameter for inactivation of influenza A virus showed that even after 1 h of exposure to these NPs >80% of HA damage, 20% of neuraminidase activities, and reduction of the infection caused by the virus in Madin-Darby Canine Kidney (MDCK) cells was observed. The Ag30-SiO_2 NPs were found to interact with viral components situated at the membrane causing them nonspecific damage resulting in virus inactivation (Park et al. 2018).

As the primary mechanism of influenza virus inhibition by AuNPs with different anionic groups blocking of viral attachment to cell surface was suggested, although viral fusion inhibition could not be excluded (Sametband et al. 2011). Li et al. (2016) proposed a new modality to inhibit viral infection by fabricating DNA-conjugated AuNPs networks on cell membranes as a protective barrier, antiviral activity of which may be attributed to steric effects, the disruption of membrane glycoproteins, and limited fusion of cell membrane bilayers. In addition, these DNA-AuNPs beside inhibition of virus attachment and entry could also inhibit viral budding and cell-to-cell spread. Multivalent sialic acid-functionalized AuNPs of 14 nm inhibited influenza virus infection. As the binding of the viral fusion protein HA to the host cell surface is mediated by sialic acid receptors, a multivalent interaction with sialic acid-functionalized AuNPs is expected to competitively inhibit viral infection (Papp et al. 2010). Intranasal immunizations with a mixture of conjugated recombinant trimetric influenza A/Aichi/2/68(H3N2) HA onto functionalized AuNPs surfaces in a repetitive, oriented configuration and AuNPs coupled with Toll-like receptor 5 agonist flagellin as particulate adjuvants resulted in improved mucosal B cell responses (increasing influenza-specific immunoglobulin (Ig) A and IgG l in nasal, tracheal, and lung washes), stimulation of antigen-specific interferon-γ-secreting $CD4^+$ cell proliferation and induced strong effector $CD8^+$ T cell activation suggesting powerful mucosal and systemic immune responses protecting hosts against lethal influenza challenges (Wang et al. 2018b).

CuI NPs of mean size of 160 nm exerted antiviral activity against an influenza A virus of swine origin (pandemic [H1N1] 2009) by generating hydroxyl radicals, suggesting that they could be applied in filters, face masks, protective clothing, and kitchen cloths as a material suitable to protect against viral attacks (Fujimori et al. 2012).

Nanocomposites, in which DNA fragments were electrostatically bound to TiO_2 NPs pre-covered with polylysine, efficiently inhibited human influenza A (subtype H3N2) (Levina et al. 2014). Composites of peptide nucleic acids with TiO_2 NP-containing DNA/peptide nucleic acid duplexes inhibited reproduction of influenza A virus (H3N2 subtype) with an efficiency of 99% and they were shown to not only penetrate through cell membranes, but also exhibit a high specific antisense activity without toxic effects on the living cells (Amirkhanov et al. 2015).

SeNPs decorated by ribavirin (RBV), a broad-spectrum antiviral drug, protected cells during H1N1 infection in vitro, while in in vivo experiments they prevented lung injury in H1N1-infected mice, significantly reduced DNA damage in lung tissue, and restrained activations of caspase-3 and proteins on the apoptosis pathway (Lin et al. 2018b).

Pieler et al. (2016) bound inactivated, concentrated, and diafiltered influenza A virus particles produced in MDCK cell suspension to magnetic sulfated cellulose microparticles (100–250 μm) and directly injected into mice for immunization and observed high anti-influenza A antibody responses and full protection against a lethal challenge with replication-competent influenza A virus, whereby 400-fold reduced number of influenza nucleoprotein gene copies in the lungs of mice immunized with antigen-loaded microparticles compared to mock-treated animals was estimated.

Hybrid inorganic-organic microcapsules obtained by encapsulating siRNA via the combination of layer-by-layer technique and in situ modification by sol-gel chemistry were characterized with high cell uptake efficiency, low toxicity, efficient intracellular delivery of siRNAs and the protection of siRNAs from premature degradation before reaching the target cells, reduced viral nucleoprotein level, and inhibited influenza A virus (H1N1) production in infected MDCK and A549 cells (Timin et al. 2017). CS NPs loaded by siRNA that were efficiently up-taken by Vero cells resulting in the inhibition of influenza virus (strain A/PR/8/34 (H1N1)) replication in vitro showed antiviral effects at nasal administration and considerably protected BALB/c mice from a lethal influenza challenge, suggesting that this nanoformulation is a promising system for controlling influenza virus infection (Jamali et al. 2018).

Figueira et al. (2018) modified the influenza fusion inhibitors by adding a cell-penetrating peptide self-assembling into 15–30 nm NPs and targeting relevant tissues infected with influenza virus A in vivo, causing reduction of viral infectivity upon interaction with the cell membrane, and it was shown that for efficacious biodistribution, fusion inhibition, and efficacy in vivo both the cell-penetrating peptide and the lipid moiety are necessary.

The amino acid sequence of influenza matrix protein 2 ectodomain (M2e) is highly conserved among human seasonal influenza A viruses (Deng et al. 2018a). Multivalent oleanolic acid protein conjugates as nonglycosylated neomucin mimics for the capture and entry inhibition of influenza A viruses (H1N1, H3N2, and H9N2) designed by Yang et al. (2018) were found to be comparable to natural glycosylated mucin, suggesting that this material could potentially be used as anti-infective barriers to prevent virus from invading host cells. Bernasconi et al. (2018) developed a fusion protein with three copies of the ectodomain of matrix protein 2, which is one of the most explored conserved influenza A virus antigens for a broadly protective vaccine and incorporated it into porous maltodextrin NPs to enhance its protective ability and the formulation resulted in broadly protective immunity against a lethal infection with heterosubtypic influenza virus, immune protection being mediated by enhanced levels of lung-resident $CD4^+$ T cells as well as anti-HA and anti-M2e serum IgG and local IgA antibodies.

The initial adhesion of most viruses is quantitatively controlled by multivalent binding of proteins to glycan receptors on the host cell. Lin et al. (2018c) to mimic virus adhesion to cell surfaces attached protein-oligomer-coated NPs to fluidic glycolipid membranes with surface glycan density varying over four orders of magnitude and found that in the binding isotherms two regions could be estimated, which could be attributed to monovalent and multivalent protein/glycan interactions at low and high glycan densities, respectively.

Lauster et al. (2017) designed multivalent peptide-polymer NPs consisting of covalent conjugates of peptidic ligands and a biocompatible polyglycerol-based hydrophilic dendritic scaffold showing binding with nanomolar affinity to the influenza A virus via its spike protein hemagglutinin. A novel seasonal recombinant HA NP influenza vaccine (NIV) formulated with a saponin-based adjuvant, Matrix-M™, induced HA inhibition (HAI) and microneutralizing antibodies against a broad range of influenza A (H3N2) subtypes. HAI-positive and HAI-negative neutralizing monoclonal antibodies derived from mice immunized with NIV were active against homologous and drifted influenza A (H3N2) strains (Smith et al. 2017). Layered protein NPs composed of structure-stabilized HA stalk domains from both HA groups, and constructed M2e, were reported to have potential to be developed into a universal influenza vaccine that could induce broad cross protection against divergent viruses (Deng et al. 2018a).

An intranasally administered biocompatible polyanhydride NP-based influenza A virus (IAV) vaccine (IAV-nanovax) that could provide protection against subsequent homologous and heterologous IAV infections in both inbred and outbred populations, when used for vaccination, resulted in promotion of the induction of germinal center B cells within the lungs, both systemic and lung local IAV-specific antibodies, and IAV-specific lung-resident memory CD4 and CD8 T cells (Zacharias et al. 2018).

Double-layered peptide NPs prepared by desolvating a composite peptide of tandem copies of nucleoprotein epitopes into NPs as cores and cross-linking another composite peptide of four tandem copies of influenza matrix protein 2 ectodomain epitopes to the core surfaces as a coating that were delivered via dissolvable microneedle patch-based skin vaccination, induced robust specific immunity and protected mice against heterosubtypic influenza A virus challenges and demonstrated a strong antigen depot effect resulting in the stronger immune responses, whereby $CD8^{+}$ T cells were involved in the protection (Deng et al. 2018b).

Intranasal vaccination with NPs formed by 3-sequential repeats of M2e on the self-assembling recombinant human heavy chain ferritin cage could induce robust immune responses such as high titers of sera M2e-specific IgG antibodies, T cell immune responses, and mucosal secretory IgA antibodies in mice in the absence of an adjuvant. These NPs also confer complete protection against a lethal infection of homo-subtypic H1N1 and hetero-subtypic H9N2 virus and could be considered as a promising, needle-free, intranasally administered cross-protective influenza vaccine as being economical and suitable for large-scale production (Qi et al. 2018).

7.4 Avian Influenza and Swine Influenza Viruses

Avian influenza is a variety of influenza caused by viruses adapted to birds. Avian influenza virus (AIV) is an A-type influenza virus belonging to the *Orthomyxoviridae* family, whereby isolated strains of "highly pathogenic avian influenza" inducing intravenous pathogenicity index >1.2 or mortality rate >75% in a defined chicken population during the specified interval of 10 days are of H5 and H7 subtypes (Chatziprodromidou et al. 2018).

Swine influenza is caused by swine influenza viruses (SwIV) that are endemic in pigs. In Europe, swine influenza is considered one of the most important primary pathogens of swine respiratory disease and infection is primarily with H1N1, H1N2, and H3N2 influenza A viruses (Brown 2013).

It was found that dietary supplementation of chromium picolinate (1500 ppb) and chromium NPs (1000 ppb) improved the performance and antibody titers against avian influenza and infectious bronchitis heat stress conditions (36 °C) in broilers (Hajializadeh et al. 2017). Serum samples collected from immunized chickens with formulation consisting of plasmid encoding HA gene of AIV, (A/Ck/Malaysia/5858/04 (H5N1)), formulated using biosynthesized AgNPs (4–18 nm) and PEG showed rapidly increasing antibody responses against H5 on day 14 after immunization and single oral administration of this formulation resulted in the induction of both the antibody and cell-mediated immune responses as well as enhanced cytokine production (Jazayeri et al. 2012). A vaccine formulation consisting of ion channel membrane matrix protein 2 of the extracellular domain of influenza A conjugated to AuNPs with CpG oligodeoxynucleotide as a soluble adjuvant, which was delivered intranasally in mice, resulted in lung B cell activation and robust serum anti-M2e IgG response, with stimulation of both IgG1 and IgG2a subtypes. Lethal challenge of vaccinated mice with A/California/04/2009 (H1N1pdm) pandemic strain, A/Victoria/3/75 (H3N2), and the highly pathogenic AIV A/Vietnam/1203/2004 (H5N1) resulted in 100, 92, and 100% protection (Tao et al. 2017). Considering a very important role of mucosal immunity in the antiviral immune response, AgNPs biofabricated using *Cinnamomum cassia* extract showed enhanced antiviral activity against highly pathogenic AIV subtype H7N3, when incubated with the virus prior to infection as well as introduced to cells after infection, whereby the tested concentrations of extract and AgNPs (up to 500 μg/mL) were found to be nontoxic to Vero cells (Fatima et al. 2016). For complexity, it is necessary to mention (as was discussed above) that Fujimori et al. (2012) described antiviral activity of CuI NPs against influenza A virus of swine origin H1N1.

Formulation using calcium phosphate NPs vaccine adjuvant and delivery platform to formulate an inactivated whole virus pandemic influenza A/CA/04/2009 (H1N1pdm) vaccine as a potential dose-sparing strategy that was intramuscularly administered to BALI/c mice at doses 0.3, 1, or 3 μg (based on HA content) resulted in higher HAI, virus neutralization, and IgG antibody titers compared to the nonadjuvanted vaccine, providing equal protection with one-third of the antigen dose as compared to the nonadjuvanted or alum (hydrated double sulfate salts of aluminum with potassium, sodium, or ammonium)-adjuvanted vaccine (Morcol et al. 2017). Solution of nanoscale scallop shell powder produced by calcination process showing a size of 500 nm inactivated AIV within 5 s, whereas the solution of greater powder particles (20 μm) could not even after 1 h incubation. Moreover, inactivation of Newcastle disease virus and goose parvovirus solution by nanopowder within 5 and 30 s, respectively, was observed as well (Thammakarn et al. 2014).

Chickens immunized with a low dose (200 μL) of bioadhesive liposomal influenza vaccine using liposomes prepared with tremella or xanthan gum as the bioadhesive polysaccharide showed considerably higher mucosal and serum antibody levels, and low-viscosity gel mixed with liposomes was found to be suitable for nasal delivery and chickens elicited higher mucosal secretory IgA and serum IgG after two vaccinations compared to application of liposome mixture with a high-viscosity gel (Chiou et al. 2009).

Vaccination of chickens with nonencapsulated AIV combined with PLGA-encapsulated CpG oligodeoxynucleotides, CpG 2007, resulted in qualitatively and quantitatively augmented antibody responses manifested as a reduction in virus shedding compared to the encapsulated AIV combined with PLGA-encapsulated CpG 2007 formulation (Singh et al. 2016a). Similarly, nonencapsulated CpG 2007 in inactivated AIV vaccines administered by the intramuscular route generated higher antibody responses compared to the CpG 2007 encapsulated in PLGA, while PLGA-encapsulated CpG 2007 in AIV vaccines administered by the aerosol route elicited higher mucosal responses compared to nonencapsulated CpG 2007 (Singh et al. 2016b). Seok et al. (2017) designed an intradermal pH1N1 DNA vaccine delivery platform using microneedles coated with a polyplex-containing PLGA/polyethyleneimine NPs inducing a greater humoral immune response due to rapid dissolution of the coated polyplex in porcine skin (within 5 min) than that observed at intramuscular polyplex delivery or naked pH1N1 DNA vaccine delivery by a dry-coated microneedles. Consequently, intradermal delivery of DNA vaccines within a cationic polyplex coated on microneedles has potential in skin immunizations. Polyanhydride NPs encapsulating subunit proteins can enhance humoral and cell-mediated immunity and provide protection upon lethal challenge. A robust immunogen, recombinant H5 hemagglutinin trimer ($H5_3$) encapsulated into polyanhydride NPs used to immunizing mice induced high neutralizing antibody titers and enhanced $CD4^+$ T cell recall responses in mice and $H5_3$-based polyanhydride nanovaccine induced protective immunity against a low-pathogenic H5N1 viral challenge (Ross et al. 2015). Similarly, intranasal delivery of nanovaccine consisting of inactivated SwIV encapsulated in polyanhydride NPs enhanced antigen-specific cellular immune response in pigs, with promise to induce cross-protective immunity (Dhakal et al. 2017a), and inactivated SwIV encapsulated in PLGA NPs (200–300 nm diameter) administered via intranasal route reduced the clinical disease and induced cross-protective cell-mediated immune response in a pig model as well (Dhakal et al. 2017b).

As a virus for challenge test Moon et al. (2012) used the HPAI H5N1 virus (A/EM/Korea/W149/06) isolated from fecal specimens collected from wild bird habitats and constructed recombinant HA antigen based on the HA1 head domain of this virus. Intranasal immunization with a mixture of recombinant influenza HA antigen or inactivated virus and poly-γ-glutamic acid (PGA)/CS NPs induced a high anti-HA IgA response in lung and IgG response in serum, including anti-HA neutralizing antibodies as well as an influenza virus-specific cell-mediated immune response in female BALB/c mice against challenge with a lethal dose of the highly pathogenic influenza A H5N1 virus (Moon et al. 2012). Pigs that were intranasally vaccinated with killed swine influenza A virus H1N2 (δ-lineage) antigens encapsulated in CS polymer-based NPs exhibited an enhanced IgG serum antibody and mucosal secretory IgA antibody responses in nasal swabs, bronchoalveolar lavage fluids, and lung lysates that were reactive against homologous (H1N2), heterologous (H1N1), and heterosubtypic (H3N2) influenza A virus strains (Dhakal et al. 2018).

Mucosal vaccination of conserved matrix protein 2, fusion peptide of hemagglutinin HA_2 and cholera toxin subunit Al (CTAI) fusion protein with poly-γ-glutamate/CS NPs induced protection against divergent influenza subtypes and was found to induce a high degree of systemic immunity (IgG and IgA) at the site of inoculation and in challenge tests in BALI/c mice with several viruses (H5N1, H1N1, H5N2, H7N3, or H9N2) provided cross protection against divergent lethal influenza subtypes up to 6 months after vaccination (Chowdhury et al. 2017). Two immunizations of modified nonreplicating mRNA encoding influenza H10 hemagglutinin and encapsulated in lipid NPs induced protective HA inhibition titers and H10-specific $CD4^+$ T cell responses after intramuscular or intradermal delivery in rhesus macaques (Liang et al. 2017). The results of hemagglutination inhibition and micro-neutralization assays showed that lipid NP-formulated modified mRNA vaccines encoding HA proteins of H10N8 (A/Jiangxi-Donghu/346/2013) or H7N9 (A/Anhui/1/2013) generated rapid and robust immune responses in mice, ferrets, and nonhuman primates, and a single dose of H7N9 mRNA was able to protect mice from a lethal challenge and reduced lung viral titers in ferrets (Bahl et al. 2017).

Coated two-layer protein nanoclusters from recombinant trimeric HA from an avian-origin H7N9 influenza that were evaluated for the virus-specific immune responses and protective efficacy in mice immunized with these nanoclusters were found to be highly immunogenic; they were able to induce protective immunity and long-lasting humoral antibody responses to this virus without the use of adjuvants, suggesting that such coated nanoclusters also have great potential for influenza vaccine production not only in response to an emerging pandemic, but also as a replacement for conventional seasonal influenza vaccines (Wang et al. 2017).

7.5 Respiratory Syncytial Virus

Respiratory syncytial virus (RSV), a negative-sense single-stranded enveloped RNA virus, is a global human pathogen responsible for lower respiratory tract infections and is considered as the major viral pathogen of the lower respiratory tract of infants. Therefore, there are urgent need to utilize convenient strategies to prevent

RSV infection, including beside of live attenuated, chimeric, and subunit vaccines also nanosized particles (Borchers et al. 2013; Clark and Guerrero-Plata 2017).

Au nanorods (AuNRs) were found to inhibit RSV in human epithelial type 2 (HEp-2) cells and in BALB/c mice by 82% and 56%, respectively, whereby the RSV inhibition correlated with marked upregulated antiviral genes due to AuNR-mediated Toll-like receptor, the nucleotide-binding oligomerization domain (NOD)-like receptor, and retinoic acid-inducible gene-I (RIG-I)-like receptor signaling pathways, whereby recruitment of immune cells to counter RSV replication was demonstrated by production of cytokines and chemokines in the lungs (Bawage et al. 2016). Treatment of human dendritic cells with AuNRs conjugated to RSV F formulated as a candidate vaccine preparation for RSV by covalent attachment of viral protein using a layer-by-layer approach induced immune responses in primary human T cells (Stone et al. 2013). Curcumin-modified AgNPs showed strong inhibitory activity against RSV infection resulting in a reduction of viral titers about two orders of magnitude at AgNPs concentration showing no toxicity to the host cells, whereby AgNPs inactivated the virus directly, which resulted in the prevention of RSV to infect the host cells (Yang et al. 2016a).

In BALB/c mice exposed to a single dose of intranasally administered TiO_2 NPs (0.5 mg/kg) that were 5 days later infected intranasally with RSV notably increased levels of interferon-γ and chemokine CCL5 in the bronchoalveolar lavage fluids were estimated compared with the control on day 5 postinfection, but not in uninfected mice. TiO_2 exposure resulted also in an increase in the infiltration of lymphocytes into the alveolar septa in lung tissues, while pulmonary viral titers were not affected. Consequently, it can be stated that the immune system was affected by a single exposure to TiO_2 NPs that exacerbated pneumonia in RSV-infected mice (Hashiguchi et al. 2015).

The curcumin-loaded β-cyclodextrin-functionalized graphene oxide (GO) composite was found to cause highly efficient inhibition of RSV infection, showed great biocompatibility to the host cells, and was able to prevent the host cells from RSV infection by directly inactivating the virus and inhibiting the viral attachment, showing prophylactic and therapeutic effects toward RSV (Yang et al. 2017a).

Recent advances in prophylactic synthetic biodegradable microparticle and nanoparticle vaccines against RSV and the multiple factors that can affect vaccine efficacy were summarized by Jorquera and Tripp (2016). Vaccination with a recombinant RSV F nanoparticle vaccine (60 or 90 g RSV F protein, with or without aluminum phosphate adjuvant) enhanced functional immunity to RSV in older adults (≥60 years), showed an acceptable safety profile, and additional immunogenicity benefit was observed with adjuvanted formulation compared to increasing antigen dose alone. Moreover, the RSV F vaccine co-administered with licensed inactivated trivalent influenza vaccine (TIV) did not impact the serum HAI antibody responses to a standard-dose TIV, and TIV did not affect the immune response to the RSV F vaccine (Fries et al. 2017).

7.6 Herpes Simplex Virus

Double-stranded DNA viruses, *Herpes simplex* virus 1 and 2 (HSV-1 and HSV-2), are members of the α-herpesvirus subfamily of *Herpesviridae* family that infect humans. HSV-1 causes cold sores, while HSV-2 causes genital herpes and these viruses can establish lifelong latent infection within peripheral nervous system (Dai and Zhou 2018).

Size-dependent interactions of AgNPs with HSV-1, HSV-2, and human parainfluenza virus type 3 resulting in reduced viral infectivity or in inhibition of the infectivity of the viruses by smaller-sized NPs that was caused probably by blocking interaction of the virus with the cell was described by Gaikwad et al. (2013). AgNPs applied at nontoxic concentrations administered prior to viral infection or soon after initial virus exposure were capable to inhibit HSV-2 replication, suggesting that the AgNPs acted during the early phases of viral replication (Hu et al. 2014). Tannic acid-modified AgNPs of 13, 33, and 46 nm showed antiviral activity and reduced both infection and inflammatory reaction in the mouse model of HSV-2 infection, when used at infection or for a postinfection treatment (Orlowski et al. 2014), and antiviral activity of such NPs with a size of 33 nm applied upon the mucosal tissues caused activation of immune response in vaginal HSV-2 (Orlowski et al. 2018a). Multifunctional tannic acid/AgNPs-based mucoadhesive hydrogel for effective vaginal treatment of HSV-2 genital infection was also reported (Szymanska et al. 2018). Investigation of the ability of tannic acid-modified Ag and AuNPs to induce dendritic cells maturation and activation in the presence of HSV-2 antigens, when used at nontoxic doses, showed that both types of these metal NPs were good activators of dendritic cells, albeit their final effect upon maturation and activation may be metal and size dependent and can help to overcome virus-induced suppression of dendritic cells activation (Orlowski et al. 2018b). AuNPs of the size of 7.86 nm surface-conjugated with gallic acid showed the antiviral efficacy against HSV infections in Vero cells with EC_{50} of 32.3 μM in HSV-1 and 38.6 μM in HSV-2 (Haider et al. 2018). AuNPs capped with mercaptoethanesulfonate (MES) strongly inhibited HSV-1, whereby they interfered with viral attachment, entry, and cell-to-cell spread, thereby preventing subsequent viral infection in a multimodal manner (Baram-Pinto et al. 2010). However, the antiviral effect of MES-capped AgNPs against HSV-1 was found to be imparted by their multivalent nature and spatially directed MES on the surface (Baram-Pinto et al. 2009). An overview related to application of AgNPs in inhibition of HSV was presented by Akbarzadeh et al. (2018).

ZnO micro-nano structures capped with multiple nanoscopic spikes mimicking cell induced filopodia and showing partially negatively charged oxygen vacancies on their nanoscopic spikes, after trapping the virions rendered them unable to enter into human corneal fibroblasts—a natural target cell for HSV-1 infection and exhibited pronouncedly enhanced anti-HSV-1 effect creating additional oxygen vacancies under UV light illumination (Mishra et al. 2011). Intravaginal ZnO tetrapod NPs with engineered oxygen vacancies, when used intravaginally, acted as a microbicide and were found to be an effective suppressor of HSV-2 genital infection in female BALB/c mice, suppressing also a reinfection, and exhibited strong adjuvant-like properties as well (Antoine et al. 2016). The antiviral effects of ZnO tetrapods can be enhanced by illuminating with UV light (Antoine et al. 2012).

SnO_2 nanowires working as a carrier of negatively charged structures can compete with HSV-1 attachment to cell-bound HS, resulting in the inhibition of virus entry and subsequent cell-to-cell spread (Trigilio et al. 2012).

Acyclovir (ACV)-loaded glycosaminoglycan-modified mesoporous SiO_2 NPs reduced the viral infection with HSV and such NPs were able to simultaneously inhibit the viral entry and DNA replication (Lee et al. 2018).

Graphene sheets uniformly anchored with spherical sulfonated magnetic NPs were able to capture and photothermally destroy HSV-1 upon irradiation with near-infrared light (808 nm, 7 min) (Deokar et al. 2017). Graphene sheets approx. 300 nm in size and with a degree of sulfation of approx. 10% were found to operate as effective viral inhibitor and inhibited HSV infection at an early stage during entry but did not affect cell-to-cell spread (Ziem et al. 2017). A multivalent 2D flexible carbon architecture fabricated using reduced GO functionalized with sulfated dendritic polyglycerol to mimic the HS-containing surface of cells and to compete with this natural binding site of viruses demonstrated excellent binding as well as efficient inhibition of the infection with orthopoxvirus possessing a HS-dependent cell entry mechanism (Ziem et al. 2016). Carbon nanodots surface-functionalized with 4-aminophenylboronic acid hydrochloride prevented HSV-1 infection on Vero and A549 cells and showed EC_{50} of 80 and 145 ng/mL, respectively, specifically acting on the early stage of virus entry (Barras et al. 2016).

C-Glycosylflavonoid-enriched fraction of *Cecropia glaziovii* encapsulated in PLGA NPs showed antiherpes properties and could be considered as a promising system for the effective drug delivery in the treatment of herpes infections (dos Santos et al. 2017). *Cymbopogon citratus* volatile oil encapsulated in PLGA NPs incorporated in carbomer hydrogels and tested against HSV using Vero cells inhibited virus at 42.2-fold lower concentration than free oil and it was 8.8- and 2.2-fold more efficient than oil-loaded NPs and hydrogel containing free oil, respectively (Almeida et al. 2018). Genistein-loaded cationic nanoemulsions with an average droplet size approx. 200–300 nm containing hydroxyethyl cellulose as a thickening agent showed considerably reduced genistein flux through excised porcine mucosa specimens compared to nanoemulsions before thickening, exhibited notable increase of genistein retention in mucosa compared to the genistein propylene glycol solution, and showed antiherpetic activity in vitro against HSV-1 (strain 29R) (Argenta et al. 2016).

Suspensions of glycyrrhizic acid NPs with particle size approx. 180 nm exhibited better anti-HSV activities compared to glycyrrhizic acid ammonium salt, especially during replication period. Morphology of HSV-1 observed by transmission electron microscopy was found to damage and shed the envelope of HSV-1 (Wang et al. 2015). The antiviral activities against HSV-2 estimated using the cytopathic effect assay showed also microemulsions with mean nanodroplet diameter of 4.7 ± 1.22 nm consisting of oil, water, surfactants, and cosurfactants that were able to destroy the HSV-2 virus at a 200-fold dilution in Dulbecco's modified eagle medium (Alkhatib et al. 2016). Pentyl gallate nanoemulsions (particle sizes of 124.8–143.7 nm; zeta potential ranging from −50.1 to −66.1 mV) demonstrated anti-HSV-1 activity and the drug reached deeper into the dermis more efficiently from the nanoemulsion as free drug, suggesting that nanoemulsions have potential to be used in topically delivering pentyl gallate in the treatment of human herpes labialis infection affecting primarily the lip (Kelmann et al. 2016).

Enhanced antiviral activity against a clinical isolate of HSV-1 was determined using ACV-loaded spherical carboxylated cyclodextrin-based nanosponges carrying carboxylic groups with the size of about 400 nm that exhibited prolonged release in comparison with that observed with nanosponges, without initial burst effect (Lembo et al. 2013). The formulation of ACV-loaded flexible membrane vesicles with particle size and zeta potential of 453.7 nm and −11.62 mV, respectively, incorporated into a hydrogel enhanced retention of drug deep inside the skin layers indicating potentially reduced need of application frequency resulting in reducing of adverse effects and suitability of such formulation for topical application against HSV-1 infection (Sharma et al. 2017b).

ACV entrapped in nanostructured lipid carriers coated with CS increased the corneal bioavailability in albino rabbits by 4.5-fold when compared to a commercially available ophthalmic ointment of drug, whereby this nanoformulation showed sustained release and improved antiviral properties of ACV through cell internalization (Seyfoddin et al. 2016). The ACV-loaded CS NPs showing a spherical shape, a size approx. 200 nm, and a zeta potential approx. −40.0 mV showed improved in vitro skin permeation and higher antiviral activity than the free drug against both the HSV-1 and the HSV-2 strains (Donalisio et al. 2018). CS as an immunomodulating adjuvant on T cells and antigen-presenting cells in HSV-1 infected mice was reported by Choi et al. (2016). In the dendritic cell-based DNA vaccine (pRSC-NLDC145.gD-IL21) carried by CS NPs, the expressed glycoprotein D in the formulation effectively targeted corneal dendritic cells and significantly alleviated the symptoms of both primary and recurrent HSV keratitis in mice via eliciting strong humoral and cellular immune response suggesting that such vaccine could be successfully used in *Herpes simplex* keratitis treatment (Tang et al. 2018).

Biomimetic supramolecular hexagonal-shaped nanoassemblies composed of chondroitin sulfate formed by mixing hydrophobically modified chondroitin sulfate with α-cyclodextrin in water showed improved antiviral activity against HSV-2 compared to hydrophobically modified chondroitin sulfate (Galus et al. 2016). ACV was found to increasingly permeate through the multilayers of human corneal epithelial cells from the drug-loaded bovine serum albumin NPs (approx. 200 nm) compared to drug solution suggesting potential of such formulation to be used as ocular drug delivery system (Suwannoi et al. 2017).

7.7 Hepatitis B and C Viruses

Hepatitis B virus (HBV), a species of the genus *Orthohepadnavirus* and a member of the *Hepadnaviridae* family of viruses, is a double-stranded DNA virus that replicates by reverse transcription, while hepatitis C virus (HCV) is an enveloped positive-sense single-stranded RNA virus of the family *Flaviviridae* (Zuckerman 1996). HBV and HBC affect the liver and can cause both acute and chronic infections, whereby cirrhosis and liver cancer may eventually develop. Viral hepatitis caused 1.34 million deaths in 2015, a number comparable to deaths caused by tuberculosis and higher than those caused by HIV and in 2013 it was the seventh leading cause of death worldwide (Stanaway et al. 2016; World Health Organization 2017). The presence of hepatitis B virus core antigen (HBcAg) that is the major structural protein of hepatitis B virus (HBV) in a blood serum indicates that a person has been exposed to HBV (Liang 2009; Inoue and Tanaka 2016).

Lu et al. (2008) estimated that AgNPs of the size 10 and 50 nm inhibited the in vitro production of HBV RNA and extracellular virions and assumed that the direct interaction between these NPs and HBV double-stranded DNA or viral particles is responsible for their antiviral mechanism. Lee et al. (2012) proposed hyaluronic acid-AuNPs/interferon-α complex for targeted treatment of HCV infection. Antiviral effect of AuNPs showing small particle size (approx. 3.5 nm) organized on the surface of larger layered double hydroxide (LDH) NPs such as MgLDH, ZnLDH, and MgFeLDH (approx. 150 nm) using HBV as a model virus and hepatoma-derived HepG2.2.215 cells for viral replication reduced the amount of viral and subviral particles released from treated cells by up to 80%; in the presence of AuNPs/LDHs the HBV particles were sequestered within the treated cells and the highest antiviral HBV response (>90% inhibition of HBV secretion) was estimated with AuNPs/MgFeLDH (Carja et al. 2015). Cu_2O NPs that were tested on antiviral activity against HCV pronouncedly inhibited the infectivity of HCVcc/Huh7.5.1 at a noncytotoxic concentration, they inhibited the entry of HCV pseudoparticles (HCVpp), including genotypes 1a, 1b, and 2a, while no effect on HCV replication was observed and they were found to stop HCV infection both at the attachment and entry stages suggesting that Cu_2O NPs could be used in the treatment of patients with chronic hepatitis C (Hang et al. 2015).

Multifunctional SeNPs with baicalin and folic acid surface-modifications designed for the targeted treatment of HBV-infected liver cancer primarily targeted lysosomes in HepG2215 cells, induced apoptosis of HepG2215 cell by downregulating the generation of reactive oxygen species (ROS) and the expression of the HBxAg protein, and showed superb ability to inhibit cancer cell migration and invasion (Fang et al. 2017).

Alum-adjuvanted HBV vaccine is considered as the most effective measure to prevent HBV infection; however, it is a frost-sensitive suspension and therefore it would be desirable to use alternative natural adjuvant system strongly immunogenic allowing for a reduction in dose and cost. Therefore, AbdelAllah et al. (2016) subcutaneously immunized mice with HBV surface antigen, HBsAg, adjuvanted with

CS and sodium alginate, either alone or combined with alum, estimated rate of seroconversion, serum HBsAg antibody, interleukin-4, and interferon-γ levels, and compared them with control mice immunized with current vaccine formula or unadjuvanted HBsAg. It was found that the solution formula with CS or sodium alginate exhibited comparable immunogenic responses to alum-adjuvanted suspension and the triple adjuvant application (alum, CS, sodium alginate) resulted in considerably higher immunogenic response than controls. Compared to traditional methods, vaccines prepared from nanoscale materials show appropriate biocompatibility and can secure effective targeting to certain tissue or cells as well as precise stimulation of immune responses (Yang et al. 2016b).

HBsAg-loaded trimethyl CS (TMC)/hydroxypropyl methylcellulose (hypromellose) phthalate (HPMCP) NPs with particle size of 158 nm showing loading capacity and loading efficiency of 76.75% and 86.29%, respectively, at 300 μg/mL concentration of the antigen exhibited improved acid stability and better protection of entrapped HBsAg from gastric destruction in vitro, whereby the antigen showed efficacious activity also after loading. Based on these findings it could be suggested that TMC/HPMCP NPs have potential to be applied in the oral delivery of HBsAg vaccine (Farhadian et al. 2015). Ndeboko et al. (2015) studied the inhibition of the replication of duck hepatitis B virus (DHBV), a reference model for human HBV infection, by peptide nucleic acid (PNA) conjugated to different cell-penetrating peptides (CPPs) and found that the PNA-CPP conjugates administered to neonatal ducklings reached the liver and inhibited DHBV replication, and in mouse model conjugation of HBV DNA vaccine to modified CS ameliorated cellular and humoral responses to plasmid-encoded antigen and plasmid DNA uptake, whereby expression could also be notably increased using gene delivery systems such as CPP-modified CS or cationic NPs.

Biodegradable NPs encapsulating RBV monophosphate prepared from the blend of poly(D,L-lactic acid) homopolymer and arabinogalactan-poly(L-lysine) conjugate were efficiently internalized in cultured HepG2 cells; they ensured sustained release of drug and could be considered as a formulation suitable for the clinical application of RBV as a therapeutic agent for chronic HCV (Ishihara et al. 2014). Liver-specific, sustained drug delivery system prepared by conjugating the liver-targeting peptide to PEGylated cyclosporine A-encapsulated PLGA NPs effectively inhibited viral replication in vitro as well as in a HCV mouse model (Jyothi et al. 2015). Adefovir encapsulated in PLGA microspheres showed sustained release and after intramuscular injection to rats considerable increase in the $t_{\max}$, $AUC_{(0-t)}$, and mean residence time, and a pronounced reduction in the $C_{\max}$ was observed, suggesting that the nanoformulation could be used for long-term treatment of chronic hepatitis B instead of the daily dose used by the patient (Ayoub et al. 2018).

Lipid nanocapsules, surface-functionalized with amphiphilic boronic acid through their postinsertion into the semirigid shell of the nanocapsules, were found to be excellent HCV entry inhibitors preventing HCV infection in the micromolar range (Khanal et al. 2015). Phenylboronic-acid-modified NPs as potential antiviral therapeutics were reported previously also by Khanal et al. (2013). siRNA-loaded lipid NPs (LNPs) modified with a hepatocyte-specific ligand, *N*-acetyl-D-galactosamine, showed pronounced improvement of hepatocyte-specificity and strong reduction in toxicity and further modification of NPs with PEG practically eliminated the LNP-associated toxicity without any detectable loss of gene silencing activity in hepatocytes, whereby a single injection of the LNPs considerably reduced HBV genomic DNA and their antigens without any sign of toxicity in chimeric mice with humanized livers that had been persistently infected with HBV (Sato et al. 2017).

7.8 Ebola and Marburg Virus

Ebola virus disease is caused by Ebola viruses (EBOVs; family *Filoviridae*), members of the group of hemorrhagic fever and it is one of the most dangerous infection diseases with mortality rates up to 90%. The EBOVs do not replicate through cell division, but instead insert their own genetic sequencing into the DNA of the host cell and subsequently hijack all cellular processes, including transcription and translation; thus, the host cell becomes a factory of viral proteins (Gebretadik et al. 2015; Murray 2015).

It was found that graphene sheets associate strongly with the EBOV matrix protein VP40 that is a potential pharmacological target for disrupting the virus life cycle, at various interfaces. Graphene can disrupt the C-terminal domain interface of VP40 hexamers being crucial in forming the Ebola viral matrix, suggesting that graphene or similar NPs-based solutions used as a disinfectant could notably reduce the spread of the disease and prevent an Ebola epidemic (Gc et al. 2017). Rodriguez-Perez et al. (2018) found that MWCNTs functionalized with glycofullerenes can be considered as potent inhibitors of Ebola infection. In two mRNA vaccines based on the EBOV envelope glycoprotein, differing by the nature of signal peptide for improved glycoprotein posttranslational translocation, the mRNAs were formulated with LNPs to facilitate delivery. Vaccination of guinea pigs induced EBOV-specific IgG and neutralizing antibody responses and 100% survival after EBOV infection (Meyer et al. 2018). Adjuvant-free dendrimer NPs vaccine platform wherein antigens are encoded by encapsulated mRNA replicons, able to generate protective immunity against many lethal pathogen challenges, including H1N1 influenza, *Toxoplasma gondii*, and EBOV, that can be formed with multiple antigen-expressing replicons and could elicit both $CD8^{+}$ T cell and antibody responses was designed by Chahal et al. (2016). Administration of siRNA encapsulated in LNPs able to target Sudan ebolavirus VP35 gene to rhesus monkeys receiving a lethal dose of the virus resulted in up to 100% survival and a reduction of viral replication (up to 4 $\log_{10}$) (Thi et al. 2016). Similar results were received with treatment of Ebola-virus-Makona-infected nonhuman primates following administration of siRNA encapsulated in LNPs suggesting the therapeutic potential of such nanoformulation in combating this lethal disease (Thi et al. 2015). Incorporation of Ebola DNA vaccine into PLGA-poly-L-lysine/poly-γ-glutamic acid increased vaccine thermostability and immunogenicity compared to free vaccine and vaccination performed to skin using a microneedle patch produced stronger immune responses than intramuscular administration (Yang et al. 2017b).

Marburg virus, similarly to Ebola virus, belongs to the family *Filoviridae* and causes severe and often fatal hemorrhagic fever in humans and nonhuman primates with very high mortality rates. The virus is transmitted from animals to humans by contact with bats or monkeys, or their bodily secretions or from person to person, although human-to-human contamination is rare (Sboui and Tabbabi 2017). Formulation of LNPs delivering siRNAs that target the anti-Marburg virus nucleoprotein administered to rhesus monkeys challenged with a lethal dose of Marburg virus-Angola showed excellent therapeutic efficacy and secured survival of infected animals (Thi et al. 2014) and protection against lethal Marburg virus infection mediated by lipid encapsulated siRNA was also observed with virus-infected guinea pigs (Ursic-Bedoya et al. 2014). Blocking of the infection of T-lymphocytes and human dendritic cells by Ebola virus using glyco-dendri-protein-NPs displaying quasi-equivalent nested polyvalency upon glycoprotein platforms, which consist of glycodendrimeric constructs (bearing up to 1620 glycans) with diameters up to 32 nm, was observed already at picomolar concentrations (Ribeiro-Viana et al. 2012).

7.9 Newcastle Disease Virus

Newcastle disease, a contagious viral bird disease, is caused by virulent strains of Newcastle disease virus (NDV), that causes substantial morbidity and mortality events worldwide in poultry resulting in devastating economic effects on global domestic poultry production. NDV, a negative-sense, single-stranded RNA virus, is capable of infecting more than 250 species of domestic and wild avian species (Brown and Bevins 2017). NDV is transmissible to humans and the exposure of humans to infected birds can cause mild conjunctivitis and influenza-like symptoms, but the NDV otherwise poses no hazard to human health (Abdisa and Tagesu 2017).

Zhao et al. (2016) designed a NDV F gene-containing DNA vaccine encapsulated in Ag@SiO_2 hollow NPs with an average diameter of 500 nm that following intranasal immunization of chickens induced high titers of serum antibody, notably stimulated lymphocyte proliferation and induced higher expression levels of interleukine-2 and interferon-γ, suggesting that Ag@SiO_2 hollow NPs could be used as an efficient and safe delivery carrier for NDV DNA vaccine to induce mucosal immunity.

The intranasal administration of quaternized CS (2-hydroxy-*N*,*N*,*N*-trimethyl propan-1-ammonium chloride CS/*N*,*O*-carboxymethyl CS) NPs loaded with the combined attenuated live vaccine against Newcastle disease and infectious bronchitis elicited immune response in chicken and induced higher titers of IgG and IgA antibodies, notably stimulated proliferation of lymphocytes and induced higher levels of interleukine-2, interleukine-4, and interferon-γ than the commercially combined attenuated live vaccine did. Induction of humoral, cellular, and mucosal immune responses protecting animals from the infection of highly virulent NDV and avian infectious bronchitis virus suggested that such CS derivative could be used as an efficient adjuvant and delivery carrier in mucosal vaccines (Zhao et al. 2017). Quaternized CS NPs loaded with the combined attenuated live vaccine against Newcastle disease and infectious bronchitis elicited immune response in chicken after intranasal administration.

El Naggar et al. (2017) designed preparation of mucosal NPs and polymer-based inactivated vaccine for Newcastle disease and H9N2 AI viruses, which after being delivered via intranasal and spray routes of administration in chickens enhanced the cell-mediated immune response and induced protection against challenge with both abovementioned viruses.

7.10 Dengue Virus and Zika Virus

Dengue is an acute viral illness caused by RNA virus, dengue virus (DENV), a member of the genus *Flavivirus* of the family *Flaviviridae*. DENV is an arthropode-borne virus that includes four different serotypes (DEN-1, DEN-2, DEN-3, and DEN-4) and spread by *Aedes* mosquitoes. Its presenting features may range from asymptomatic fever to dreaded complications such as hemorrhagic fever and shock and it is considered as a major global public health challenge in the tropic and subtropic nations (Hasan et al. 2016). To achieve durable protective immunity against all four serotypes DENV vaccines must induce balanced, serotype-specific neutralizing antibodies. A tetravalent DENV protein subunit vaccine, based on recombinant envelope protein (rE) adsorbed to the surface of PLGA NPs, was designed. Particulate rE induced higher neutralizing antibody titers compared to the soluble rE antigen alone and stimulated a more balanced serotype-specific antibody response to each DENV serotype compared to soluble antigens, suggesting that such vaccines might overcome unbalanced immunity observed for leading live attenuated vaccine candidates (Metz et al. 2018). AgNPs showed in vitro antiviral activity against dengue serotype DEN-2 infecting Vero cells and the activity of AgNPs against the dengue vector *Aedes aegypti* expressed by IC_{50} values ranged from 10.24 ppm (I instar larvae) to 21.17 ppm (pupae) (Sujitha et al. 2015). AgNPs biosynthesized using the aqueous extract of *Bruguiera cylindrica* leaves significantly inhibited the production of dengue viral envelope E protein in Vero cells and downregulated the expression of dengue viral E gene and they were found to be suitable for application at low doses to reduce larval and pupal population of *Ae. aegypti*, without detrimental effects of predation rates of mosquito predators, such as *Carassius auratus* (Murugan et al. 2015). Stabile AgNPs (particle size from 7 to 32 nm and zeta potential −15.58 mV) biosynthesized using silver nitrate and *Carica papaya* leaf extract demonstrated good binding affinity against dengue type 2 virus nonstructural protein 1 (Renganathan et al. 2019). Treatment with a hybrid consisting of AuNPs functionalized with domain III of envelope glycoprotein derived from serotype 2 of DENV (EDIII) resulted in a high level of antibody, which mediates serotype-specific neutralization of DENV in BALB/c mice, was found to be size-dependent and according to researchers the hybrid concept could also be adopted for the development of a tetravalent vaccine against four serotypes of DENV (Quach et al. 2018). It is important to note that good antiviral activity against dengue virus exhibited also SeNPs (Ramya et al. 2015).

Zika virus (ZIKV) is a mosquito-borne flavivirus that is the focus of an ongoing pandemic and public health emergency and represents a public health threat due to its teratogenic nature causing microcephaly in babies born to infected mothers and association with the serious neurological condition Guillain-Barre syndrome in adults (Plourde and Bloch 2016). Haque et al. (2018) proposed strategies to design effective and safe vaccines against ZIKV, including Toll-like receptors based NPs vaccines. The development of biomimetic nanodecoy (ND) that traps ZIKV and inhibits ZIKV infection was suggested by Rao et al. (2019). The ND, which is composed of a gelatin nanoparticle core camouflaged by mosquito medium host cell membranes, effectively adsorbs ZIKV and inhibits ZIKV replication in ZIKV-susceptible cells. AgNPs fabricated using leaf extracts of *Cleistanthus collinus* (triangular and pentagonal shape with sizes 66.27–75.09 nm) and *Strychnos nux-vomica* (spherical and round shape with sizes 54.45–60.84 nm) can be applied as a natural biolarvicidal agent to vector control strategy as an eco-friendly approach to prohibit Zika, chikungunya, and dengue fever in the future (Jinu et al. 2018). Insecticidal AgNPs prepared using *Suaeda maritima* were found to be effective against the dengue vector *Ae. aegypti* as well (Suresh et al. 2018). Bacterial exopolysaccharide-coated ZnO NPs nanoparticles showed high antibiofilm activity and larvicidal toxicity against malaria and ZIKV vectors *Anopheles stephensi* and *Ae. aegypti* with 100% mortality against third instars mosquito larvae at very low doses, whereby in the midgut of treated mosquito larvae presence of damaged cells and tissues was observed (Abinaya et al. 2018). Promising toxic and repellent activity against ZIKV mosquito vectors showed also TiO_2 NPs prepared using an extract of *Argemone mexicana* that were capped with poly(styrenesulfonate)/poly(allylamine hydrochloride) (Murugan et al. 2018). Nanoscale silicate platelets modified with anionic sodium dodecyl sulfate significantly suppressed the plaque-forming ability of Japanese encephalitis virus (JEV) at noncytotoxic concentrations and blocked infection with DENV and influenza A virus and also reduced the lethality of JEV and DENV infection in mouse challenge models (Liang et al. 2014).

Efficiency of green-synthesized metal AgNPs and AuNPs used as biopesticides against *Ae. aegypti* that can spread dengue virus and *Culex quinquefasciatus* or *An. stephensi* that transmit Zika virus were presented in several papers (e.g., Jampílek and Kráľová 2019d; Govindarajan and Benelli 2017; Lallawmawma et al. 2015; Pavunraj et al. 2017; Ishwarya et al. 2017; Suganya et al. 2017).

7.11 Pseudorabies Virus

Rabies virus is a neurotropic virus that causes rabies in humans and animals, while pseudorabies virus (PRV) is a herpesvirus of swine, a member of the *Alphaherpesvirinae* subfamily, and the etiological agent of Aujeszky's disease. PRV infection progresses from acute infection of the respiratory epithelium to latent infection in the peripheral nervous system, whereby sporadic reactivation from latency can transmit PRV to new hosts (Pomeranz et al. 2005).

Au nanoclusters surface-stabilized with histidine that strongly inhibited the proliferation of PRV were found to function via blockage of the viral replication process rather than the processes of attachment, penetration, or release and they were observed to be mainly localized to nucleus (Feng et al. 2018). Both GO and reduced GO showing the nanosheet structure tested against a DNA virus, PRV, and a RNA virus, porcine epidemic diarrhea virus (PEDV), suppressed the infection of these viruses for a 2 log reduction, and their potent antiviral activity can be attributed to the unique single-layer structure and negative charge of GO and reduced GO, whereby GO inactivated both viruses by structural destruction prior to viral entry (Ye et al. 2015).

A nonlinear globular G2 dendrimer comprising citric acid and PEG-600, adjuvanticity effect of which was investigated in veterinary rabies vaccine, did not show significant toxic effect in J774A.1 cells and ensured higher survival rate in the mice after virus challenge in vivo due to adjuvanticity effect of dendrimer resulting in rising of neutralizing antibodies against rabies virus and thus enhancing immune responses (Asgary et al. 2018).

7.12 Other Viruses

AgNPs of 25 nm were found to prevent viral entry of *Vaccinia virus*, an enveloped virus belonging to the poxvirus family, by macropinocytosis-dependent mechanism, which resulted in inhibition of *Vaccinia virus* infection (Trefry and Wooley 2013).

Broglie et al. (2015) tested antiviral activity of Au/CuS core/shell NPs against human Norovirus GI.1 (Norwalk) virus-like particles as a model viral system and found that virucidal efficacy significantly increased with increasing NPs concentration and/or contact time of virus-like particles with NPs.

CuI NPs demonstrated high antiviral activity against the nonenveloped virus feline calicivirus (FCV) as a surrogate for human norovirus (the most common etiological agent of gastroenteritis) that was attributed to Cu^{+} ions, followed by generation of ROS and subsequent capsid protein oxidation (Shionoiri et al. 2012).

The investigation of in vitro antiviral effects of MgO NPs in the foot-and-mouth disease (FMD), an extremely contagious viral disease of cloven-hoofed animals, on Razi Bovine kidney cell line showed that the MgO NPs exhibited virucidal and antiviral activities and they inhibited FMD virus by more than 90% at the early stages of infection such as attachment and penetration but not after penetration (Rafiei et al. 2015).

McGill et al. (2018) developed a mucosal nanovaccine with the post-fusion F and G glycoproteins from bovine respiratory syncytial virus (BRSV) encapsulated in polyanhydride NPs and tested it against BRSV infection using a neonatal calf model. They observed reduced pathology in the lungs, reduced viral burden, and decreased virus shedding compared to unvaccinated control calves showing correlation with BRSV-specific immune responses in the respiratory tract and peripheral blood.

Glutathione-capped Ag_2S nanoclusters showed strong antiviral activity against PEDV used as a model of coronavirus and pronounced reduction of the infection of PEDV by about three orders of magnitude at the noncytotoxic concentration at 12 h post-infection was observed. The Ag_2S nanoclusters inhibited the synthesis of viral negative-strand RNA and viral budding and positively regulated the generation of interferon-stimulating genes and the expression of proinflammation cytokines (Du et al. 2018).

Khandelwal et al. (2014) evaluated antiviral efficacy of AgNPs (5–30 nm) against *Ovine rinderpest* (*Peste des petits ruminants virus*, PPRV), a prototype *Morbillivirus*, causing disease in small ruminants, such as goats and sheep, and estimated significant inhibition of PPRV replication by AgNPs in an experiment using Vero cells already at noncytotoxic concentration, whereby AgNPs blocked the viral entry into the target cells due to interaction of AgNPs with the virion surface as well as with the virion core.

7.13 Conclusions

Nanoscale science and nanotechnology have unambiguously demonstrated to have a great potential in providing novel and improved solutions. Nano-size materials change their physical and chemical properties in comparison with bulk materials and have helped to improve and innovate a variety of pharmaceutical, medical, industrial, and agricultural products. Thus, nanoformulations of antivirotics and antiviral vaccines have become an important tool in the fight against various types of viruses due to modified bioavailability, ability to target viral or cellular proteins and sustainable release of drugs. Also, other nanosized materials were found to exhibit antiviral activity, and their combinations with antivirotics thus could provide remarkable medicines, especially against resistant viral pathogens. However, in spite of these significant benefits of nanomaterials in drug development, an increased attention should be devoted to the potential "intrinsic" toxicity of these nanomedicines caused by particle size that is able, within side effects, to induce various pathological processes in cells/tissues, which can result in various adverse/hazardous effects on animals and humans.

Acknowledgements This study was supported by the Slovak Research and Development Agency (projects APVV-17-0373 and APVV-17-0318).

References

AbdelAllah NH, Abdeltawab NF, Boseila AA, Amin MA. Chitosan and sodium alginate combinations are alternative, efficient, and safe natural adjuvant systems for hepatitis B vaccine in mouse model. Evid Based Complement Alternat Med. 2016;2016:7659684.

Abdisa T, Tagesu T. Review on Newcastle disease of poultry and its public health importance. J Veter Sci Technol. 2017;8:441.

Abinaya M, Vaseeharan B, Divya M, Sharmili A, Govindarajan M, Alharbi NS, Kadaikunnan S, Khaled JM, Benelli G. Bacterial exopolysaccharide (EPS)-coated ZnO nanoparticles showed high antibiofilm activity and larvicidal toxicity against malaria and Zika virus vectors. J Trace Elem Med Biol. 2018;45:93–103.

Akbarzadeh A, Kafshdooz L, Razban Z, Tbrizi AD, Rasoulpour S, Khalilov R, Kavetskyy T, Saghfi S, Nasibova AN, Kaamyabi S, Kafshdooz T. An overview application of silver nanoparticles in inhibition of herpes simplex virus. Artif Cells Nanomed Biotechnol. 2018;46:263–7.

Al-Ghananeem AM, Smith M, Coronel ML, Tran H. Advances in brain targeting and drug delivery of anti-HIV therapeutic agents. Expert Opin Drug Deliv. 2013;10:973–85.

Alkhatib MH, Aly MM, Rahbeni RA, Balamash KS. Antimicrobial activity of biocompatible microemulsions against *Aspergillus niger* and herpes simplex virus type 2. Jundishapur J Microbiol. 2016;9:e37437.

Almeida KB, Araujo JL, Cavalcanti JF, Romanos MTV, Mourao SC, Amaral ACF, Falcao DQ. In vitro release and anti-herpetic activity of *Cymbopogon citratus* volatile oil-loaded nanogel. Rev Bras. 2018;28:495–502.

Amirkhanov RN, Mazurkova NA, Amirkhanov NV, Zarytova VF. Composites of peptide nucleic acids with titanium dioxide nanoparticles IV. Antiviral activity of nanocomposites containing DNA/PNA duplexes. Russ J Bioorg Chem. 2015;41:140–6.

Antoine TE, Hadigal SR, Yakoub AM, Mishra YK, Bhattacharya P, Haddad C, Valyi-Nagy T, Adelung R, Prabhakar BS, Shukla D. Intravaginal zinc oxide tetrapod nanoparticles as novel immunoprotective agents against genital herpes. J Immunol. 2016;196:4566–75.

Antoine TE, Mishra YK, Trigilio J, Tiwari V, Adelung R, Shukla D. Prophylactic, therapeutic and neutralizing effects of zinc oxide tetrapod structures against herpes simplex virus type-2 infection. Antivir Res. 2012;96:363–75.

Ardestani MS, Fordoei AS, Abdoli A, Cohan RA, Bahramali G, Sadat SM, Siadat SD, Moloudian H, Koopaei NN, Bolhasani A, Rahimi P, Hekmat S, Davari N, Aghasadeghi MR. Nanosilver based anionic linear globular dendrimer with a special significant antiretroviral activity. J Mater Sci Mater Med. 2015;26:179.

Argenta DF, Bidone J, Misturini FD, Koester LS, Bassani VL, Simoes CMO, Teixeira HF. In vitro evaluation of mucosa permeation/retention and antiherpes activity of genistein from cationic nanoemulsions. J Nanosci Nanotechnol. 2016;16:1282–90.

Ariza-Saenz M, Espina M, Bolanos N, Calpena AC, Gomara MJ, Haro I, Garcia ML. Penetration of polymeric nanoparticles loaded with an HIV-1 inhibitor peptide derived from GB virus C in a vaginal mucosa model. Eur J Pharm Biopharm. 2017;120:98–106.

Ariza-Saenz M, Espina M, Calpena A, Gomara MJ, Perez-Pomeda I, Haro I, Garcia ML. Design, characterization, and biopharmaceutical behavior of nanoparticles loaded with an HIV-1 fusion inhibitor peptide. Mol Pharm. 2018;15:5005–18.

Asgary V, Shoari A, Moayad MA, Ardestani MS, Bigdeli R, Ghazizadeh L, Khosravy MS, Panahnejad E, Janani A, Bashar R, Abedi M, Ahangri Cohan R. Evaluation of G2 citric acid-based dendrimer as an adjuvant in veterinary rabies vaccine. Viral Immunol. 2018;31:47–54.

Augustine R, Ashkenazi DL, Arzi RS, Zlobin V, Shofti R, Sosnik A. Nanoparticle-in-microparticle oral drug delivery system of a clinically relevant darunavir/ritonavir antiretroviral combination. Acta Biomater. 2018;74:344–59.

Ayoub MM, Elantouny NG, El-Nahas HM, Ghazy FED. Injectable PLGA Adefovir microspheres; the way for long term therapy of chronic hepatitis-B. Eur J Pharm Sci. 2018;118:24–31.

Bahl K, Senn JJ, Yuzhakov O, Bulychev A, Brito LA, Hassett KJ, Laska ME, Smith M, Almarsson O, Thompson J, Ribeiro AM, Watson M, Zaks T, Ciaramella G. Preclinical and clinical demonstration of immunogenicity by mRNA vaccines against H10N8 and H7N9 influenza viruses. Mol Ther. 2017;25:1316–27.

Baram-Pinto D, Shukla S, Gedanken A, Sarid R. Inhibition of HSV-1 attachment, entry, and cell-to-cell spread by functionalized multivalent gold nanoparticles. Small. 2010;6:1044–50.

Baram-Pinto D, Shukla S, Perkas N, Gedanken A, Sarid R. Inhibition of herpes simplex virus type 1 infection by silver nanoparticles capped with mercaptoethane sulfonate. Bioconjug Chem. 2009;20:1497–502.

Barras A, Pagneux Q, Sane F, Wang Q, Boukherroub R, Hober D, Szunerits S. High efficiency of functional carbon nanodots as entry inhibitors of herpes simplex virus type 1. ACS Appl Mater Interfaces. 2016;8:9004–13.

Bawage SS, Tiwari PM, Singh A, Dixit S, Pillai SR, Dennis VA, Singh SR. Gold nanorods inhibit respiratory syncytial virus by stimulating the innate immune response. Nanomedicine. 2016;12:2299–310.

Bayon E, Morlieras J, Dereuddre-Bosquet N, Gonon A, Gosse L, Courant T, Le Grand R, Marche PN, Navarro FP. Overcoming immunogenicity issues of HIV p 24 antigen by the use of innovative nanostructured lipid carriers as delivery systems: evidences in mice and non-human primates. NPJ Vaccines. 2018;3:46.

Belgamwar A, Khan S, Yeole P. Intranasal chitosan-g-HPβCD nanoparticles of efavirenz for the CNS targeting. Artif Cells Nanomed Biotechnol. 2018;46:374–86.

Bernasconi V, Bernocchi B, Ye L, Le MQ, Omokanye A, Carpentier R, Schon K, Saelens X, Staeheli P, Betbeder D, Lycke N. Porous nanoparticles with self-adjuvanting M2e-fusion protein and recombinant hemagglutinin provide strong and broadly protective immunity against influenza virus infections. Front Immunol. 2018;9:2060.

Borchers AT, Chang C, Gershwin ME, Gershwin LJ. Respiratory syncytial virus—a comprehensive review. Clin Rev Allergy Immunol. 2013;45:331–79.

Broglie JJ, Alston B, Yang C, Ma L, Adcock AF, Chen W, Yang LJ. Antiviral activity of gold/copper sulfide core/shell nanoparticles against human norovirus virus-like particles. PLoS One. 2015;10:e0141050.

Brown IH. History and epidemiology of Swine influenza in Europe. Curr Top Microbiol Immunol. 2013;370:133–46.

Brown VR, Bevins SN. A review of virulent Newcastle disease viruses in the United States and the role of wild birds in viral persistence and spread. Vet Res. 2017;48:68.

Cagno V, Andreozzi P, D'Alicarnasso M, Silva PJ, Mueller M, Galloux M, Le Goffic R, Jones ST, Vallino M, Hodek J, Weber J, Sen S, Janeček ER, Bekdemir A, Sanavio B, Martinelli C, Donalisio M, Rameix Welti MA, Eleouet JF, Han YX, Kaiser L, Vukovic L, Tapparel C, Král P, Krol S, Lembo D, Stellacci F. Broad-spectrum non-toxic antiviral nanoparticles with a virucidal inhibition mechanism. Nat Mater. 2018;17:195–203.

Capek I. Viral nanoparticles, noble metal decorated viruses and their nanoconjugates. Adv Colloid Interf Sci. 2015;222:119–34.

Carja G, Grosu EF, Petrarean C, Nichita N. Self-assemblies of plasmonic gold/layered double hydroxides with highly efficient antiviral effect against the hepatitis B virus. Nano Res. 2015;8:3512–23.

Chahal JS, Khan OF, Cooper CL, McPartlan JS, Tsosie JK, Tilley LD, Sidik SM, Lourido S, Langer R, Bavari S, Ploegh H, Anderson DG. Dendrimer-RNA nanoparticles generate protective immunity against lethal Ebola, H1N1 influenza, and *Toxoplasma gondii* challenges with a single dose. Proc Natl Acad Sci U S A. 2016;113:E4133–42.

Chaowanachan T, Krogstad E, Ball C, Woodrow KA. Drug synergy of tenofovir and nanoparticle-based antiretrovirals for HIV prophylaxis. PLoS One. 2013;8:e61416.

Chatziprodromidou IP, Arvanitidou M, Guitian J, Apostolou T, Vantarakis G, Apostolos Vantarakis A. Global avian influenza outbreaks 2010–2016: a systematic review of their distribution, avian species and virus subtype. Syst Rev. 2018;7:17.

Chiou CJ, Tseng LP, Deng MC, Jiang PR, Tasi SL, Chung TW, Huang YY, Liu DZ. Mucoadhesive liposomes for intranasal immunization with an avian influenza virus vaccine in chickens. Biomaterials. 2009;30:5862–8.

Choi B, Jo DH, Anower AKMM, Islam SMS, Sohn S. Chitosan as an immunomodulating adjuvant on T-cells and antigen-presenting cells in herpes simplex virus type 1 infection. Mediat Inflamm. 2016;2016:4374375.

Chowdhury MYE, Kim TH, Uddin MB, Kim JH, Hewawaduge CY, Ferdowshi Z, Sung MH, Kim CJ, Lee JS. Mucosal vaccination of conserved sM2, HA2 and cholera toxin subunit Al (CTAI) fusion protein with poly γ-glutamate/chitosan nanoparticles (PC NPs) induces protection against divergent influenza subtypes. Vet Microbiol. 2017;201:240–51.

Clark CM, Guerrero-Plata A. Respiratory syncytial virus vaccine approaches: a current overview. Curr Clin Microbiol Rep. 2017;4:202–7.

Climent N, Garcia I, Marradi M, Chiodo F, Miralles L, Maleno MJ, Gatell JM, Garcia F, Penades S, Plana M. Loading dendritic cells with gold nanoparticles (GNPs) bearing HIV-peptides and mannosides enhance HIV-specific T cell responses. Nanomedicine. 2018;14:339–51.

Coulibaly FS, Ezoulin MJM, Purohit SS, Ayon NJ, Oyler NA, Youan BBC. Layer-by-layer engineered microbicide drug delivery system targeting HIV-1 gp120: physicochemical and biological properties. Mol Pharm. 2017;14:3512–27.

Cunha-Reis C, Machado A, Barreiros L, Araújo F, Nunes R, Seabra V, Ferreira D, Segundo MA, Sarmento B, Neves JD. Nanoparticles-in-film for the combined vaginal delivery of anti-HIVmicrobicide drugs. J Control Release. 2016;243:43–53.

Dai X, Zhou ZH. Structure of the herpes simplex virus 1 capsid with associated tegument protein complexes. Science. 2018;360:eaao7298.

De Clercq E, Li G. Approved antiviral drugs over the past 50 years. Clin Microbiol Rev. 2016;29:695–747.

Deng L, Chang TZ, Wang Y, Li S, Wang S, Matsuyama S, Yu GY, Compans RW, Li JD, Prausnitz MR, Champion JA, Wang BZ. Heterosubtypic influenza protection elicited by double-layered polypeptide nanoparticles in mice. Proc Natl Acad Sci U S A. 2018b;115:E7758–67.

Deng L, Mohan T, Chang TZ, Gonzalez GX, Wang Y, Kwon YM, Kang SM, Compans RW, Champion JA, Wang BZ. Double-layered protein nanoparticles induce broad protection against divergent influenza A viruses. Nat Commun. 2018a;9:359.

Deng L, Wang BZ. A perspective on nanoparticle universal influenza vaccines. ACS Infect Dis. 2018;4:1656–65.

Deokar AR, Nagvenkar AP, Kalt I, Shani L, Yeshurun Y, Gedanken A, Sarid R. Graphene-based "hot plate" for the capture and destruction of the herpes simplex virus type 1. Bioconjug Chem. 2017;28:1115–22.

Desai J, Thakkar H. Enhanced oral bioavailability and brain uptake of darunavir using lipid nanoemulsion formulation. Colloids Surf B: Biointerfaces. 2019;175:143–9.

Dey P, Bergmann T, Cuellar-Camacho JL, Ehrmann S, Chowdhury MS, Zhang MZ, Dahmani I, Haag R, Azad W. Multivalent flexible nanogels exhibit broad-spectrum antiviral activity by blocking virus entry. ACS Nano. 2018;12:6429–42.

Dhakal S, Goodman J, Bondra K, Lakshmanappa YS, Hiremath J, Shyu DL, Ouyang K, Kang KI, Krakowka S, Wannemuehler M, Won Lee C, Narasimhan B, Renukaradhya GJ. Polyanhydride nanovaccine against swine influenza virus in pigs. Vaccine. 2017a;35:1124–31.

Dhakal S, Hiremath J, Bondra K, Lakshmanappa YS, Shyu DL, Ouyang K, Kanga KI, Binjawadagi B, Goodman J, Tabynov K, Krakowka S, Narasimhan B, Won Lee C, Gourapura J, Renukaradhya GJ. Biodegradable nanoparticle delivery of inactivated swine influenza virus vaccine provides heterologous cell-mediated immune response in pigs. J Control Release. 2017b;247:194–205.

Dhakal S, Renu S, Ghimire S, Lakshmanappa YS, Hogshead BT, Feliciano-Ruiz N, Lu FJ, HogenEsch H, Krakowka S, Lee CW, Renukaradhya GJ. Mucosal immunity and protective efficacy of intranasal inactivated influenza vaccine is improved by chitosan nanoparticle delivery in pigs. Front Immunol. 2018;9:934.

Di Gianvincenzo P, Marradi M, Martinez-Avila OM, Bedoya LM, Alcami J, Penades S. Gold nanoparticles capped with sulfate-ended ligands as anti-HIV agents. Bioorg Med Chem Lett. 2010;20:2718–21.

Dolez PI. Nanoengineering: global approaches to health and safety issues. Amsterdam: Elsevier; 2015.

Donalisio M, Leone F, Civra A, Spagnolo R, Ozer O, Lembo D, Cavalli R. Acyclovir-loaded chitosan nanospheres from nano-emulsion templating for the topical treatment of herpesviruses infections. Pharmaceutics. 2018;10:46.

dos Santos TC, Rescignano N, Boff L, Reginatto FH, Simoes CMO, de Campos AM, Mijangos C. In vitro antiherpes effect of C-glycosyl flavonoid enriched fraction of *Cecropia glaziovii* encapsulated in PLGA nanoparticles. Mater Sci Eng C Mater Biol Appl. 2017;75:1214–20.

Du T, Liang JG, Dong N, Lu J, Fu YY, Fang LR, Xiao S, Han HY. Glutathione-capped Ag_2S nanoclusters inhibit coronavirus proliferation through blockage of viral RNA synthesis and budding. ACS Appl Mater Interfaces. 2018;10:4369–78.

El Naggar HM, Madkour MS, Hussein AH. Preparation of mucosal nanoparticles and polymer-based inactivated vaccine for Newcastle disease and H9N2 AI viruses. Vet World. 2017;10:187–93.

Endsley AN, Ho RJY. Enhanced anti-HIV efficacy of indinavir after inclusion in CD4-targeted lipid nanoparticles. J Acquir Immune Defic Syndr. 2012;61:417–24.

Fang XY, Wu XL, Li CE, Zhou BW, Chen XY, Chen TF, Yang F. Targeting selenium nanoparticles combined with baicalin to treat HBV-infected liver cancer. RSC Adv. 2017;7:8178–85.

Farhadian A, Dounighi NM, Avadi M. Enteric trimethyl chitosan nanoparticles containing hepatitis B surface antigen for oral delivery. Hum Vaccin Immunother. 2015;11:2811–8.

Fatima M, Zaidi NUSS, Amraiz D, Afzal F. In vitro antiviral activity of *Cinnamomum cassia* and its nanoparticles against H7N3 influenza A virus. J Microbiol Biotechnol. 2016;26:151–9.

Fayaz AM, Ao ZJ, Girilal M, Chen LY, Xiao XZ, Kalaichelvan PT, Yao XJ. Inactivation of microbial infectiousness by silver nanoparticles-coated condom: a new approach to inhibit HIV- and HSV-transmitted infection. Int J Nanomedicine. 2012;7:5007–18.

Feng CC, Fang PX, Zhou YR, Liu LZ, Fang LR, Xiao SB, Liang JG. Different effects of His-Au NCs and MES-Au NCs on the propagation of pseudorabies virus. Global Chall. 2018;2:1800030.

Feng F, Sakoda Y, Ohyanagi T, Nagahori N, Shibuya H, Okamastu M, Miura N, Kida H, Nishimura S. Novel thiosialosides tethered to metal nanoparticles as potent influenza A virus haemagglutinin blockers. Antivir Chem Chemother. 2013;23:59–65.

Figueira TN, Augusto MT, Rybkina K, Stelitano D, Noval MG, Harder OE, Veiga AS, Huey D, Alabi CA, Biswas S, Niewiesk S, Moscona A, Santos NC, Castanho MARB, Porotto M. Effective in vivo targeting of influenza virus through a cell-penetrating/fusion inhibitor tandem peptide anchored to the plasma membrane. Bioconjug Chem. 2018;29:3362–76.

Fries L, Shinde V, Stoddard JJ, Thomas DN, Kpamegan E, Lu HX, Smith G, Hickman SP, Piedra P, Glenn GM. Immunogenicity and safety of a respiratory syncytial virus fusion protein (RSV F) nanoparticle vaccine in older adults. Immun Ageing. 2017;14:8.

Fujimori Y, Sato T, Hayata T, Nagao T, Nakayama M, Nakayama T, Sugamata R, Suzuki K. Novel antiviral characteristics of nanosized copper(I) iodide particles showing inactivation activity against 2009 pandemic H1N1 influenza virus. Appl Environ Microbiol. 2012;78:951–5.

Gaikwad S, Ingle A, Gade A, Rai M, Falanga A, Incoronato N, Russo L, Galdiero S, Galdiero M. Antiviral activity of mycosynthesized silver nanoparticles against herpes simplex virus and human parainfluenza virus type 3. Int J Nanomedicine. 2013;8:4303–14.

Galdiero S, Falanga A, Vitiello M, Cantisani M, Marra V, Galdiero M. Silver nanoparticles as potential antiviral agents. Molecules. 2011;16:8894–918.

Galus A, Mallet JM, Lembo D, Cagno V, Djabourov M, Lortat-Jacob H, Bouchemal K. Hexagonal-shaped chondroitin sulfate self-assemblies have exalted anti-HSV-2 activity. Carbohydr Polym. 2016;136:113–20.

Gao Y, Kraft JC, Yu D, Ho RJY. Recent developments of nanotherapeutics for targeted and long-acting, combination HIV chemotherapy. Eur J Pharm Biopharm. 2018;138:75–91.

Gc JB, Pokhrel R, Bhattarai N, Johnson KA, Gerstman BS, Stahelin RV, Chapagain PP. Graphene-VP40 interactions and potential disruption of the Ebola virus matrix filaments. Biochem Biophys Res Commun. 2017;493:176–81.

Gebretadik FA, Seifu MF, Gelaw BK. Review on Ebola virus disease: its outbreak and current status. Epidemiology (Sunnyvale). 2015;5:204.

German Advisory Committee Blood (Arbeitskreis Blut), Subgroup 'Assessment of Pathogens Transmissible by Blood'. Human immunodeficiency virus (HIV). Transf Med Chemother. 2016;43:203–22.

Gerson T, Makarov E, Senanayake TH, Gorantla S, Poluektova LY, Vinogradov SV. Nano-NRTIs demonstrate low neurotoxicity and high antiviral activity against HIV infection in the brain. Nanomedicine. 2014;10:177–85.

Gong Y, Chowdhury P, Midde NM, Rahman MA, Yallapu MM, Kumar S. Novel elvitegravir nanoformulation approach to suppress the viral load in HIV-infected macrophages. Biochem Biophys Rep. 2017;12:214–9.

Govindarajan M, Benelli G. Ovicidal and larvicidal potential on malaria, dengue and filariasis mosquito vectors. J Clust Sci. 2017;28:15–36.

Gupta S, Kesarla R, Chotai N, Misra A, Omri A. Systematic approach for the formulation and optimization of solid lipid nanoparticles of efavirenz by high pressure homogenization using design of experiments for brain targeting and enhanced bioavailability. Biomed Res Int. 2017;2017:5984014.

Haider A, Das S, Ojha D, Chattopadhyay D, Mukherjee A. Highly monodispersed gold nanoparticles synthesis and inhibition of herpes simplex virus infections. Mater Sci Eng C Mater Biol Appl. 2018;89:413–21.

Hajializadeh F, Ghahri H, Talebi A. Effects of supplemental chromium picolinate and chromium nanoparticles on performance and antibody titers of infectious bronchitis and avian influenza of broiler chickens under heat stress condition. Vet Res Forum. 2017;8:259–64.

Hang XF, Peng HR, Song HY, Qi ZT, Miao XH, Xu WS. Antiviral activity of cuprous oxide nanoparticles against Hepatitis C virus in vitro. J Virol Methods. 2015;222:150–7.

Haque A, Akcesme FB, Pant AB. A review of Zika virus: hurdles toward vaccine development and the way forward. Antivir Ther. 2018;23:285–93.

Hasan S, Jamdar SF, Alalowi M, Al Ageel Al Beaiji SM. Dengue virus: a global human threat: review of literature. J Int Soc Prevent Communit Dent. 2016;6:1–6.

Hashiguchi S, Yoshida H, Akashi T, Komemoto K, Ueda T, Ikarashi Y, Miyauchi A, Konno K, Yamanaka S, Hirose A, Kurokawa M, Watanabe W. Titanium dioxide nanoparticles exacerbate pneumonia in respiratory syncytial virus (RSV)-infected mice. Environ Toxicol Pharmacol. 2015;39:879–86.

Hillaireau H, Dereuddre-Bosquet N, Skanji R, Bekkara-Aounallah F, Caron J, Lepetre S, Argote S, Bauduin L, Yousfi R, Rogez-Kreuz C, Desmaële D, Rousseau B, Gref R, Andrieux K, Clayette P, Couvreur P. Anti-HIV efficacy and biodistribution of nucleoside reverse transcriptase inhibitors delivered as squalenoylated prodrug nanoassemblies. Biomaterials. 2013;34:4831–8.

Hu RL, Li SR, Kong FJ, Hou RJ, Guan XL, Guo F. Inhibition effect of silver nanoparticles on herpes simplex virus 2. Genet Mol Res. 2014;13:7022–8.

Iannazzo D, Pistone A, Galvagno S, Ferro S, De Luca L, Monforte AM, Da Ros T, Hadad C, Prato M, Pannecouque C. Synthesis and anti-HIV activity of carboxylated and drug-conjugated multi-walled carbon nanotubes. Carbon. 2015;82:548–61.

Inoue T, Tanaka Y. Hepatitis B virus and its sexually transmitted infection—an update. Microbial Cell. 2016;3:420–37.

Ishihara T, Kaneko K, Ishihara T, Mizushima T. Development of biodegradable nanoparticles for liver-specific ribavirin delivery. J Pharm Sci. 2014;103:4005–11.

Ishwarya R, Vaseeharan B, Anuradha R, Rekha R, Govindarajan M, Alharbi NS, Kadaikunnan S, Khaled JM, Benelli G. Eco-friendly fabrication of Ag nanostructures using the seed extract of *Pedalium murex*, an ancient Indian medicinal plant: histopathological effects on the *Zika virus* vector *Aedes aegypti* and inhibition of biofilm-forming pathogenic bacteria. J Photochem Photobiol B. 2017;174:133–43.

Jackman JA, Lee J, Cho NJ. Nanomedicine for infectious disease applications: innovation towards broad-spectrum treatment of viral infections. Small. 2016;12:1133–9.

Jamali A, Mottaghitalab F, Abdoli A, Dinarvand M, Esmailie A, Kheiri MT, Atyabi F. Inhibiting influenza virus replication and inducing protection against lethal influenza virus challenge through chitosan nanoparticles loaded by siRNA. Drug Deliv Transl Res. 2018;8:12–20.

Jampílek J. Potential of agricultural fungicides for antifungal drug discovery. Expert Opin Drug Discovery. 2016a;11:1–9.

Jampílek J. How can we bolster the antifungal drug discovery pipeline? Fut Med Chem. 2016b;8:1393–7.

Jampílek J. Design and discovery of new antibacterial agents: advances, perspectives, challenges. Curr Med Chem. 2018;25:4972–5006.

Jampílek J, Kráľová K. Nano-antimicrobials: activity, benefits and weaknesses. In: Ficai A, Grumezescu AM, editors. Nanostructures for antimicrobial therapy. Amsterdam: Elsevier; 2017. p. 23–54.

Jampílek J, Kráľová K. Application of nanobioformulations for controlled release and targeted biodistribution of drugs. In: Sharma AK, Keservani RK, Kesharwani RK, editors. Nanobiomaterials: applications in drug delivery. Warentown: CRC Press; 2018. p. 131–208.
Jampílek J, Kráľová K. Recent advances in lipid nanocarriers applicable in the fight against cancer. In: Grumezescu AM, editor. Nanoarchitectonics in biomedicine—recent progress of nanoarchitectonics in biomedical science. Amsterdam: Elsevier; 2019a. p. 219–94.
Jampílek J, Kráľová K. Nanotechnology based formulations for drug targeting to central nervous system. In: Keservani RK, Sharma AK, editors. Nanoparticulate drug delivery systems. Warentown: Apple Academic Press & CRC Press; 2019b. p. 151–220.
Jampílek J, Kráľová K. Natural biopolymeric nanoformulations for brain drug delivery. In: Keservani RK, Sharma AK, Kesharwani RK, editors. Nanocariers for brain targetting: principles and applications. Warentown: CRC Press; 2019c. p. 131–204.
Jampílek J, Kráľová K. Nano-biopesticides in agriculture: state of art and future opportunities. In: Koul O, editor. Nano-biopesticides today and future perspectives. Amsterdam: Academic Press & Elsevier; 2019d. p. 397–447.
Jampílek J, Záruba K, Oravec M, Kuneš M, Babula P, Ulbrich P, Brezaniová I, Tříska J, Suchý P. Preparation of silica nanoparticles loaded with nootropics and their in vivo permeation through blood–brain barrier. Biomed Res Int. 2015;2015:812673.
Jazayeri SD, Ideris A, Zakaria Z, Shameli K, Moeini H, Omar AR. Cytotoxicity and immunological responses following oral vaccination of nanoencapsulated avian influenza virus H5 DNA vaccine with green synthesis silver nanoparticles. J Control Release. 2012;161:116–23.
Jiang Y, Cao S, Bright DK, Bever AM, Blakney AK, Suydam IT, Woodrow KA. Nanoparticle-based ARV drug combinations for synergistic inhibition of cell-free and cell-cell HIV transmission. Mol Pharm. 2015;12:4363–74.
Jinu U, Rajakumaran S, Senthil-Nathan S, Geetha N, Venkatachalam P. Potential larvicidal activity of silver nanohybrids synthesized using leaf extracts of *Cleistanthus collinus* (Roxb.) Benth. ex Hook. f. and *Strychnos nux-vomica* L. *nux-vomica* against dengue, Chikungunya and Zika vectors. Physiol Mol Plant Pathol. 2018;101:163–71.
Jorquera PA, Tripp RA. Synthetic biodegradable microparticle and nanoparticle vaccines against the respiratory syncytial virus. Vaccine. 2016;4:45.
Joshy KS, Alex SM, Snigdha S, Kalarikkal N, Pothen LA, Thomas S. Encapsulation of zidovudine in PF-68 coated alginate conjugate nanoparticles for anti-HIV drug delivery. Int J Biol Macromol. 2018a;107:929–37.
Joshy KS, Snigdha S, George A, Kalarikkal N, Pothen LA, Thomas S. Poly(vinylpyrrolidone)-lipid based hybrid nanoparticles for anti viral drug delivery. Chem Phys Lipids. 2018b;210:82–9.
Joshy KS, George A, Snigdha S, Joseph B, Kalarikkal N, Pothen LA, Thomas S. Novel core-shell dextran hybrid nanosystem for anti-viral drug delivery. Mater Sci Eng C Mater Biol Appl. 2018c;93:864–72.
Joshy KS, Snigdha S, George A, Kalarikkal N, Pothen LA, Thomas S. Core-shell nanoparticles of carboxy methyl cellulose and compritol-PEG for antiretroviral drug delivery. Cellulose. 2017;24:4759–71.
Jyothi KR, Beloor J, Jo A, Minh NN, Choi TG, Kim JH, Akter S, Lee SK, Maeng CH, Baik HH, Kang I, Ha J, Kim SS. Liver-targeted cyclosporine A-encapsulated poly (lactic-co-glycolic) acid nanoparticles inhibit hepatitis C virus replication. Int J Nanomedicine. 2015;10:903–21.
Kelmann RG, Colombo M, Lopes SCD, Nunes RJ, Pistore M, Agnol DD, Rigotto C, Silva IT, Roman SS, Teixeira HF, Oliveira Simões CM, Koester LS. Pentyl gallate nanoemulsions as potential topical treatment of herpes labialis. J Pharm Sci. 2016;105:2194–203.
Khanal M, Barras A, Vausselin T, Feneant L, Boukherroub R, Siriwardena A, Dubuisson J, Szunerits S. Boronic acid-modified lipid nanocapsules: a novel platform for the highly efficient inhibition of hepatitis C viral entry. Nanoscale. 2015;7:1392–402.
Khanal M, Vausselin T, Barras A, Bande O, Turcheniuk K, Benazza M, Zaitsev V, Teodorescu CM, Boukherroub R, Siriwardena A, Dubuisson J, Szunerits S. Phenylboronic-acid-modified nanoparticles: potential antiviral therapeutics. ACS Appl Mater Interfaces. 2013;5:12488–98.

Khandelwal N, Kaur G, Chaubey KK, Singh P, Sharma S, Tiwari A, Singh SV, Kumar N. Silver nanoparticles impair Peste des petits ruminants virus replication. Virus Res. 2014;190:1–7.

Kos J, Ku CF, Kapustíková I, Oravec M, Zhang HJ, Jampílek J. 8-Hydroxyquinoline-2-carboxanilides as antiviral agents against avian influenza virus. ChemistrySelect. 2019;4:4582–7.

Krishnaraj RN, Chandran S, Pal P, Berchmans S. Investigations on the antiretroviral activity of carbon manotubes using computational molecular approach. Comb Chem High Throughput Screen. 2014;17:531–5.

Kuchipudi SV, Niessly RH. Novel flu viruses in bats and cattle: "pushing the envelope" of influenza infection. Vet Sci. 2018;5:71.

Kumar P, Lakshmi YS, Bhaskar C, Golla K, Kondapi AK. Improved safety, bioavailability and pharmacokinetics of zidovudine through lactoferrin nanoparticles during oral administration in rats. PLoS One. 2015;10:e0140399.

Kumar SD, Singaravelu G, Ajithkumar S, Murugan K, Nicoletti M, Benelli G. Mangrove-mediated green synthesis of silver nanoparticles with high HIV-1 reverse transcriptase inhibitory potential. J Clust Sci. 2017a;28:359–67.

Kumar P, Lakshmi YS, Kondapi AK. Triple drug combination of zidovudine, efavirenz and lamivudine loaded lactoferrin nanoparticles: an effective nano first-line regimen for HIV therapy. Pharm Res. 2017b;34:257–68.

Lallawmawma H, Sathishkumar G, Sarathbabu S, Ghatak S, Sivaramakrishnan S, Gurusubramanian G, Kumar NS. Synthesis of silver and gold nanoparticles using *Jasminum nervosum* leaf extract and its larvicidal activity against filarial and arboviral vector *Culex quinquefasciatus* say (Diptera: *Culicidae*). Environ Sci Pollut Res. 2015;22:17753–68.

Lara HH, Ixtepan-Turrent L, Garza-Trevino EN, Rodriguez-Padilla C. PVP-coated silver nanoparticles block the transmission of cell-free and cell-associated HIV-1 in human cervical culture. J Nanobiotechnol. 2010;8:15.

Lauster D, Glanz M, Bardua M, Ludwig K, Hellmund M, Hoffmann U, Hamann A, Boettcher C, Haag R, Hackenberger CPR, Herrmann A. Multivalent peptide-nanoparticle conjugates for influenza-virus inhibition. Angew Chem Int Ed. 2017;56:5931–6.

Lee EC, Nguyen CTH, Strounina E, Davis-Poynter N, Ross BP. Structure-activity relationships of GAG mimetic-functionalized mesoporous silica nanoparticles and evaluation of acyclovir-loaded antiviral nanoparticles with dual mechanisms of action. ACS Omega. 2018;3:1689–99.

Lee IH, Palombo MS, Zhang XP, Szekely Z, Sinko PJ. Design and evaluation of a CXCR4 targeting peptide 4DV3 as an HIV entry inhibitor and a ligand for targeted drug delivery. Eur J Pharm Biopharm. 2019;138:11–22.

Lee KL, Twyman RM, Fiering S, Steinmetz NF. Virus-based nanoparticles as platform technologies for modern vaccines. Wiley Interdiscip Rev Nanomed Nanobiotechnol. 2016;8:554–78.

Lee MY, Yang JA, Jung HS, Beack S, Choi JE, Hur W, Koo H, Kim K, Yoon SK, Hahn SK. Hyaluronic acid-gold nanoparticle/interferon α complex for targeted treatment of hepatitis C virus infection. ACS Nano. 2012;6:9522–31.

Lembo D, Swaminathan S, Donalisio M, Civra A, Pastero L, Aquilano D, Vavia P, Trotta F, Cavalli R. Encapsulation of Acyclovir in new carboxylated cyclodextrin-based nanosponges improves the agent's antiviral efficacy. Int J Pharm. 2013;443:262–72.

Levina AS, Repkova MN, Ismagilov ZR, Shikina NV, Mazurkova NA, Zarytova VF. Efficient inhibition of human influenza a virus by oligonucleotides electrostatically fixed on polylysine-containing TiO_2 nanoparticles. Russ J Bioorg Chem. 2014;40:179–84.

Li CM, Zheng LL, Yang XX, Wan XY, Wu WB, Zhen SJ, Li YF, Luo LF, Huang CZ. DNA-AuNP networks on cell membranes as a protective barrier to inhibit viral attachment, entry and budding. Biomaterials. 2016;77:216–26.

Liang F, Lindgren G, Lin A, Thompson EA, Ols S, Rohss J, John S, Hassett K, Yuzhakov O, Bahl K, Brito LA, Salter H, Ciaramella G, Loré K. Efficient targeting and activation of antigen presenting cells in vivo after modified mRNA vaccine administration in rhesus macaques. Mol Ther. 2017;25:2635–47.

Liang JJ, Wei JC, Lee YL, Hsu SH, Lin JJ, Lin YL. Surfactant-modified nanoclay exhibits an antiviral activity with high potency and broad spectrum. J Virol. 2014;88:4218–28.

Liang TJ. Hepatitis B: the virus and disease. Hepatology. 2009;49:13–21.

Lin GM, Li L, Panwar N, Wang J, Tjin SC, Wang XM, Yong KT. Non-viral gene therapy using multifunctional nanoparticles: status, challenges, and opportunities. Coord Chem Rev. 2018a;374:133–52.

Lin JK, Wang K, Xia XY, Shen L. Quantification of multivalency in protein-oligomer-coated nanoparticles targeting dynamic membrane glycan receptors. Langmuir. 2018c;34:8415–21.

Lin ZF, Li YH, Gong GF, Xia Y, Wang CB, Chen Y, Hua L, Zhong JY, Tang Y, Liu XM, Zhu B. Restriction of H1N1 influenza virus infection by selenium nanoparticles loaded with ribavirin via resisting caspase-3 apoptotic pathway. Int J Nanomedicine. 2018b;13:5787–97.

Lu L, Sun RWY, Chen R, Hui CK, Ho CM, Luk JM, Lau GKK, Che CM. Silver nanoparticles inhibit hepatitis B virus replication. Antivir Ther. 2008;13:253–62.

Mandal S, Belshan M, Holec A, Zhou Y, Destache C. An enhanced emtricitabine-loaded long-acting nanoformulation for prevention or treatment of HIV infection. Antimicrob Agents Chemother. 2017b;61:e01475–16.

Mandal S, Prathipati PK, Kang GB, Zhou Y, Yuan Z, Fan WJ, Li QS, Destache CJ. Tenofovir alafenamide and and elvitegravir loaded nanoparticles for long-acting prevention of HIV-1 vaginal transmission. AIDS. 2017a;31:469–76.

Marcos-Almaraz MT, Gref R, Agostoni V, Kreuz C, Clayette P, Serre C, Couvreur P, Horcajada P. Towards improved HIV-microbicide activity through the co-encapsulation of NRTI drugs in biocompatible metal organic framework nanocarriers. J Mater Chem B. 2017;5:8563–9.

Martins C, Araujo F, Gomes MJ, Fernandes C, Nunes R, Li W, Santos HA, Borges F, Sarmento B. Using microfluidic platforms to develop CNS-targeted polymeric nanoparticles for HIV therapy. Eur J Pharm Biopharm. 2019;138:111–24.

McConnachie LA, Kinman LM, Koehn J, Kraft JC, Lane S, Lee W, Collier AC, Ho RJY. Long-acting profile of 4 drugs in 1 anti-HIV nanosuspension in nonhuman primates for 5 weeks after a single subcutaneous injection. J Pharm Sci. 2018;107:1787–90.

McGill JL, Kelly SM, Kumar P, Speckhart S, Haughney SL, Henningson J, Narasimhan B, Sacco RE. Efficacy of mucosal polyanhydride nanovaccine against respiratory syncytial virus infection in the neonatal calf. Sci Rep. 2018;8:3021.

McKenzie Z, Kendall M, Mackay RM, Tetley TD, Morgan C, Griffiths M, Clark HW, Madsen J. Nanoparticles modulate surfactant protein A and D mediated protection against influenza A infection in vitro. Philos Trans R Soc Lond B Biol Sci. 2015;370:20140049.

Metz SW, Thomas A, Bracjbill A, Xianwen Y, Stone M, Horvath K, Miley M, Luft C, DeSimone JM, Tian S, de Silva AM. Nanoparticle delivery of a tetravalent E protein subunit vaccine induces balanced, type-specific neutralizing antibodies to each dengue virus serotype. PLoS Negl Trop Dis. 2018;12:e0006793.

Meyer M, Huang E, Yuzhakov O, Ramanathan P, Ciaramella G, Bukreyev A. Modified mRNA-based vaccines elicit robust immune responses and protect guinea pigs from Ebola virus disease. J Infect Dis. 2018;217:451–5.

Milovanovic M, Arsenijevic A, Milovanovic J, Kanjevac T, Arsenijevic N. Nanoparticles in antiviral therapy. In: Grumezescu AM, editor. Antimicrobial nanoarchitectonics: from synthesis to applications. Amsterdam: Elsevier; 2017. p. 383–410.

Mishra YK, Adelung R, Roehl C, Shukla D, Spors F, Tiwari V. Virostatic potential of micro-nano filopodia-like ZnO structures against herpes simplex virus-1. Antivir Res. 2011;92:305–12.

Mohideen M, Quijano E, Song E, Deng Y, Panse G, Zhang W, Clark MR, Saltzman WM. Degradable bioadhesive nanoparticles for prolonged intravaginal delivery and retention of elvitegravir. Biomaterials. 2017;144:144–54.

Moon HJ, Lee JS, Talactac MR, Chowdhury MYE, Kim JH, Park ME, Choi YK, Sung MH, Kim CJ. Mucosal immunization with recombinant influenza hemagglutinin protein and poly γ-glutamate/chitosan nanoparticles induces protection against highly pathogenic influenza A virus. Vet Microbiol. 2012;160:277–89.

Morcol T, Hurst BL, Tarbet EB. Calcium phosphate nanoparticle (CaPNP) for dose-sparing of inactivated whole virus pandemic influenza A (H1N1) 2009 vaccine in mice. Vaccine. 2017;35:4569–77.

Mori Y, Ono T, Miyahira Y, Nguyen VQ, Matsui T, Ishihara M. Antiviral activity of silver nanoparticle/chitosan composites against H1N1 influenza A virus. Nanoscale Res Lett. 2013;8:93.

Murray M. Ebola virus disease: a review of its past and present. Anesth Analg. 2015;121:798–809.

Murugan K, Dinesh D, Paulpandi M, Althbyani ADM, Subramaniam J, Madhiyazhagan P, Wang L, Suresh U, Kumar PM, Mohan J, Rajaganesh R, Wei H, Kalimuthu K, Parajulee MN, Mehlhorn H, Benelli G. Nanoparticles in the fight against mosquito-borne diseases: bioactivity of *Bruguiera cylindrica*-synthesized nanoparticles against dengue virus DEN-2 (*in vitro*) and its mosquito vector *Aedes aegypti* (Diptera: Culicidae). Parasitol Res. 2015;114:4349–61.

Murugan K, Jaganathan A, Rajaganesh R, Suresh U, Madhavan J, Senthil-Nathan S, Rajasekar A, Higuchi A, Kumar SS, Alarfaj AA, Nicoletti M, Petrelli R, Cappellacci L, Maggi F, Benelli G. Poly(styrene sulfonate)/poly(allylamine hydrochloride) encapsulation of TiO_2 nanoparticles boosts their toxic and repellent activity against Zika virus mosquito vectors. J Clust Sci. 2018;29:27–39.

National Nanotechnology Initiative. Big things from a tiny world. Arlington, VA: National Nanotechnology Initiative; 2008.

Ndeboko B, Lemamy GJ, Nielsen PE, Cova L. Therapeutic potential of cell penetrating peptides (CPPs) and cationic polymers for chronic hepatitis B. Int J Mol Sci. 2015;16:28230–41.

Nowacek AS, Balkundi S, McMillan J, Roy U, Martinez-Skinner A, Mosley RL, Kanmogne G, Kabanov AV, Bronich T, Gendelman HE. Analyses of nanoformulated antiretroviral drug charge, size, shape and content for uptake, drug release and antiviral activities in human monocyte-derived macrophages. J Control Release. 2011;150:204–11.

Nunes R, Araujo F, Barreiros L, Bartolo I, Segundo MA, Taveira N, Sarmento B, das Neves J. Noncovalent PEG coating of nanoparticle drug carriers improves the local pharmacokinetics of rectal anti-HIV microbicides. ACS Appl Mater Interfaces. 2018;10:34942–53.

Orlowski P, Kowalczyk A, Tomaszewska E, Ranoszek-Soliwoda K, Wegrzyn A, Grzesiak J, Celichowski G, Grobelny J, Eriksson K, Krzyzowska M. Antiviral activity of tannic acid modified silver nanoparticles: potential to activate immune response in herpes genitalis. Viruses. 2018a;10:524.

Orlowski P, Tomaszewska E, Gniadek M, Baska P, Nowakowska J, Sokolowska J, Nowak Z, Donten M, Celichowski G, Grobelny J, Krzyzowska M. Tannic acid modified silver nanoparticles show antiviral activity in herpes simplex virus type 2 infection. PLoS One. 2014;9:e104113.

Orlowski P, Tomaszewska E, Ranoszek-Soliwoda K, Gniadek M, Labedz O, Malewski T, Nowakowska J, Chodaczek G, Celichowski G, Grobelny J, Krzyzowska M. Tannic acid-modified silver and gold nanoparticles as novel stimulators of dendritic cells activation. Front Immunol. 2018b;9:1115.

Osminkina LA, Timoshenko VY, Shilovsky IP, Kornilaeva GV, Shevchenko SN, Gongalsky MB, Tamarov KP, Abramchuk SS, Nikiforov VN, Khaitov MR, Karamov EV. Porous silicon nanoparticles as scavengers of hazardous viruses. J Nanopart Res. 2014;16:2430.

Papp I, Sieben C, Ludwig K, Roskamp M, Boettcher C, Schlecht S, Herrmann A, Haag R. Inhibition of influenza virus infection by multivalent sialic-acid-functionalized gold nanoparticles. Small. 2010;6:2900–6.

Parboosing R, Maguire GEM, Govender P, Kruger HG. Nanotechnology and the treatment of HIV infection. Viruses. 2012;4:488–520.

Park SJ, Ko YS, Lee SJ, Lee C, Woo K, Ko G. Inactivation of influenza A virus via exposure to silver nanoparticle-decorated silica hybrid composites. Environ Sci Pollut Res. 2018;25:27021–30.

Patel BK, Parikh RH, Patel N. Targeted delivery of mannosylated-PLGA nanoparticles of antiretroviral drug to brain. Int J Nanomedicine. 2018;13:97–100.

Patra JK, Das G, Fraceto LF, Campos EVR, Rodriguez-Torres MDP, Acosta-Torres LS, Diaz-Torres LA, Grillo R, Swamy MK, Sharma S, Habtemariam S, Shin HS. Nano based drug delivery systems: recent developments and future prospects. J Nanobiotechnol. 2018;16:71.

Pavunraj M, Baskar K, Duraipandiyan V, Al-Dhabi NA, Rajendran V, Benelli G. Toxicity of Ag nanoparticles synthesized using stearic acid from *Catharanthus roseus* leaf extract against *Earias vittella* and mosquito vectors (*Culex quinquefasciatus* and *Aedes aegypti*). J Clust Sci. 2017;28:2477–92.

Pentak D, Kozik V, Bąk A, Dybał P, Sochanik A, Jampílek J. Methotrexate and cytarabine-loaded nanocarriers for multidrug cancer therapy. Spectroscopic study. Molecules. 2016;21:1689.

Pieler MM, Frentzel S, Bruder D, Wolff MW, Reichl U. A cell culture-derived whole virus influenza A vaccine based on magnetic sulfated cellulose particles confers protection in mice against lethal influenza A virus infection. Vaccine. 2016;34:6367–74.

Pisárčik M, Jampílek J, Lukáč M, Horáková R, Devínsky F, Bukovský M, Kalina M, Tkacz J, Opravil T. Silver nanoparticles stabilised by cationic gemini surfactants with variable spacer length. Molecules. 2017;22:1794.

Pisárčik M, Lukáč M, Jampílek J, Bilka F, Bilková A, Pašková Ľ, Devínsky F, Horáková R, Opravil T. Silver nanoparticles stabilised with cationic single-chain surfactants. Structure-physical properties-biological activity relationship study. J Mol Liq. 2018;272:60–72.

Plourde AR, Bloch EN. A literature review of Zika virus. Emerg Infect Dis. 2016;22:1185–92.

Pomeranz LE, Reynolds AE, Hengartner CJ. Molecular biology of pseudorabies virus: impact on neurovirology and veterinary medicine. Microbiol Mol Biol Rev. 2005;69:462–500.

Prathipati PK, Mandal S, Pon G, Vivekanandan R, Destache CJ. Pharmacokinetic and tissue distribution profile of long acting tenofovir alafenamide and elvitegravir loaded nanoparticles in humanized mice model. Pharm Res. 2017;34:2749–55.

Qi M, Zhang XE, Sun XX, Zhang XW, Yao YF, Liu SL, Chen Z, Li W, Zhang ZP, Chen JJ, Cui Z. Intranasal nanovaccine confers homo- and hetero-subtypic influenza protection. Small. 2018;14:1703207.

Quach QH, Ang SK, Chu JHJ, Kah JCY. Size-dependent neutralizing activity of gold nanoparticle-based subunit vaccine against dengue virus. Acta Biomater. 2018;78:224–35.

Rafiei S, Rezatofighi SE, Ardakani MR, Madadgar O. In vitro anti-foot-and-mouth disease virus activity of magnesium oxide nanoparticles. IET Nanobiotechnol. 2015;9:247–51.

Rai M, Deshmukh SD, Ingle AP, Gupta IR, Galdiero M, Galdiero S. Metal nanoparticles: the protective nanoshield against virus infection. Crit Rev Microbiol. 2016;42:46–56.

Ramya S, Shanmugasundaram T, Balagurunathan R. Biomedical potential of actinobacterially synthesized selenium nanoparticles with special reference to anti-biofilm, anti-oxidant, wound healing, cytotoxic and anti-viral activities. J Trace Elem Med Biol. 2015;32:30–9.

Rao L, Wang WB, Meng QF, Tian MF, Cai B, Wang YC, Li AX, Zan MH, Xiao F, Bu LL, Li G, Li A, Liu Y, Guo SS, Zhao XZ, Wang TH, Liu W, Wu J. A biomimetic nanodecoy traps Zika virus to prevent viral infection and fetal microcephaly development. Nano Lett. 2019;19:2215–22.

Renganathan S, Aroulmoji V, Shanmugam G, Devarajad G, Rao KV, Rajendar V, Park SH. Silver nanoparticle synthesis from carica papaya and virtual screening for anti-dengue activity using molecular docking. Mater Res Express. 2019;6:035028.

Ribeiro-Viana R, Sanchez-Navarro M, Luczkowiak J, Koeppe JR, Delgado R, Rojo J, Davis BG. Virus-like glycodendrinanoparticles displaying quasi-equivalent nested polyvalency upon glycoprotein platforms potently block viral infection. Nat Commun. 2012;3:1303.

Rodriguez-Perez L, Ramos-Soriano J, Perez-Sanchez A, Illescas BM, Munoz A, Luczkowiak J, Lasala F, Rojo J, Delgado R, Martin N. Nanocarbon-based glycoconjugates as multivalent inhibitors of Ebola virus infection. J Am Chem Soc. 2018;140:9891–8.

Ross KA, Loyd H, Wu WW, Huntimer L, Ahmed S, Sambol A, Broderick S, Flickinger Z, Rajan K, Bronich T, Mallapragada S, Wannemuehler MJ, Carpenter S, Narasimhan B. Hemagglutinin-based polyanhydride nanovaccines against H5NI influenza elicit protective virus neutralizing titers and cell-mediated immunity. Int J Nanomedicine. 2015;10:229–43.

Sametband M, Shukla S, Meningher T, Hirsh S, Mendelson E, Sarid R, Gedanken A, Mandelboim M. Effective multi-strain inhibition of influenza virus by anionic gold nanoparticles. Med Chem Commun. 2011;2:421–3.

Sato Y, Matsui H, Yamamoto N, Sato R, Munakata T, Kohara M, Harashima H. Highly specific delivery of siRNA to hepatocytes circumvents endothelial cell-mediated lipid nanoparticle-

associated toxicity leading to the safe and efficacious decrease in the hepatitis B virus. J Control Release. 2017;266:216–25.
Sboui S, Tabbabi A. Marburg virus disease: a review literature. J Genes Proteins. 2017;1:1.
Scherliess R. Future of nanomedicines for treating respiratory diseases. Expert Opin Drug Deliv. 2019;16:59–68.
Senanayake TH, Gorantla S, Makarov E, Lu Y, Warren G, Vinogradov SV. Nanogel-conjugated reverse transcriptase inhibitors and their combinations as novel antiviral agents with increased efficacy against HIV-1 infection. Mol Pharm. 2015;12:4226–36.
Seok H, Noh JY, Lee DY, Kim SJ, Song CS, Kim YC. Effective humoral immune response from a H1N1 DNA vaccine delivered to the skin by microneedles coated with PLGA-based cationic nanoparticles. J Control Release. 2017;265:66–74.
Seyfoddin A, Sherwin T, Patel DV, McGhee CN, Rupenthal ID, Taylor JA, Al-Kassas R. Ex vivo and in vivo evaluation of chitosan coated nanostructured lipid carriers for ocular delivery of Acyclovir. Curr Drug Deliv. 2016;13:923–34.
Sharma G, Thakur K, Setia A, Amarji B, Singh MP, Raza K, Katare OP. Fabrication of acyclovir-loaded flexible membrane vesicles (FMVs): evidence of preclinical efficacy of antiviral activity in murine model of cutaneous HSV-1 infection. Drug Deliv Transl Res. 2017b;7:683–94.
Sharma RK, Cwiklinski K, Aalinkeel R, Reynolds JL, Sykes DE, Quaye E, Oh J, Mahajan SD, Schwartz SA. Immunomodulatory activities of curcumin-stabilized silver nanoparticles: efficacy as an antiretroviral therapeutic. Immunol Investig. 2017a;46:833–46.
Shionoiri N, Sato T, Fujimori Y, Nakayama T, Nemoto M, Matsunaga T, Tanaka T. Investigation of the antiviral properties of copper iodide nanoparticles against feline calicivirus. J Biosci Bioeng. 2012;113:580–6.
Shohani S, Mondanizadeh M, Abdoli A, Khansarinejad B, Salimi-Asl M, Ardestani M, Ghanbari M, Haj MS, Zabihollahi R. Trimethyl chitosan improves anti-HIV effects of atripla as a new nano-formulated drug. Curr HIV Res. 2017;15:56–65.
Siddiq A, Younus I, Shamim A, Badar S. Nanostructures for antiviral therapy: in the last two decades. Curr Nanosci. 2017;13:229–46.
Sillman B, Bade AN, Dash PK, Bhargavan B, Kocher T, Mathews S, Su H, Kanmogne GD, Poluektova LY, Gorantla S, McMillan JE, Gautam N, Alnouti Z, Edawa H, Gendelman HE. Creation of a long-acting nanoformulated dolutegravir. Nat Commun. 2018;9:443.
Singh L, Kruger HG, Maguire GEM, Govender T, Parboosing R. The role of nanotechnology in the treatment of viral infections. Ther Adv Infect Dis. 2017;4:105–31.
Singh SM, Alkie TN, Abdelaziz KT, Hodgins DC, Novy A, Nagy E, Sharif S. Characterization of immune responses to an inactivated avian influenza virus vaccine adjuvanted with nanoparticles containing CpG ODN. Viral Immunol. 2016b;29:269–75.
Singh SM, Alkie TN, Nagy E, Kulkarni RR, Hodgins DC, Sharif S. Delivery of an inactivated avian influenza virus vaccine adjuvanted with poly(D,L-lactic-co-glycolic acid) encapsulated CpG ODN induces protective immune responses in chickens. Vaccine. 2016a;34:4807–13.
Smith G, Liu Y, Flyer D, Massare MJ, Zhou B, Patel N, Ellingsworth L, Lewis M, Cummings JF, Glenn G. Novel hemagglutinin nanoparticle influenza vaccine with Matrix-M™ adjuvant induces hemagglutination inhibition, neutralizing, and protective responses in ferrets against homologous and drifted A(H3N2) subtypes. Vaccine. 2017;35:5366–72.
Stanaway JD, Flaxman AD, Naghavi M, Fitymaurice C, Vos T, Abubakar I, Abu-Raddad LJ, Assadi R, Bhala N, Cowie B, Forouzanfour MH, Groeger J, Hanafiah KM, Jacobsen KH, James SL, MacLachlan J, Malekzadeh R, Martin NK, Mokdad AA, Mokdad AH, Murray CJL, Plass D, Rana S, Rein DB, Richardus JH, Sanabria J, Saylan M, Shahraz S, So S, Vlassov VV, Weiderpass E, Wiersma ST, Younis M, Yu C, El Sayed ZM, Cooke GS. The global burden of viral hepatitis from 1990 to 2013: findings from the Global Burden of Disease study. Lancet. 2016;388:1081–8.
Stone J, Thornburg NJ, Blum DL, Kuhn SJ, Wright DW, Crowe JE. Gold nanorod vaccine for respiratory syncytial virus. Nanotechnology. 2013;24:295102.
Suganya P, Vaseeharan B, Vijayakumar S, Balan B, Govindarajan M, Alharbi NS, Kadaikunnan S, Khaled JM, Benelli G. Biopolymer zein-coated gold nanoparticles: synthesis, antibacterial

potential, toxicity and histopathological effects against the *Zika virus* vector *Aedes aegypti*. J Photochem Photobiol B. 2017;173:404–11.
Sujitha V, Murugan K, Paulpandi M, Panneerselvam C, Suresh U, Roni M, Nicoletti M, Higuchi A, Madhiyazhagan P, Subramaniam J, Dinesh D, Vadivalagan C, Chandramohan B, Alarfaj AA, Munusamy MA, Barnard DR, Benelli G. Green-synthesized silver nanoparticles as a novel control tool against dengue virus (DEN-2) and its primary vector *Aedes aegypti*. Parasitol Res. 2015;114:3315–25.
Suresh U, Murugan K, Panneerselvam C, Rajaganesh R, Roni M, Aziz A, Al-Aoh HAN, Trivedi S, Rehman H, Kumar S, Higuchi A, Canale A, Benelli G. *Suaeda maritima*-based herbal coils and green nanoparticles as potential biopesticides against the dengue vector *Aedes aegypti* and the tobacco cutworm *Spodoptera litura*. Physiol Mol Plant Pathol. 2018;101:225–35.
Suwannoi P, Chomnawang M, Sarisuta N, Reichl S, Mueller-Goymann CC. Development of Acyclovir-loaded albumin nanoparticles and improvement of Acyclovir permeation across human corneal epithelial T cells. J Ocul Pharmacol Ther. 2017;33:743–52.
Szunerits S, Barras A, Khanal M, Pagneux Q, Boukherroub R. Nanostructures for the inhibition of viral infections. Molecules. 2015;20:14051–1408.
Szymanska E, Orlowski P, Winnicka K, Tomaszewska E, Baska P, Celichowski G, Grobelny J, Basa A, Krzyzowska M. Multifunctional tannic acid/silver nanoparticle-based mucoadhesive hydrogel for improved local treatment of HSV infection: in vitro and in vivo studies. Int J Mol Sci. 2018;19:387.
Takizawa N, Yamasaki M. Current landscape and future prospects of antiviral drugs derived from microbial products. J Antibiot. 2018;71:45–52.
Tang R, Zhai YJ, Dong LL, Malla T, Hu K. Immunization with dendritic cell-based DNA vaccine pRSC-NLDC145.gD-IL21 protects mice against herpes simplex virus keratitis. Immunotherapy. 2018;10:189–200.
Tao W, Hurst BL, Shakya AK, Uddin MJ, Ingrole RS, Hernandez-Sanabria M, Arya RP, Bimler L, Paust S, Tarbet EB, Gill HS. Consensus M2e peptide conjugated to gold nanoparticles confers protection against H1N1, H3N2 and H5N1 influenza A viruses. Antivir Res. 2017;141:62–72.
Taubenberger JK, Morens DM. Influenza: the once and future pandemic. Public Health Rep. 2010;125:16–26.
Thammakarn C, Satoh K, Suguro A, Hakim H, Ruenphet S, Takehara K. Inactivation of avian influenza virus, Newcastle disease virus and goose parvovirus using solution of nano-sized scallop shell powder. J Vet Med Sci. 2014;76:1277–80.
Thi EP, Lee ACH, Geisbert JB, Ursic-Bedoya R, Agans KN, Robbins M, Deer DJ, Fenton KA, Kondratowicz AS, MacLachlan I, Geisbert TW, Mire CE. Rescue of non-human primates from advanced Sudan ebolavirus infection with lipid encapsulated siRNA. Nat Microbiol. 2016;1:16142.
Thi EP, Mire CE, Lee ACH, Geisbert JB, Zhou JZ, Agans KN, Snead NM, Deer DJ, Barnard TR, Fenton KA, MacLachlan I, Geisbert TW. Lipid nanoparticle siRNA treatment of Ebola-virus-Makona-infected nonhuman primates. Nature. 2015;521:362–5.
Thi EP, Mire CE, Ursic-Bedoya R, Geisbert JB, Lee ACH, Agans KN, Robbins M, Deer DJ, Fenton KA, MacLachlan I, Geisbert TW. Marburg virus infection in nonhuman primates: therapeutic treatment by lipid-encapsulated siRNA. Sci Transl Med. 2014;6:250ra116.
Timin AS, Muslimov AR, Petrova AV, Lepik KV, Okilova MV, Vasin AV, Afanasyev BV, Sukhorukov GB. Hybrid inorganic-organic capsules for efficient intracellular delivery of novel siRNAs against influenza A (H1N1) virus infection. Sci Rep. 2017;7:102.
Tomitaka A, Arami H, Huang ZH, Raymond A, Rodriguez E, Cai Y, Febo M, Takemura Y, Nair M. Hybrid magneto-plasmonic liposomes for multimodal image-guided and brain-targeted HIV treatment. Nanoscale. 2018;10:184–94.
Trefry JC, Wooley DP. Silver nanoparticles inhibit vaccinia virus infection by preventing viral entry through a macropinocytosis-dependent mechanism. J Biomed Nanotechnol. 2013;9:1624–35.
Trigilio J, Antoine TE, Paulowicz I, Mishra YK, Adelung R, Shukla D. Tin oxide nanowires suppress herpes simplex virus-1 entry and cell-to-cell membrane fusion. PLoS One. 2012;7:e48147.

Ursic-Bedoya R, Mire CE, Robbins M, Geisbert JB, Judge A, MacLachlan I, Geisbert TW. Protection against lethal Marburg virus infection mediated by lipid encapsulated small interfering RNA. J Infect Dis. 2014;209:562–70.

Vijayakumar S, Ganesan S. Gold nanoparticles as an HIV entry inhibitor. Curr HIV Res. 2012;10:643–6.

Wang C, Zhu WD, Luo Y, Wang BZ. Gold nanoparticles conjugating recombinant influenza hemagglutinin trimers and flagellin enhanced mucosal cellular imunity. Nanomedicine. 2018b;14:1349–60.

Wang L, Chang TZ, He Y, Kim JR, Wang S, Mohan T, Berman Z, Tompkins SM, Tripp RA, Compans RW, Champion JA, Wang BZ. Coated protein nanoclusters from influenza H7N9 HA are highly immunogenic and induce robust protective imunity. Nanomedicine. 2017;13:253–62.

Wang W, Zhao YL, Zhao XH, Zu YG, Fu YJ. Anti-HSV-1 activity of glycyrrhizic acid nanoparticles prepared by supercritical antisolvent process. Curr Nanosci. 2015;11:366–70.

Wang Y, Deng L, Kang SM, Wang BZ. Universal influenza vaccines: from viruses to nanoparticles. Expert Rev Vaccines. 2018a;17:967–76.

Warren G, Makarov E, Lu Y, Senanayake T, Rivera K, Gorantla S, Poluektova LY, Vinogradov SV. Amphiphilic cationic nanogels as brain-targeted carriers for activated nucleoside reverse transcriptase inhibitors. J Neuroimmune Pharmacol. 2015;10:88–101.

World Health Organization. Global hepatitis report. 2017. http://apps.who.int/iris/bitstream/10665/255016/1/9789241565455-eng.pdf?ua=1.

Xiang DX, Chen Q, Pang L, Zheng CL. Inhibitory effects of silver nanoparticles on H1N1 influenza A virus in vitro. J Virol Methods. 2011;178:137–42.

Xiang DX, Zheng Y, Duan W, Li XJ, Yin JJ, Shigdar S, O'Connor ML, Marappan M, Zhao XJ, Miao YQ, Xiang B, Zheng CLL. Inhibition of A/Human/Hubei/3/2005 (H3N2) influenza virus infection by silver nanoparticles in vitro and in vivo. Int J Nanomedicine. 2013;8:4103–13.

Yadavalli T, Shukla D. Role of metal and metal oxide nanoparticles as diagnostic and therapeutic tools for highly prevalent viral infections. Nanomedicine. 2017;13:219–30.

Yang HW, Ye L, Guo XD, Yang CL, Compans RW, Prausnitz MR. Ebola vaccination using a DNA vaccine coated on PLGA-PLL/γPGA nanoparticles administered using a microneedle patch. Adv Healthc Mater. 2017b;6:1600750.

Yang L, Li W, Kirberger M, Liao WZ, Ren JY. Design of nanomaterial based systems for novel vaccine development. Biomater Sci. 2016b;4:785–802.

Yang XX, Li CM, Huang CZ. Curcumin modified silver nanoparticles for highly efficient inhibition of respiratory syncytial virus infection. Nanoscale. 2016a;8:3040–8.

Yang XX, Li CM, Li YF, Wang J, Huang CZ. Synergistic antiviral effect of curcumin functionalized graphene oxide against respiratory syncytial virus infection. Nanoscale. 2017a;9:16086–92.

Yang Y, He HJ, Chang H, Yu Y, Yang MB, He Y, Fan ZC, Iyer SS, Yu P. Multivalent oleanolic acid human serum albumin conjugate as nonglycosylated neomucin for influenza virus capture and entry inhibition. Eur J Med Chem. 2018;143:1723–31.

Ye SY, Shao K, Li ZH, Guo N, Zuo YP, Li Q, Lu ZC, Chen L, He QG, Han HY. Antiviral activity of graphene oxide: how sharp edged structure and charge matter. ACS Appl Mater Interfaces. 2015;7:21571–9.

Zacharias ZR, Ross KA, Hornick EE, Goodman JT, Narasimhan B, Waldschmidt TJ, Legge KL. Polyanhydride nanovaccine induces robust pulmonary B and T cell immunity and confers protection against homologous and heterologous influenza A virus infections. Front Immunol. 2018;9:1953.

Zazo H, Millan CG, Colino CI, Lanao JM. Applications of metallic nanoparticles in antimicrobial therapy. In: Grumezescu AM, editor. Antimicrobial nanoarchitectonics: from synthesis to applications. Amsterdam: Elsevier; 2017. p. 411–44.

Zhao K, Li SS, Li W, Yu L, Duan XT, Han JY, Wang XH, Jin Z. Quaternized chitosan nanoparticles loaded with the combined attenuated live vaccine against Newcastle disease and infectious bronchitis elicit immune response in chicken after intranasal administration. Drug Deliv. 2017;24:1574–86.

Zhao K, Rong GY, Hao Y, Yu L, Kang H, Wang X, Wang XH, Jin Z, Ren ZY, Li ZJ. IgA response and protection following nasal vaccination of chickens with Newcastle disease virus DNA vaccine nanoencapsulated with Ag@SiO_2 hollow nanoparticles. Sci Rep. 2016;6:25720.

Zhou T, Su H, Dash P, Lin Z, Shetty BLD, Kocher T, Szlachetka A, Lamberty B, Fox HS, Poluektova L, Gorantia S, McMillan J, Gautam N, Mosley RL, Alnouti Y, Edgawa B, Gendelman HE. Creation of a nanoformulated cabotegravir prodrug with improved antiretroviral profiles. Biomaterials. 2018;151:53–65.

Ziem B, Azab W, Gholami MF, Rabe JP, Osterrieder N, Haag R. Size-dependent inhibition of herpesvirus cellular entry by polyvalent nanoarchitectures. Nanoscale. 2017;9:3774–83.

Ziem B, Thien H, Achazi K, Yue C, Stern D, Silberreis K, Gholami MF, Beckert F, Groeger D, Muelhaupt R, Rabe JP, Nitsche A, Haag R. Highly efficient multivalent 2D nanosystems for inhibition of orthopoxvirus particles. Adv Healthc Mater. 2016;5:2922–30.

Zoppe JO, Ruottinen V, Ruotsalainen J, Ronkko S, Johansson LS, Hinkkanen A, Jarvinen K, Seppala J. Synthesis of cellulose nanocrystals carrying tyrosine sulfate mimetic ligands and inhibition of alphavirus infection. Biomacromolecules. 2014;15:1534–42.

Zuckerman AJ. Hepatitis viruses. In: Baron S, editor. Medical microbiology. 4th ed. Galveston: University of Texas Medical Branch at Galveston; 1996. p. 1–169.

Chapter 8
The Potential of Gold and Silver Antimicrobials: Nanotherapeutic Approach and Applications

Heejeong Lee and Dong Gun Lee

Abstract Nanoparticles are promising antimicrobial agents for use and its potential as therapy for skin and soft tissue microbial infection has been investigated. Metallic ion and metal nanoparticles demonstrated efficacy against microbial pathogen, causing cutaneous and life-threatening systemic infections. Among various nanoparticles, silver and gold nanoparticles have been focused in this chapter. In recent researches, dual mechanisms of gold and silver nanoparticles were determined against pathogenic microorganisms. They are similar to cell death, but in a different pathway, including potent antimicrobial activity. The obvious difference is the presence or absence of reactive oxygen species level. Silver nanoparticles change cell death process depending on reactive oxygen species concentration, but gold nanoparticles do not change or influence. Due to nontoxicity to human and their antimicrobial effects, nanoparticles have been applied to diverse fields and possess clinical utility as a novel wound healing and antimicrobial agent for microbial infection.

Keywords Gold nanoparticle · Silver nanoparticle · Skin infection · Microbial infection

8.1 Introduction

The skin is a first nonspecific defense barrier to protect body organs and tissues and the most exposed organ to impairment and injury, burns, and scratches. Damaging the epithelium and connective structures is intolerable to protect from the external environment. A functional epidermis or even other skin layers should be refabricated, known as wound repair or wound healing happened by a cascade of

---

H. Lee · D. G. Lee (✉)
School of Life Sciences, College of Natural Sciences, Kyungpook National University, Daegu, Republic of Korea
e-mail: dglee222@knu.ac.kr

M. Rai, B. Jamil (eds.), *Nanotheranostics*,
https://doi.org/10.1007/978-3-030-29768-8_8

intersecting phases. The repair is continued until the damaged skin structure replaces by the formation of a scar (Negut et al. 2018). Human skin diseases can commonly cause complications if not treated properly. Although many antifungal agents are used to treat skin infections, most of the fungal pathogen have been reported to be resistant to antifungal agents (Rai et al. 2017a). Anatomically, bacteria, viruses, yeast, dermatophytes, and parasites can infect the skin and subcutaneous tissue. For example, subcutaneous infections in hospitals are multilayer infections. The infection of the skin and subcutaneous tissue is called cellulitis and the infection of deep connective tissue layer is called as fasciitis (Kujath and Kujath 2010). Acute purulent inflammation in the dermis and subcutaneous tissue is referred to as cellulitis. It is characterized by erythema, fever, swelling, and pain in the multilayer of skin and subcutaneous tissues where the bacteria are involved (Kujath and Kujath 2010). Mostly, it is caused by group A hemolytic streptococci or yellow staphylococci (Kujath and Kujath 2010). Common pathogens include group A Streptococci, *Staphylococcus aureus*, and—depending on the mechanism of injury—gram-negative pathogens such as *Pseudomonas aeruginosa, Escherichia coli*, and *Enterobacter cloacae* (Kujath and Kujath 2010). Side effects of systemic and topical antifungal drugs may limit its use from time to time (Rai et al. 2017a).

Nanoparticle-based formulations are becoming increasingly important for the treatment of skin infections due to increased skin permeability, target delivery, and release control. They also provide benefits by increasing the bioavailability of the active ingredient, which continues to be effective at the site of infection (Rai et al. 2017a). Nanoparticles are defined as promising materials ranging in size from 10 to 100 nm due to the variety of potential catalytic, optoelectronic, biological, pharmaceutical and biomedical applications (Mohanraj and Chen 2006; DeLouise 2012). Nanoparticles reveal some advantages in relation to other materials, including their storage stability in the biological fluid, easy and varied manufacturing technology, and controlled release. Furthermore, they also fill the objective of delivering and encapsulating active substances known as vectors or carriers to the target site (Bermúdez et al. 2017). Not all antimicrobial agents can exhibit the desired antimicrobial effect and can reach the pathogens at an effective concentration through the plasma membranes of the host. Nanoparticles can be used as materials for targeted local delivery of drugs in the case of infections inside the body (Zhang et al. 2010). Nanoscale materials have been used effectively as antimicrobial agents for many infectious diseases due to their ability to interact with microorganisms (Huh and Kwon 2011; Albanese et al. 2012). Therefore, nanoparticles for treating microbial infection have potential applications in the economics of pharmaceutical development, personal hygiene industries, and food (Zhang et al. 2013; Kulshreshtha et al. 2017).

Various metal nanoparticles are known to suffer the most demanding processing conditions and have proven their value as antimicrobials (Sondi and Salopek-Sondi 2004; Espitia et al. 2012; Beyth et al. 2015). The inherent advantage of such a nanostructure is that it is relatively difficult to develop resistance because it acts by more than one mode of action (Mühling et al. 2009). In clinical conditions, metal-based nanoparticles are utilized in the coatings to block attachment of microbial pathogen and infection to their safety instead of being applied directly as a therapeutic agent (Karlsson et al. 2008; Schrand et al. 2010). Also, for treating skin infections, there is less risk of systemic adverse effects. Nanoparticles such as titanium dioxide, sulfur, copper, silver, zinc oxide, sulfur, and gold have been used to manage skin diseases such as pityriasis, skin candidiasis, folliculitis, tinea, and seborrheic dermatitis (Rai et al. 2017a). The study of gold nanoparticles (AuNPs) has been steadily continuing and their unique physical and chemical properties make them promising therapeutics (Alkilany et al. 2012). For cancer treatments, AuNPs has been widely used as a thermal therapy and drug delivery system because of their photophysical properties that allow the particles to bind drugs and then trigger release in a remote site (Huang et al. 2008; Brown et al. 2010). Silver nanoparticles (AgNPs) can be utilized to develop and prepare nanomedicine, a new generation of biosensors, drug delivery systems, antimicrobials, silver-based dressings, nanogels, silver-coated medical devices, and nano-lotions (Sahu et al. 2013). Medical applications in the form of silver-based dressings, nano-lotions, nanogels, and silver-coated medical devices have been contained (Oberdorster et al. 2005). This chapter focuses on the AuNPs and AgNPs used in the treatment of various skin infections by microbial pathogens and their mechanism of action.

8.2 Requirements for Nanoparticles to Be Effective Antimicrobial Agents

Nanoparticles are applied as a mediator for all types of drugs and provide an alternative to supplement the limitations of antimicrobial treatment, increase effectiveness, and minimize undesirable side effects (Bermúdez et al. 2017). Most of the drug delivery areas can be reached through microcirculation by blood capillaries or through pores in various surfaces and membranes. One of the primary requirements is the ability to move freely on available blood capillaries or pores in membranes and to cross various boundaries that can come in the way. Most of the gates, openings, and apertures at the cytoplasmic or subcellular level have nanometer size. Therefore, nanoparticles are fitted to reach the subcellular level. Indeed, nanoparticles should be smaller than 300 nm for efficient transport (Bermúdez et al. 2017). The characters of nanostructures depend on their size, so they can be controlled and manipulated according to requirements (Kulshreshtha et al. 2017). The ability to move freely is not helpful for drug delivery purposes. The delivery system has to reach the destination site and cross the blood capillary walls, reaching the extracellular fluid of the tissue. On the other hand, the nanoparticles cross the other cells when they enter the target cell. These are the major barriers to drug delivery. The nanoparticles do many things through the capillaries or vessels and across the barrier while the carrier stays (Bermúdez et al. 2017).

During storage, nanoparticles have to be stable for size, size distribution, surface morphology, and other important chemical and physical properties. The formulation must be non-immunogenic and biocompatible. Nanoparticle should be able to regenerate the nanoparticles after administration. The by-product of nanoparticle formulation has to be non-toxic, stable, easy to administer, and safe. It also has to be achieving good circulation and bioavailability (Kulshreshtha et al. 2017). The antimicrobial properties of nanoparticles depend strongly on their stability in the media. The nanoparticles must be stored in the media to allow sufficient time for interaction between the pathogenic microorganisms and the nanoparticles (Shrivastava et al. 2007). Stability ultimately determines the biological consequences of nanoparticles and their in vivo collapse, targeting ability, and toxicity (Panyam and Labhasetwar 2003). The property can be improved by changing the manufacturing method. The stability of the nanoparticles allows the drug to be administered via the oral route, thus extending the shelf life of the drug (Gelperina et al. 2005; Kulshreshtha et al. 2017). Compatibility with biosafety and biological systems is the most serious parameter to estimate the applicability of drugs. Quantitative biocompatibility analysis is important to confirm the interaction of drugs with biological machinery. Safety concerns related to both the environment and humans should be markedly assessed to evaluate the long-term impact of nanoparticle-based antimicrobials. These two factors are essential for the public acceptance of nanoparticle-based antimicrobials (Debnath et al. 2010; Kulshreshtha et al. 2017).

8.3 Mode of Actions of AuNPs in Microbial Infection

The toxic effects of metallic nanoparticles vary depending upon their dose and configuration. Smaller nanoparticles have higher toxicities. AuNPs is examined to be comparatively safer than other metallic nanoparticles owing to the nontoxic and inert nature of gold. AuNPs is able to be conjugated to a variety of macromolecules including antimicrobial agents through the surface synthesis techniques for nanoparticles owing to inherent elemental properties (Lee and Lee 2018). AuNPs can enter cells through the pinocytosis pathway and can be placed in lysosomes without entering the nucleus, which potentially helps to minimize toxicity. In addition, during exposure to AuNPs, the redox properties are beneficial in decreasing the level of reactive oxygen species (ROS) generated. Colloidal gold has been applied for therapeutic purposes, indicating that gold has high biocompatibility for centuries (Rajchakit and Sarojini 2017). AuNPs can be applied for cell and tissue imaging, cancer treatment, and drug delivery. AuNPs exhibits potent antifungal activity through inhibition of the H^+-ATPase-mediated proton pump to *Candida albicans* (Wani and Ahmad 2013). Proton pump can be modulated by exogenous or endogenous molecules (Lagadic-Gossmann et al. 2004). Malfunction of proton pump leads to intracellular changes that have been implicated in intracellular damage (Lagadic-Gossmann et al. 2004; Seong and Lee 2018). In respect of intracellular networks, physical changes in DNA prevent cell division from completing, leading to apoptotic cell death (Roos and Kaina 2006; Henry et al. 2013). AuNPs-treated *C. albicans* cells undergo nucleic acid damage such as DNA fragmentation and nucleus condensation

(Seong and Lee 2018). It has been reported that the nanoparticles may bind directly into DNA, causing cell damage (Lecoeur et al. 2001). Direct interaction with DNA causes DNA fragmentation and nucleus condensation. The backup and breakdown on intracellular maintenance result from mitochondrial dysfunction, including increase in mitochondrial mass, collapse of mitochondrial Ca^{2+} homeostasis, and membrane depolarization of the mitochondria (Nugent et al. 2007; Seong and Lee 2018). Mitochondria plays an important role in cell survival and provide cellular energy, adenosine triphosphate (ATP), and the stimulation of apoptosis process (Akbar et al. 2016). Imbalance of mitochondrial ion homeostasis has been shown to help understand an unusual status. Mitochondrial Ca^{2+} effect on cell function which is several steps of energy metabolism to synchronize ATP (Hajnóczky et al. 2006). Mitochondrial Ca^{2+} uptake may induce mitochondrial permeability pore openings, leading to stimulate in apoptotic signals by mitochondrial mass changes (Rizzuto et al. 2012). The disturbance of Ca^{2+} homeostasis in mitochondria typically associates with a decrease of membrane potential (Carraro and Bernardi 2016). Depolarization of mitochondrial membrane potential is known to constitute a failure of matrix configuration and energy metabolism, and cytochrome *c* release during apoptotic cell death (Gottlieb et al. 2003; Carraro and Bernardi 2016) (Fig. 8.1).

AuNPs in *C. albicans* cells exerts ROS-independent apoptosis through organelle disruption, including attenuation of mitochondrial homeostasis and destruction of nucleic acid and nucleus (Seong and Lee 2018). ROS is produced as by-products or part of cell collapse in many cellular processes, including aging, inflammation,

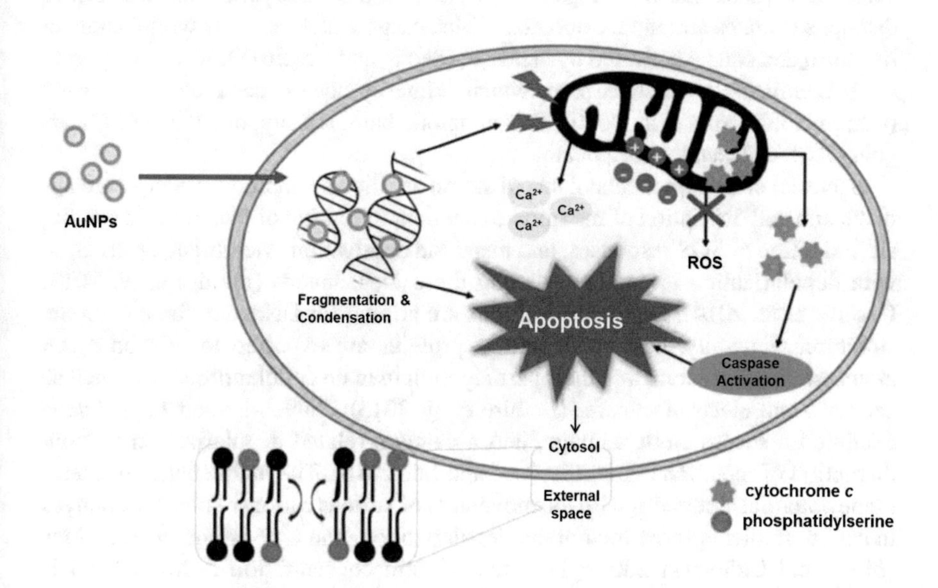

Fig. 8.1 AuNPs exerts ROS-independent apoptosis effects against *Candida albicans* through organelles disruption, such as attenuation of mitochondrial homeostasis and destruction of nucleic acid and nucleus (Seong and Lee 2018)

proliferation, and death (Carmona-Gutierrez et al. 2010; Li et al. 2011; Schieber and Chandel 2014). Their excess causes oxidative stress and macromolecular oxidative damage, resulting in loss of cell function, necrosis, and apoptosis (Nordberg and Arner 2001; Mittler et al. 2011). Whereas, the moderate concentration of ROS also participates in physiological responses as part of the defense mechanism and signal transduction process (Espinosa-Diez et al. 2015). Approximately, the apoptotic processes were prevented by ROS scavenger in apoptosis (Simon et al. 2000). When a shift in the redox state of a cell occurs, the interference that disrupts with this balance can play a regulating role in induction of apoptosis. The proper balance between the production of free radicals and the production of its eliminator antioxidants helps to maintain a homeostatic state (Mukherjee et al. 2014a). In this case, antioxidant has no protective effect against ROS-independent apoptosis response (Ko et al. 2005; Seong and Lee 2018). AuNPs triggers metacaspase-activated apoptotic signaling unrelated to ROS levels (Seong and Lee 2018). Cytochrome *c* in the mitochondria is a constituent of the electron transport chain and it transfers electrons from complex III to IV. The release of cytochrome *c* from mitochondria to cytosol is a hallmark of dysfunction of electron transport chain (Huttemann et al. 2011). Cytochrome *c* is released from the cytoplasm after treatment with AuNPs. The function as an electron transfer transmitter is lost. Cytochrome *c* has been exhibited to have dual functions in regulating energetic metabolism and cell death (Di Lisa et al. 2001). The cytochrome *c* released from the mitochondria is an essential stimulus in the apoptotic pathway because it can activate metacaspase followed by the cleavage of metacaspase substrates (Barbu et al. 2013). Metacaspases in yeast are a functional homologue of caspases in the eukaryotic cells that cleave distinct substrates and induce apoptosis. Metacaspase always exists within an inactive form and can be activated by cleavage (Seong and Lee 2018). In yeast, apoptosis is dominated by metacaspase, which primarily acts to devastate cells inside (Carmona-Gutierrez et al. 2010). This apoptotic hallmark supports the cells undergoing ROS-independent cell death.

Bacterial cell death mechanisms of action are diverse, including energy metabolism and cell disruption of membrane function, inhibition of nucleic acid synthesis, induction of SOS responses, and metal ion deprivation. Membrane destruction with depolarization is common among these mechanisms (Lemire et al. 2013; Cushnie et al. 2014). Distinct metal ions are critical for DNA and the membrane structure. Generally, half of all known proteins are expected to rely on metal atoms for their structure and their participation in main cellular processes, such as catalysis and electron transfer (Lemire et al. 2013). AuNPs-treated *E. coli* cells exhibited a severe DNA damage, and a calcium-related depolarization without disruption of membrane integrity (Lee and Lee 2018). The maintenance of membrane potential associates with the movement of various cations. Thus, the changes in cation gradients show membrane depolarization, and Ca^{2+} is one of these ions (Mann and Cidlowski 2001). The free calcium concentration in the cytosol is enhanced in the AuNPs-treated cells, indicating that the calcium signal process is related to bacterial cell death (Lee and Lee 2018). Cytoplasmic calcium level is correlated with processes, including sporulation, bacterial pathogenicity, chemotaxis, and differentiation (Bruni et al. 2017). The Ca^{2+} gradient is delicately preserved by a system of transporters and channels (Yun and Lee 2016). Depolarization of cytoplasmic membrane is also involved with a change in the Ca^{2+} gradient due to the regulation of bacterial function: cell division, ATP generation, and persister formation have all been restricted to membrane potential (Yun and Lee 2016; Bruni et al. 2017). During membrane voltage depolarization, *E. coli* also reveal cytoplasmic calcium influx (Bruni et al. 2017). Depolarization of the cytoplasmic membrane potential occurs, and then the imbalance between the plasma membrane Ca^{2+} export and Ca^{2+} influx leads to a sustained elevation in cytosolic Ca^{2+} (Hajnoczky et al. 2006). From the membrane depolarization with calcium fluctuations it can be realized that AuNPs exposed to cells induces an increase in intracellular Ca^{2+} export and influx across the plasma membrane and release Ca^{2+} from the intracellular stores (Bruni et al. 2017).

Similar to eukaryotes, programmed cell death in bacteria is a regulated process that contributes to multiple aspects of bacterial growth (Allocati et al. 2015). Severe DNA damage and membrane depolarization may lead to cell death following apoptotic-like death (Koksharova 2013). Enzyme-catalyzed DNA fragmentation ultimately initiates a series of cascade reactions and events that lead to apoptosis in the eukaryotic cells. Similarly, in prokaryotic cells, DNA damage causes decrease in the complex I and II activities in the respiratory chain, the expression of Edin and SOS Edin genes, leading to the formation of hydroxyl radicals, leading to membrane depolarization and cell destruction (Nagamalleswari et al. 2017). Its role as a caspase-like protein was performed as a major regulator in both apoptosis-like death and SOS repair system in *E. coli*. AuNPs activates caspase-like protein as RecA protein expression (Lee and Lee 2018). In *E. coli* cells, RecA proposed as caspase-like protein (Dwyer et al. 2012). RecA-induced ROS response would have caused the DNA damage and cell death-induced SOS response, leading to RecA activation. AuNPs in *E. coli* induces overexpression of the RecA protein as the stimulation of bacterial caspase-like protein (Lee and Lee 2018). ROS is required for normal cellular function by all aerobic organisms. They play mediators of intracellular signaling pathways and serve an important role in maintaining homeostasis from prokaryotic cells to eukaryotic cells (Mittler et al. 2011). In common with eukaryotic cells, accumulation of intracellular ROS was not observed in the presence of AuNPs. However, changes in intracellular glutathione level have occurred, although no accumulation of ROS has been shown. Glutathione, which is an antioxidant tripeptide, possesses two free carboxyl and one sulfhydryl groups. The sulfhydryl group plays a variety of cellular functions in optimization of the intracellular redox potential. After AuNPs is exposed to cells, the GSH level substantially decreases regardless of boost of intracellular ROS level. Balance between antioxidative forces and oxidative stress helps in maintaining cell homoeostasis (Lee and Lee 2018). AuNPs penetrates through cell membranes and exhibits apoptosis-like death indicating several hallmarks like caspase-like protein activation, severe DNA damage, and membrane depolarization. However, they induced imbalance of redox status

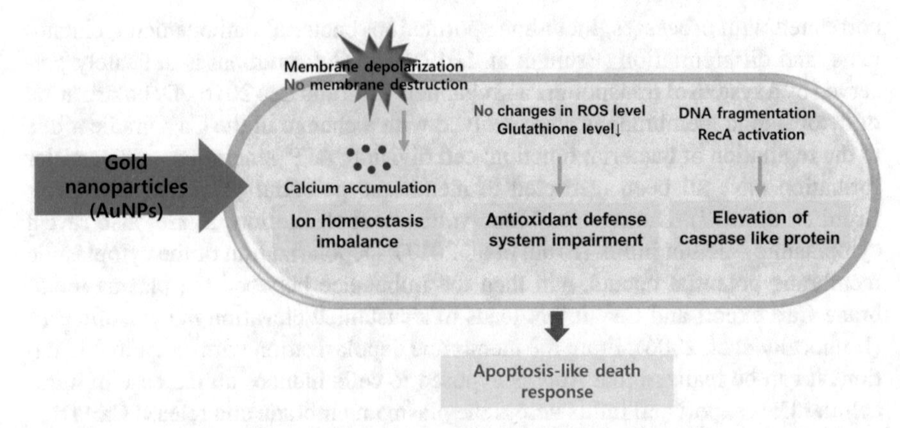

Fig. 8.2 AuNPs induces apoptotic-like cell death on *Escherichia coli*. AuNPs triggers a redox imbalance through changes in the glutathione level. AuNPs penetrates cell membrane and damages DNA and nucleic acid (Lee and Lee 2018)

without elevation of ROS level (Lee and Lee 2018). Furthermore, many modified AuNPs have been reported to enter the cytoplasm and penetrate the cell wall (Zhou et al. 2012) (Fig. 8.2).

8.4 Mode of Actions of Silver Nanoparticles (AgNPs) in Microbial Infection

Metal ions can be leaked from each nanoparticle. These ions can target other protein groups. Also, ions can react with intracellular components to produce soluble salts that can interfere with biological function (Niskanen et al. 2010). Silver ions can reduce chloride ions to precipitate silver chloride to inhibit DNA replication and inhibit cell respiration at relatively low concentrations (Niskanen et al. 2010). Antimicrobial activity of AgNPs is determined through many types of research and their mechanisms are discovered on pathogenic microorganisms. AgNPs has dual mechanisms such as membrane disturbance and programmed cell death (Rai et al. 2017b). Maintaining membrane integrity for cell survival is a significant function. It regulates intracellular responses and protects cells from the external harsh environment (Benyagoub et al. 1996; Portet et al. 2009). AgNPs interacts with the microbial membrane and changes membrane integrity. In case of fungal cells, the effect of AgNPs on cytoplasmic membranes tends to increase with increasing concentration. To maintain membrane potential, fungal cells form an ionic gradient across the membrane. The ion gradient is maintained by the limited membrane permeability for ions and small solutes (Yu 2003). Nonetheless, if the membrane loses their stability due to extreme membrane damage, various ions leak from the cell and membrane permeability enhances. Undesired ion leaks then lead to depolarize the microbial membrane potential (Bolintineanu et al. 2010). After the cells treated with

AgNPs, the membrane depolarization is induced via increasing membrane permeability. Under stress conditions such as drying, oxidation, dehydration, and toxicants, the cells undergo protein denaturation and inactivation. To stabilize or restore this condition, the fungal cells contain trehalose, a non-reducing disaccharide of trehalose, which acts as a compatible solute in the stabilization of the biological structure. AgNPs induces cellular stress on the membrane and releases some intracellular components such as trehalose and glucose (Alvarez-Peral et al. 2002; Garg et al. 2002). A cell cycle is an assemblage of highly ordered processes that consequence in duplication of cells. The cells process the cell cycle through checkpoints to distinguish the disordered situation. The location of arrest within the cell cycle depends on the phase at which the damage is recognized. Inhibition of cell division associates with cell cycle arrest and membrane damage can be caused by the disruption of membrane integrity. The DNA content represents the G1, S, and G2/M phase of the cell cycle. AgNPs in *C. albicans* provokes physiological damage leading to the G2/M phase cell cycle arrest (Rai et al. 2017b).

In contrast to mechanisms of AuNPs, AgNPs induces ROS accumulation in fungal cells and produces intracellular oxidative damage. Hydroxyl radicals, which are neutral in the form of hydroxyl ions, promote oxidative damage of the cells and have high reactivity (Haruna et al. 2002). When metal ions and hydrogen peroxide are present, hydroxyl radicals are generated through the Fenton reaction, and substances in the cytosol and nucleus such as proteins, lipids, and DNA are damaged, resulting in cell death (Rollet-Labelle et al. 1998). Significant formation of strong oxidant, hydroxyl radical, is caused by AgNPs in *C. albicans* cells. Hydroxyl radicals perform a significant role in AgNPs-induced death and activate caspases that promote apoptotic signals (Hwang et al. 2012). Cytochrome *c* released from mitochondria binds to protease that activates apoptotic signals in the cytosol (Pereira et al. 2007). AgNPs produces the cytochrome *c* release from the mitochondria into the cytoplasm and the formation of hydroxyl radicals affects release of cytochrome *c*. Mostly, the cytoplasmic membrane is asymmetrically composed of phosphate lipids, such as phosphatidylserine, phosphatidylethanolamine, phosphatidylcholine, and sphingomyelin (Rai et al. 2017b). Sphingomyelin and phosphatidylcholine are set on the outer leaflets of the plasma membrane whereas phosphatidylserine and phosphatidylethanolamine are intensified on the inner leaflets of the plasma membrane (Li et al. 2008). In the presence of hydroxyl radicals, phosphatidylserine externalization is confirmed in early stage of apoptotic cell death. Also, cause of DNA damage is cleavage of selected groups of substrates by caspases and oxidative stress by ROS formation (Rai et al. 2017b). In particular, DNA damage by oxidative stress induces apoptosis, and during apoptosis, these DNAs are broken into short fragments by endonucleases activated. It is regarded as a characteristic of apoptotic death (Wadskog et al. 2004). AgNPs promotes oxidative state and caspase activity of *C. albicans* cells, resulting in DNA fragmentation (Fig. 8.3).

Antimicrobial agents must approach effective concentrations that can be applied to the molecular target. Their effective operation should avoid pathogen defense mechanisms such as pump-mediated efflux and enzymatic modification in the process. The agents can act through various mechanisms to prevent the growth of bacteria or kill (Li et al. 2008).

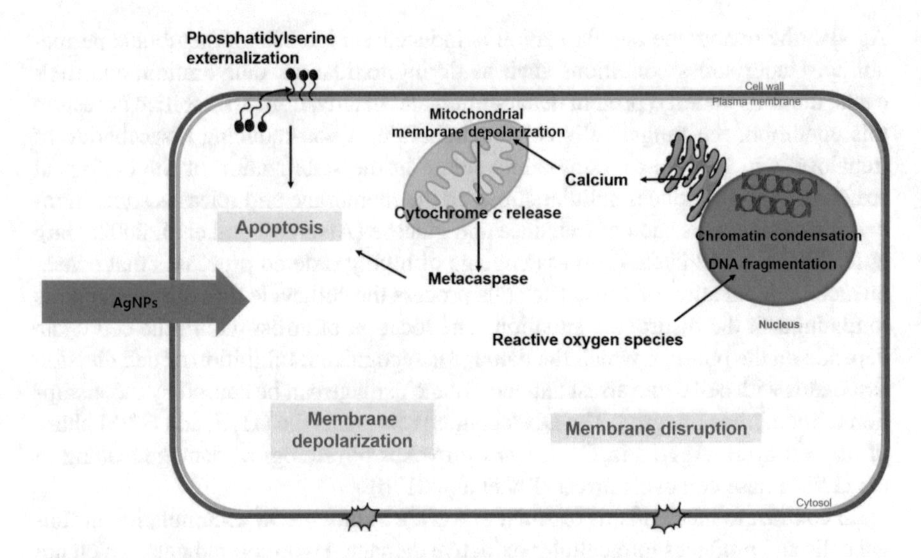

Fig. 8.3 Antifungal mechanisms of AgNPs. Cell cycle arrest is caused by AgNPs and membrane disruption leads to leakage of intracellular components. Furthermore, according to ROS generation, AgNPs promotes apoptosis through DNA damage and mitochondria dysfunction

Bacteria can be classified as gram-negative bacteria or gram-positive bacteria and their structural differences make the virulence (Durán et al. 2016). Due to the lack of peptidoglycan between the cytoplasmic membrane and the outer membrane, gram-negative bacteria are significant for defending bacteria from harmful substances such as degrading enzymes, drugs, detergents, and toxin and are composed of lipopolysaccharides (LPS) (Amro et al. 2000; Morones et al. 2005). Contrary to gram-negative bacteria, gram-positive bacteria have a peptidoglycan layer about 30 nm thick instead of outer membrane (Morones et al. 2005). AgNPs increases the membrane leakage of reducing sugar and enhances protein leakage through the membranes of *E. coli*. Additionally, AgNPs reduces the level of intracellular ATP, resulting in the death of bacterial cells (Mukherjee et al. 2014b). Thus, AgNPs permeabilizes the barrier of the outer membrane, peptidoglycan layer, periplasm, and the inner membrane, resulting in the release of cellular substances, such as reduction of proteins and sugars (Lee et al. 2014; Mukherjee et al. 2014b; Rai et al. 2017b). The membrane structure of *E. coli* cell-treated AgNPs was observed by using transmission electron microscopy (TEM) and scanning electron microscopy (SEM). In micrographs by SEM, AgNPs-treated cells show large leakage and others are misshapen and fragmentary. In the TEM micrograph, the surface of the untreated cells was intact and smooth, and some filaments around the cells were clear and distinct. On the other hand, the membrane of cells treated with AgNPs was severely damaged. Many gaps and pits appeared and the membrane was fragmentary. Moreover, membrane fragments could spontaneously form vesicles and are similar in size to spheroid that is between 1/5 and 1/20 of the size of *E. coli* cells (Li et al. 2010). In AgNPs-treated cells, the membrane vesicles were dispersed and dissolved. Membrane components are scattered and disorganized from their original ordered and close arrangement. This phenomenon suggests that AgNPs affects some phospholipids and proteins, leading to membrane collapse leading to cell disintegration and death (Rai et al. 2017b). Silver inhibits the function of proteins that interfere with the growth of *E. coli* by interacting with thiol groups of cysteines by inhibiting respiration and substituting hydrogen atoms (Holt and Bard 2005; Kim et al. 2008). Withal, AgNPs inactivates the respiratory chain dehydrogenase to inhibit cell respiration and growth (Rai et al. 2017b). The main factor of antimicrobial property is a positive charge to nanostructures. The positive charge figures out the changes of respiratory enzyme present in the bacterial membrane. Likewise, an ionic efflux pump located in the membrane is also affected and accumulates a lethal ion concentration inside the cell, leading to cell death (Allaker 2010). ROS, which results from oxidation of respiratory enzymes, facilitates DNA degradation and reacts with various cellular components (Xia et al. 2010).

A bactericidal agent can induce apoptosis-like death responses, which has similar hallmarks with eukaryotic apoptosis (Dwyer et al. 2012). AgNPs has bactericidal activity and the potential for stimulating an apoptosis-like death. Like *E. coli* cells treated with AgNPs, induction of an apoptosis-like death is proceeded by the production of ROS (Dwyer et al. 2012). ROS-induced AgNPs in bacteria is involved in apoptosis-like death response (Lee et al. 2014). Furthermore, AgNPs exerts change in membrane potential and calcium accumulation (Lee et al. 2014). Calcium ions are involved in several biological activities such as maintenance of motility, cell structure, and signaling pathways including transport and regulation of cellular processes and cell death (Yoon et al. 2012; Lee et al. 2014). AgNPs causes an imbalance in ion homeostasis, in particular, the accumulation of cytoplasmic calcium ions (Lee et al. 2014). Phosphatidylserine externalization is concerned by stimulation of calcium signaling of enzymes that dissociate membrane arrangement (Rai et al. 2017b). AgNPs induces the phosphatidylserine externalization and DNA fragmentation of *E. coli* cells, showing apoptosis-like death (Lee et al. 2014). During the repair of the damaged DNA, SulA binds to FtsZ and stops cell division, resulting in cell elongation (Lee et al. 2014). Phenomena of apoptosis-like death including cell filamentation and DNA fragmentation were observed in AgNPs-treated cells. Also, the halt of cell division progression relates prolonged cell size (Lee et al. 2014). When extreme DNA damage occurs in bacterial cells, the SOS response activates RecA protein to repair DNA damage. The RecA indicates a capacity of coprotease in the catalytic cleavage of LexA repressor that inhibits the expression of genes involving DNA damage repair (Maul and Sutton 2005). The relationship between caspase and RecA speculated that RecA may act as a major regulator in both bacterial apoptosis-like death response and SOS repair system (Dwyer et al. 2012). Cell death by AgNPs involves the DNA repair system and severe DNA damage and it is activated by the RecA (Fig. 8.4).

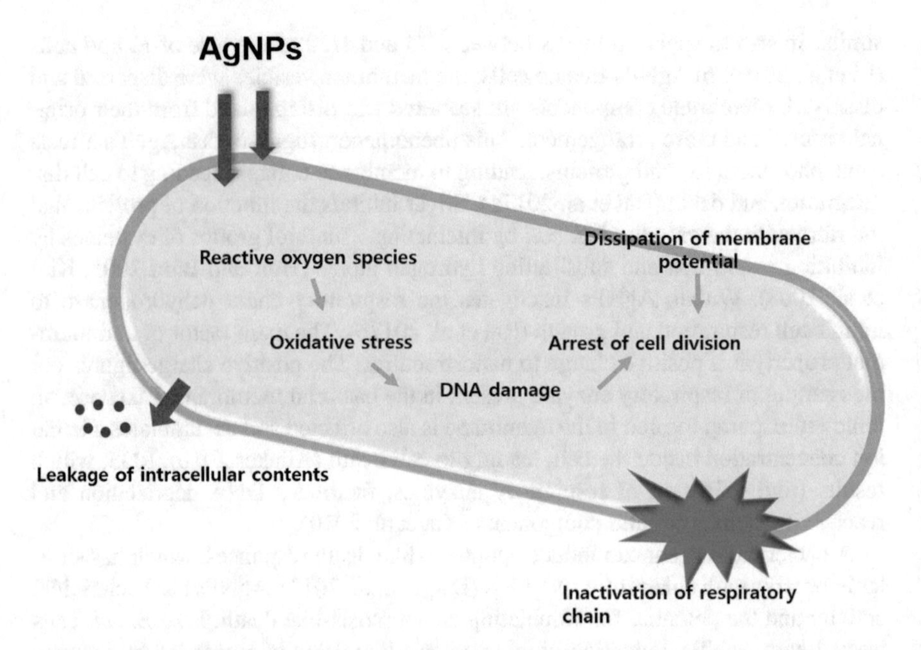

Fig. 8.4 Antibacterial mechanisms of AgNPs. According to oxidative stress, AgNPs exerts the antibacterial effect through membrane damage and apoptosis-like death process. Also, AgNPs shows inactivation of respiratory chain

8.5 Use of Antimicrobials for Treatment of Skin and Soft Tissue Infection

Nanotechnology is overcoming barriers to treating skin infections (Ikoba et al. 2015). Furthermore, applying nanotechnology to skin cancer has been devoted to designing new imaging and therapeutic approaches (DeLouise 2012). Nanomaterials are constructed to release the therapeutic drug for a period of time without releasing the damaged skin after drug release (Ikoba et al. 2015). Nanoparticles are used against a wide variety of microorganisms, including biofilm-forming bacteria and multidrug-resistant strains. Serving as drug delivery agents or conjugates counteract to increase the efficiency of existing antimicrobial therapy (Kulshreshtha et al. 2017). They avoid metastatic tumors and the common side effects of radiation therapy and chemotherapy, such as hair loss, nausea, infertility, and diarrhea. The nanoparticles can also be used as a component of an antimicrobial surface coating to avoid nosocomial infection and even as a vaccine component to prevent microbial infection (Kulshreshtha et al. 2017). The gold standard for tissue infection remains systemic antibiotics, with or without topical antiseptics (Englander and Friedman 2010). With reduced production of proinflammatory cytokines, smaller size and lower cytotoxicity promoted the use of AuNPs for drug delivery (Ikoba et al. 2015). The antimicrobial properties of AgNPs are beneficial to skin infection and can be fabricated into various protocols (Ikoba et al. 2015). Recently produced wound dressing inhibits microbial infections and demonstrates exceptional wound care. Silver nanocrystalline dressings decrease chronic wounds by bacterial infections (Ip et al. 2006; Leaper 2006).

8.6 Future Prospects and Concluding Remarks

The future of nanotechnology is expected to bring a lot of convergence technologies. Different research areas and technologies are closely linked. Different technologies can be fused together to develop more useful technologies. Nanotechnology will play a key role in all convergence technologies. Drug delivery may seem like a simple technique, but it requires a complex adaptation of a variety of clinically useful fusion techniques (Bermúdez et al. 2017). Nanoparticles offer unique properties when compared to micro- or macroparticles. Notable features include easy suspension in liquids, small size, deep access to cells and cell organs, high surface area, easy sterilization of particles, and various optical and magnetic properties (Bermúdez et al. 2017). The nonpolar properties of these biological agents can interfere with entry into cells and can be addressed by binding with appropriate nanoparticles (Kulshreshtha et al. 2017). Using nanoparticles with other antimicrobial effects can diminish the inherent toxicity of nanoparticles to mammalian cells, enhance the damage mechanism, increase bactericidal effects, and reduce resistance development (Lee and Lee 2018). In this chapter, we introduced AuNPs and AgNPs, which are representative nanoparticles. These nanoparticles have potential as antimicrobial agents and it would have a wide range of applications in numerous fields. Nanoparticles is going to play an important role in the development of next-generation therapeutics.

References

Akbar M, Essa MM, Daradkeh G, Abdelmegeed MA, Choi Y, Mahmood L, Song BJ. Mitochondrial dysfunction and cell death in neurodegenerative diseases through nitroxidative stress. Brain Res. 2016;1637:34–55.

Albanese A, Tang PS, Chan WC. The effect of nanoparticle size, shape, and surface chemistry on biological systems. Annu Rev Biomed Eng. 2012;14:1–16.

Alkilany AM, Lohse SE, Murphy CJ. The gold standard: gold nanoparticle libraries to understand the nano–bio interface. Acc Chem Res. 2012;46(3):650–61.

Allaker R. The use of nanoparticles to control oral biofilm formation. J Dent Res. 2010;89(11):1175–86.

Allocati N, Masulli M, Di Ilio C, De Laurenzi V. Die for the community: an overview of programmed cell death in bacteria. Cell Death Dis. 2015;6:e1609.

Alvarez-Peral FJ, Zaragoza O, Pedreno Y, Argüelles J-C. Protective role of trehalose during severe oxidative stress caused by hydrogen peroxide and the adaptive oxidative stress response in Candida albicans. Microbiology. 2002;148(8):2599–606.

Amro NA, Kotra LP, Wadu-Mesthrige K, Bulychev A, Mobashery S, Liu G-y. High-resolution atomic force microscopy studies of the Escherichia coli outer membrane: structural basis for permeability. Langmuir. 2000;16(6):2789–96.

Barbu EM, Shirazi F, McGrath DM, Albert N, Sidman RL, Pasqualini R, Arap W, Kontoyiannis DP. An antimicrobial peptidomimetic induces Mucorales cell death through mitochondria-mediated apoptosis. PLoS One. 2013;8(10):e76981.

Benyagoub M, Willemot C, Belanger RR. Influence of a subinhibitory dose of antifungal fatty acids from Sporothrix flocculosa on cellular lipid composition in fungi. Lipids. 1996;31(10):1077–82.

Bermúdez JM, Cid AG, Romero AI, Villegas M, Villegas NA, Palma SD. New trends in the antimicrobial agents delivery using nanoparticles. In: Antimicrobial nanoarchitectonics. Amsterdam: Elsevier; 2017. p. 1–28.

Beyth N, Houri-Haddad Y, Domb A, Khan W, Hazan R. Alternative antimicrobial approach: nano-antimicrobial materials. Evid Based Complement Alternat Med. 2015;2015:246012.

Bolintineanu D, Hazrati E, Davis HT, Lehrer RI, Kaznessis YN. Antimicrobial mechanism of pore-forming protegrin peptides: 100 pores to kill E. coli. Peptides. 2010;31(1):1–8.

Brown SD, Nativo P, Smith J-A, Stirling D, Edwards PR, Venugopal B, Flint DJ, Plumb JA, Graham D, Wheate NJ. Gold nanoparticles for the improved anticancer drug delivery of the active component of oxaliplatin. J Am Chem Soc. 2010;132(13):4678–84.

Bruni GN, Weekley RA, Dodd BJT, Kralj JM. Voltage-gated calcium flux mediates Escherichia coli mechanosensation. Proc Natl Acad Sci U S A. 2017;114(35):9445–50.

Carmona-Gutierrez D, Eisenberg T, Buttner S, Meisinger C, Kroemer G, Madeo F. Apoptosis in yeast: triggers, pathways, subroutines. Cell Death Differ. 2010;17(5):763–73.

Carraro M, Bernardi P. Calcium and reactive oxygen species in regulation of the mitochondrial permeability transition and of programmed cell death in yeast. Cell Calcium. 2016;60(2):102–7.

Cushnie TP, Cushnie B, Lamb AJ. Alkaloids: an overview of their antibacterial, antibiotic-enhancing and antivirulence activities. Int J Antimicrob Agents. 2014;44(5):377–86.

Debnath M, Prasad GB, Bisen PS. Molecular diagnostics: promises and possibilities. New York: Springer Science & Business Media; 2010.

DeLouise LA. Applications of nanotechnology in dermatology. J Invest Dermatol. 2012;132(3 Pt 2):964–75.

Di Lisa F, Menabo R, Canton M, Barile M, Bernardi P. Opening of the mitochondrial permeability transition pore causes depletion of mitochondrial and cytosolic NAD+ and is a causative event in the death of myocytes in postischemic reperfusion of the heart. J Biol Chem. 2001;276(4):2571–5.

Durán N, Durán M, de Jesus MB, Seabra AB, Fávaro WJ, Nakazato G. Silver nanoparticles: a new view on mechanistic aspects on antimicrobial activity. Nanomedicine. 2016;12(3):789–99.

Dwyer DJ, Camacho DM, Kohanski MA, Callura JM, Collins JJ. Antibiotic-induced bacterial cell death exhibits physiological and biochemical hallmarks of apoptosis. Mol Cell. 2012;46(5):561–72.

Englander L, Friedman A. Nitric oxide nanoparticle technology: a novel antimicrobial agent in the context of current treatment of skin and soft tissue infection. J Clin Aesthet Dermatol. 2010;3(6):45–50.

Espinosa-Diez C, Miguel V, Mennerich D, Kietzmann T, Sánchez-Pérez P, Cadenas S, Lamas S. Antioxidant responses and cellular adjustments to oxidative stress. Redox Biol. 2015;6:183–97.

Espitia PJP, Soares N d FF, dos Reis Coimbra JS, de Andrade NJ, Cruz RS, Medeiros EAA. Zinc oxide nanoparticles: synthesis, antimicrobial activity and food packaging applications. Food Bioprocess Technol. 2012;5(5):1447–64.

Garg AK, Kim J-K, Owens TG, Ranwala AP, Do Choi Y, Kochian LV, Wu RJ. Trehalose accumulation in rice plants confers high tolerance levels to different abiotic stresses. Proc Natl Acad Sci. 2002;99(25):15898–903.

Gelperina S, Kisich K, Iseman MD, Heifets L. The potential advantages of nanoparticle drug delivery systems in chemotherapy of tuberculosis. Am J Respir Crit Care Med. 2005;172(12):1487–90.

Gottlieb E, Armour S, Harris M, Thompson C. Mitochondrial membrane potential regulates matrix configuration and cytochrome c release during apoptosis. Cell Death Differ. 2003;10(6):709.
Hajnóczky G, Csordás G, Das S, Garcia-Perez C, Saotome M, Roy SS, Yi M. Mitochondrial calcium signalling and cell death: approaches for assessing the role of mitochondrial Ca^{2+} uptake in apoptosis. Cell Calcium. 2006;40(5):553–60.
Hajnoczky G, Csordas G, Das S, Garcia-Perez C, Saotome M, Sinha Roy S, Yi M. Mitochondrial calcium signalling and cell death: approaches for assessing the role of mitochondrial Ca^{2+} uptake in apoptosis. Cell Calcium. 2006;40(5–6):553–60.
Haruna S, Kuroi R, Kajiwara K, Hashimoto R, Matsugo S, Tokumaru S, Kojo S. Induction of apoptosis in HL-60 cells by photochemically generated hydroxyl radicals. Bioorg Med Chem Lett. 2002;12(4):675–6.
Henry CM, Hollville E, Martin SJ. Measuring apoptosis by microscopy and flow cytometry. Methods. 2013;61(2):90–7.
Holt KB, Bard AJ. Interaction of silver (I) ions with the respiratory chain of *Escherichia coli*: an electrochemical and scanning electrochemical microscopy study of the antimicrobial mechanism of micromolar Ag+. Biochemistry. 2005;44(39):13214–23.
Huang X, Jain PK, El-Sayed IH, El-Sayed MA. Plasmonic photothermal therapy (PPTT) using gold nanoparticles. Lasers Med Sci. 2008;23(3):217.
Huh AJ, Kwon YJ. "Nanoantibiotics": a new paradigm for treating infectious diseases using nanomaterials in the antibiotics resistant era. J Control Release. 2011;156(2):128–45.
Huttemann M, Pecina P, Rainbolt M, Sanderson TH, Kagan VE, Samavati L, Doan JW, Lee I. The multiple functions of cytochrome c and their regulation in life and death decisions of the mammalian cell: from respiration to apoptosis. Mitochondrion. 2011;11(3):369–81.
Hwang IS, Lee J, Hwang JH, Kim KJ, Lee DG. Silver nanoparticles induce apoptotic cell death in Candida albicans through the increase of hydroxyl radicals. FEBS J. 2012;279(7):1327–38.
Ikoba U, Peng H, Li H, Miller C, Yu C, Wang Q. Nanocarriers in therapy of infectious and inflammatory diseases. Nanoscale. 2015;7(10):4291–305.
Ip M, Lui SL, Poon VK, Lung I, Burd A. Antimicrobial activities of silver dressings: an in vitro comparison. J Med Microbiol. 2006;55(Pt 1):59–63.
Karlsson HL, Cronholm P, Gustafsson J, Moller L. Copper oxide nanoparticles are highly toxic: a comparison between metal oxide nanoparticles and carbon nanotubes. Chem Res Toxicol. 2008;21(9):1726–32.
Kim K-J, Sung WS, Moon S-K, Choi J-S, Kim JG, Lee DG. Antifungal effect of silver nanoparticles on dermatophytes. J Microbiol Biotechnol. 2008;18(8):1482–4.
Ko CH, Shen S-C, Hsu C-S, Chen Y-C. Mitochondrial-dependent, reactive oxygen species-independent apoptosis by myricetin: roles of protein kinase C, cytochrome c, and caspase cascade. Biochem Pharmacol. 2005;69(6):913–27.
Koksharova OA. Bacteria and phenoptosis. Biochemistry (Mosc). 2013;78(9):963–70.
Kujath P, Kujath C. Complicated skin, skin structure and soft tissue infections - are we threatened by multi-resistant pathogens? Eur J Med Res. 2010;15(12):544–53.
Kulshreshtha NM, Jadhav I, Dixit M, Sinha N, Shrivastava D, Bisen PS. Nanostructures as antimicrobial therapeutics. In: Antimicrobial nanoarchitectonics. Amsterdam: Elsevier; 2017. p. 29–59.
Lagadic-Gossmann D, Huc L, Lecureur V. Alterations of intracellular pH homeostasis in apoptosis: origins and roles. Cell Death Differ. 2004;11(9):953.
Leaper DJ. Silver dressings: their role in wound management. Int Wound J. 2006;3(4):282–94.
Lecoeur H, Prevost MC, Gougeon ML. Oncosis is associated with exposure of phosphatidylserine residues on the outside layer of the plasma membrane: a reconsideration of the specificity of the annexin V/propidium iodide assay. Cytometry. 2001;44(1):65–72.
Lee H, Lee DG. Gold nanoparticles induce a reactive oxygen species-independent apoptotic pathway in Escherichia coli. Colloids Surf B: Biointerfaces. 2018;167:1–7.
Lee W, Kim KJ, Lee DG. A novel mechanism for the antibacterial effect of silver nanoparticles on Escherichia coli. Biometals. 2014;27(6):1191–201.

Lemire JA, Harrison JJ, Turner RJ. Antimicrobial activity of metals: mechanisms, molecular targets and applications. Nat Rev Microbiol. 2013;11(6):371–84.

Li Q, Mahendra S, Lyon DY, Brunet L, Liga MV, Li D, Alvarez PJ. Antimicrobial nanomaterials for water disinfection and microbial control: potential applications and implications. Water Res. 2008;42(18):4591–602.

Li W-R, Xie X-B, Shi Q-S, Zeng H-Y, You-Sheng O-Y, Chen Y-B. Antibacterial activity and mechanism of silver nanoparticles on Escherichia coli. Appl Microbiol Biotechnol. 2010;85(4):1115–22.

Li ZY, Yang Y, Ming M, Liu B. Mitochondrial ROS generation for regulation of autophagic pathways in cancer. Biochem Biophys Res Commun. 2011;414(1):5–8.

Mann CL, Cidlowski JA. Glucocorticoids regulate plasma membrane potential during rat thymocyte apoptosis in vivo and in vitro. Endocrinology. 2001;142(1):421–9.

Maul RW, Sutton MD. Roles of the Escherichia coli RecA protein and the global SOS response in effecting DNA polymerase selection in vivo. J Bacteriol. 2005;187(22):7607–18.

Mittler R, Vanderauwera S, Suzuki N, Miller G, Tognetti VB, Vandepoele K, Gollery M, Shulaev V, Van Breusegem F. ROS signaling: the new wave? Trends Plant Sci. 2011;16(6):300–9.

Mohanraj V, Chen Y. Nanoparticles-a review. Trop J Pharm Res. 2006;5(1):561–73.

Morones JR, Elechiguerra JL, Camacho A, Holt K, Kouri JB, Ramírez JT, Yacaman MJ. The bactericidal effect of silver nanoparticles. Nanotechnology. 2005;16(10):2346.

Mühling M, Bradford A, Readman JW, Somerfield PJ, Handy RD. An investigation into the effects of silver nanoparticles on antibiotic resistance of naturally occurring bacteria in an estuarine sediment. Mar Environ Res. 2009;68(5):278–83.

Mukherjee A, Sikdar S, Bishayee K, Boujedaini N, Khuda-Bukhsh AR. Flavonol isolated from ethanolic leaf extract of Thuja occidentalis arrests the cell cycle at G2-M and induces ROS-independent apoptosis in A549 cells, targeting nuclear DNA. Cell Prolif. 2014a;47(1):56–71.

Mukherjee S, Chowdhury D, Kotcherlakota R, Patra S. Potential theranostics application of biosynthesized silver nanoparticles (4-in-1 system). Theranostics. 2014b;4(3):316.

Nagamalleswari E, Rao S, Vasu K, Nagaraja V. Restriction endonuclease triggered bacterial apoptosis as a mechanism for long time survival. Nucleic Acids Res. 2017;45(14):8423–34.

Negut I, Grumezescu V, Grumezescu AM. Treatment strategies for infected wounds. Molecules. 2018;23(9):2392.

Niskanen J, Shan J, Tenhu H, Jiang H, Kauppinen E, Barranco V, Picó F, Yliniemi K, Kontturi K. Synthesis of copolymer-stabilized silver nanoparticles for coating materials. Colloid Polym Sci. 2010;288(5):543–53.

Nordberg J, Arner ES. Reactive oxygen species, antioxidants, and the mammalian thioredoxin system. Free Radic Biol Med. 2001;31(11):1287–312.

Nugent SM, Mothersill CE, Seymour C, McClean B, Lyng FM, Murphy JE. Increased mitochondrial mass in cells with functionally compromised mitochondria after exposure to both direct gamma radiation and bystander factors. Radiat Res. 2007;168(1):134–42.

Oberdorster G, Oberdorster E, Oberdorster J. Nanotoxicology: an emerging discipline evolving from studies of ultrafine particles. Environ Health Perspect. 2005;113(7):823–39.

Panyam J, Labhasetwar V. Biodegradable nanoparticles for drug and gene delivery to cells and tissue. Adv Drug Deliv Rev. 2003;55(3):329–47.

Pereira C, Camougrand N, Manon S, Sousa MJ, Côrte-Real M. ADP/ATP carrier is required for mitochondrial outer membrane permeabilization and cytochrome c release in yeast apoptosis. Mol Microbiol. 2007;66(3):571–82.

Portet T, Camps i Febrer F, Escoffre JM, Favard C, Rols MP, Dean DS. Visualization of membrane loss during the shrinkage of giant vesicles under electropulsation. Biophys J. 2009;96(10):4109–21.

Rai M, Ingle A, Pandit R, Paralikar P, Gupta I, Anasane N, Dolenc-Voljč M. Nanotechnology for the treatment of fungal infections on human skin. In: The microbiology of skin, soft tissue, bone and joint infections. Amsterdam: Elsevier; 2017a. p. 169–84.

Rai M, Zacchino S, Derita M. Nano-Ag particles and pathogenic microorganisms: antimicrobial mechanism and its application. In: Essential oils and nanotechnology for treatment of microbial diseases. Boca Raton: CRC Press; 2017b. p. 187–200.

Rajchakit U, Sarojini V. Recent developments in antimicrobial-peptide-conjugated gold nanoparticles. Bioconjug Chem. 2017;28(11):2673–86.

Rizzuto R, De Stefani D, Raffaello A, Mammucari C. Mitochondria as sensors and regulators of calcium signalling. Nat Rev Mol Cell Biol. 2012;13(9):566–78.

Rollet-Labelle E, Grange M-J, Elbim C, Marquetty C, Gougerot-Pocidalo M-A, Pasquier C. Hydroxyl radical as a potential intracellular mediator of polymorphonuclear neutrophil apoptosis. Free Radic Biol Med. 1998;24(4):563–72.

Roos WP, Kaina B. DNA damage-induced cell death by apoptosis. Trends Mol Med. 2006;12(9):440–50.

Sahu N, Soni D, Chandrashekhar B, Sarangi BK, Satpute D, Pandey RA. Synthesis and characterization of silver nanoparticles using Cynodon dactylon leaves and assessment of their antibacterial activity. Bioprocess Biosyst Eng. 2013;36(7):999–1004.

Schieber M, Chandel NS. ROS function in redox signaling and oxidative stress. Curr Biol. 2014;24(10):R453–62.

Schrand AM, Rahman MF, Hussain SM, Schlager JJ, Smith DA, Syed AF. Metal-based nanoparticles and their toxicity assessment. Wiley Interdiscip Rev Nanomed Nanobiotechnol. 2010;2(5):544–68.

Seong M, Lee DG. Reactive oxygen species-independent apoptotic pathway by gold nanoparticles in Candida albicans. Microbiol Res. 2018;207:33–40.

Shrivastava S, Bera T, Roy A, Singh G, Ramachandrarao P, Dash D. Characterization of enhanced antibacterial effects of novel silver nanoparticles. Nanotechnology. 2007;18(22):225103.

Simon HU, Haj-Yehia A, Levi-Schaffer F. Role of reactive oxygen species (ROS) in apoptosis induction. Apoptosis. 2000;5(5):415–8.

Sondi I, Salopek-Sondi B. Silver nanoparticles as antimicrobial agent: a case study on E. coli as a model for Gram-negative bacteria. J Colloid Interface Sci. 2004;275(1):177–82.

Wadskog I, Maldener C, Proksch A, Madeo F, Adler L. Yeast lacking the SRO7/SOP1-encoded tumor suppressor homologue show increased susceptibility to apoptosis-like cell death on exposure to NaCl stress. Mol Biol Cell. 2004;15(3):1436–44.

Wani IA, Ahmad T. Size and shape dependant antifungal activity of gold nanoparticles: a case study of Candida. Colloids Surf B: Biointerfaces. 2013;101:162–70.

Xia XR, Monteiro-Riviere NA, Riviere JE. Intrinsic biological property of colloidal fullerene nanoparticles (nC60): lack of lethality after high dose exposure to human epidermal and bacterial cells. Toxicol Lett. 2010;197(2):128–34.

Yoon MJ, Kim EH, Kwon TK, Park SA, Choi KS. Simultaneous mitochondrial Ca^{2+} overload and proteasomal inhibition are responsible for the induction of paraptosis in malignant breast cancer cells. Cancer Lett. 2012;324(2):197–209.

Yu SP. Regulation and critical role of potassium homeostasis in apoptosis. Prog Neurobiol. 2003;70(4):363–86.

Yun DG, Lee DG. Antibacterial activity of curcumin via apoptosis-like response in Escherichia coli. Appl Microbiol Biotechnol. 2016;100(12):5505–14.

Zhang L, Pornpattananangkul D, Hu C-M, Huang C-M. Development of nanoparticles for antimicrobial drug delivery. Curr Med Chem. 2010;17(6):585–94.

Zhang Y, Chan HF, Leong KW. Advanced materials and processing for drug delivery: the past and the future. Adv Drug Deliv Rev. 2013;65(1):104–20.

Zhou Y, Kong Y, Kundu S, Cirillo JD, Liang H. Antibacterial activities of gold and silver nanoparticles against Escherichia coli and bacillus Calmette-Guerin. J Nanobiotechnol. 2012;10:19.

Chapter 9
Theranostic Potential of Aptamers in Antimicrobial Chemotherapy

Bushra Jamil, Nagina Atlas, Asma Qazi, and Bushra Uzair

Abstract Worldwide the infectious diseases are a major threat to the human population. However, the most important parameter is the prompt and sensitive diagnosis of pathogens. Therefore, tremendous efforts have been made for the development of rapid and portable detection techniques. The identification and treatment of infectious diseases at the nano and molecular levels is a hard task to achieve because of the scarcity of effective probes for characterization and recognition of biomarkers of these pathogens. Nonetheless, if it made possible simultaneous diagnosis and treatment at the specific spot, i.e., theranostics can be beneficial for treating the disease at a cellular level and can be helpful to understand the disease system. However, for theranostics a sensing system should be able to detect and measure biomarkers quickly. In this regard, aptamers (oligonucleotide polymers consist of single-stranded DNA (ssDNA) or RNA) have displayed the ability to be used as probes for the recognition of various targets at molecular level. DNA and RNA have the capacity of doing much lot than just keeping genetic information and therefore are also known as functional nucleic acids. Aptamer-based biosensors would be an attractive format because they can be developed for various molecules using the same sensing format. Therefore, the aptasensors utilizing aptamers for various bacterial infections have stimulating theranostic potential as well. In this chapter, the potential of aptamers as theranostic agent for bacterial infections has been discussed. The advantages and limitations of aptamer-based theranostics for the development of personalized medicine are also discussed.

Keywords SELEX · Aptamers · Microbial recognition elements · Advanced microbial detection

B. Jamil (✉)
Department of DMLS, University of Lahore, Islamabad, Pakistan

A. Qazi
Department of Biogenetics, National University of Medical Sciences, Rawalpindi, Pakistan

N. Atlas · B. Uzair
Department of Biological Sciences, International Islamic University, Islamabad, Pakistan

M. Rai, B. Jamil (eds.), *Nanotheranostics*,
https://doi.org/10.1007/978-3-030-29768-8_9

Nomenclature

ABSA	Aptamer-based sandwich assay
AuNPs	Gold nanoparticles
Bap	Biofilm-associated protein
ELASA	Enzyme-linked aptamer sorbent assay
ELISA	Enzyme-linked immunosorbent assay
FDA	Food and Drug Administration
LNA	Locked nucleic acid
mAbs	Monoclonal antibodies
MRE	Molecular recognition elements
PEG	Polyethylene glycol
SELEX	Systematic evolution of ligands by exponential enrichment
SEs	Staphylococcal enterotoxins
SiNPs	Silica nanoparticles
ssDNA	Single-stranded DNA
WB-SELEX	Whole-bacteria SELEX

9.1 Introduction

Infectious diseases pose major threat to the human health. Despite the availability of numerous anti-infective agents and disinfectants, control and prevention of infectious diseases is still a major problem in health care settings because of the complexities of pathogens. The emergences of resistant pathogens have made the situation more daunting. It is essential to expedite the exploration of anti-infective agents. Exploration of new molecular elements for the characterization of specific pathogens is the step forward for anti-infective agent discovery. Complete genome sequence of some pathogens and human as well as novel findings from genome, transcriptome, and proteome projects helped in identifying some new targets for battle against human infectious diseases (Pan et al. 2018).

Most therapeutics agents do not have the target specificity rather they act in the whole body. Therefore the therapeutic agents used today in chemotherapy, radiotherapy, and immunotherapy have numerous side effects. It is therefore required that scientists should discover site-specific agents for treatment. This will help in the reduction of dosage of the therapeutic agent and will have maximum therapeutic ability with minimum side effects (Langer 1998). Site-specific therapy would also help in mitigating the emergence of resistance. It is important in all areas of medicine, especially in areas where microorganisms are the causative agents, e.g., diseases due to microbes, water and food quality, monitoring of environment, drugs control quality, research, and many other aspects.

Particularly for bacterial diseases, fast and accurate diagnosis and treatment is needed for the best possible patient management. For this purpose, initially antibodies were envisioned as the excellent molecule (Syed 2014; Zimbres et al. 2013).

Antibodies are providing a source of innate or acquired immunity. They have the ability to capture antigens present on the surface of pathogens and subject them to lysis. Success of vaccination in preventive therapy was possible just because of antibodies. One promising alternative to antibiotics is pathogen-specific monoclonal antibodies (mAbs). The basic concept of antibodies originates from serum therapy. Serum therapy (or passive therapy) was very popular before the invention of antibiotics. Antibiotics have ruled the medical field for more than 50 years and it is now the emergence of multidrug-resistant pathogens that have created the havoc for searching and developing new agents for addressing this problem (Saylor et al. 2009).

DNA and RNA are the naturally occurring polymers that not only encode and transmit genetic information but have enormous other. They can be used as affinity probes for identifying DNA, RNA, lipids, sugars, and whole cells as well.

9.2 Aptamers

Aptamers were first discovered in 1990 and can be defined on the basis of their characteristics:

- Aptamers are short nucleic acid sequence of 30–70 nucleotides in length.
- Have unique three-dimensional structures because nucleic acid has the capacity to fold into secondary, tertiary, and quaternary structures.
- Can identify and bind to a variety of targets because of their unique 3D structure.
- Have specificity and affinity.

The aptamer word is a combination of Latin and Greek words.

- "aptus" is a Latin word meaning to fit into.
- "meros" is Greek word meaning region (Ku et al. 2015; Syed and Pervaiz 2010).

Before binding at a specific target oligonucleotide aptamers adapt three-dimensional structures (having stems, loops, bulges, hairpins, pseudoknots, triplexes, or quadruplexes) and then they tend to non-covalently attach to its specific biological targets with high accuracy (Fig. 9.1).

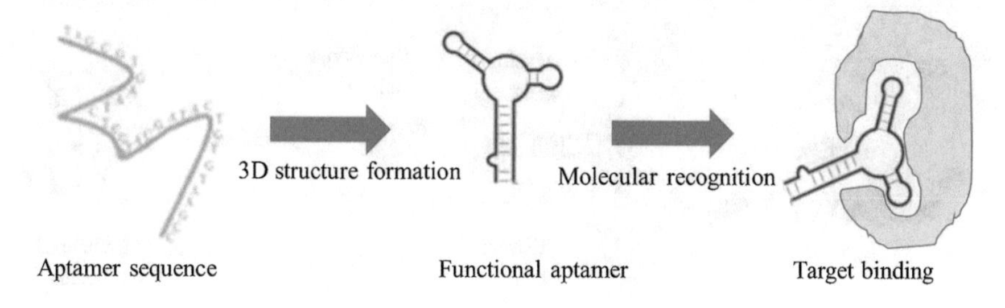

Fig. 9.1 Aptamers target binding mechanism

Aptamers are functionally utilized as agonist, antagonists, or direct targeting ligands. They are also given the name of chemical antibodies. Aptamers have evolved as a most popular type of biomolecules because of their promising potential as a highly target specificity and promising results in diagnosis, therapeutics, biosensing, and targeted drug delivery (Blind and Blank 2015; Santosh and Yadava, 2014; Song et al. 2012). Aptamers are actually single-stranded DNA molecular recognition elements (MRE).

Aptamers' most satisfactory application is the advancements in diagnostic assays for the recognition of bacterial agents or their exotoxins (Cheon et al. 2019; Das et al. 2019; Torres-chavolla and Alocilja 2009; Zimbres et al. 2013). In 2005, the United Sates Food and Drug Administration (FDA) approved the first aptamer known as anti-VEGF. It is an RNA aptamer (Macugen) to be used as therapeutic agent. Until now many new aptamer-based recognition of microbes and therapeutics have been in the progress of clinical evaluation (Sundaram et al. 2013).

9.3 Antibodies Applications in Therapeutics

Antibodies play a major role in our immune system. These are the source of natural defense against pathogens. Initially antibodies were considered as the ultimate candidate because of their specificity toward antigen on the surface of bacteria. They have accurate specificity to identify bacteria and other foreign substances and can distinguish the self from non-self-entities.

Human body has multiple types of antibodies like IgG, IgM, IgA, IgD, and IgE. Each type of antibody has Fab (fragment antigen binding) portion and it varies for each antigen. On the basis of diversity in Fab regions there are thousands of varieties of antibodies. Another type of antibody, known as monoclonal antibodies (mAbs), were discovered in 1975 by Georges Kohler and Cesar Milstein. mAbs are induced by clones obtained from a single parent cell (Fig. 9.2). Recent advances in molecular biology, technology, and antibody engineering make it possible to generate defined, homogenous, fully human and/or humanized mAbs with a single antigen-specificity to target a pathogen of interest (Babb and Pirofski 2017). After the discovery of mAbs, Ronald Levy in 1976 found that a monoclonal antibody can

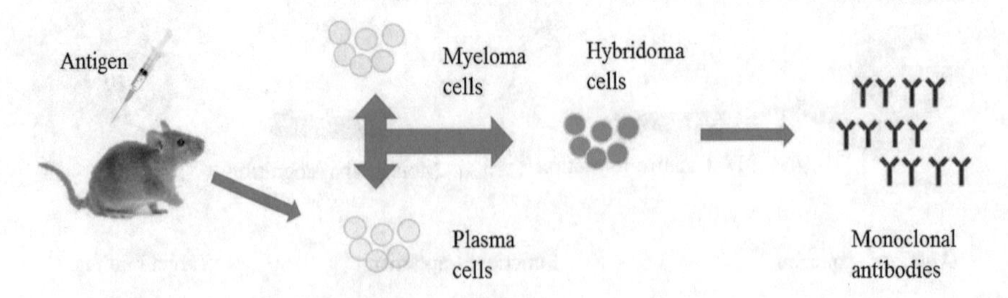

Fig. 9.2 Production process of monoclonal antibodies

particularly distinguish cancer cells (Levy and Miller 1983). Four years later, Lee Nadler found that these antibodies were able to recognize non-Hodgkin lymphoma cell. After these discoveries it was assumed that future for antibody therapeutics is promising and presumptions were high but unfortunately the results of monoclonal antibody-based therapeutics were not satisfactory because of certain other factor, including easy and economical current availability of antimicrobial drugs, small markets, high costs, and microbial antigenic variation.

There were several problems that limit the approval of antibody therapeutics. In 1992, the first monoclonal antibody got approval for therapeutic purpose. Several different researches have been conducted on mAbs in order to maximize their properties and efficiency. FDA approved ten antibodies in the year 2015 and 2016, and until now 68 monoclonal antibodies have been approved. For almost quarter a century the monoclonal antibodies have showed promising results in research, diagnosis as well as in therapeutics (Peruski and Peruski 2003).

In order to improve the performance of monoclonal antibodies huge progress has been done particularly in the therapeutics and now mAbs have also been showing direct bactericidal effect (Saylor et al. 2009). Monoclonal antibodies adopt indirect mechanisms of killing bacterial pathogens by the following methods:

- Fc-mediated functions, e.g., inflammatory response modulation
- Stimulating opsonic phagocytosis
- Increasing complement-mediated consequences

Despite their huge applications, mAbs still have certain limitations, e.g., for the synthesis of antibodies animal host is required, so they are prepared at the cost of animal lives (Bruno 2015). They require cold storage because they get altered at room temperature. One of the most important limitations is that antibodies can't be synthesized against some toxins and chemicals which are toxic to the host animal (Syed and Pervaiz 2010). Although antibodies were once considered as an ultimate option for many clinical applications, many problems regarding mAbs are not yet resolved and these "magic bullet" are still not an ideal choice for each disease.

9.4 Theranostic Potential of Aptamers

Recently, scientists have diverted their attention toward aptamers instead of using antibodies. Various different studies conducted globally have successfully utilized aptamers in the progression of various kinds of nano-diagnostic assays, biosensors as well as usual affinity assays, e.g., enzyme-linked immunosorbent assay (ELISA). The DNA or RNA probes having 3D structure can be functionalized with various chemical groups for covalent bonding with the transduced surface (Baptista 2014; Kim et al. 2014; Stoltenburg et al. 2007; Sun et al. 2016). Table 9.1 describes the differences between aptamers and antibodies.

Visible and prompt detection of specific pathogen in a complex biological environment is the biggest challenge in the field of microbial diagnosis. Microbes are

Table 9.1 Comparison of antibodies to aptamers (Adapted from Syed and Jamil 2018)

Advantages	Antibodies	Aptamers
Size	150–170 kDa	12–30 kDa. Aptamers are about ten times smaller than antibodies, giving them better access to tissues and cells
Selectivity		Aptamers are easily selected against small, non-immunogenic molecules and molecules that are toxic to antibody-producing cells
Development time	4–6 months	1–3 months
Labeling		Facile
Stability	Less stable	More stable
Immunogenicity	May cause	Non-immunogenic
Toxin recognition	No	Yes
Refrigeration	Required	Not required
Reproducibility	Batch to batch variability	Reproducible, less to no variation among batches.
Cost	Costly	Cost effective
Mass production	Produced in animals	Mass produced chemically or enzymatically
Shelf life	Limited	Prolonged
Production process	Immunization of animal or tissue culture. Purification required	Chemically produced. Do not require immunization, no tissue culture. No purification from serum
Target molecule	>600 Da	<60 Da (Target small molecules)

microscopic and are extremely diverse in nature. Strains of the same bacterial species differ from each other on the basis of very small molecular patterns. These differences are important to classify them into various strains or biotypes or for the segregation of virulent strains from those avirulent strains. Therefore their identification is not facile. Aptamers have received considerable attention in this regard because of their unique three-dimensional structures and because of this they can differentiate between closely related molecules with similar structures as well. Many groups have developed aptamers that can differentiate sensitive and resistant bacteria, and virulent and avirulent bacteria. Likewise, many probes have been developed that can differentiate different species and strains of bacteria. An important advantage of aptamers is that they can equally identify bacterial toxins. An aptasensor was developed by Sharma et al. for the detection of aflatoxins. In this model the structure-switching signaling aptasensing platform transduces signal from aptamer target recognition into a measureable signal (Fig. 9.3).

There are two groups of aptamers that are divided on the basis of its target. One group of aptamer targets bacterial antigen on the surface of the cell or virulence factors of the bacteria while other group targets the whole cells with identified or unidentified molecular targets. Until now various studies on antimicrobial aptamers have been mostly done on *Staphylococcus aureus*, *Mycobacterium tuberculosis*,

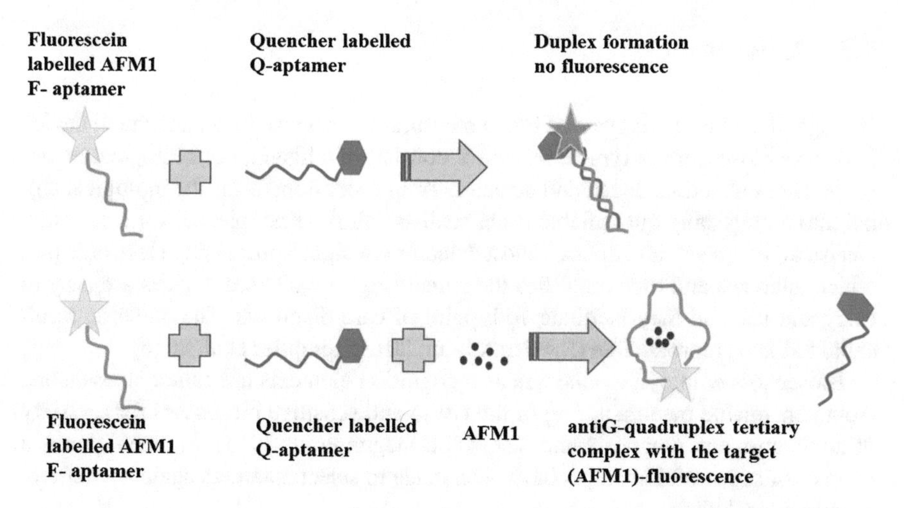

Fig. 9.3 Aptamers detecting bacterial toxins (Adapted with permission from Syed and Jamil 2018)

Salmonella typhi, and *Listeria monocytogenes* (Baig et al. 2015; Chang et al. 2013; Chen et al, 2007; Pan et al. 2005, 2014, 2018; Qin et al. 2009; Vivekananda et al. 2014; Wang et al. 2016; Yeom et al. 2016; Zelada-Guillén et al. 2009, 2012).

Staphylococcus aureus, particularly the MRSA, is a common pathogen responsible for many foodborne, community, and hospital-acquired diseases (Cao et al. 2009; Hong et al. 2015). Many researchers have developed aptamers for targeting *Staphylococcus aureus* directly or for targeting staphylococcal enterotoxins (SEs) that may cause food poisoning (Cao et al. 2009; Chang et al. 2013; DeGrasse 2012; Huang et al. 2015). Zhao et al. (2017) developed a sensitive theranostic probe from silica nanoparticles (SiNPs). SiNPs were coated with vancomycin-modified polyelectrolyte-cypate complexes (SiO_2-Cy-Van). In this complex cypate is a dye that gives signals in NIR region (700–900 nm). MRSA can attract vancomycin-modified polyelectrolyte-cypate complexes from SiNPs. The MRSA-associated complex can be further detected by near-infrared fluorescence.

Likewise, Dai et al. (2013) developed the method for detection of multidrug-resistant *Salmonella*. They designed "multifunctional core shell nanoplatforms." It consists of magnetic core–plasmonic shell nanoparticle, a methylene blue-bound aptamer, and an MDRB Salmonella DT94-specific antibody. By this technique magnetic separation and fluorescence imaging can be done simultaneously.

9.5 Aptasensor

Biological signals are converted into a measurable response by an analytical device known as biosensors. A typical biosensor consists of a ligand, i.e., biological recognition element, a transducer, and detector. Transducer transforms the biological signal into a physically quantifiable event such as calorimetric, piezoelectric acoustic, electrical, magnetic, and optical, and a detector is a signal-processing electronic part which analyzes and then amplifies the signal (Fig. 9.4). These devices are easy to carry and use and they facilitate with point-of-care diagnosis. The outcome result should also be reproducible (Syed and Jamil 2018; Templier et al. 2016).

Biosensors employing aptamers as recognition elements are called aptasensors. Aptamers enable manufacturing of the cheap and sensitive biosensors for a variety of applications in diagnosis and research (Duzgun et al. 2013). In recent years, a number of successful attempts have been made to select aptamers against a number of targets, including

- Proteins
- Cellular components
- Microbial antigens
- Whole cells (Niu et al. 2014; Qureshi et al. 2015)

Furthermore, aptamers may be modified chemically for surface immobilization and to undergo analyte-dependent conformational changes (Wu et al. 2012). Many groups have reported their use in different types of biosensors as biological recognition element (Hong et al. 2012). Aptamer-conjugated gold nanoparticles (AuNPs) have been developed for the colorimetric identification of microbial cells or their products.

SELEX (systematic evolution of ligands by exponential enrichment) is a highly sensitive method of selecting DNA sequences from large oligonucleotide libraries containing up to 9^{16} sequences that have greater specificity for their targets (Fig. 9.5).

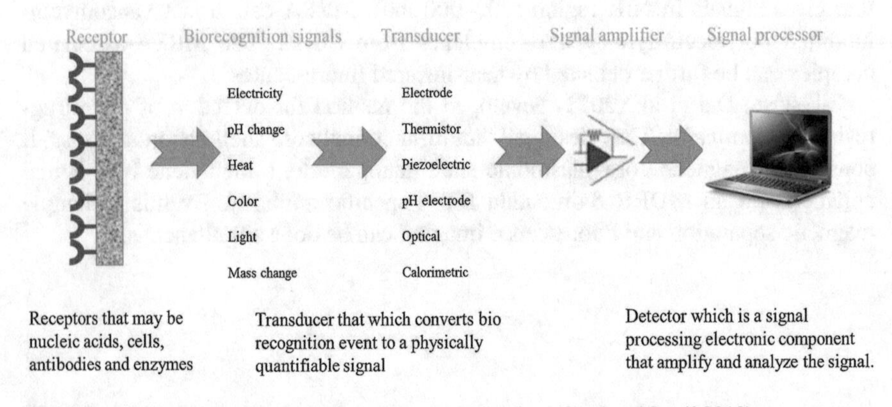

Fig. 9.4 A typical biosensor (Adapted with permission from Syed and Jamil 2018)

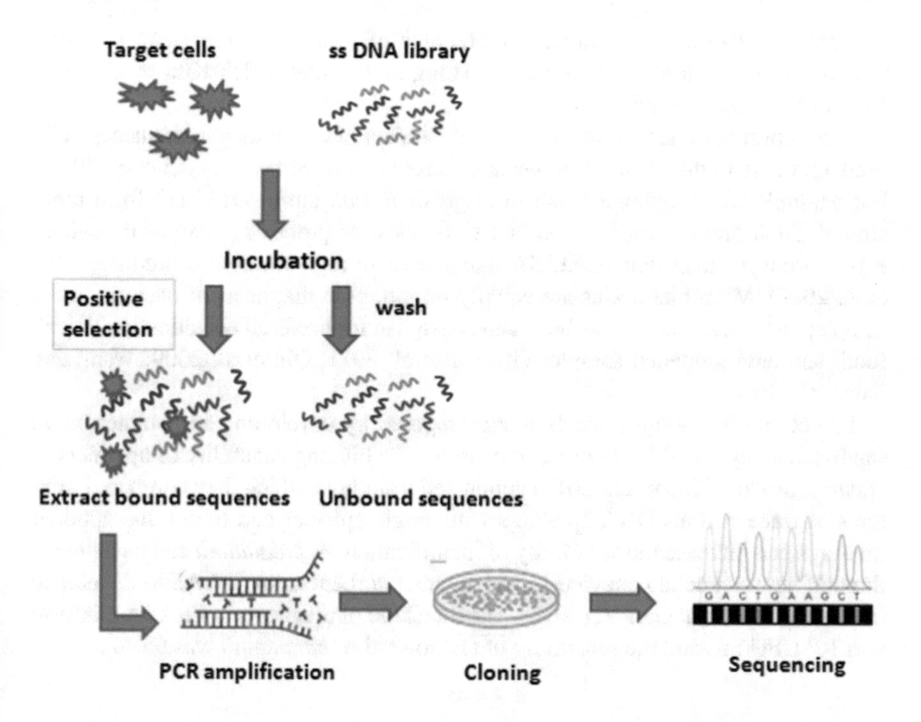

Fig. 9.5 Five steps of SELEX (Adapted with permission from Syed and Jamil 2018)

Thus, SELEX have following four steps:

1. Incubation of nucleic acid library (ssDNA) of about 10^{16} sequences with the target cells
2. Removal of free nucleic acids
3. Extraction of bounded nucleic acids from the target cells
4. Amplification of the desired collected nucleic acids through PCR (Hjalmarsson et al. 2004; Szeto et al. 2013).

SELEX can be used for the selection of aptamers that binds to different targets; these target peptides, proteins, toxins, and antigens on the cells that are unable to generate immune responses in animals host for the production of antibodies. Aptamers mostly identify small target molecules with unique 3D (three-dimensional) structures and some aptamers can even distinguish slight changes in its specific target. By combining this property with negative selection against other molecules or cells, we can select specific aptamers that can distinguish slight changes in the target. Different versions of SELEX protocols are available for DNA and RNA aptamers, such as Classical SELEX, Counter SELEX, Negative SELEX, Cell-specific SELEX, Microfluidic SELEX and MonoLEX SELEX (Germer et al. 2013; Syed and Jamil 2018).

Aptamers can even distinguish molecules of minor differences in structure ranging from picomole to nanomoles (Hong and Sooter 2015; Kim et al. 2013; Marton et al. 2016; Syed 2014).

Different researchers conducted various studies in which they have successfully used aptamers to differentiate between different strains of bacteria (Li et al. 2008). For example, RNA aptamers can distinguish *E. coli* strain O157:H7 from other strains. Such recognition is important to be used as probes in analytical devices, e.g., biosensors, later flow or ELISA-like assays for high affinity toward targets (Li et al. 2008). Microbial toxins are equally important in diagnosis of bacteria in the sample; therefore, aptasensors have been designed for bacterial detection in clinical, food, and environmental samples (Kumar et al. 2011; Qin et al. 2009; Wang and Salazar 2016).

Lee et al. (2015a) detected *L. monocytogenes* by developing an aptamer-based sandwich assay (ABSA). In order to examine the binding capability of aptamers in aptamer mixture Kim et al. (2014) conducted a study in which they compared mixtures of three various DNA aptamers with single aptamer and found that aptamer mixtures can increase the sensitivity of identification. *A. baumannii* can be detected through new test called enzyme-linked aptamer sorbent assay (ELASA) developed by Rasoulinejad and Gargari (2016). The threshold of diagnosis in ELASA platform was 10^3 CFU/mL and the sensitivity of test toward *A. baumannii* was 95.47%.

9.6 Drawbacks of Aptamer-Based Theranostics Potential

Although aptamers showed promising results in detecting bacterial species and in therapeutic, but still there are some drawbacks of aptamers. Few of them are discussed below.

9.6.1 In Vivo Efficiency of Aptamers

Mostly the process of aptamers selection is in vitro by SELEX whereas the original structures in vivo are very complex and therefore may compromise the efficiency of aptamers. Aptamers can sometimes skip its target and bind to nontarget molecules. Thus, the aptamers may be unable to detect the exact bacterial species in complex environment, e.g., food matrices and biological fluids. Therefore, aptamers should be carefully designed to improve its affinity toward target and be able to successfully distinguish structurally similar molecules in low concentration of sample. Huge variation in the strains of bacteria and the complex structures of targets molecules is an additional challenge for the performance of aptamers (Pan et al. 2018).

9.6.2 Degradation by Nuclease

Cellular nucleases can digest unmodified aptamers, specially RNA aptamers, making their half-life very short. Cellular nucleases are present in the fluids of body or cells. Therefore, chemical changes to stabilize them against the action of nucleases are required. Following modifications may be helpful in this regard:

- The activity of exonuclease usually goes from 3′ → 5′. If the 3′-end is capping with some capping agent, it would block the action of exonucleases. 3′ capping could be done with inverted thymidine and biotin (Diener et al. 2009; Dougan et al. 2000; Fine et al. 2005; Ni et al. 2017). Molecular capping in this way would introduce another 5′ end and no 3′ end. Thus, it automatically increases the resistance against 3′ exonucleases. Whereas, capping at 5′ end protects the action of 5′ → 3′ exonucleases. Capping agents in this regard may be cholesterol, PEG, proteins, fatty acids, and polycations.
- Phosphate modification also aid in augmenting stability. Phosphorothioate/methylphosphonate replaces non-bridging phosphodiester oxygen with sulfur in the phosphate backbone (Gao et al. 2016).
- Likewise, base modifications have also been recommended by many scientists to render stability. For example alterations at 2′-sugar with a fluoro (F), *O*-methyl (OCH_3) or amino (NH_2) groups induces stability. 2′ modifications in duplexes give them more stability as compared to modification made in DNA and RNA.
- Locked nucleic acid (LNA) is the most effective stabilization technique. LNA is basically the intramolecular bridge formation between 2'-oxygen and 4'-carbon with a methylene group. LNAs displayed more thermodynamic stability and make aptamers more resistant toward the action of ribozymes and serum nucleases (Karlsen and Wengel 2012; Pan et al. 2018; Shigdar et al. 2013).

9.6.3 Extraction from Body

Aptamers have a very minute size (<5 nm) and mass (6–30 kDa). Due to their extremely small size they get excreted through kidneys. In order to extend the aptamer circulation in human serum they should be conjugated to compounds having higher molecular mass, e.g., cholesterol, polyethylene glycol (PEG) (Lee et al. 2015b), liposome, an antibody (Heo et al. 2016) and other nanomaterials (Liao et al. 2015). Conjugation of PEG not only increases half-life of aptamers in serum but also can increase drug solubility and stability. The first aptamer-based drug called Macugen was modified through conjugation of PEG (Swierczewska et al. 2015). The pharmacokinetic experiments performed on rhesus monkeys showed that this aptamer has a half-life of 9.3 h (Nimjee et al. 2017).

9.7 Conclusion

In the recent two decades tremendous research work has been done on utilization of aptamers that can target bacteria. Until now a lot of aptamers have been produced that specifically not only target virulence factors of bacteria but also target the whole bacterial cells. A lot of progress has also been made in the development of aptasensors. In the diverse field of aptasensors, DNA or RNA aptamers act as biological recognition elements, and an enzyme-linked oligonucleotide assay, fluorescence, colorimetry, electrochemistry, and mass sensitivity have gained importance as read-out formats. Researchers are still trying to optimize aptasensor usage in diagnosing and treating diseases. Both aptamers and antibodies can be potentially used in diagnosis and therapeutics; however, aptamers have smaller size, can be easily generated, have low immunogenicity, are more stable, and can be easily labeled and modified as compared to antibodies. Therapeutic aptamers are more accurate than antibodies in the treatment of antitoxin, antivenom and in treating MDR bacteria but their therapeutic effects and safety should be studied further. The broad-spectrum usage of aptamers in the medical field and its long-term effects need to be studied further (Lange et al. 2017). Researchers are trying to optimize the SELEX methods, bringing modifications in aptamer, combining aptamers and targeted delivery of drug. The aptamer technology continues to reveal its promising feature in the field of theranostics.

References

Babb R, Pirofski LA. Help is on the way: monoclonal antibody therapy for multi-drug resistant bacteria. Virulence. 2017;8:1055–8.

Baig IA, Moon JY, Lee SC, Ryoo SW, Yoon MY. Development of ssDNA aptamers as potent inhibitors of Mycobacterium tuberculosis acetohydroxyacid synthase. Biochem Biophys Acta. 2015;1854:1338–50.

Baptista PV. Nanodiagnostics: leaving the research lab to enter the clinics? Diagnosis. 2014;1:305–9.

Blind M, Blank M. Aptamer selection technology and recent advances. Mol Ther Nucleic Acids. 2015;4(1–7):e223.

Bruno JG. Predicting the uncertain future of aptamer-based diagnostics and therapeutics. Molecules. 2015;20:6866–87.

Cao X, Li S, Chen L, Ding H, Xu H, Huang Y, Li J, Liu N, Cao W, Zhu Y, Shen B. Combining use of a panel of ssDNA aptamers in the detection of Staphylococcus aureus. Nucleic Acids Res. 2009;37:4621–8.

Chang YC, Yang CY, Sun RL, Cheng YF, Kao WC, Yang PC. Rapid single cell detection of Staphylococcus aureus by aptamer-conjugated gold nanoparticles. Sci Rep. 2013;3:1863.

Chen F, Zhou J, Luo F, Mohammed AB, Zhang XL. Aptamer from whole-bacterium SELEX as new therapeutic reagent against virulent Mycobacterium tuberculosis. Biochem Biophys Res Commun. 2007;357:743–8.

Cheon HJ, Lee SM, Kim SR, Shin HY, Seo YH, Cho YK, Lee SP, Kim MI. Colorimetric detection of MPT64 antibody based on an aptamer adsorbed magnetic nanoparticles for diagnosis of tuberculosis. J Nanosci Nanotechnol. 2019;19(2):622–6.

Dai X, Fan Z, Lu Y, Ray PC. Multifunctional nanoplatforms for targeted multidrug-resistant-bacteria theranostic applications. ACS Appl Mater Interfaces. 2013;5(21):11348–54.
Das R, Dhiman A, Kapil A, Bansal V, Sharma TK. Aptamer-mediated colorimetric and electrochemical detection of Pseudomonas aeruginosa utilizing peroxidase-mimic activity of gold NanoZyme. Anal Bioanal Chem. 2019;411:1229–38.
DeGrasse JA. A single-stranded DNA aptamer that selectively binds to Staphylococcus aureus enterotoxin B. PLoS One. 2012;7(3):e33410.
Diener JL, Daniel Lagasse HA, Duerschmied D, Merhi Y, Tanguay JF, Hutabarat R, Gilbert J, Wag ner DD, Wchaub R. Inhibition of von Willebrand factor-mediated platelet activation and thrombosis by the anti-von Willebrand factor A1-domain aptamer ARC1779. J Thromb Haemost. 2009;7:1155–62.
Dougan H, Lyster DM, Vo CV, Stafford A, Weitz JI, Hobbs JB. Extending the lifetime of anticoagulant oligodeoxynucleotide aptamers in blood. Nucl Med Biol. 2000;27:289–97.
Duzgun A, Imran H, Levon K, Rius FX. Protein detection with potentiometric aptasensors: a comparative study between polyaniline and single-walled carbon nanotubes transducers. Sci World J. 2013;2:1–8.
Fine SL, Martin DF, Kirkpatrick P. Pegaptanib sodium. Nat Rev Drug Discov. 2005;4:1878–87.
Gao S, Zheng X, Jiao B, Wang L. Post-SELEX optimization of aptamers. Anal Bioanal Chem. 2016;408:4567–73.
Germer K, Leonard M, Zhang X. RNA aptamers and their therapeutic and diagnostic applications. Int J Biochem Mol Biol. 2013;4:27–40.
Heo K, Min SW, Sung HJ, Kim HG, Kim HJ, Kim YH, Choi BK, Han S, Chung S, Lee ES, Chung J, Kim IH. An aptamer-antibody complex (oligobody) as a novel delivery platform for targeted cancer therapies. J Control Release. 2016;229:1–9.
Hjalmarsson K, Marcellaro A, Norlander L. Aptamers: future tools for diagnostics and therapy. Umeå: Swedish Defence Research Agency; 2004.
Hong KL, Sooter LJ. Single-stranded DNA aptamers against pathogens and toxins: identification and biosensing applications. Biomed Res Int. 2015;2015:1–31.
Hong P, Li W, Li J. Applications of aptasensors in clinical diagnostics. Sensors. 2012;12:1181–93.
Hong KL, Battistella L, Salva AD, Williams RM, Sooter LJ. In vitro selection of single-stranded DNA molecular recognition elements against S. aureus alpha toxin and sensitive detection in human serum. Int J Mol Sci. 2015;16(2):2794–809.
Huang Y, Chen X, Duan N, Wu S, Wang Z, Wei X, Wang Y. Selection and characterization of DNA aptamers against Staphylococcus aureus enterotoxin C1. Food Chem. 2015;166:623–9.
Karlsen KK, Wengel J. Locked nucleic acid and aptamers. Nucleic Acid Ther. 2012;22:366–70.
Kim YS, Song MY, Jurng J, Kim BC. Isolation and characterization of DNA aptamers against Escherichia coli using a bacterial cell–systematic evolution of ligands by exponential enrichment approach. Anal Biochem. 2013;436(1):22–8.
Kim YS, Chung J, Song MY, Jurng J, Kim BC. Aptamer cocktails: enhancement of sensing signals compared to single use of aptamers for detection of bacteria. Biosens Bioelectron. 2014;54:195–8.
Ku TH, Zhang T, Luo H, Yen TM, Chen PW, Han V, Lo YH. Nucleic acid aptamers: an emerging tool for biotechnology and biomedical sensing. Sensors. 2015;15:16281–313.
Kumar VGS, Urs TA, Ranganath R. MPT 64 Antigen detection for rapid confirmation of *M. tuberculosis* isolates. BMC Res Notes. 2011;4:1–4.
Lange MJ, Nguyen PDM, Callaway MK, Johnson MC, Burke DH. RNA-protein interactions govern antiviral specificity and encapsidation of broad spectrum anti-HIV reverse transcriptase aptamers. Nucleic Acids Res. 2017;45:6087–97.
Langer R. Drug delivery and targeting. Nature. 1998;392:5–10.
Lee SH, Ahn JY, Lee KA, Um HJ, Sekhon SS, Park TS, Min J, Kim YH. Analytical bioconjugates, aptamers, enable specific quantitative detection of Listeria monocytogenes. Biosens Bioelectron. 2015a;68:272–80.

Lee CH, Lee SH, Kim JH, Noh YH, Noh GJ, Lee SW. Pharmacokinetics of a cholesterol-conjugated aptamer against the hepatitis C virus (HCV) NS5B protein. Mol Ther Nucleic Acids. 2015b;4:e254.
Levy R, Miller RA (1983) Biological and clinical implications of lymphocyte hybridomas: tumor therapy with monoclonal antibodies. Annu Rev Med. 34(1):107–16.
Li N, Wang Y, Pothukucky A, Syrett A, Husain N, Gopalakrisha S, Kosaraju P, Ellington AD. Aptamers that recognize drug-resistant HIV-1 reverse transcriptase. Nucleic Acids Res. 2008;36:6739–51.
Liao J, Liu B, Liu J, Zhang J, Chen K, Liu H. Cell-specific aptamers and their conjugation with nanomaterials for targeted drug delivery. Expert Opin Drug Deliv. 2015;12:493–506.
Marton S, Cleto F, Krieger MA, Cardoso J. Isolation of an aptamer that binds specifically to *E. coli*. PLoS One. 2016;11:e0153637.
Ni S, Yao H, Wang L, Lu J, Jiang F, Lu A, Zhang G. Chemical modifications of nucleic acid aptamers for therapeutic purposes. Int J Mol Sci. 2017;18(8):1683.
Nimjee SM, White RR, Becker RC, Sullenger BA. Aptamers as therapeutics. Annu Rev Pharmacol Toxicol. 2017;57:61–79.
Niu S, Lv Z, Liu J, Bai W, Yang S, Chen A. Colorimetric aptasensor using unmodified gold nanoparticles for homogeneous multiplex detection. PLoS One. 2014;9:1–6.
Pan Q, Zhang XL, Wu HY, He PW, Wang F, Zhang MS, Hu JM, Xia B, Wu J. Aptamers that preferentially bind type IVB pili and inhibit human monocytic-cell invasion by Salmonella enterica serovar typhi. Antimicrob Agents Chemother. 2005;49:4052–60.
Pan Q, Wang Q, Sun X, Xia X, Wu S, Luo F, Zhang XL, Wu S, Luo F, Zhang XL. Aptamer against mannose-capped lipoarabinomannan inhibits virulent Mycobacterium tuberculosis infection in mice and rhesus TE monkeys. Mol Ther. 2014;22:940–51.
Pan Q, Luo F, Liu M, Zhang XL. Oligonucleotide aptamers: promising and powerful diagnostic and therapeutic tools for infectious diseases. J Infect. 2018;77:83–98.
Peruski AH, Peruski LF. Immunological methods for detection and identification of infectious disease and biological warfare agents. Clin Diagn Lab Immunol. 2003;10:506–13.
Qin L, Zheng R, Ma Z, Feng Y, Liu Z, Yang H, Wang J, Jin R, Lu J, Ding Y, Hu Z. The selection and application of ssDNA aptamers against MPT64 protein in *Mycobacterium tuberculosis*. Clin Chem Lab Med. 2009;47:405–11.
Qureshi A, Gurbuz Y, Niazi JH. Label-free capacitance based aptasensor platform for the detection of HER2/ErbB2 cancer biomarker in serum. Sens Actuat B Chem. 2015;220:1145–51.
Rasoulinejad S, Gargari SLM. Aptamer-nanobody based ELASA for specific detection of Acinetobacter baumannii isolates. J Biotechnol. 2016;231:46–54.
Santosh B, Yadava PK. Nucleic acid aptamers: research tools in disease diagnostics and therapeutics. Biomed Res Int. 2014;2014:1–13.
Saylor C, Dadachova E, Casadevall A. Monoclonal antibody-based therapies for microbial diseases. Vaccine. 2009;27:38–46.
Shigdar S, Macdonald J, O'Connor M, Wang T, Xiang D, Al Shamaileh H, Qiao L, Wei M, Zhou SF, Zhu Y, Kong L, Bhattacharya S, Li C, Duan W. Aptamers as theranostic agents: modifications, serum stability and functionalisation. Sensors. 2013;13:13624–37.
Song KM, Lee S, Ban C. Aptamers and their biological applications. Sensors. 2012;12:612–31.
Stoltenburg R, Reinemann C, Strehlitz B. SELEX—a (r)evolutionary method to generate high-affinity nucleic acid ligands. Biomol Eng. 2007;24:381–403.
Sun X, Pan Q, Yuan C, Wang Q, Tang XL, Ding K, Zhou X, Zhang XL. A single ssDNA aptamer binding to ManLAM of BCG enhances immunoprotective effects against Tuberculosis. J Am Chem Soc. 2016;138:11680–9.
Sundaram P, Kurniawan H, Byrne ME, Wower J. Therapeutic RNA aptamers in clinical trials. Eur J Pharm Sci. 2013;48:259–71.
Swierczewska M, Lee KC, Lee S. What is the future of PEGylated therapies? Expert Opin Emerg Drugs. 2015;20:531–6.

Syed MA. Advances in nanodiagnostic techniques for microbial agents. Biosens Bioelectron. 2014;51:391–400.

Syed MA, Jamil B. Aptamers and aptasensors as novel approach for microbial detection and identification: an appraisal. Curr Drug Targets. 2018;19:1560–72.

Syed MA, Pervaiz S. Advances in Aptamers. Oligonucleotides. 2010;20:215–24.

Szeto K, Latulippe DR, Ozer A, Pagano JM, White BS, Shalloway D, Lis JT, Craighead HG. Rapid-SELEX for RNA Aptamers. PLoS One. 2013;8:1–11.

Templier V, Roux A, Roupioz Y, Livache T. Ligands for label-free detection of whole bacteria on biosensors: a review. TrAC Trends Anal Chem. 2016;79:71–9.

Torres-chavolla E, Alocilja EC. Aptasensors for detection of microbial and viral pathogens. Biosens Bioelectron. 2009;24:3175–82.

Vivekananda J, Salgado C, Millenbaugh NJ. DNA aptamers as a novel approach to neutralize Staphylococcus aureus alpha-toxin. Biochem Biophys Res Commun. 2014;444:433–8.

Wang Y, Salazar K. Culture-independent rapid detection methods for bacterial pathogens and toxins in food matrices. Compr Rev Food Sci Food Saf. 2016;15:183–205.

Wang K, Wu D, Chen Z, Zhang X, Yang X, Yang CJ, Lan X. Inhibition of the superantigenic activities of Staphylococcal enterotoxin A by an aptamer antagonist. Toxicon. 2016;119:21–7.

Wu W, Li M, Wang Y, Ouyang H, Wang L, Li C, Cao Y, Meng Q, Lu J. Aptasensors for rapid detection of *Escherichia coli* O157:H7 and *Salmonella typhimurium*. Nanoscale Res Lett. 2012;7:1–7.

Yeom JH, Lee B, Kim D, Lee JK, Kim S, Bae J, Park Y, Lee K. Gold nanoparticle-DNA aptamer conjugate-assisted delivery of antimicrobial peptide effectively eliminates intracellular Salmonella enterica serovar Typhimurium. Biomaterials. 2016;104:43–51.

Zelada-Guillén GA, Riu J, Duzgun A, Rius FX. Immediate detection of living bacteria at ultra-low concentrations using a carbon nanotube based potentiometric aptasensor. Angew Chem Int Ed Engl. 2009;48:7334–7.

Zelada-Guillén GA, Sebastián-Avila JL, Blondeau P, Riu J, Rius FX. Label-free detection of Staphylococcus aureus in skin using real-time potentiometric biosensors based on carbon nanotubes and aptamers. Biosens Bioelectron. 2012;31:226–32.

Zhao Z, Yan R, Yi X, Li J, Rao J, Guo Z, Yang Y, Li W, Li YQ, Chen C. Bacteria-activated theranostic nanoprobes against methicillin-resistant Staphylococcus aureus infection. ACS Nano. 2017;11(5):4428–38.

Zimbres FM, Tarnok A, Ulrich H, Wregner C. Aptamers: novel molecules as diagnostic markers in bacterial and viral infections? Biomed Res Int. 2013;2:1–7.

Chapter 10
Current and Future Aspects of Smart Nanotheranostic Agents in Cancer Therapeutics

Qurrat Ul Ain

Abstract Despite the wide range of knowledge and information about cancer and advances in its treatment, still it is among the leading cause of mortality. Scientists around the globe are working on developing the new strategies to combat this fatal disease, and fortunately significant advances have already been achieved. In this regard, nanomedicines can play a vital role by improving the bio-distribution and the target site delivery of chemotherapeutics. Along with therapeutic applications, nanomedicine formulations have been used for imaging purposes as well. Nanotheranostics is a relatively new but flourishing field, which combines the diagnosis and therapy for personalized treatment. Combining the nanomaterials of diverse origins, e.g., polymers, liposomes, micelles, and antibodies, scientists have successfully developed the smart nanoparticles for both diagnostics and therapeutics at the same time in vivo. In addition, theranostics can be conjugated with bioligands for targeted drug delivery to treat and monitor the treatment response at molecular level. Potential applications of nanotheranostic medicines are assessment of drug biodistribution, site-targeted drug delivery, and visualization of drug release at the delivery site. These applications help to optimize the strategies based on triggered drug release and the prediction of therapeutic responses. In the near future, nanotheranostics are the practical solution for cancer and other lethal diseases to cure or at least treat them in the early stage.

This chapter summarizes the smart nanoparticles, developed for the simultaneous imaging and therapy, approaches for their targeted delivery, current applications and the challenges in their development and future perspectives for cancer therapy.

Keywords Cancer · Nanotheranostics · Chemotherapy · Nanoparticles

---

Q. Ul Ain (✉)
Department of Molecular Medicine, National University of Medical Sciences, Rawalpindi, Pakistan

M. Rai, B. Jamil (eds.), *Nanotheranostics*,
https://doi.org/10.1007/978-3-030-29768-8_10

10.1 Introduction

Despite the intense research on cancer diagnosis and therapeutics, it remains a substantial global threat. Cancer is usually developed because of genetic damage in normal cells. The damage can be either congenital or caused by some external factors. Cancer develops when a threat of mutations take place within the genes that control the cell's growth and division. These mutations cause an overstimulation of cell division and an abnormal cell metabolism. Cancer is a composite of multiple neoplastic diseases characterized by uncontrolled cell growth capable of intruding any part of the body (Esteva et al. 2017). The normal body cells contain tumor suppressor genes as well, but in cancer, the expression of these genes get slow down and this lead to uncontrolled cell division. Normally it takes a long time and enough mutations to transform a cell into a cancerous cell.

At present, cancer incidence is the second foremost cause of death in the United States after cardiac diseases. However, it is predicted to exceed cardiac diseases in terms of mortality rate in upcoming years (Robinson et al. 2015; Siegel et al. 2015; WHO 2015). Breast cancer is the leading cause of deaths in women (0.52 million in 2012) and pulmonary cancer is the prime cause of deaths (1.1 million in 2012) in men (Siegel et al. 2017). About 13.59 million cancer deaths were reported in the European Union alone in 2016 (Malvezzi et al. 2018). It is expected that total number of cancer patients would cross 21 million by 2030 (Bhakta-Guha et al. 2015). Around 60% of new cancer cases are from Asia, Africa, and Central and South American countries and almost 70% of cancer deaths are reported from these states.

This chapter summarizes the smart nanoparticles, developed for the simultaneous imaging and therapy, approaches for their targeted delivery, current applications and the challenges in their development and future perspectives for cancer therapy.

10.2 Therapeutic Modality and Application Analysis of Oncology Drugs

Therapeutic modality analysis of global oncology drugs market has divided it into chemotherapy, targeted therapy, and immunotherapy. In 2017, chemotherapy occupied the major share. Nonetheless, it is expected that immunotherapy would assume the major role in upcoming years due to better efficiency and increased targeted therapy with lesser side effects. Applications analysis of global oncology drugs market by applications is divided broadly into breast cancer, prostate cancer, blood cancer, lung cancer, gastrointestinal cancer, and others. In 2017, blood cancer occupied the major share, and lung cancer applications were found to be the fastest growing segment due to the high incidence of lung cancer (World Health Organization 2017).

The overwhelming acceptance of the concept of personalized medicine and immunology has driven a shift in cancer treatment from chemotherapy in the past decade. From 2011 to 2017, about 84 new drugs have been approved globally. Especially in immuno-oncology PD-1 and PD-L1 inhibitors have seen a quick endorsement due to their noteworthy clinical profile and approval for various cancers. Additionally, there are around 630 unique molecules in the development pipeline including 278 biological therapies and 82 vaccines (World Health Organization 2017).

10.2.1 Current Cancer Diagnostics and Treatment Strategies

Conventional strategies that are commonly used to combat cancer include chemotherapy, immunotherapy, radiotherapy, targeted therapies, stem cell transplant therapy, and surgery (Lyman et al. 2015). Surgical interventions have always been a popular and successful mode of cancer treatment. However, post-surgery relapse often requires taking on adjuvant therapies, such as chemotherapy and radiotherapy (or combination of both) (Howell and Valle 2015). Numerous other techniques such as laparoscopic, thoracoscopic, endoscopic, laser surgeries and cryo-based surgeries have found success in managing different types of cancer (DeSantis et al. 2014). Chemotherapy is effective in controlling various types of cancer, including breast cancer, lymphoma, myeloma, sarcoma, lung cancer, leukemia, etc., by administration of drugs such as antitumor antibiotics (daunorubicin, epirubicin, mitomycin C), corticosteroids (prednisone, dexamethasone, methylprednisolon, etc.), topoisomerase inhibitors (topotecan, teniposide, etc.), alkylating agents (mechlorethamine, streptozocin, busulfan), antimetabolites (floxuridine, 5-fluorouracil, etc.), and many mitotic inhibitors (epothilones, taxanes). Various noticeable immunotherapeutic cancer drugs have been developed in the field of nanotheranostics. These drugs boost the immune system to fight the tumor cells (such as PD-1, CD244, CD160, CTLA-4, VISTA, and BTLA) (Baksh and Weber 2015). Anti-PD-1 antibodies like nivolumab, pembrolizumab, and pidilizumab are used to treat head and neck carcinoma, renal cell carcinoma, melanoma, and lymphoma (Moreno et al. 2015). Recent advancements in radiotherapy, such as image guided radiotherapy (IGRT), intensity modulated radiotherapy (IMRT), and four-dimensional conformal radiotherapy (4D CRT), boost their activities in the treatment of the progression of breast, prostate, neck and head cancers (Moreno et al. 2015). Recently, another development in cancer therapeutics is the use of the RNA interference (RNAi) technology. In RNAi technique cellular proteins that are responsible for neoplasticity are targeted to prevent malignancy. Different approaches are introduced in this field including micro RNAs (miRNA), small interfering RNA (siRNA), and short hairpin RNAs (shRNA). Micro RNAs (miRNAs) are also used for both prognosis and therapy in cancer (Bucci et al. 2005). For instance, antisense miRNAs antagomiRs are used to disrupt the RNA-induced silencing complex (RISC) (Bertoli et al. 2015) and miRNA sponges are used to silence miR-9 to inhibit metastasis (Conde et al. 2015). Small interfering RNA (siRNA) is used to treat myeloid leukemia by blocking mRNA translation (Ma et al. 2010). Short hairpin RNA (shRNA) is used to treat myeloma in inhibiting the translation of overexpressed proteins, which are responsible for drug resistance (Lee et al. 2015). Despite the significant progress in cancer therapeutics, the side effects caused by chemotherapy are not mitigating. It includes hair loss, diarrhea, constipation, osteoporosis, premature ovarian failure, infertility, and certain viral infections caused by herpesviridae (Landry et al. 2015; Gibson and Keefe 2006; Keidan et al. 1989; Elad et al. 2010; Lee et al. 2012a). Likewise, radiotherapy also has its adverse effects including nausea, gastric pain, heart diseases, gastrointestinal lesions, and other conditions (Brydøy et al. 2007; Carretero et al. 2007; Lee et al. 2012b). However, RNAi-based therapies are less problematic, yet their stability and functional efficiency are more compromised as compared to other conventional therapies.

For developing a targeted chemotherapy, effective diagnosis of tumors is required. Usually, a computed tomography (CT) scan, magnetic resonance imaging (MRI), ultrasound, nuclear scan, X-ray, and most essentially biopsy (Jagasia et al. 2015) are performed for this purpose. In recent years, different kinds of tumor markers are used for cancer diagnosis. Many kinds of tumor markers are available including carbohydrate antigen 19-9 (CA19-9) and carcinoembryonic antigen (CEA) for cholangiocarcinoma diagnosis (Ramage et al. 1995), uPA/PAI1 for breast cancer (Weigelt et al. 2005), b2-microglobulin (B2M) for chronic lymphocytic leukemia, multiple myeloma, ALK gene rearrangements in non-small cell lung carcinoma, human chorionic gonadotropin, (HCG), a-fetoprotein, OCAA/OCAA-1, and pregnancy-zone protein (PZP) for ovarian cancer (Kulke et al. 2011).

There are still numerous limitations to these diagnostic strategies. For example, the CT scans are quite unresponsive to smaller tumors (Whitlock et al. 2008), and the use of ultrasounds in early cancer detection is restricted due to low-contrast resolution (Moon et al. 2015). MRI also has some limitations, such as high costs, interval in quantitation, deficient delivery of contrast agents, and lack of intraoperative image acquisition (Koo et al. 2006). Though the molecular biomarkers have high potential in cancer diagnosis, yet they are still not reliable in determining the complications at the molecular level that drive tumor growth (Zhang et al. 2014).

The abovementioned treatments and diagnostic modalities have helped in improving (to some extent) the current dreadful scenario of cancer treatment. Latest clinical practices are generating a considerable shift in their functional paradigms—from conventional therapies to a personalized therapy model based on molecular-level diagnosis. Approaches that are more sophisticated are in need to be developed to selectively target tumor cells with more effective therapeutic efficacy and less toxicity. One such arena that has come to the forefront in this regard is the field of nanotheranostics.

10.3 Smart Nanotheranostics in Cancer Therapeutics

"Theranostics" refers to concurrent combination of diagnosis and therapy. The term is derived from thera(py) + (diag)nostics to combine the two fields for advanced applications (Funkhouser 2002). Nanotheranostic is the term that complies to develop new nanomedicine strategies using smart nanomaterials such as polymer conjugations, carbon nanotubes, dendrimers, liposomes, and micelles, and biodegradable polymers for sustained, controlled, and targeted co-delivery of diagnostic and therapeutic agents. The whole idea behind the development of nanotheranostics is to diagnose and treat the diseases as early as possible (Fig. 10.1)

Conclusively, nanotheranostics are considered as highly suitable systems for better understanding of various important aspects of the drug development and delivery systems. These nanotheranostic systems also have the potential to contribute toward "personalized medicine," with more effective and less toxic therapies for individual patients.

Smart nanotheranostic systems are composed of different types of smart nanomaterial agents such as smart magnetic nanotheranostic agents, smart gold nanotheranostic agents, smart graphene nanotheranostic agents, smart silica nanotheranostic agents, smart lipid and polymer nanotheranostic agents, and smart protein-based nanotheranostic agents (Fig. 10.2).

10.3.1 Smart Magnetic Nanotheranostic Agents

Magnetic NP-based smart nanotheranostics have their magnetic and biocompatible properties and it makes them ideal candidate for MRI. Superparamagnetic iron oxide nanoparticles (SPIONPs), commonly called magnetite and maghemite, are most frequently used iron oxide nanomaterial agents (Xie et al. 2011). The main limitation in the use of magnetic nanoparticles is their poor hydrophilicity and intracellular aggregation. To overcome this issue, hydrophilic polymers are usually added to the nanocrystal surface. Various polymers have been studied well for this purpose, including polyaniline, dendrimer, polyvinyl pyrrolidone, and dextran.

Poly (acryl amide) (PAA) has been used to co-encapsulate a lipophilic dye and taxol within hydrophobic pouches, resulting in a smart theranostic nanocarrier for both fluorescence and MR-based imaging of drug delivery in cancer patients. Furthermore, in a study foliate was conjugated onto PAA-iron oxide nanoparticles (IONPs) to directly target cancer cells with overexpressed foliate receptors. Later on, their work showed that incorporating polymer could decrease the water solubility problem of magnetic nanoparticles (Santra et al. 2009).

In another study, three polymers, namely *N*-isopropylacrylamide (NIPAAM), acrylic acid, and PEG methacrylate were used to surround the superparamagnetic IONPs to generate smart thermosensitive nanotheranostic agents.

Fig. 10.1 Schematic illustration of smart nanotheranostics for cancer treatment. Almost all smart nanotheranostic agents carry nano-size, a therapeutic agent, a diagnostic agent, targeting ligand, and an anti-fouling agent

N-Isopropylacrylamide (NIPAAM) worked as a temperature-sensitive polymer; acrylic acid enhanced the conjugation to iron surface and PEG methacrylate provided a stealth coating and increased the circulation time and enabled the reactive groups for folic acid coupling (Rastogi et al. 2011).

Amphiphilic polymers that have both hydrophilic and hydrophobic characteristics and are easier to use instead of using three polymers for individual functionalities are also available. For instance, pluronic F127, nonionic triblockcopolymer, is composed of a central hydrophobic chain of polyoxypropylene (poly [propylene oxide]) flanked

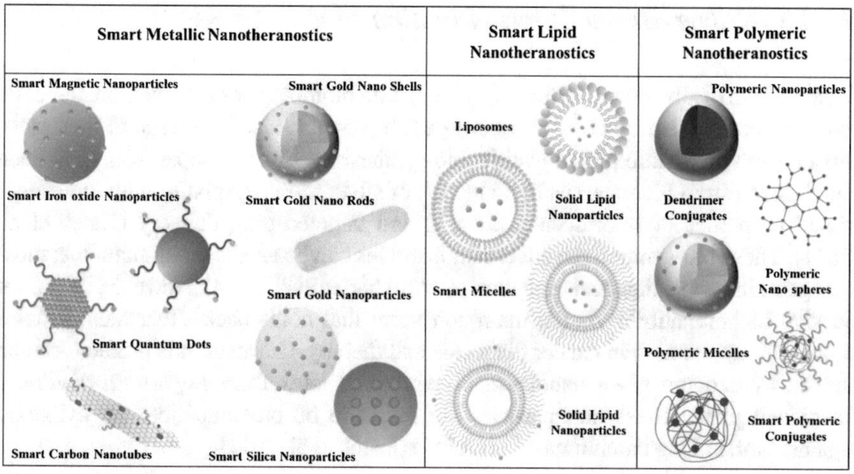

Fig. 10.2 Schematic illustration of chemical-based classification of nanocarriers used in smart nanotheranostic delivery systems

by two hydrophilic chains of polyoxyethylene (poly [ethylene oxide]). Along with β-cyclodextrin it was coated onto the magnetic NPs and hence yielded an efficient encapsulation with smaller particle size, higher drug-loading efficacy, lower protein binding, and superior uptake of particles in cancer cells (Yallapu et al. 2011).

10.3.2 Smart Gold Nanotheranostic Agents

Smart gold nanotheranostic agents have both diagnostic and treatment potential for cancer (Wang et al. 2012). The most attractive characteristic of gold nanomaterials (GNMs) is their tunable optical property that facilitates the localized surface plasmon resonance (LSPR). By modifying the morphology of gold nanomaterials, the LSPR of gold nanomaterials can be adjusted. Gold (Au) NPs, nanoshells, nanorods (AuNR), and nanocages display unique optical and thermal properties, which turns the gold nanomaterials to be used as potential theranostic agents (Choi et al. 2012). In a study, layered double hydroxide-gadolinium/gold (LDH-Gd/Au) nanocomposite has been developed for CT/MRI dual-modality imaging and anticancer therapy (Wang et al. 2013). These nanocomposites displayed high loading capacity for non-anionic anticancer drug DOX and the loaded DOX had pH-responsive drug release at acidic tumor microenvironment. Even though these properties of gold nanomaterials are quite fascinating to tune them as smart theranostics, they also have disadvantages, such as high production cost and stability in natural physiological conditions (Xie et al. 2010). More stable surface chemistry of gold nanomaterials is critically required for their clinical translation and applications.

10.3.3 Silver-Based Smart Nanotheranostic Agents

Conjugation of silver nanoparticles (AgNPs) with biomolecules and chemotherapeutic drug molecules via non-covalent or covalent bonds has raised the issue of their cytotoxicity in vitro. More biocompatible bio-synthesized silver nanoparticles have been synthesized from biomaterials by reduction of silver nitrate ($AgNO_3$) with multifunctional properties of fluorescence imaging and targeted drug delivery (Kateb et al. 2011). These bio-synthesized silver nanoparticles can be used as smart nanotheranostic agents in cancer diagnosis and therapy (Mukherjee et al. 2014). However, their poor in vivo biocompatibility is the main constraint that holds back silver nanoparticles from their application in cancer diagnosis and therapy. However, this problem can be solved by capping silver nanoparticles with stem latex from *Euphorbia nivulia*, a medicinal plant. These nanoparticles are found to be biocompatible yet cytotoxic against human lung carcinoma cells (A549) (Kateb et al. 2011).

10.3.4 Smart Graphene Nanotheranostic Agents

Graphene is a two-dimensional layer of sp2-bonded carbon that has tremendous applications in nanotechnology (Novoselov et al. 2005). In recent years, graphene oxide (GO)-based smart nanotheranostic agents have gained substantial attention due to its unique physical properties such as colloidal stability, large surface area, easy surface modification, as well as better electrical and mechanical properties that are not very common in other nanotheranostic materials (Draz et al. 2014). Graphene-based smart nanotheranostic agents can do image-guided removal of tumor by synergistic photothermal therapy (PTT). Graphene nanosheets can enhance apoptosis in CD44+ KB carcinoma cell lines by using NIR imaging (Miao et al. 2015). Dendrimer-grafted nano-graphene oxide conjugated with gadolinium (Gd-NGO) with positive charge on their surface can carry the negatively charged miRNA, thereby forming Gd-NGO/miRNA complexes for anti-neoplastic action. Gd-NGO can also effectively carry the chemotherapeutic agent, epirubicin (EPI). Simultaneous conjugation of both miRNA and epirubicin with Gd-NGO (GdNGO-miRNA-EPI) also worked synergistically for the inhibition of glioblastoma growth in comparison with the individual conjugates (Cao et al. 2017). GdNGO-miRNA-EPI were also quite effective as contrasting agent for tracking the site of drug delivery and quantitative analysis through MRI. Graphene oxide-iron oxide nanoparticles' (GO-IONPs) nanocomposites thus constitute well-reported agents for multimodal imaging, photothermal therapy, and drug delivery (Yang et al. 2012; Ma et al. 2012). Graphene oxide-gold nanoparticle's GO-AuNP composites also showed significant results in phototherapy (Zedan et al. 2012).

Furthermore, graphene oxide-gold-iron oxide nanoparticle's (GO-Au-IONP) assemblies enhanced the optical absorbance and superparamagnetic and photothermal therapeutic potential in NIR laser irradiation therapies observed by both

X-ray imaging and MRI (Shi et al. 2013). Phthalocyanine (Pc)-dendrimer-based low-oxygen graphene nanoparticles (Pc-LOG-NPs) have shown both functionalities of being photodynamic (Pc-dendrimer) and photothermal (LOG) in the treatment of ovarian cancer. The Pc-LOG nanoparticles can visualize the unresected cancer cell's margins and ablate the cancerous cells via PTT (Taratula et al. 2015).

10.3.5 Smart Silica Nanotheranostic Agents

Higher surface area, firm siloxane chemistry, and distinct tunable nanostructures allow effective fabrication of SiNPs according to the desired surface for theranostic applications (Wang et al. 2012). Silica nanoparticles are commonly classified as solid silica nanoparticles and mesoporous silica nanoparticles (MSNs). Sol-gel synthesis and microemulsion techniques have been employed to prepare silica-based nanoparticles for their theranostic applications (Vivero-Escoto and Huang 2011). Mesoporous silica-based smart nanotheranostics have been loaded or encapsulated with a wide range of imaging molecules, targeting agents (such as superparamagnetic iron oxide nanoparticles, and Gd complexes for MRI imaging), and chemotherapeutic drugs like Camptothecin, Doxorubicin, and Paclitaxel (He et al. 2012). Development of trifunctionalized MSNs for smart nanotheranostic application was synthesized by combining the imaging, targeting, and therapeutic agents as one single-particle platform, which showed exceptional targeting of human glioblastoma cells and negligible collateral damage with strong therapeutic effects (Chen et al. 2013).

10.3.6 Smart Lipid-Based Nanotheranostic Agents

Liposomes are one of the most extensively studied nanomaterials for cancer therapeutics. These liposomes are developed from lipids containing a lipophilic tail and hydrophilic head group that at once form spheres at their critical concentrations (CMC) (Kirschbaum and Baeumner 2015; Kumar et al. 2012). Liposomes are spherical vesicles with outer phospholipid bilayers encircling aqueous compartment (Cheng et al. 2010).

Currently smart nanotheranostic liposomes are developed by multimodal imaging agents like radioisotopes, fluorescent probes and magnetic nanoparticles or quantum dots (QDs). Applications of smart nanotheranostic liposomes in the cancer diagnosis have been reportedly done by utilizing positron emission tomography (PET) imaging, magnetic resonance imaging (MRI), near-infrared resonance (NIR) fluorescent imaging, and single-photon emission computed tomography (SPECT) (Sen and Mandal 2013). The imaging agents can be conjugated on the surface covalently, loaded within the hydrophobic core.

The therapeutic agents can be embedded in the lipophilic bilayer shell or encapsulated in the hydrophilic core and there is always the option of conjugating the molecular probe for targeting on the surface of liposomes. These multifunctional liposomes can have the high circulation time in the blood and can evade host defenses and can concurrently facilitate in vitro or in vivo imaging (Muthu et al. 2012). Gadolinium-based liposomes worked quite efficiently as smart nanotheranostics and delivered promising results (Li et al. 2012).

10.3.7 Smart Micelle-Based Smart Nanotheranostic Agents

Micelles are evolving as powerful and multifunctional platforms to be used as smart theranostic delivery systems in cancer treatment (Kumar et al. 2012). Micelles are nanosized spherical structures, composed of self-assembled amphiphilic block copolymers in the form of a core/shell structure in aqueous media. A multifunctional micellar (cRGD-DOX-SPIO micelles) was developed by conjugating a targeting ligand, doxorubicin (DOX), and an MRI-visible agent. Polyethyleneglycol-polylacticacid (PEG-PLA) was used to develop the core of micelle, and a cluster of superparamagnetic iron oxide nanoparticles (SPIO) were loaded into these cores where as cRGD ligand was used as targeting agent on the micelle surface. This multifunctional smart nanotheranostic micelles showed higher targeted uptake of drug in vitro in endothelial tumor cells (Sailor and Park 2012).

10.3.8 Smart Polymer and Dendrimer-Based Nanotheranostic Agents

Polymeric nanomaterials have well-studied physiochemical prosperities which make them biocompatible, versatile, and multifunctional for theranostic applications in cancer treatment. Many polymeric materials have been used in cancer treatment to enhance their circulation time, anticancer efficacy, increased stimuli-responsive drug release and targeted delivery. Currently, some of these polymer-based smart nanotheranostic agents are in different stages of clinical development.

A multifunctional polymer-based nanotheranostic platform has been developed in which the hydrophobic therapeutic agent (doxorubicin) was co-encapsulated along with hydrophobic superparamagnetic nanocrystals or hydrophobic quantum dots and a folate group was attached on the surface of the polymeric nanoparticles to target folate receptors on cancer cells (Sailor and Park 2012). Another example of polymeric nanotheranostics was presented by loading poly hydroxypropyl methacrylamide (HPMA) copolymers with Cu-64 (an intrinsic theranostic agent) and cRGD as targeting ligand to target tumor angiogenesis. These smart nanotheranostic particles showed much better accumulation of Poly(HPMA)-c(RGD)-64Cu in tumor sites after systemic injection (Yuan et al. 2013).

Like polymeric nanotheranostics, dendrimer-based nanotheranostics have also been developed due to their distinctive characteristics such as a single dendrimer can be used as a dais for both imaging and targeting agents to identify cancer cells (Wolinsky and Grinstaff 2008; Fernandez-Fernandez et al. 2011). These exclusive characteristic advantages have been employed in different studies. For example, poly (amido amine) (PAMAM) generation five dendrimer (with a diameter of around 5 nm and more than 100 functional primary amines) was conjugated on the surface of ethylenediamine core. These dendrimer-based smart nanotheranostics have the potential to be used for imaging, targeting, intracellular drug delivery, and covalent attachment to folic acid. This multifunctional smart nanotheranostics showed 100-fold higher cytotoxicity than free anticancer drug (Zhang et al. 2008).

10.4 Conclusion and Future Perspectives

Nanotheranostics represents the advancement of multidisciplinary nanoscience including chemistry, biology, material science, medical physics, electromagnetics, and oncology. Nanotheranostics have been developed so that diagnostics and therapeutics can work side by side. High impact development is conceivable in this field due to their multiple functionalities (Bardhan et al. 2011). Hopefully in near future nanotheranostics will enter in the clinical trials and soon will become the norm rather than the fiction (Janib et al. 2010). Further innovation of smart nanotheranostics by using multiple imaging agents is expected in the future with more trustworthy and reproducible procedures to attain better therapeutic efficacy and to be scaled up to production levels (Mitra et al. 2012).

Smart nanotheranostic platforms with their exclusive ability for simultaneous imaging and treatment are the ray of hope toward the development of effective cancer therapy. Nevertheless, a number of challenges confront the developers before the translation of these nanotheranostics to the clinical application. These limitations are quite difficult due to the lack of availability of physiologically relevant test-beds for designers. For the clinical translation of these smart nanotheranostics, biological aspects are much needed to be addressed and combined with their engineering. The biological issues are very complex and can only be replicated by an in vivo set up. Majority of smart nanotheranostics are evaluated in vitro and very few have been tested for their in vivo efficacy. This shows either the lack of access to the relevant animal models or healthy collaborations with biological experts. This concludes the inactive broadcasting of the facts to the nanomedicine community, which can influence future of nanotheranostics. The complexity of cancer requires more interdisciplinary collaboration and experts of different specialties such as chemists and bioengineers, biologists and biochemists, pharmacologists and pharmacists, experts on drug safety, statisticians, and physicians have to work together closely for clinical translation of smart nanotheranostics to achieve the solution for this fatal disease. At the same time, it is also vital that patients participate in clinical trials so that the findings could be translated into the medical standard of care.

References

Baksh K, Weber J. Immune checkpoint protein inhibition for cancer: preclinical justification for CTLA-4 and PD-1 blockade and new combinations. Semin Oncol. 2015;42(3):363–77.

Bardhan R, Lal S, Joshi A, Halas NJ. Theranostic nanoshells: from probe design to imaging and treatment of cancer. Acc Chem Res. 2011;44(10):936–46.

Bertoli G, Cava C, Castiglioni I. MicroRNAs: new biomarkers for diagnosis, prognosis, therapy prediction and therapeutic tools for breast cancer. Theranostics. 2015;5(10):1122.

Bhakta-Guha D, Saeed M, Greten H, Efferth T. Dis-organizing centrosomal clusters: specific cancer therapy for a generic spread? Curr Med Chem. 2015;22(6):685–94.

Brydøy M, Fosså SD, Dahl O, Bjøro T. Gonadal dysfunction and fertility problems in cancer survivors. Acta Oncol. 2007;46(4):480–9.

Bucci MK, Bevan A, Roach M. Advances in radiation therapy: conventional to 3D, to IMRT, to 4D, and beyond. CA Cancer J Clin. 2005;55(2):117–34.

Cao Y, Xu L, Kuang Y, Xiong D, Pei R. Gadolinium-based nanoscale MRI contrast agents for tumor imaging. J Mater Chem B. 2017;5(19):3431–61.

Carretero C, Munoz-Navas M, Betes M, Angos R, Subtil JC, Fernandez-Urien I, De la Riva S, Sola J, Bilbao JI, De Luis E. Gastroduodenal injury after radioembolization of hepatic tumors. Am J Gastroenterol. 2007;102(6):1216.

Chen N-T, Cheng S-H, Souris JS, Chen C-T, Mou C-Y, Lo L-W. Theranostic applications of mesoporous silica nanoparticles and their organic/inorganic hybrids. J Mater Chem B. 2013;1(25):3128–35.

Cheng S-H, Lee C-H, Chen M-C, Souris JS, Tseng F-G, Yang C-S, Mou C-Y, Chen C-T, Lo L-W. Tri-functionalization of mesoporous silica nanoparticles for comprehensive cancer theranostics—the trio of imaging, targeting and therapy. J Mater Chem. 2010;20(29):6149–57.

Choi KY, Liu G, Lee S, Chen X. Theranostic nanoplatforms for simultaneous cancer imaging and therapy: current approaches and future perspectives. Nanoscale. 2012;4(2):330–42.

Conde J, Edelman ER, Artzi N. Target-responsive DNA/RNA nanomaterials for microRNA sensing and inhibition: The jack-of-all-trades in cancer nanotheranostics? Adv Drug Deliv Rev. 2015;81:169–83.

DeSantis CE, Lin CC, Mariotto AB, Siegel RL, Stein KD, Kramer JL, Alteri R, Robbins AS, Jemal A. Cancer treatment and survivorship statistics, 2014. CA Cancer J Clin. 2014;64(4):252–71.

Draz MS, Fang BA, Zhang P, Hu Z, Gu S, Weng KC, Gray JW, Chen FF. Nanoparticle-mediated systemic delivery of siRNA for treatment of cancers and viral infections. Theranostics. 2014;4(9):872.

Elad S, Zadik Y, Hewson I, Hovan A, Correa MEP, Logan R, Elting LS, Spijkervet FK, Brennan MT. A systematic review of viral infections associated with oral involvement in cancer patients: a spotlight on Herpesviridea. Support Care Cancer. 2010;18(8):993–1006.

Esteva A, Kuprel B, Novoa RA, Ko J, Swetter SM, Blau HM, Thrun S. Dermatologist-level classification of skin cancer with deep neural networks. Nature. 2017;542(7639):115.

Fernandez-Fernandez A, Manchanda R, McGoron AJ. Theranostic applications of nanomaterials in cancer: drug delivery, image-guided therapy, and multifunctional platforms. Appl Biochem Biotechnol. 2011;165(7–8):1628–51.

Funkhouser J. Reinventing pharma: the theranostic revolution. Curr Drug Discov. 2002;2:17–9.

Gibson RJ, Keefe DM. Cancer chemotherapy-induced diarrhoea and constipation: mechanisms of damage and prevention strategies. Support Care Cancer. 2006;14(9):890.

He Q, Ma M, Wei C, Shi J. Mesoporous carbon@ silicon-silica nanotheranostics for synchronous delivery of insoluble drugs and luminescence imaging. Biomaterials. 2012;33(17):4392–402.

Howell M, Valle JW. The role of adjuvant chemotherapy and radiotherapy for cholangiocarcinoma. Best Pract Res Clin Gastroenterol. 2015;29(2):333–43.

Jagasia MH, Greinix HT, Arora M, Williams KM, Wolff D, Cowen EW, Palmer J, Weisdorf D, Treister NS, Cheng G-S. National Institutes of Health consensus development project on criteria for clinical trials in chronic graft-versus-host disease: I. The 2014 Diagnosis and Staging Working Group report. Biol Blood Marrow Transpl. 2015;21(3):389–401.e381.

Janib SM, Moses AS, MacKay JA. Imaging and drug delivery using theranostic nanoparticles. Adv Drug Deliv Rev. 2010;62(11):1052–63.

Kateb B, Chiu K, Black KL, Yamamoto V, Khalsa B, Ljubimova JY, Ding H, Patil R, Portilla-Arias JA, Modo M. Nanoplatforms for constructing new approaches to cancer treatment, imaging, and drug delivery: what should be the policy? NeuroImage. 2011;54:S106–24.

Keidan RD, Fanning J, Gatenby RA, Weese JL. Recurrent typhlitis. Dis Colon Rectum. 1989;32(3):206–9.

Kirschbaum SE, Baeumner AJ. A review of electrochemiluminescence (ECL) in and for microfluidic analytical devices. Anal Bioanal Chem. 2015;407(14):3911–26.

Koo Y-EL, Reddy GR, Bhojani M, Schneider R, Philbert MA, Rehemtulla A, Ross BD, Kopelman R. Brain cancer diagnosis and therapy with nanoplatforms. Adv Drug Deliv Rev. 2006;58(14):1556–77.

Kulke MH, Siu LL, Tepper JE, Fisher G, Jaffe D, Haller DG, Ellis LM, Benedetti JK, Bergsland EK, Hobday TJ. Future directions in the treatment of neuroendocrine tumors: consensus report of the National Cancer Institute Neuroendocrine Tumor clinical trials planning meeting. J Clin Oncol. 2011;29(7):934.

Kumar R, Kulkarni A, Nagesha DK, Sridhar S. In vitro evaluation of theranostic polymeric micelles for imaging and drug delivery in cancer. Theranostics. 2012;2(7):714.

Landry B, Valencia-Serna J, Gul-Uludag H, Jiang X, Janowska-Wieczorek A, Brandwein J, Uludag H. Progress in RNAi-mediated molecular therapy of acute and chronic myeloid leukemia. Mol Ther Nucleic Acids. 2015;4:e240.

Lee HS, Park JY, Shin SH, Kim SB, Lee JS, Lee A, Ye BJ, Kim YS. Herpesviridae viral infections after chemotherapy without antiviral prophylaxis in patients with malignant lymphoma: incidence and risk factors. Am J Clin Oncol. 2012a;35(2):146–50.

Lee VH, Ng SC, Leung T, Au GK, Kwong DL. Dosimetric predictors of radiation-induced acute nausea and vomiting in IMRT for nasopharyngeal cancer. Int J Radiat Oncol Biol Phys. 2012b;84(1):176–82.

Lee J-H, Chae J-W, Kim JK, Kim HJ, Chung JY, Kim Y-H. Inhibition of cisplatin-resistance by RNA interference targeting metallothionein using reducible oligo-peptoplex. J Control Release. 2015;215:82–90.

Li S, Goins B, Zhang L, Bao A. Novel multifunctional theranostic liposome drug delivery system: construction, characterization, and multimodality MR, near-infrared fluorescent, and nuclear imaging. Bioconjug Chem. 2012;23(6):1322–32.

Lyman GH, Bohlke K, Khorana AA, Kuderer NM, Lee AY, Arcelus JI, Balaban EP, Clarke JM, Flowers CR, Francis CW. Venous thromboembolism prophylaxis and treatment in patients with cancer: American Society of Clinical Oncology clinical practice guideline update 2014. J Clin Oncol. 2015;33(6):654.

Ma L, Young J, Prabhala H, Pan E, Mestdagh P, Muth D, Teruya-Feldstein J, Reinhardt F, Onder TT, Valastyan S. miR-9, a MYC/MYCN-activated microRNA, regulates E-cadherin and cancer metastasis. Nat Cell Biol. 2010;12(3):247.

Ma X, Tao H, Yang K, Feng L, Cheng L, Shi X, Li Y, Guo L, Liu Z. A functionalized graphene oxide-iron oxide nanocomposite for magnetically targeted drug delivery, photothermal therapy, and magnetic resonance imaging. Nano Res. 2012;5(3):199–212.

Malvezzi M, Carioli G, Bertuccio P, Negri E, La Vecchia C. Relation between mortality trends of cardiovascular diseases and selected cancers in the European Union, in 1970–2017. Focus on cohort and period effects. Eur J Cancer. 2018;103:341–55.

Miao W, Shim G, Kim G, Lee S, Lee H-J, Kim YB, Byun Y, Oh Y-K. Image-guided synergistic photothermal therapy using photoresponsive imaging agent-loaded graphene-based nanosheets. J Control Release. 2015;211:28–36.

Mitra RN, Doshi M, Zhang X, Tyus JC, Bengtsson N, Fletcher S, Page BD, Turkson J, Gesquiere AJ, Gunning PT. An activatable multimodal/multifunctional nanoprobe for direct imaging of intracellular drug delivery. Biomaterials. 2012;33(5):1500–8.
Moon H, Yoon C, Lee TW, Ha K-S, Chang JH, Song T-K, Kim K, Kim H. Therapeutic ultrasound contrast agents for the enhancement of tumor diagnosis and tumor therapy. J Biomed Nanotechnol. 2015;11(7):1183–92.
Moreno BH, Parisi G, Robert L, Ribas A. Anti-PD-1 therapy in melanoma. Semin Oncol. 2015;42(3):466–73.
Mukherjee S, Chowdhury D, Kotcherlakota R, Patra S. Potential theranostics application of biosynthesized silver nanoparticles (4-in-1 system). Theranostics. 2014;4(3):316.
Muthu MS, Kulkarni SA, Raju A, Feng S-S. Theranostic liposomes of TPGS coating for targeted co-delivery of docetaxel and quantum dots. Biomaterials. 2012;33(12):3494–501.
Novoselov K, Jiang D, Schedin F, Booth T, Khotkevich V, Morozov S, Geim A. Two-dimensional atomic crystals. Proc Natl Acad Sci. 2005;102(30):10451–3.
Ramage JK, Donaghy A, Farrant JM, Iorns R, Williams R. Serum tumor markers for the diagnosis of cholangiocarcinoma in primary sclerosing cholangitis. Gastroenterology. 1995;108(3):865–9.
Rastogi R, Gulati N, Kotnala RK, Sharma U, Jayasundar R, Koul V. Evaluation of folate conjugated pegylated thermosensitive magnetic nanocomposites for tumor imaging and therapy. Colloids Surf B Biointerfaces. 2011;82(1):160–7.
Robinson D, Van Allen EM, Wu Y-M, Schultz N, Lonigro RJ, Mosquera J-M, Montgomery B, Taplin M-E, Pritchard CC, Attard G. Integrative clinical genomics of advanced prostate cancer. Cell. 2015;161(5):1215–28.
Sailor MJ, Park JH. Hybrid nanoparticles for detection and treatment of cancer. Adv Mater. 2012;24(28):3779–802.
Santra S, Kaittanis C, Grimm J, Perez JM. Drug/dye-loaded, multifunctional iron oxide nanoparticles for combined targeted cancer therapy and dual optical/magnetic resonance imaging. Small. 2009;5(16):1862–8.
Sen K, Mandal M. Second generation liposomal cancer therapeutics: transition from laboratory to clinic. Int J Pharm. 2013;448(1):28–43.
Shi X, Gong H, Li Y, Wang C, Cheng L, Liu Z. Graphene-based magnetic plasmonic nanocomposite for dual bioimaging and photothermal therapy. Biomaterials. 2013;34(20):4786–93.
Siegel RL, Miller KD, Jemal A. Cancer statistics, 2015. CA Cancer J Clin. 2015;65(1):5–29.
Siegel RL, Miller KD, Jemal A. Cancer statistics, 2017. CA Cancer J Clin. 2017;67(1):7–30.
Taratula O, Patel M, Schumann C, Naleway MA, Pang AJ, He H, Taratula O. Phthalocyanine-loaded graphene nanoplatform for imaging-guided combinatorial phototherapy. Int J Nanomed. 2015;10:2347.
Vivero-Escoto JL, Huang Y-T. Inorganic-organic hybrid nanomaterials for therapeutic and diagnostic imaging applications. Int J Mol Sci. 2011;12(6):3888–927.
Wang L-S, Chuang M-C, Ho J-aA. Nanotheranostics—a review of recent publications. Int J Nanomed. 2012;7:4679.
Wang L, Xing H, Zhang S, Ren Q, Pan L, Zhang K, Bu W, Zheng X, Zhou L, Peng W. A Gd-doped Mg-Al-LDH/Au nanocomposite for CT/MR bimodal imagings and simultaneous drug delivery. Biomaterials. 2013;34(13):3390–401.
Weigelt B, Peterse JL, Van't Veer LJ. Breast cancer metastasis: markers and models. Nat Rev Cancer. 2005;5(8):591.
Whitlock EP, Lin JS, Liles E, Beil TL, Fu R. Screening for colorectal cancer: a targeted, updated systematic review for the US Preventive Services Task Force. Ann Intern Med. 2008;149(9):638–58.
WHO. Cancer fact sheet N 297. 2015.
Wolinsky JB, Grinstaff MW. Therapeutic and diagnostic applications of dendrimers for cancer treatment. Adv Drug Deliv Rev. 2008;60(9):1037–55.
World Health Organization. Cancer fact sheet. Updated Feb 2017.

Xie J, Lee S, Chen X. Nanoparticle-based theranostic agents. Adv Drug Deliv Rev. 2010;62(11):1064–79.

Xie J, Liu G, Eden HS, Ai H, Chen X. Surface-engineered magnetic nanoparticle platforms for cancer imaging and therapy. Acc Chem Res. 2011;44(10):883–92.

Yallapu MM, Othman SF, Curtis ET, Gupta BK, Jaggi M, Chauhan SC. Multi-functional magnetic nanoparticles for magnetic resonance imaging and cancer therapy. Biomaterials. 2011;32(7):1890–905.

Yang K, Hu L, Ma X, Ye S, Cheng L, Shi X, Li C, Li Y, Liu Z. Multimodal imaging guided photothermal therapy using functionalized graphene nanosheets anchored with magnetic nanoparticles. Adv Mater. 2012;24(14):1868–72.

Yuan J, Zhang H, Kaur H, Oupicky D, Peng F. Synthesis and characterization of theranostic poly (HPMA)-c (RGDyK)-DOTA-64Cu copolymer targeting tumor angiogenesis: tumor localization visualized by positron emission tomography. Mol Imaging. 2013;12(3):203–12.

Zedan AF, Moussa S, Terner J, Atkinson G, El-Shall MS. Ultrasmall gold nanoparticles anchored to graphene and enhanced photothermal effects by laser irradiation of gold nanostructures in graphene oxide solutions. ACS Nano. 2012;7(1):627–36.

Zhang L, Gu F, Chan J, Wang A, Langer R, Farokhzad O. Nanoparticles in medicine: therapeutic applications and developments. Clin Pharmacol Ther. 2008;83(5):761–9.

Zhang MH, Man HT, Zhao XD, Dong N, Ma SL. Estrogen receptor-positive breast cancer molecular signatures and therapeutic potentials. Biomed Rep. 2014;2(1):41–52.

Chapter 11
Biosynthesized Metallic Nanoparticles as Emerging Cancer Theranostics Agents

Muhammad Ovais, Ali Talha Khalil, Muhammad Ayaz, and Irshad Ahmad

Abstract Cancer is considered as a great health challenge liable for outstripped demises worldwide. Currently it is treated mainly by chemotherapy and radiotherapy. However, there is a perpetual demand for the development of novel therapeutic drugs to combat this devastating disease. In this regard nanomedicine can provide an alternative platform for its diagnosis and treatment but its conventional synthesis through physiochemical methods has several shortcomings like high cost, energy intensive, and toxicity concerns. Consequently, the green synthesis of biogenic metallic nanoparticles (MNPs) from plants provides an alternate paradigm which has been proved safer, eco-friendly, energy proficient, inexpensive, and less toxic in nature. Additionally, the green MNPs have multipurpose biomedical applications like drug delivery agents, anticancerous mediators, photothermal therapy, and bio-imaging. This chapter will provide ample information on the current status of green MNPs, its anticancerous mechanisms, and efficiency in cancer diagnosis. Other issues like polydispersity and toxicity are also highlighted. Keeping in view all of the challenges, the authors anticipate biogenic MNPs may contribute to shift the paradigm toward development of novel nanomedicine that can prove as biocompatible theranostic agents in near future.

M. Ovais (✉)
CAS Key Laboratory for Biomedical Effects of Nanomaterials and Nanosafety, CAS Center for Excellence in Nanoscience, National Center for Nanoscience and Technology (NCNST), Beijing, People's Republic of China

University of Chinese Academy of Sciences, Beijing, People's Republic of China
e-mail: movais@bs.qau.edu.pk; movais@nanoctr.cn

A. T. Khalil
Department of Eastern Medicine and Surgery, Qarshi University, Lahore, Pakistan

M. Ayaz
Department of Pharmacy, University of Malakand, Chakdara, Khyber Pakhtunkhwa, Pakistan

I. Ahmad
Department of Life Sciences, King Fahd University of Petroleum and Minerals (KFUPM), Dhahran, Saudi Arabia

M. Rai, B. Jamil (eds.), *Nanotheranostics*,
https://doi.org/10.1007/978-3-030-29768-8_11

Keywords Metal nanoparticles · Bionanomaterials · Cancer theranostics · Green synthesis · Biocompatibility

11.1 Introduction

Cancer is a deadly disease initiated by mutations in genes that activate a sequence of events at molecular level, ultimately progressed to tumor formation. The hallmarks of cancer development are a multistep process involving proliferative signaling, circumvent growth suppressors, escaping cell death, facilitating replicative immortality, stimulate angiogenesis, invasion, and metastasis (Hanahan and Weinberg 2011; Hollstein et al. 2017). According to the current global statistics, GLOBOCAN has estimated the cancer incidence and mortality because of 36 different types of cancers among 185 countries. 18.1 million people have been diagnosed with cancer and 9.6 million cancer deaths with an emphasis on geographical inconsistency crossways 20 world constituencies (Bray et al. 2018). The global incidence of cancer is expected to rise to 27.5 million by 2040 and 16.3 million cancer deaths due to diverse risk factors for instance smoking, unnatural diet, lack of exercise, and less pregnancies (WHO 2017). According to a current report on cancer prevalence and death patterns in Europe 3.91 million new cases of cancer have been projected with 1.93 million demises. Lung, colorectal, breast, and pancreatic cancers are the major causes of death in EU-28 with estimated new cases of cancer in males (1.6 million) and females (1.4 million) (Ferlay et al. 2013).

The iMShealth Institute for Healthcare Informatics has published a report on the global market for cancer which is tremendously increased to a highest level of US$107 billion in 2015, and is expected to reach US$150 billion by 2020. Highest prevalence of cancer can be observed in high-income countries (HIC) (Hao et al. 2010) as compared to low- and middle-income countries (LMIC). The lung, colorectal, breast, and prostate cancer incidence rate is very high in HIC whereas stomach, liver, esophageal, and cervical cancer is prevalent in LMIC. Though cancer prevalence is high in HICs, the mortality rate is plateauing or declining now in many cancers due to diminishing identified risk factors, early detection and screening, and better treatment process (Torre et al. 2016). Generally, the cancer incidence is caused by some external or internal factors. The external factors comprise an exposure to chemicals, radiations, or viruses. For instance, the workers who are unprotected from the toxic chemicals and ionizing radiations have a more chance of getting cancer (Manzoor et al. 2016). The internal factors responsible for cancer development include hormonal disturbance, mutagenesis, and weak immune conditions which might work in a chronological order to activate or promote the process of carcinogenesis (Anand et al. 2008). For cancer treatment different types of therapeutic approaches can be adopted like radiotherapy, chemotherapy, immunotherapy, photodynamic therapy, cancer vaccinations, surgery, and stem cell transformation; however, these therapies alone or in combination are accompanied by severe side effects. These side effects comprise toxicity, non-specificity, constrained bioavailability, debauched clearance, and constraint metastasis (Lim et al. 2011;

Patra et al. 2014; Mukherjee and Patra 2016). Many chemotherapeutic drugs have been reported to cause different types of toxicities; e.g., 5-fluorouracil is a common chemotherapeutic agent but allied with myelotoxicity, cardiotoxicity, gastrointestinal disorders, mucositis, myelosuppression, hand–foot syndrome, and the shrinking of blood vessels (Macdonald 1999; Boilève et al. 2019). Similarly, another anticancer drug (Doxorubicin) is concomitant to renal toxicity, cardiotoxicity, and myelotoxicity (Avilés et al. 1993; Farzanegi et al. 2019). The bleomycin and cyclophosphamide have been reported to cause cutaneous toxicity, and lungs toxicity and bladder toxicity, respectively (Adamson 1976; Fraiser et al. 1991; Andersen et al. 2019).

Keeping in view the limitations of currently available anticancer agents, in this chapter, we have provided the readers new outlook on the biosynthesis of metallic nanoparticles. We believe that this innovative approach will contribute to explore new avenues in order to develop potential biogenic agents for the effective cancer treatment with trivial side effects.

11.2 The Interface of Nanotechnology, Biological Constituents and Cancer

Nanotechnology is considered as the sixth revolutionary technology of the millennium after industrial, nuclear energy, green, information technology, and biotechnology revolution (Manimaran 2015). During the last several years tremendous developments have been accomplished in the synthesis of metallic nanoparticles (MNPs) and their valuable applications in different arenas of biological sciences including food, agriculture, engineering, electronics, biomedical instruments, cosmetics, and medication. The MNPs have obtained significant position due to their explicit physicochemical properties and diverse industrial uses (Slavin et al. 2017; Khan et al. 2017; Ovais et al. 2018a; Khalil et al. 2018a).

Previously, MNPs have been synthesized by the classical physicochemical methods which are facing the problems of low yield, costly, and unsafe due to the association of harmful chemical compounds coated by the outer surface of MNPs which possess adverse side effects in the biomedical applications. Considering these apprehensions, various research groups have now focused on the synthesis of MNPs by using biological constituents which are safe, cost-effective, biocompatible, and environment friendly (Mukherjee et al. 2013, 2015; Patra et al. 2015; Ovais et al. 2018b; Hameed et al. 2019). Moreover, the bioinspired MNPs are synthesized and scaled up easily with proper morphologies and higher biocompatibility that compelled researchers to utilize such resources as nanofactories (Baker et al. 2013; Singh et al. 2015). Different types of microorganisms (bacteria, yeasts, and fungi) and plant extracts have been used for the synthesis of MNPs (Patra et al. 2014; Kumari et al. 2017). The synthesis of these MNPs does not require any additional capping or stabilizing agents as the biomolecules of the microorganisms and plants can accomplish this task themselves (Shah et al. 2018).

Due to the adverse side effects and other limitations of the currently available anticancer agents, novel strategies need to be developed for better cancer treatment. Nanomedicines are considered as an exciting area for the cancer therapy (Iqbal et al. 2018). Due to the significant progress in research on medicinal plants and nanotechnology, there is a hope for the cancer patients to use safe and economical nanomedicine in the near future (Mukherjee et al. 2012; Ovais et al. 2016, 2017). Many research groups are now actively involved in the cutting edge research on the synthesis of multifunctional green MNPs for cancer treatment (Bhaumik et al. 2015; Singh et al. 2015). These green MNPs will take place in the classical cancer therapies due to their targeted and site-specific activity. These characteristics enhance the efficacy of the drug, as the MNPs can dodge immune responses and cross the impermeable membranes (Burda et al. 2005), thus considered as valuable tool to combat cancer. Due to the rational existence of free electrons in the transmission band, MNPs carry surface plasmon resonance in UV-visible regions determined by dielectric constant, size and particle surrounds the band shift (Tessier et al. 2001; Burda et al. 2005). Additionally, MNPs possess distinctive characteristic, i.e., the absorbance of the wavelengths which provides significant information regarding their sizes, shapes, and interparticle features (Mulvaney 1996). Due to the dynamic features, MNPs are considered to play a significant role by providing enhanced targeting, gene silencing, and as drug delivery agent in cancer therapy. The functionalized MNPs carrying targeted molecules comprise control energy released at the tumors site. MNPs can also be used as a diagnostic tool for cancer cell imaging. Within a short time, the biocompatible and functionalized MNPs will modernize cancer treatment and management (Sharma et al. 2018).

11.3 Synthesis of MNPs via Biological Resources

Various physical and chemical methods are employed to synthesize NPs. The chemical methods involving the use of chemicals as reducing agents are often associated with undesirable effects, including biocompatibility issues and environmental hazards. A recent approach is green or biogenic synthesis, whereby various natural sources including plants, algae, fungi, yeast, and other microorganisms are used as reducing agents (Rahman et al. 2019). Among these, plant-mediated synthesis of biogenic NPS has got more significance and emerged as a separate discipline called phytonanotechnology. Plant-mediated synthesis of NPs has got several advantages over the other techniques since it is rapid, less expansive, biocompatible, and devoid of environmental hazards. A general schematic diagram is presented in Fig 11.1.

Synthesis of biogenic NPs using plants involves the incubation of plant extracts with solution of metal salts like silver nitrate. In the first step, reduction of metals take place (i.e., Ag^{+} to Ag^{0}) followed by agglomeration and stabilization leading to formation of colloidal NPs as clusters (Park 2014; Duan et al. 2015). The first evidence of the formation of AgNPs is the change in the color of metal salt and

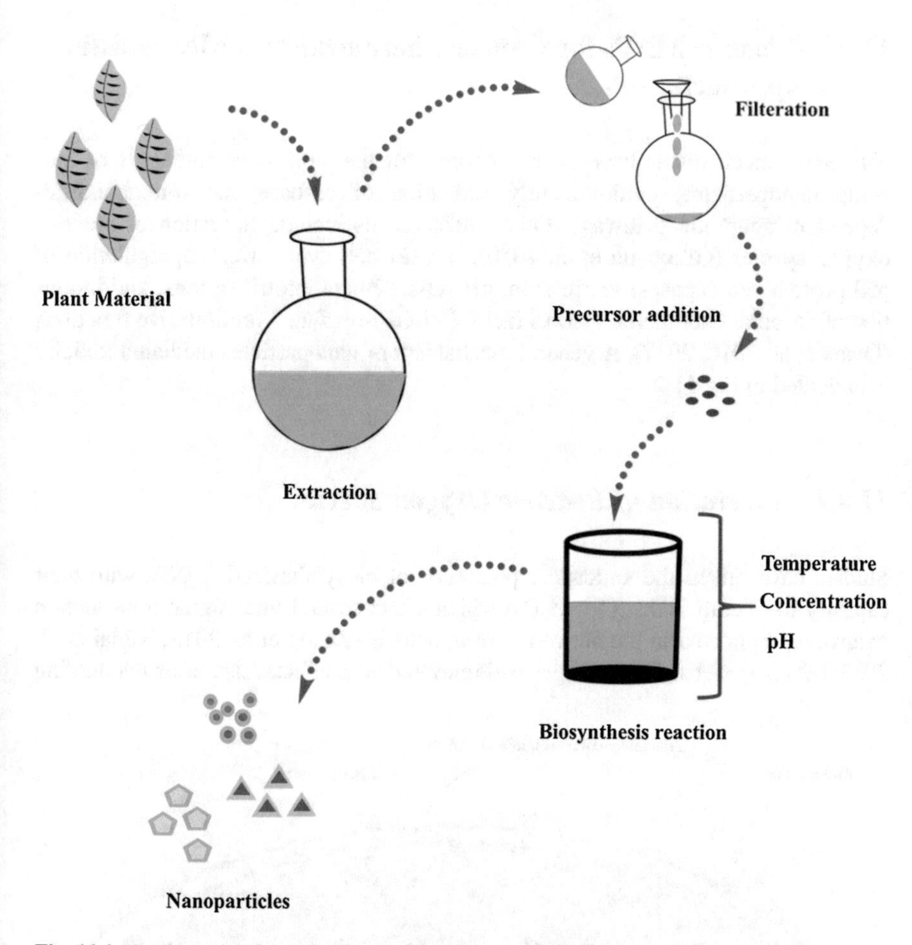

Fig. 11.1 A schematic of generic steps in biosynthesis of metal nanoparticles

extract mixture from colorless to dark brown. During synthesis of biogenic NPs, several parameters need optimization, including concentration of metal salt and extract solutions, temperature, pH, and incubation time. These factors have major effect on physicochemical properties of the NPs, including resultant size, shape, and stability (Kumar et al. 2012; Ovais et al. 2016). For instance, synthesis of AgNPs is possible at various pH and temperature ranges, yet pH 7 and room temperature (25 °C) are considered as the optimum pH-temperature conditions for spherical and small size AgNPs with ideal biological properties (Iravani and Zolfaghari 2013).

Among other natural reducing agents are included isolated chloroplasts, microalgae, and other microbes. The NPs synthesized via microalgae were eco-friendlier and scalable with prolog fabrication of metallic NPs (Dahoumane et al. 2017). The exact mechanism involved in these processes is not understood, but microbial enzymes are known to be implicated in the biogenic synthesis (Ovais et al. 2018a).

11.4 Biogenic MNPs for Cancer Theranostics: A Mechanistic Approach

Numerous mechanisms have been proposed for the anticancer potentials of biogenic nanoparticles, predominantly activation of caspase and mitochondrial-dependent apoptotic pathways. Other mechanisms include liberation of reactive oxygen species (Giljohann et al. 2010), sub-G1 cell cycle arrest, upregulation of p53 protein and capase-3 expression, pH-reliant liberation of Ag ions, and inhibition of vascular endothelial growth factor (VEGF)-mediated proliferative functions (Ovais et al. 2016, 2017). A general mechanism of nanoparticles mediated toxicity is indicated in Fig 11.2.

11.4.1 Liberation of Reactive Oxygen Species

Studies have linked the anticancer potentials of biosynthesized AgNPs with their capacity to liberate ROS (O_2^-, H_2O_2) which effect several vital signal transduction pathways implicated in the activation of apoptosis (Zhang et al. 2013; Minai et al. 2013; Mukherjee et al. 2014). Superoxide elevated level effects respiration uncoupling

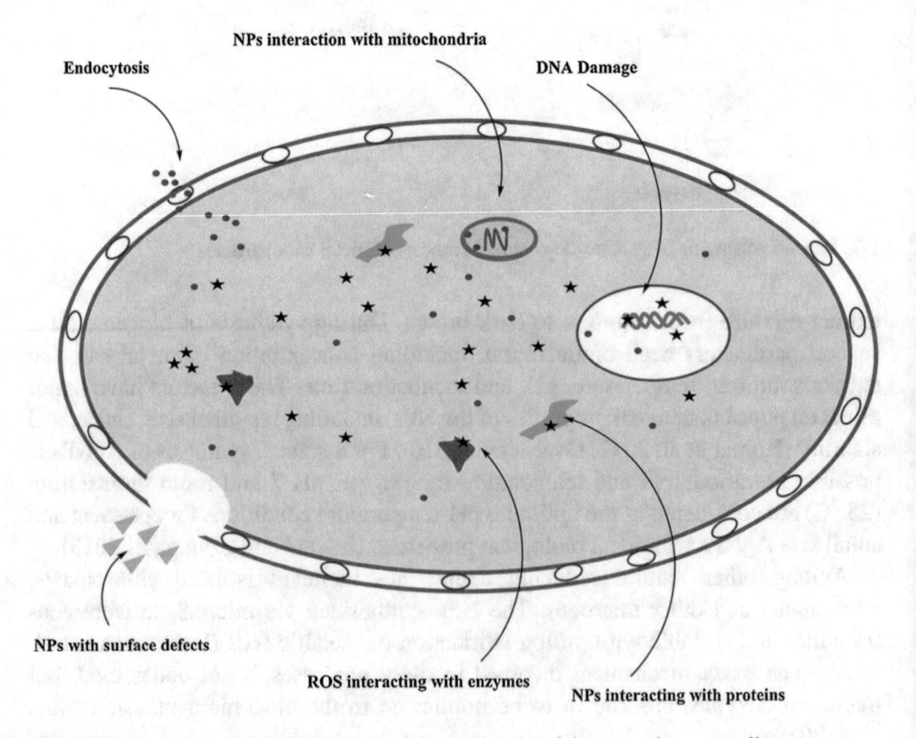

Fig. 11.2 General mechanism of nanoparticle-mediated toxicity toward cancer cells

and mitochondrial transmembrane potentials (Garrido et al. 2006). Various drugs by the virtue of their ability to liberate superoxide radicals exhibit strong anticancer properties (Velayutham et al. 2005). In summary, biogenic AgNPs generate ROS, thus causing apoptosis and cell death.

11.4.2 Upregulation of p53 Protein and Caspase-3 Expression

B16 cells treated with biogenic AgNPs have been shown to upregulate p53 protein and active caspase-3 expression using western blotting (Mukherjee et al. 2014). The potentials of AgNPs to trigger p53 upregulation via activation of apoptotic pathways and subsequent cellular death are well established now (Amaral et al. 2010; Mei et al. 2012). In a study, Gurunathan and coworkers (2013) reported that Ag ions released from AgNPs cause activation of caspase-3 and ultimate increase in oxidative stress and cellular damages (Gurunathan et al. 2013). Furthermore, the acidic tumor microenvironment aids in the liberation of phytoconstituents from AgNPs that augment its anticancer potential (Mukherjee et al. 2014).

11.4.3 Sub-G1 Arrest and Cancer Suppression

The cell cycle assay using fluorescence-activated B16 melanoma cancer cells treated with biogenic AgNPs show that the treated cells undergo apoptosis, potentially mediated by sub-G1 arrest (Mukherjee et al. 2014). The induction of apoptosis in cancer cells accumulating sub-G1 DNA was confirmed by Beach and coworkers (Beach et al. 2011). Further studies conformed the association between enhanced malignant cell population in sub-G1 phase and activation of caspase-3, an apoptotic protease responsible for initiation of apoptosis (Mao et al. 2004). Further, curcumin treatment of cancer cells causes sub-G1 phase arrest, indicating a direct relation between sub-G1 phase arrest in cancer cells and apoptosis (Chang et al. 2011). These evidences suggest that death in biogenic NP-treated cancer cells might be linked with rise in sub-G1 phase cancer cells, a phenomenon highly related to induction of apoptosis.

11.4.4 pH-Reliant Liberation of Ag Ions and Cancer Cells Death

Studies suggests that liberation of Ag ions from biogenic AgNPs is responsible for the cancer cell death, and cell death is dependent on concentration of Ag ions in various cell type of cell lines (Gurunathan et al. 2013; Mukherjee et al. 2014).

As release of Ag from NPs is pH dependent, high concentrations of silver ions are released in acidic pH with consequent high anticancer activity of these biogenic AgNPs at acidic environment (Tannock and Rotin 1989). A positive correlation between releases of Ag ions with acidic pH has been reported. Biogenic AgNPs incubated overnight at pH 5 (acidic) and at 7.4 (physiological buffer) suggested that Ag ions' release was twice high at acidic pH as compared to buffer environment. Percent lethality of cancerous cells were high in comparison to normal cells, which further strengthen the argument of pH-dependent Ag release and selective death (Mukherjee et al. 2014). Moreover, ROS generation by Ag ions at acidic environment might be another possible mechanism of pH-dependent cellular death (Asharani et al. 2008).

11.4.5 Inhibition of VEGF-Mediated Functions

Biogenic AgNPs have been reported to possess significant anti-angiogenic potentials in some major pathways for arresting tumor growth. Biogenic AgNPs exhibited anti-angiogenic potential by inhibiting vascular endothelial growth factor (VEGF)-mediated cell proliferation and inhibition of new blood vessel formation in tumor microenvironment. It was reported that using porcine retinal endothelial cells inhibits the VEGF-prompted vascular permeability mediated via Src pathway or proto-oncogene tyrosine-protein kinase. Another study reported the inhibition of VEGF-mediated cell proliferation via PI3K/Akt pathway. It is an intracellular signaling pathway important in regulating the cell cycle. *Saliva officinalis*-mediated biogenic AgNPs were reported to possess considerable anti-angiogenic properties

11.5 Challenges for Biogenic MNPs as Future Cancer Nanomedicine

Nanomaterials are in use of the human beings unknowingly since ancient times, but it's the recent developments which are considered a hallmark in nanotechnology research (Rao and Gan 2015; Singh et al. 2016). For example, gold nanoparticles were used in staining glasses for decorative purposes by ancient civilizations (Giljohann et al. 2010), and now gold nanoparticles have demonstrated good results in the targeted delivery of methotrexate, doxorubicin, and paclitaxel, while also been used for detection of tumors, imaging, photothermal therapies, etc. (Rai et al. 2016). The use of eco-friendly and green methods has become a cornerstone in the fabrication of metal nanoparticles because of several advantages. Among green methods, the use of medicinal plants for nanomaterials synthesis is highly preferred (Ovais et al. 2017). Usually biogenic methods are easy to perform, and the aqueous extracts of medicinal plants are biocompatible and devoid of generating toxic wastes

(Dauthal and Mukhopadhyay 2016; Ovais et al. 2017; Khalil et al. 2018b). The process does not require high temperature or pressure while the phytochemicals perform a dual role of chelating as well as stabilizing the nanoparticles (Ovais et al. 2016; Khalil et al. 2017). Albeit their potential advantages, there remain some concerns in the use of biogenic nanomaterials for biomedical applications, which needs to be addressed. Lack of homogeneity and reproducibility remains the most important questions to be solved in green chemistry methods (Dauthal and Mukhopadhyay 2016).

11.5.1 Polydispersity to Monodispersity

Obtaining monodispersed metal nanoparticles always remained a challenge and a major hindrance in the large-scale applications of the biosynthesized nanoparticles. The involvement of the diverse nature of phytochemicals in plant extracts is the reason for the polydispersity of biogenic metal nanoparticles (Ovais et al. 2016). Furthermore, variation in phytochemical content and nature of secondary metabolites can be the subject of geographical and seasonal variations. Plant systems are exposed to diverse sorts of harsh conditions, which can alter the phytochemistry. Production of chemical entities are upregulated or downregulated depending on the conditions and environment (Khalil et al. 2018b). Such variations not only led to polydispersity but also indicate the need of obtaining a standardized raw material for the fabrication of nanoparticles. There are different optimization strategies that can be adopted to increase the monodispersity. One such method is increasing the concentration of the plant extracts used. In a recent report, *Citrus paradise* extracts were used for the biosynthesis of Au nanoparticles. Authors reported that the increasing concentration of extracts leads to more monodisperse nature of AuNPs (Silva-De Hoyos et al. 2018). With reference to the standardization issues and availability of raw materials, tissue culturing techniques can be useful, in which the plants can be propagated under constant conditions and their automation will lead to more uniform and standardized raw materials.

Considering advanced medical applications in which biogenic nanoparticles can be applied, it's imperative to elucidate the mechanistic aspects as well as to extensively characterize the nature of nanoparticles. While using aqueous extracts of plants as a reducing and stabilizing agent, it's difficult to identify a particular chemical entity that has reduced and capped the nanoparticles. It's difficult to use such nanoparticles for therapeutic applications such as drug delivery to specific site. The concept of uni-capped nanoparticles is getting popular and recently utilized. This concept extends from extracts to the specific isolated pure phytochemicals for the biosynthesis of nanoparticles. This method is already being reported to have improved the monodispersity of the nanoparticles. While synthesizing the Ag and Au nanoparticles through the aqueous extracts of *Indian propolis*, phytochemicals obtained from extracts (pinocembrin and galangin) indicated relatively monodisperse nature with uniform distribution for the pure phytochemical-based nanoparticles as compared to extracts (Roy et al. 2010). One another article reports higher stability for the Ag nanoparticles synthesized through reduction by pure phytochemical (β-sitosterol-D-glucopyranoside) isolated from *Desmostachya bipinnata*, as compared to the *Desmostachya bipinnata* extracts reduced Ag nanoparticles (Ahmed et al. 2014). Another method that can contribute in improving the uniformity is diafiltration. In conventional washing and centrifugation of the biogenic nanomaterials, large nanoclusters and impurities are often retained, which compromises the uniformity of the nanoparticles. Diafiltration has been recently described as an effective size-based separation technique which can be applied in biogenic synthesis (Sweeney et al. 2006).

11.5.2 Toxicological and Accumulation

Toxicity of the metal nanoparticles is a long-standing concern in the scientific community. Understanding the risks associated with the use of metal nanoparticles is important for deriving full-fledged advantages of nanotechnology. Although some of the metal nanoparticles like TiO_2 and ZnO are considered being safe, but significant knowledge gaps exist, when it comes to evaluate the nature and toxicity of the metal nanoparticles. Mostly, the toxicity of nanomaterial is evaluated using cell culture-based experiments, in which cellular density and number of preceding cell divisions can affect the toxicity. Cytotoxicity and DNA damaging potential is reported for a number of metal oxide nanoparticles, while their also exist contradictory reports as the method of synthesis, solubility, and concentration affect the behavior of nanoparticles, and hence the results (Suresh et al. 2013; Papavlassopoulos et al. 2014). Various metal nanoparticles (CuO, TiO_2, Fe_2O_3 and Fe_3O_4) when assessed for toxicity in human A549 cells revealed more toxicity than their macroscale counterparts (Kim et al. 2010). Increased membrane permeability and elevated intracellular concentration of metal nanoparticles may progress to dysfunctional mitochondria in case of nanoparticles. Recently, biogenic selenium nanoparticles were reported ten times less toxic than the chemogenic selenium nanoparticles to zebra fish (Mal et al. 2017). Similarly, de Lima et al. (2012) reviewed different studies on the toxicological aspects of silver nanoparticles. Different literature indicated that the biogenically capped silver nanoparticles have reduced toxicity as compared to the AgNPs synthesized through chemical method. In a recent article, silver nanoparticles were fabricated using polysaccharide (Galactomannan) isolated from *Punica granatum* and their toxicity was evaluated in cancer cells. The biogenically stabilized silver nanoparticles showed excellent toxicity toward cancer cells, but indicated more compatibility toward normal cells as compared to $AgNO_3$, which revealed drastic toxicity toward normal cells (Padinjarathil et al. 2018). In another study, the rhamnogalacturonan gum obtained from *Cochlospermum gossypium* was used for the synthesis of silver nanoparticles and their compatibility studies on HeLa cells revealed cytotoxicity beyond 2.5 μg/mL (Kora and Sashidhar 2018). Similarly the biogenic selenium nanoparticles are less toxic than other forms of selenium like selenite (Benko et al. 2012; Shakibaie et al. 2013; Wadhwani et al. 2016). In general, the smaller size of nanoparticles can lead to significant reactivity and enhanced toxicity (Kim et al. 2011), but the size of the particles is not the sole contributor to the toxicity of nanoparticles. Reduction in size increases the surface area and enhances the oxidation and DNA-damaging abilities (Karlsson et al. 2009). Generally, nanomaterials with size <100 nm possess the ability to penetrate the cell membrane, <40 nm can penetrate nuclei, and <35 nm can cross the blood brain barrier (Gliga et al. 2014). Such penetrations have the capacity to induce oxidative stress, genomic and mitochondrial DNA damage, and apoptosis. There is evidence of accumulation of the nanoparticles in the body tissues. Asharani et al. (2011) studied the effects of different metal nanoparticles capped with PVP on the development of Zebra fish embryos. It was found that the silver and platinum nanoparticles caused delayed hatching, dropped heart rate, and also manifested phenotypic changes like malformation of eyes or their absence while gold nanoparticles did not show any toxicity. The accumulation of nanoparticles in embryos was confirmed through the ICP-OES (inductively coupled plasma optical emission spectroscopy). In one of the studies, silver nanoparticles (10 nm) were found to be accumulated in the liver and gill tissues (Scown et al. 2010). In a recent work of Lacave et al. (2018), zebra fish was studied after exposure to the silver nanoparticles and ions. It was observed that the silver ions were accumulated all over the body, while the silver nanoparticles were accumulated in gills, liver, and intestine. The hepatic transcriptome analysis initially indicated regulation of larger number of transcripts by the silver ions as compared to the silver nanoparticles; however, later on, silver nanoparticles induced significantly high number of transcripts. Hyperplasia and inflammation was also found in the gills of zebra fish after 6 months of exposure to the silver nanoparticles, which indicates the long-term adverse effects. There are other factors including synthesis methods, absence or presence of capping agents, nature of the capping agent, shape, and solubility (Asharani et al. 2011; Panda et al. 2011; Sufian et al. 2017). Most of the metal nanoparticles have tendency to dissolve at body pH that results in increase of the intracellular metal ion concentration causing stressed condition (Sufian et al. 2017).

11.6 Conclusions

Cancer is a dreadful disease and among a leading cause of mortalities in the world. The present anticancer therapies like chemotherapy and radiotherapy are not only expensive but also possess side effects. In this regard, nanotechnology is considered as an array of hope for effective and selective killing of the cancerous cells from the body. The newly emerging area of nanobiotechnology has yielded impressive results in different studies. Extracts of medicinal plants can be used as a low-cost chelating and capping agents for the biosynthesis of metal nanoparticles, providing an eco-friendly, cheap, and safe route for the synthesis of nanoparticles. The field is still in its infancy, and needs extensive research and unique strategies to cope with issues like polydispersity in green chemistry-based synthesis. Uni-capped nanoparticles seem to be the next big leap in green chemistry-based biosynthesis of metal nanoparticles. Considering the impressive results in vitro, these nanomaterials must be subjected to extensive in vivo studies to determine their toxicity, accumulation adsorption, etc. in the body. Once their safety is ensured, only then one can reap the therapeutic benefits of nanomaterials.

References

Adamson IY. Pulmonary toxicity of bleomycin. Environ Health Perspect. 1976;16:119–25.

Ahmed KBA, Subramaniam S, Veerappan G, Hari N, Sivasubramanian A, Veerappan A. β-Sitosterol-d-glucopyranoside isolated from Desmostachya bipinnata mediates photoinduced rapid green synthesis of silver nanoparticles. RSC Adv. 2014;4(103):59130–6.

Amaral JD, Xavier JM, Steer CJ, Rodrigues CM. The role of p53 in apoptosis. Discov Med. 2010;9(45):145–52.

Anand P, Kunnumakkara AB, Sundaram C, Harikumar KB, Tharakan ST, Lai OS, Sung B, Aggarwal BB. Cancer is a preventable disease that requires major lifestyle changes. Pharm Res. 2008;25(9):2097–116.

Andersen MD, Kamper P, d'Amore A, Clausen M, Bentzen H, d'Amore F. The incidence of bleomycin induced lung toxicity is increased in Hodgkin lymphoma patients over 45 years exposed to granulocyte-colony stimulating growth factor. Leuk Lymphoma. 2019;60(4):927–33.

Asharani PV, Lian Wu Y, Gong Z, Valiyaveettil S. Toxicity of silver nanoparticles in zebrafish models. Nanotechnology. 2008;19(25):255102.

Asharani PV, Lianwu Y, Gong Z, Valiyaveettil S. Comparison of the toxicity of silver, gold and platinum nanoparticles in developing zebrafish embryos. Nanotoxicology. 2011;5(1):43–54.

Avilés A, Arévila N, Díaz Maqueo JC, Gómez T, García R, Nambo MJ. Late cardiac toxicity of doxorubicin, epirubicin, and mitoxantrone therapy for Hodgkin's disease in adults. Leuk Lymphoma. 1993;11(3-4):275–9.

Baker S, Rakshith D, Kavitha KS, Santosh P, Kavitha HU, Rao Y, Satish S. Plants: emerging as nanofactories towards facile route in synthesis of nanoparticles. BioImpacts. 2013;3(3):111–7.

Beach JA, Nary LJ, Hirakawa Y, Holland E, Hovanessian R, Medh RD. E4BP4 facilitates glucocorticoid-evoked apoptosis of human leukemic CEM cells via upregulation of Bim. J Mol Signal. 2011;6(1):13.

Benko I, Nagy G, Tanczos B, Ungvari E, Sztrik A, Eszenyi P, Prokisch J, Banfalvi G. Subacute toxicity of nano-selenium compared to other selenium species in mice. Environ Toxicol Chem. 2012;31(12):2812–20.

Bhaumik J, Thakur NS, Aili PK, Ghanghoriya A, Mittal AK, Banerjee UC. Bioinspired nanotheranostic agents: synthesis, surface functionalization, and antioxidant potential. ACS Biomater Sci Eng. 2015;1(6):382–92.

Boilève A, Wicker C, Verret B, Leroy F, Malka D, Jozwiak M, Pontoizeau C, Ottolenghi C, De Lonlay P, Ducreux M, Hollebecque A. 5-Fluorouracil rechallenge after 5-fluorouracil-induced hyperammonemic encephalopathy. Anti-Cancer Drugs. 2019;30(3):313–7.

Bray F, Ferlay J, Soerjomataram I, Siegel RL, Torre LA, Jemal A. Global cancer statistics 2018: GLOBOCAN estimates of incidence and mortality worldwide for 36 cancers in 185 countries. CA Cancer J Clin. 2018;68(6):394–424.

Burda C, Chen X, Narayanan R, El-Sayed MA. Chemistry and properties of nanocrystals of different shapes. Chem Rev. 2005;105(4):1025–102.

Chang YJ, Tai CJ, Kuo LJ, Wei PL, Liang HH, Liu TZ, Wang W, Tai CJ, Ho YS, Wu CH, Huang MT. Glucose-regulated protein 78 (GRP78) mediated the efficacy to curcumin treatment on hepatocellular carcinoma. Ann Surg Oncol. 2011;18(8):2395–403.

Dahoumane SA, Mechouet M, Wijesekera K, Filipe CDM, Sicard C, Bazylinski DA, Jeffryes C. Algae-mediated biosynthesis of inorganic nanomaterials as a promising route in nanobiotechnology—a review. Green Chem. 2017;19(3):552–87.

Dauthal P, Mukhopadhyay M. Noble metal nanoparticles: plant-mediated synthesis, mechanistic aspects of synthesis, and applications. Ind Eng Chem Res. 2016;55(36):9557–77.

de Lima R, Seabra AB, Durán N. Silver nanoparticles: a brief review of cytotoxicity and genotoxicity of chemically and biogenically synthesized nanoparticles. J Appl Toxicol. 2012;32(11):867–79.

Duan H, Wang D, Li Y. Green chemistry for nanoparticle synthesis. Chem Soc Rev. 2015;44(16):5778–92.

Farzanegi P, Asadi M, Abdi A, Etemadian M, Amani M, Amrollah V, Shahri F, Gholami V, Abdi Z, Moradi L, Ghorbani S. Swimming exercise in combination with garlic extract administration as a therapy against doxorubicin-induced hepatic, heart and renal toxicity to rats. Toxin Rev. 2019:1–10. https://doi.org/10.1080/15569543.2018.1559194.

Ferlay J, Steliarova-Foucher E, Lortet-Tieulent J, Rosso S, Coebergh JW, Comber H, Forman D, Bray F. Cancer incidence and mortality patterns in Europe: estimates for 40 countries in 2012. Eur J Cancer. 2013;49(6):1374–403.

Fraiser LH, Kanekal S, Kehrer JP. Cyclophosphamide toxicity. Characterising and avoiding the problem. Drugs. 1991;42(5):781–95.

Garrido C, Galluzzi L, Brunet M, Puig PE, Didelot C, Kroemer G. Mechanisms of cytochrome c release from mitochondria. Cell Death Differ. 2006;13(9):1423–33.

Giljohann DA, Seferos DS, Daniel WL, Massich MD, Patel PC, Mirkin CA. Gold nanoparticles for biology and medicine. Angew Chem Int Ed. 2010;49(19):3280–94.

Gliga AR, Skoglund S, Wallinder IO, Fadeel B, Karlsson HL. Size-dependent cytotoxicity of silver nanoparticles in human lung cells: the role of cellular uptake, agglomeration and Ag release. Part Fibre Toxicol. 2014;11(1):11.

Gurunathan S, Han JW, Eppakayala V, Jeyaraj M, Kim JH. Cytotoxicity of biologically synthesized silver nanoparticles in MDA-MB-231 human breast cancer cells. BioMed Res Int. 2013;2013:535796.

Hameed S, Khalil AT, Ali M, Numan M, Khamlich S, Shinwari ZK, Maaza M. Greener synthesis of ZnO and Ag–ZnO nanoparticles using Silybum marianum for diverse biomedical applications. Nanomedicine. 2019;14(6):655–73.

Hanahan D, Weinberg RA. Hallmarks of cancer: the next generation. Cell. 2011;144(5):646–74.

Hao R, Xing R, Xu Z, Hou Y, Gao S, Sun S. Synthesis, functionalization, and biomedical applications of multifunctional magnetic nanoparticles. Adv Mater. 2010;22(25):2729–42.

Hollstein M, Alexandrov LB, Wild CP, Ardin M, Zavadil J. Base changes in tumour DNA have the power to reveal the causes and evolution of cancer. Oncogene. 2017;36(2):158–67.

Iqbal J, Abbasi BA, Ahmad R, Mahmood T, Ali B, Khalil AT, Kanwal S, Shah SA, Alam MM, Badshah H, Munir A. Nanomedicines for developing cancer nanotherapeutics: from benchtop to bedside and beyond. Appl Microbiol Biotechnol. 2018;102(22):9449–70.

Iravani S, Zolfaghari B. Green synthesis of silver nanoparticles using Pinus eldarica bark extract. BioMed Res Int. 2013;2013:1–5.

Karlsson HL, Gustafsson J, Cronholm P, Möller L. Size-dependent toxicity of metal oxide particles—a comparison between nano-and micrometer size. Toxicol Lett. 2009;188(2):112–8.

Khalil AT, Ovais M, Ullah I, Ali M, Jan SA, Shinwari ZK, Maaza M. Bioinspired synthesis of pure massicot phase lead oxide nanoparticles and assessment of their biocompatibility, cytotoxicity and in-vitro biological properties. Arabian J Chem. 2017. https://doi.org/10.1016/j.arabjc.2017.08.009.

Khalil AT, Ayaz M, Ovais M, Wadood A, Ali M, Shinwari ZK, Maaza M. In vitro cholinesterase enzymes inhibitory potential and in silico molecular docking studies of biogenic metal oxides nanoparticles. Inorg Nano-Metal Chem. 2018a;48(9):441–8.

Khalil AT, Khalil AT, Ovais M, Ullah I, Ali M, Shinwari ZK, Hassan D, Maaza M. Sageretia thea (Osbeck.) modulated biosynthesis of NiO nanoparticles and their in vitro pharmacognostic, antioxidant and cytotoxic potential. Artif Cells Nanomed Biotechnol. 2018b;46(4):838–52.

Khan I, Saeed K, Khan I. Nanoparticles: properties, applications and toxicities. Arabian J Chem. 2017. https://doi.org/10.1016/j.arabjc.2017.05.011.

Kim JK, Seo SJ, Kim KH, Kim TJ, Chung MH, Kim KR, Yang TK. Therapeutic application of metallic nanoparticles combined with particle-induced X-ray emission effect. Nanotechnology. 2010;21(42):425102.

Kim JS, Sung JH, Ji JH, Song KS, Lee JH, Kang CS, Yu IJ. In vivo genotoxicity of silver nanoparticles after 90-day silver nanoparticle inhalation exposure. Saf Health Work. 2011;2(1):34–8.

Kora AJ, Sashidhar RB. Biogenic silver nanoparticles synthesized with rhamnogalacturonan gum: antibacterial activity, cytotoxicity and its mode of action. Arabian J Chem. 2018;11(3):313–23.

Kumar R, Roopan SM, Prabhakarn A, Khanna VG, Chakroborty S. Agricultural waste Annona squamosa peel extract: biosynthesis of silver nanoparticles. Spectrochim Acta A Mol Biomol Spectrosc. 2012;90:173–6.

Kumari R, Barsainya M, Singh DP. Biogenic synthesis of silver nanoparticle by using secondary metabolites from Pseudomonas aeruginosa DM1 and its anti-algal effect on *Chlorella vulgaris* and *Chlorella pyrenoidosa*. Environ Sci Pollut Res. 2017;24(5):4645–54.

Lacave JM, Vicario-Parés U, Bilbao E, Gilliland D, Mura F, Dini L, Cajaraville MP, Orbea A. Waterborne exposure of adult zebrafish to silver nanoparticles and to ionic silver results in differential silver accumulation and effects at cellular and molecular levels. Sci Total Environ. 2018;642:1209–20.

Lim Z-ZJ, Li J-EJ, Ng C-T, Yung L-YL, Bay BH. Gold nanoparticles in cancer therapy. Acta Pharmacol Sin. 2011;32(8):983–90.

Macdonald JS. Toxicity of 5-fluorouracil. Oncology. 1999;13(7 Suppl 3):33–4.

Mal J, Veneman WJ, Nancharaiah YV, van Hullebusch ED, Peijnenburg WJ, Vijver MG, Lens PN. A comparison of fate and toxicity of selenite, biogenically, and chemically synthesized selenium nanoparticles to zebrafish (Danio rerio) embryogenesis. Nanotoxicology. 2017;11(1):87–97.

Manimaran M. A review on nanotechnology and its implications in agriculture and food industry. Asian J Plant Sci Res. 2015;5(7):13–5.

Manzoor M, Khan AHA, Ullah R, Khan MZ, Ahmad I. Environmental epidemiology of cancer in South Asian population: risk assessment against exposure to polycyclic aromatic hydrocarbons and volatile organic compounds. Arabian J Sci Eng. 2016;41(6):2031–43.

Mao X, Seidlitz E, Truant R, Hitt M, Ghosh HP. Re-expression of TSLC1 in a non-small-cell lung cancer cell line induces apoptosis and inhibits tumor growth. Oncogene. 2004;23(33):5632–42.

Mei N, Zhang Y, Chen Y, Guo X, Ding W, Ali SF, Biris AS, Rice P, Moore MM, Chen T. Silver nanoparticle-induced mutations and oxidative stress in mouse lymphoma cells. Environ Mol Mutagen. 2012;53(6):409–19.

Minai L, Yeheskely-Hayon D, Yelin D. High levels of reactive oxygen species in gold nanoparticle-targeted cancer cells following femtosecond pulse irradiation. Scient Rep. 2013;3:2146.

Mukherjee S, Patra CR. Therapeutic application of anti-angiogenic nanomaterials in cancers. Nanoscale. 2016;8(25):12444–70.

Mukherjee S, Sushma V, Patra S, Barui AK, Bhadra MP, Sreedhar B, Patra CR. Green chemistry approach for the synthesis and stabilization of biocompatible gold nanoparticles and their potential applications in cancer therapy. Nanotechnology. 2012;23(45):455103.

Mukherjee S, Vinothkumar B, Prashanthi S, Bangal PR, Sreedhar B, Patra CR. Potential therapeutic and diagnostic applications of one-step in situ biosynthesized gold nanoconjugates (2-in-1 system) in cancer treatment. RSC Adv. 2013;3(7):2318–29.

Mukherjee S, Chowdhury D, Kotcherlakota R, Patra S, Vinothkumar B, Bhadra MP, Sreedhar B, Patra CR. Potential theranostics application of bio-synthesized silver nanoparticles (4-in-1 system). Theranostics. 2014;4(3):316–35.

Mukherjee S, Dasari M, Priyamvada S, Kotcherlakota R, Bollu VS, Patra CR. A green chemistry approach for the synthesis of gold nanoconjugates that induce the inhibition of cancer cell proliferation through induction of oxidative stress and their in vivo toxicity study. J Mater Chem B. 2015;3(18):3820–30.

Mulvaney P. Surface plasmon spectroscopy of nanosized metal particles. Langmuir. 1996;12(3): 788–800.

Ovais M, Khalil AT, Raza A, Khan MA, Ahmad I, Islam NU, Saravanan M, Ubaid MF, Ali M, Shinwari ZK. Green synthesis of silver nanoparticles via plant extracts: beginning a new era in cancer theranostics. Nanomedicine. 2016;12(23):3157–77.

Ovais M, Raza A, Naz S, Islam NU, Khalil AT, Ali S, Khan MA, Shinwari ZK. Current state and prospects of the phytosynthesized colloidal gold nanoparticles and their applications in cancer theranostics. Appl Microbiol Biotechnol. 2017;101(9):3551–65.

Ovais M, Zia N, Ahmad I, Khalil AT, Raza A, Ayaz M, Sadiq A, Ullah F, Shinwari ZK. Phytotherapeutic and nanomedicinal approaches to cure Alzheimer's disease: present status and future opportunities. Front Aging Neurosci. 2018a;10:284.

Ovais M, Khalil AT, Ayaz M, Ahmad I, Nethi SK, Mukherjee S. Biosynthesis of metal nanoparticles via microbial enzymes: a mechanistic approach. Int J Mol Sci. 2018b;19(12):4100.

Padinjarathil H, Joseph MM, Unnikrishnan BS, Preethi GU, Shiji R, Archana MG, Maya S, Syama HP, Sreelekha TT. Galactomannan endowed biogenic silver nanoparticles exposed enhanced cancer cytotoxicity with excellent biocompatibility. Int J Biol Macromol. 2018;118:1174–82.

Panda KK, Achary VM, Krishnaveni R, Padhi BK, Sarangi SN, Sahu SN, Panda BB. In vitro biosynthesis and genotoxicity bioassay of silver nanoparticles using plants. Toxicol In Vitro. 2011;25(5):1097–105.

Papavlassopoulos H, Mishra YK, Kaps S, Paulowicz I, Abdelaziz R, Elbahri M, Maser E, Adelung R, Röhl C. Toxicity of functional nano-micro zinc oxide tetrapods: impact of cell culture conditions, cellular age and material properties. PLoS One. 2014;9(1):e84983.

Park Y. New paradigm shift for the green synthesis of antibacterial silver nanoparticles utilizing plant extracts. Toxicol Res. 2014;30(3):169.

Patra CR, Mukherjee S, Kotcherlakota R. Biosynthesized silver nanoparticles: a step forward for cancer theranostics? Nanomedicine. 2014;9(10):1445–8.

Patra S, Mukherjee S, Barui AK, Ganguly A, Sreedhar B, Patra CR. Green synthesis, characterization of gold and silver nanoparticles and their potential application for cancer therapeutics. Mater Sci Eng C. 2015;53:298–309.

Rahman S, Rahman L, Khalil AT, Ali N, Zia D, Ali M, Shinwari ZK. Endophyte-mediated synthesis of silver nanoparticles and their biological applications. Appl Microbiol Biotechnol. 2019;103:1–19.

Rai M, Ingle AP, Birla S, Yadav A, Santos CA. Strategic role of selected noble metal nanoparticles in medicine. Crit Rev Microbiol. 2016;42(5):696–719.

Rao PV, Gan SH. Recent advances in nanotechnology-based diagnosis and treatments of diabetes. Curr Drug Metab. 2015;16(5):371–5.

Roy N, Mondal S, Laskar RA, Basu S, Begum NA. Biogenic synthesis of Au and Ag nanoparticles by Indian propolis and its constituents. Colloids Surf B Biointerfaces. 2010;76(1):317–25.

Scown TM, Santos EM, Johnston BD, Gaiser B, Baalousha M, Mitov S, Lead JR, Stone V, Fernandes TF, Jepson M, van Aerle R, Tyler CR. Effects of aqueous exposure to silver nanoparticles of different sizes in rainbow trout. Toxicol Sci. 2010;115(2):521–34.

Shah A, Lutfullah G, Ahmad K, Khalil AT, Maaza M. Daphne mucronata-mediated phytosynthesis of silver nanoparticles and their novel biological applications, compatibility and toxicity studies. Green Chem Lett Rev. 2018;11(3):318–33.

Shakibaie M, Shahverdi AR, Faramarzi MA, Hassanzadeh GR, Rahimi HR, Sabzevari O. Acute and subacute toxicity of novel biogenic selenium nanoparticles in mice. Pharm Biol. 2013;51(1):58–63.

Sharma A, Goyal AK, Rath G. Recent advances in metal nanoparticles in cancer therapy. J Drug Target. 2018;26(8):617–32.

Silva-De Hoyos LE, Sánchez-Mendieta V, Camacho-López MA, Trujillo-Reyes J, Vilchis-Nestor AR. Plasmonic and fluorescent sensors of metal ions in water based on biogenic gold nanoparticles. Arabian J Chem. 2018. https://doi.org/10.1016/j.arabjc.2018.02.016.

Singh P, Kim YJ, Singh H, Wang C, Hwang KH, Farh ME, Yang DC. Biosynthesis, characterization, and antimicrobial applications of silver nanoparticles. Int J Nanomed. 2015;10:2567–77.

Singh P, Kim YJ, Zhang D, Yang DC. Biological synthesis of nanoparticles from plants and microorganisms. Trends Biotechnol. 2016;34(7):588–99.

Slavin YN, Asnis J, Häfeli UO, Bach H. Metal nanoparticles: understanding the mechanisms behind antibacterial activity. J Nanobiotechnol. 2017;15(1):65.

Sufian MM, Khattak JZK, Yousaf S, Rana MS. Safety issues associated with the use of nanoparticles in human body. Photodiagn Photodyn Ther. 2017;19:67–72.

Suresh AK, Pelletier DA, Doktycz MJ. Relating nanomaterial properties and microbial toxicity. Nanoscale. 2013;5(2):463–74.

Sweeney SF, Woehrle GH, Hutchison JE. Rapid purification and size separation of gold nanoparticles via diafiltration. J Am Chem Soc. 2006;128(10):3190–7.

Tannock IF, Rotin D. Acid pH in tumors and its potential for therapeutic exploitation. Cancer Res. 1989;49(16):4373–84.

Tessier PM, Velev OD, Kalambur AT, Lenhoff AM, Rabolt J, Kaler EW. Structured metallic films for optical and spectroscopic applications via colloidal crystal templating. Adv Mater. 2001;13(6):396–400.

Torre LA, Siegel RL, Ward EM, Jemal A. Global cancer incidence and mortality rates and trends—an update. Cancer Epidemiol Prev Biomark. 2016;25(1):16–27.

Velayutham M, Villamena FA, Fishbein JC, Zweier JL. Cancer chemopreventive oltipraz generates superoxide anion radical. Arch Biochem Biophys. 2005;435(1):83–8.

Wadhwani SA, Shedbalkar UU, Singh R, Chopade BA. Biogenic selenium nanoparticles: current status and future prospects. Appl Microbiol Biotechnol. 2016;100(6):2555–66.

WHO. Obesity and overweight fact sheet. 2017. Available from: http://www.who.int/mediacentre/factsheets/fs311/en/. Accessed 1 Jul 2018.

Zhang D, Zhao YX, Gao YJ, Gao FP, Fan YS, Li XJ, Duan ZY, Wang H. Anti-bacterial and in vivo tumor treatment by reactive oxygen species generated by magnetic nanoparticles. J Mater Chem B. 2013;1(38):5100–7.

Chapter 12
Superparamagnetic Iron Oxide Nanoparticles for Cancer Theranostic Applications

Dipak Maity, Ganeshlenin Kandasamy, and Atul Sudame

Abstract In the last few decades, superparamagnetic iron oxide nanoparticles (SPIONs—particularly magnetite (Fe_3O_4)/maghemite (Fe_2O_3) nanoparticles) have gained a great deal of attention in many biomedical applications, including magnetic targeting based cell isolation/sorting, tissue engineering, gene delivery, and magnetofection, due to their unique magnetic properties, excellent chemical stability, biodegradability, and low toxicity as compared to other magnetic materials (for instance, Co, Mn, and Ni). But recently, SPIONs (in the form of ferrofluids—i.e., SPIONs dispersed in a carrier fluid) have become a highly promising candidate for their use as therapeutic and diagnostic (theranostic) agents in cancer treatment applications such as magnetic fluid hyperthermia (MFH) and magnetic resonance imaging (MRI), respectively. However, the theranostic efficacies of the SPIONs (or ferrofluids) might alter due to the differences in their physicochemical/dispersibility/magnetic properties that are significantly impacted by their synthesis methods and their stabilization process. In this chapter, we have initially discussed the crystal structure/composition and different synthesis methods of the SPIONs. Then, we have described the role of the SPIONs in the formation of the ferrofluids along with their stabilization process via diverse interactions. Finally, we have discussed about their (1) intrinsic cancer theranostic applications of SPIONs such as magnetic fluid hyperthermia, magnetic resonance imaging, and magnetic nanoparticle-based drug delivery and (2) combined cancer theranostics applications including MRI as an adjuvant to fluorescence imaging, thermo-chemotherapy, thermo-radiotherapy, and thermo-immunotherapy.

D. Maity (✉)
Department of Chemical Engineering, Institute of Chemical Technology Mumbai, IOC Campus, Bhubaneswar, OD, India

G. Kandasamy
Department of Biomedical Engineering, Vel Tech Rangarajan Dr. Sagunthala R&D Institute of Science and Technology, Chennai, TN, India

A. Sudame
Department of Mechanical Engineering, Shiv Nadar University, Dadri, UP, India

M. Rai, B. Jamil (eds.), *Nanotheranostics*,
https://doi.org/10.1007/978-3-030-29768-8_12

Keywords SPIONs · Magnetic nanoparticles · Biomedical applications · Nanomedicine · Magnetic fluid hyperthermia · Magnetic resonance imaging · Ferrofluids · Theranostics · Cancer treatment

12.1 Introduction

Superparamagnetic iron oxide nanoparticles (SPIONs) are one of the most commonly used superparamagnetic nanoparticles (SPNs) and they are extensively investigated in various biomedical applications including drug delivery, magnetic fluid hyperthermia (MFH), magnetic resonance imaging (MRI), cell isolation and/or sorting, gene delivery, and tissue engineering due to their unique magnetic properties, excellent chemical stability, biodegradability, and low toxicity as compared to other magnetic materials (for instance, Co, Mn, and Ni) (Odenbach 2002; Prashant et al. 2010; Merbach et al. 2013; Wang et al. 2013; Demirer et al. 2015; Li et al. 2016; Ali et al. 2016). Generally, the SPIONs have core-shell structures which are composed of the magnetite (Fe_3O_4) and/or maghemite (γ-Fe_2O_3) cores, and the non-magnetic organic/inorganic surface coatings (or surfactants) (Kumar and Leuschner 2005; Maity and Agrawal 2007; Issa et al. 2013; Kandasamy and Maity 2015). The surfactants/surface coating molecules play an important role (along with the reactants during the synthesis process) in determining physicochemical properties (i.e., size, shape, surface charge, colloidal stability), and magnetic properties (magnetic susceptibility, saturation magnetization, superparamagnetic behavior) beside the purpose to protect the SPIONs from their aggregation/agglomeration. Moreover, the surface coating molecules enable them for effective surface functionalization or bio-conjugation (by bearing suitable surface functional groups), to improve biocompatibility (by reducing toxicity) and also to enhance hydrophilicity (water dispersibility) so that the SPIONs could be efficiently used for their instantaneous biomedical applications (Liu et al. 2009; Mahmoudi et al. 2011).

Iron oxide-based magnetic nanoparticles are usually synthesized in the nanodimensional regime—i.e., 1–100 nanometers (nm). In general, the large-sized magnetic particles display coercivity (H_c) and remanent magnetization (M_r) values due to their multi-domain structure ascribed to different crystallite orientations (as shown in Fig. 12.1). However, when the sizes of these particles are reduced to submicron (i.e., nanometer) regime, the multi-domain structure will get modified into a single-domain structure, and the coercivity value increases to maximum. The reduced nanometer-size at which these particles possess a single-domain structure is determined as single-domain size with a specific critical radius (r_c). For example, the r_c values of single-domain Fe_3O_4 and γ-Fe_2O_3 nanoparticles are, respectively, calculated as ~30 and ~60 nm (Trohidou 2014; Li et al. 2017). Also, when the size of the magnetic particles is reduced further, these particles might possess "super-

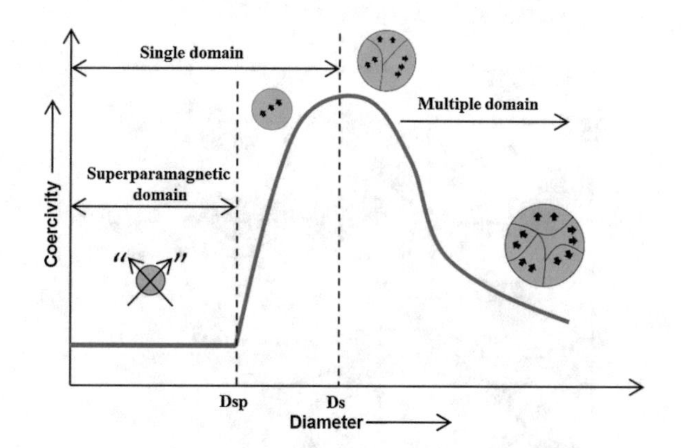

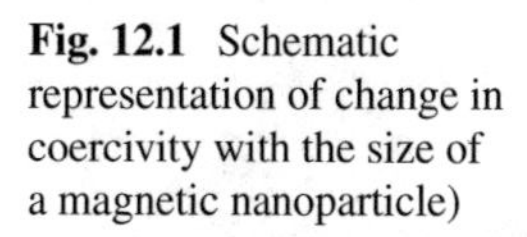
Fig. 12.1 Schematic representation of change in coercivity with the size of a magnetic nanoparticle)

paramagnetism" and this reduced size is called as superparamagnetic size, usually in the range of 4–20 nm for Fe_3O_4/γ-Fe_2O_3 nanoparticles at room temperature (Ortega and Giorgio 2012). Herein, "superparamagnetism" indicates that the size-reduced magnetic nanoparticles display a robust paramagnetic nature with high saturation magnetization (M_S) and magnetic susceptibility (χ) under the influence of an externally applied magnetic field. Moreover, these nanoparticles might lose their magnetization completely once the magnetic field is removed, which results in zero H_c and M_r.

Superparamagnetic iron oxide nanoparticles (SPIONs), especially magnetite (Fe_3O_4) nanoparticles, have an inverse spinal crystal structure composed of (1) both divalent iron (Fe^{2+}) and trivalent iron (Fe^{3+}) ions at octahedral sites and (2) one trivalent iron (Fe^{3+}) ions at the tetrahedral sites—as shown in Fig. 12.2 (Bastow and Trinchi 2009). Herein, the total stoichiometric ratio of Fe^{2+} to Fe^{3+} ions is 0.5. Moreover, the crystal structure of maghemite (Fe_2O_3) nanoparticles is similar to that of magnetite nanoparticles (i.e., spinel structure); however, the only difference is that all the iron ions are in the trivalent state (i.e., Fe^{3+} ions). Besides, the oxygen anions (O^{2-}) are arranged among the iron ions to form a close-packed array with cubic structure in both Fe_3O_4 and γ-Fe_2O_3 nanoparticles (Cornell and Schwertmann 2004; Qiao et al. 2009). Usually, the magnetic moments in the SPIONs (magnetite/maghemite) originate from the presence of unpaired 3d electrons in Fe^{3+}/Fe^{2+} cations in their crystal structure. However, these cations are located far apart from each other to hinder their interaction (for magnetic moment formation). Nevertheless, an exchange coupling between the cations (Fe^{3+} and Fe^{2+} ions) is possible through the non-magnetic oxygen anions (O^{2-}) which helps in the formation of the magnetic moments (Moskowitz 1991; Spaldin 2003; Tartaj et al. 2003; Liu et al. 2009; Thanh 2012; Wu et al. 2015).

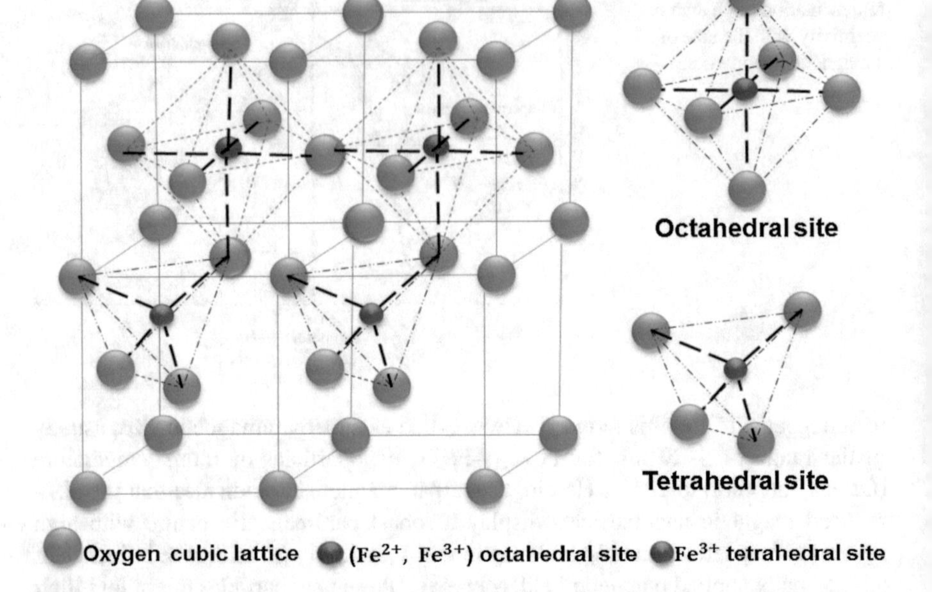

Fig. 12.2 Crystal structure representation of magnetite (Fe_3O_4) unit cell through ball-and-stick model. (Adapted from Bastow and Trinchi 2009)

12.2 Synthesis Methods

SPIONs are one of the common magnetic nanoparticles approved by Food and Drug Administration (FDA) for usage in biomedical applications such as cancer therapeutics and/or diagnostics (theranostics) (Revia and Zhang 2016; Stephen et al. 2012). However, in-depth studies are required to use these SPIONs effectively in theranostic applications under clinical scenarios. Therefore, the researchers are fine-tuning the synthesis methods to obtain high-quality SPIONs with good colloidal stability, high magnetization, and narrow size distribution. The following are the major hydrolytic and non-hydrolytic synthetic chemical routes that are widely utilized to synthesize high-quality SPIONs.

12.2.1 Hydrolytic Synthetic Routes

Hydrolytic synthetic routes are utilized as conventional routes to directly synthesize hydrophilic SPIONs, based on the chemical reactions among iron precursors in aqueous conditions. Besides, the SPIONs synthesized via hydrolytic methods are more appropriate for their instant biomedical applications. The major synthetic routes including coprecipitation, hydrothermal, microemulsion, and sonochemical methods are discussed for the synthesis of SPIONs as follows.

12.2.1.1 Coprecipitation Method

Coprecipitation method is an extensively used hydrolytic route to synthesize SPIONs, where the precipitation of iron oxide nanoparticles is made via chemical reactions between ferric/ferrous salts (nitrates/sulfates/chlorides/perchlorates) and a base (NaOH/NH_4OH) under aqueous condition at slightly elevated temperatures (i.e., 40–80 °C). The main reaction mechanism involved in the formation of the SPIONs (for e.g., Fe_3O_4) is as follows (Ahn et al. 2012):

$$Fe^{2+} + 2Fe^{3+} + 8OH^{-} \leftrightarrows Fe(OH)_2 + 2Fe(OH)_3 \rightarrow Fe_3O_4 \downarrow + 4H_2O$$

This reaction is usually performed in an inert atmosphere (for e.g., nitrogen (N_2) or argon (Ar)) to avoid the formation of unwanted iron oxide phases (such as α-Fe_2O_3) in the as-synthesized SPIONs. Moreover, the reaction mechanism in the formation of the SPIONs usually passes through a topotactic transition (structural change to crystalline solid) phase in either one of the following routes: (1) nucleation → akaganeite phase → goethite → hematite/maghemite → magnetite (Fe_3O_4); (2) nucleation → ferrous hydroxide → lepidocrocite → maghemite → magnetite (Fe_3O_4). Furthermore, the path of this topotactic transition is majorly dependent on the variations in the pH of the aqueous reaction mixture. In addition, the physicochemical properties (such as shape, size, colloidal stability, and morphology) of the SPIONs can be tuned by altering reaction temperature, time of reaction, concentration of reactants, type of base, stabilizing agents, and reactant molarity (Mahmoudi et al. 2011; Ahn et al. 2012; Fu and Ravindra 2012; Mojica Pisciotti et al. 2014; Wu et al. 2015).

12.2.1.2 Microemulsion Method

Microemulsion is an optically transparent and thermodynamically stable solution and is classified into three major types: (1) water-in-oil, (2) oil-in-water, and (3) bicontinuous microemulsions. Out of these types, water-in-oil microemulsion is mostly used to synthesize the SPIONs, where the reverse micelles (containing the aqueous droplets of reactants—surrounded by a surfactant monolayer) are formed in a continuous oil phase, which might react with each other to form the SPIONs. Additionally, the synthesis of the SPIONs can also be carried out in either of the following two routes: (1) mixing of two or more microemulsions that contain different iron precursors; and (2) adding a precipitating agent (i.e., for example, ammonia) dropwise into the microemulsion containing the iron precursors. The reactions might take place inside the droplets (that mainly act as a nanoscale reactor) and the final nanoparticles can be collected by removing the excessive surfactants/solvents (Boutonnet et al. 2008; Okoli et al. 2011). In this method, the physiochemical/magnetic properties of SPIONs are majorly dependent on the droplet size, concentration of the precursors, and type of surfactants/solvent.

12.2.1.3 Hydrothermal Method

Hydrothermal method is another conventional method used to synthesize the SPIONs (Kim et al. 2013). In this method, the nanoparticles are synthesized by performing the aqueous chemical reactions among the iron precursors in the presence/absence of the surfactants in a sealed container (inside an autoclave) which provide high temperature/vapor pressure (up to 250 °C/4 MPa) for chemical reactions. After the reaction, the mixture of aqueous solution is cooled down to the room temperature and the SPIONs are obtained by removing the residual surfactants, unreacted precursors, and other impurities. In this method, the physicochemical/magnetic properties of the SPIONs can be modified by tuning the reaction temperature, reaction time, amount of surfactant, and precursors (Kim et al. 2013; Piñeiro et al. 2015).

12.2.1.4 Sonochemical Method

Sonochemical method is based on inducing the reaction among mixture of iron precursors (for example, ferric or ferrous salts) via ultrasound irradiation having frequency ranging from 20 to 60 kHz to synthesize the SPIONs (Wu et al. 2008). This ultrasound irradiation (containing alternating expansive and compressive acoustic waves) generates cavitation microbubbles (i.e., cavities) in iron precursor solution, which induces nano-crystal nucleation by accumulating the ultrasonic energy. Finally, the microbubbles might collapse and subject to release the stored concentrated energy (with a heating and cooling rate of $>10^{10}$ K/s) that tends to increase the temperatures within the cavitation bubbles in a very short time (~1 ns) (Morel et al. 2008). Because of this, H^+ and OH^- radicals are produced through the decomposition of water, which further react with iron precursor mixtures to form the SPIONs (Yoffe et al. 2013). This method is beneficial to reduce unwanted growth of nano-crystals. However, the SPIONs with controlled physicochemical/magnetic properties are difficult to synthesize via this method (Pinkas et al. 2008; Wu et al. 2015).

Nonetheless, more research works are essential to overcome the drawbacks associated with these hydrolytic synthetic routes, including low crystallinity, broad particle size distribution, and/or complicated surface characteristics (Qiao et al. 2009).

12.2.2 Non-hydrolytic Synthetic Routes

The non-hydrolytic synthetic routes used in the preparation of the SPIONs are based on the chemical reactions of the iron precursors (in the presence/absence of surfactants) dissolved in organic solvents, which consequently results in hydrophobic SPIONs. Herein, thermal decomposition method is a major non-hydrolytic synthetic route used for the synthesis of the hydrophobic SPIONs with controlled physicochemical properties including size/shape/composition/crystal structure and magnetic properties (for instance, saturation magnetization) (Mutin and Vioux 2009; Qiao et al. 2009).

12.2.2.1 Thermal Decomposition Method

Thermal decomposition method (thermolysis) is based on the decomposition of the iron precursors (without/with the presence of surfactants) in high boiling point organic solvent by heating them at very high temperatures ranging from 200 to 350 °C. Herein, decomposition of the precursor occurs due to the breakage of their chemical bonds (via endothermic reaction) in an inert atmosphere by supplying continuous nitrogen/argon gas to avoid the formation of unnecessary iron oxide phases. The precursor materials and capping agents frequently used in the synthesis of the SPIONs via thermolysis are as follows: (1) iron precursors—ferric acetylacetonate ($[Fe(acac)_5]$ (acac = acetylacetonate)), iron pentacarbonyl ($Fe(CO)_5$), and iron cupferon ($[Fe(cup)_3]$ (cup = *N*-nitrosophenylhydroxylamine)); and (2) surfactants—oleylamine, hexadecylamine, oleic acid, linolenic acid, steric acid, and other fatty amines/acids (Demirer et al. 2015; Piñeiro et al. 2015). The physicochemical/magnetic properties of the SPIONs can be controlled by optimizing the reaction temperature, amount of precursors/surfactants, and reaction time (Maity et al. 2008).

Nevertheless, SPIONs prepared by this method are hydrophobic in nature, and cannot be directly used for cancer theranostic application. Therefore, the additional surface modification procedures like bilayer surfactant stabilization/ligand exchange methods are required to modify the hydrophobic nature of the surface of the SPIONs into hydrophilic nature. Recently, one-pot thermal decomposition method using polyol-based surfactants/solvents are extensively used to directly synthesize the hydrophilic SPIONs for instant use in cancer theranostic applications (Maity et al. 2008, 2009, 2010a, 2011a; Turcheniuk et al. 2013; Li et al. 2015; Piñeiro et al. 2015; Kandasamy et al. 2018a, 2019a, b).

12.3 Ferrofluids or Magnetic Fluids

12.3.1 Ferrofluid Formation: SPIONs as Building Blocks

Ferrofluids or magnetic fluids are homogeneous stable/colloidal suspensions of the SPIONs (coated with suitable molecules on their surfaces) that are dispersed in an appropriate carrier liquid (which can be polar/nonpolar). Hence, a typical ferrofluid or magnetic fluid consists of the following components (by volume): (1) magnetic solids (~5%), (2) surfactants (~10%), and (3) a carrier fluid (~85%) (Gupta and Gupta 2015). Moreover, the individual components that form the ferrofluids are explained as follows:

1. *Magnetic Solids*: Herein, the magnetic cores of the magnetic nanoparticles/solids act as a main source in the formation of the ferrofluids. The sizes of the magnetic cores inside the nanoparticles should be adequately (a) small enough for their uniform suspension in the carrier liquid via Brownian motion (which is the random motion of particles in a liquid due to multiple collisions among them) and (b) large enough to make considerable contribution in the magnetic response of the ferrofluids, while applying a magnetic field. Therefore, magnetic cores of the nanoparticles should have sizes $\leq D_{sp}$ or the individual nanoparticles should be superparamagnetic for their effective usage as magnetic solids. In general, SPIONs (such as Fe_3O_4 and/or γ-Fe_2O_3 nanoparticles) having sizes of 5–20 nm mainly act as the magnetic solids (Araki et al. 2009; Raj et al. 1995; Kalikmanov 2001).
2. *Surfactants*: The surfactants are mainly used to avoid clumping/aggregation and oxidation of the magnetic solids/SPIONs during the formation of the ferrofluids since (a) the SPIONs have very high dipole–dipole magnetic interactions and (b) it is difficult to maintain the dispersibility of the SPIONs in the ferrofluid suspensions by only Brownian motion—ascribed to their heavy weight (Raj and Boulton 1987). The surfactants mainly prevent the agglomeration/oxidation problems (even when exposed to the strong magnetic/gravitation fields and atmosphere) by creating strong electrostatic and/or steric repulsions around them. Generally, the surfactants consist of a polar head and/or a nonpolar tail (or vice versa), where one of their ends might adsorb onto the surface of the SPIONs and the other ends are exposed to the carrier liquid (to create repulsions) while forming ferrofluid suspensions. In addition, the surfactants play also important role in reducing the viscosity by decrementing the packing density of the SPIONs (Raj and Boulton 1987; Odenbach 2003; Scherer and Neto 2005; Gupta and Gupta 2015). The most commonly used surfactants in the ferrofluid formation are phosphonic acid, carboxylic acid, silane, catechol, polymers, and gold—as shown in Fig. 12.3 (Turcheniuk et al. 2013).
3. *Carrier Liquid*: The selection of a carrier liquid is significant in order to govern the overall physical properties of the ferrofluids. The carrier liquid is generally selected based on the following categories: (a) the surface nature (either hydro-

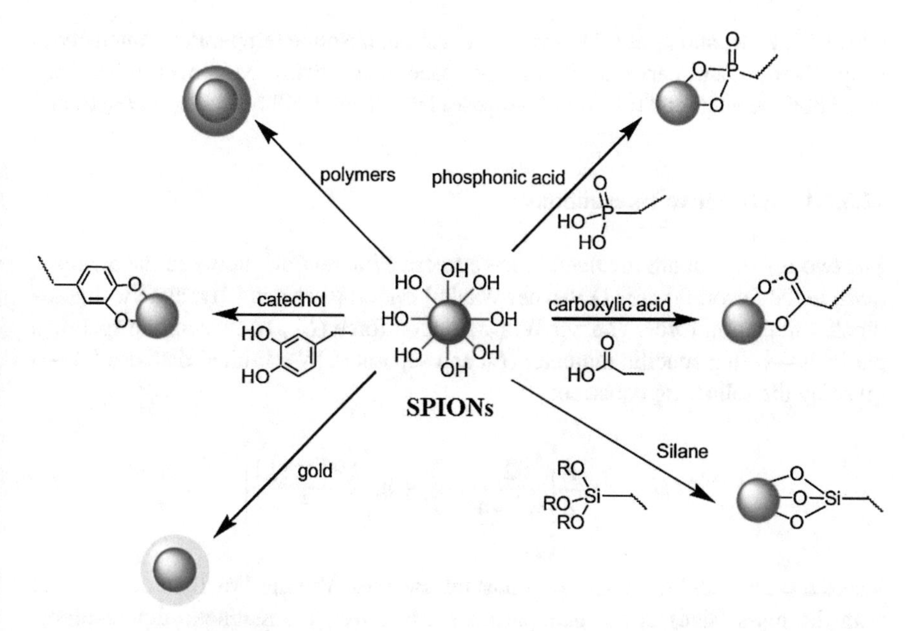

Fig. 12.3 Common surfactants used for the protection/stabilization of SPIONs. (Adapted from Turcheniuk et al. 2013)

phobic or hydrophilic) of the SPIONs, (b) field of application (water for biological applications), and compatibility of the SPIONs with the carrier liquid. The common carrier liquids (or solvent) used in ferrofluid formation include organic (hexane/toluene/tetrahydrofuran) or aqueous (water/biologic media including phosphate buffer saline (PBS), fetal bovine serum (FBS), and Dulbecco's modified eagle medium (DMEM)) based solvents (Raj and Boulton 1987; Raj et al. 1995; Odenbach 2003; Scherer and Neto 2005; Gupta and Gupta 2015).

12.3.2 Ferrofluid Stabilization

The stabilization of the ferrofluids is chiefly dependent on (1) the balance between the attractive and repulsive interactions, and (2) contribution from thermal energy. The typical nanoparticle diameter (*D*) to avoid the agglomeration can be evaluated by using the following equation that compares the thermal energy with the dipole–dipole pair energy.

$$D \le \left(\frac{72 K_B T}{\pi u_0 M^2} \right)^{1/3}$$

where K_B, T, M, and u_0 are Boltzmann's constant, absolute temperature, intensity of magnetization, and permeability of free space, respectively. Moreover, it has been noted that the magnetic (i.e., dipole–dipole) interactions will be lower, if $D \leq 10$ nm.

12.3.2.1 Attractive Interactions

The two forces that are involved in the attractive interactions between the nanoparticles in the ferrofluid are (1) Van der Waals-London force and (2) magnetic dipole–dipole interaction force. Van der Waals-London force (U_{AW}) between two spherical particles—with a specific diameter (D) and separated by a limited distance (r)—is given by the following equation:

$$U_{AW} = -\frac{A}{6}\left[\frac{2}{\alpha^2 - 4} + \frac{2}{\alpha^2} + \ln\left(\frac{\left(\alpha^2 - 4\right)}{\alpha^2}\right)\right]$$

where $\alpha = 2r/D$ and A—Hamaker constant. Besides, Van der Waals force increases with the mass (size) of the nanoparticles. Moreover, the magnetic dipole–dipole interaction force (U_{Ad}) between two magnetic dipoles (μ_1) and (μ_2) separated by a specific distance (r) is given by the following equation:

$$U_{Ad} = -\frac{u_o}{4\pi^3}\left[u_1 \cdot u_2 - 3\left(u_1 \cdot \frac{\overline{r}}{r}\right)\left(u_1 \cdot \frac{\overline{r}}{r}\right)\right]$$

where $\overline{r}$ is the relative position of the nanoparticles. The total attractive force between two nanoparticles is the sum of their Van der Waals-London and magnetic dipole–dipole interaction forces.

12.3.2.2 Repulsive Interactions

The major two forces involved in the repulsive interactions between the nanoparticles inside the ferrofluid are (1) electrostatic repulsion (long range) and (2) steric repulsion (short range) forces. In ionic ferrofluids (i.e., the SPIONs are surface-coated with ionic surfactants), the electrostatic repulsive forces keep the nanoparticles apart to avoid their agglomeration for assuring the colloidal stability. Moreover, the electrostatic repulsive force (U_R) between two electrically charged spherical particles—with diameter (D) and separation by a distance (r)—is given by the following equation:

$$U_R = -\frac{D\pi\sigma^2}{\varepsilon_0 \varepsilon_r k^2}\exp\left[-k\left(r - D\right)\right]$$

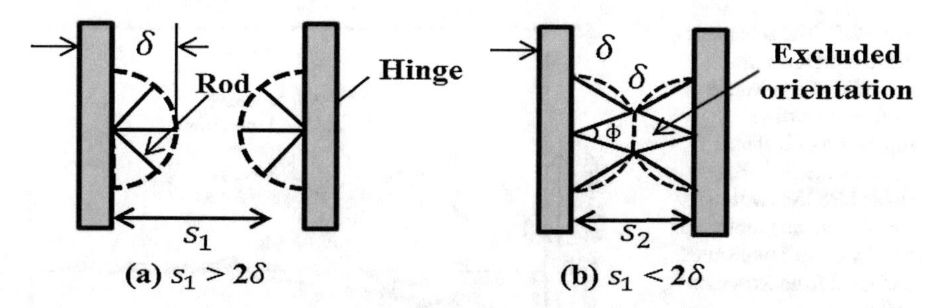

Fig. 12.4 Geometric illustration of steric repulsion energy using rigid rods attached onto universal hinge. (**a**) $s_1 > 2\delta$. (**b**) $s_1 < 2\delta$. (Adapted from Rosensweig 1997)

where σ is surface charge density and given by the equation: $\sigma = \varepsilon_0\varepsilon_r\varphi_0$ and ε ($\varepsilon_0\varepsilon_r$) is the electric permittivity of the fluid carrier. Moreover, φ_0 is surface potential of the charged nanoparticle at Helmholtz plane (in a double-layer model).

Besides, the steric repulsive force is linearly dependent on the temperature and can be understood by the geometric illustration of rigid rods attached onto a universal hinge—as shown in Fig. 12.4 (Rosensweig 1997). Here, the head pole groups of the surfactant are assumed to be adsorbed onto the surface, and the tail groups (rod) are assumed to form a specific hemisphere orientation under the thermal motion. The equation for the steric repulsive force between the spherical particles—having a specific diameter (D) with surfactants shell thickness (δ) and density (ξ molecular per nm^2), at temperature T—is given below:

$$\frac{U_t}{kT} = -\frac{\pi d^2 \xi}{2}\left[2 - \frac{l+2}{t}\ln\left(\frac{1+t}{1+l/2}\right) - \frac{l}{t}\right]$$

where $l = 2s/D$, and $s = r - D$ is the separation between the surfaces and $t = 2\delta/D$.

Thus, the colloidal stability of the nanoparticles in ferrofluid suspension is mainly dependent on the net content between the attractive and repulsive interactions among them. Moreover, the net interaction curve in the ferrofluid stabilization is given in Fig. 12.5 (Araki et al. 2009; Rosensweig 1997; Kalikmanov 2001; Scherer and Neto 2005).

12.4 Intrinsic Cancer Theranostic Applications

The most prominent cancer theranostic applications of the SPIONs include (1) magnetic fluid hyperthermia (MFH) therapy, (2) magnetic resonance imaging (MRI), and (3) magnetic drug delivery, where these SPIONs act as heating agents (HEA), contrast enhancing agents (CEA), and drug carriers, respectively (Kudr et al. 2017).

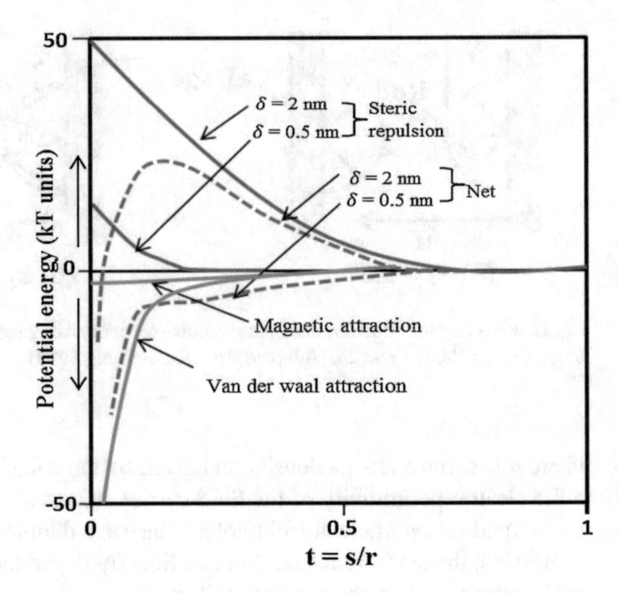

Fig. 12.5 The schematic curve represents the potential energy versus surface-to-surface separation of/distance between the colloidal stable SPIONs (with $d = \sim 10$ nm and molecular density of 10^{18} molecules). (Adapted from Rosensweig 1997)

12.4.1 Magnetic Fluid Hyperthermia Therapy

Magnetic fluid hyperthermia (MFH) therapy is a localized thermal therapy used for the treatment of malignant tumors/cancer, in which superparamagnetic nanoparticles (especially SPIONs) act as heating generating agents (Kandasamy and Maity 2015). Figure 12.6 gives a schematic representation of the application of the SPIONs (in the form of ferrofluids) in MFH application.

In brief, the SPIONs are normally delivered and localized near to the tumor site via passive/magnetic/active targeting, and subsequently exposed to an alternating magnetic field (AMF) for a certain time period (for e.g., ~1–2 h). This process generates heat which raises the tumor temperature to about 42–45 °C that is used for treating the cancer cells. Herein, the induced heat causes obstructions or ceases many cellular functions including cell proliferation and gene expressions in cancer cells which tends to induce cell death via apoptosis (Laurent et al. 2011; Revia and Zhang 2016). However, the heat produced during this therapy create very minimal damage to the nearby normal cells/organs as compared to other conventional treatment methods (for instance, chemo-radiation-therapies) (Hervault and Thanh 2014; Abadeer and Murphy 2016). Besides, the generated heat via this method could be controlled by tuning the physicochemical/magnetic properties of the SPIONs (including the size, shape, saturation magnetization, surfactants, and dispersion medium) and also the applied AMF (Kandasamy and Maity 2015; Kandasamy et al. 2018a, b).

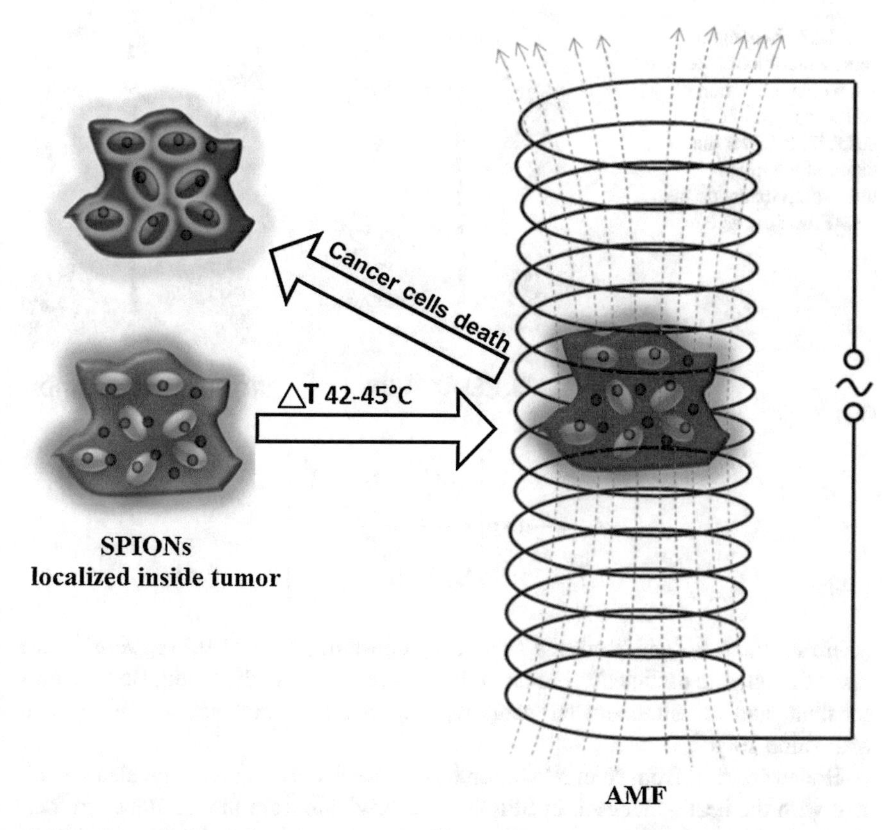

Fig. 12.6 Schematic representation of magnetic fluid hyperthermia (MFH) based cancer treatment by using ferrofluids containing the SPIONs

12.4.1.1 Heating Generation Mechanism

The heat generated by the SPIONs is due to the susceptibility loss, hysteresis loss, viscous heating or either one of their combinations, under the influence of the AMF. However, the susceptibility loss is a leading phenomenon for the heat generation in SPIONs, which majorly works on the Neel and Brownian relaxation losses. Generally, the magnetic moments of the SPIONs are fluctuated in the nanosecond timescale at their easy axis due to superparamagnetic nature (i.e., in the absence of an externally applied AMF). However, under the influence of an external AMF, Neel's relaxation time lag behind the reversal time of AMF (as shown in Fig. 12.7), which is caused by the energy of anisotropy that resists the magnetic moments to orient in the direction of applied AMF (Laurent et al. 2008). As a result, the energy that consumes to overcome the anisotropy is dissipated in the form of heat and known as "Neel relaxation loss." Moreover, the time for relaxation of magnetic moments of the nanoparticles (i.e., Neel relaxation time) is expressed by the following equation:

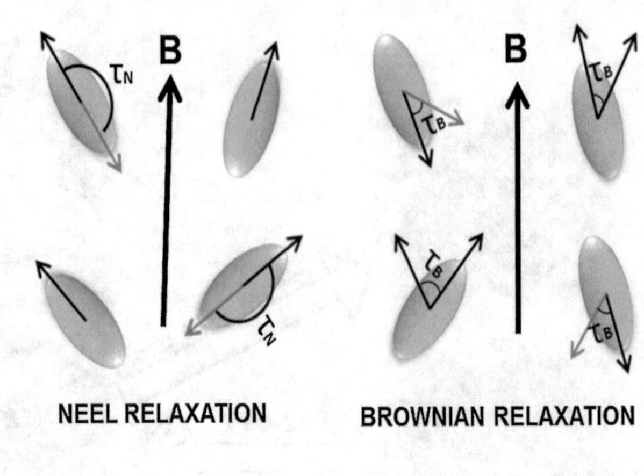

Fig. 12.7 Schematic representation of Néel and Brownian relaxation of SPIONs in colloidal suspension under the externally applied alternating magnetic field (AMF, marked as B)

$$\tau_N = \frac{1}{f_0} \exp\left[\frac{KV\left(1 - H/H_K\right)^2}{k_B T} \right]$$

where K, V, H, k_B, and T are the anisotropy constant (i.e., 3 × 10^5 ergs/cm^3 in the case of magnetite particles), volume of the nanoparticles, applied field, Boltzmann's constant, and measurement temperature, respectively. Moreover, f_0 is a constant with value 10^9 s^{-1}.

However, apart from Neel relaxation losses, the Brownian losses are also associated with the heat generation in SPIONs that originate from the hindrance to their physical rotational mobility (on application of an external AMF) by the viscosity of the carrier liquid (as shown in Fig. 12.7). Herein, the energy—that is consumed to overcome the viscosity—is dissipated in the form of heat. Moreover, the Brownian relaxation time is calculated by the following equation:

$$\tau_B = \frac{3VV_H\eta}{k_B T}$$

where V_H is the hydrodynamic volume of the SPIONs (by considering total effective diameter of the magnetic core and non-magnetic shell/coating), and η is viscosity of medium.

So the hindrances in the Néel and Brownian relaxations lead to a phase lag (between the AMF and the orientation direction of the magnetic moments/particles) that tends to generate heat in the form of susceptibility loss. Moreover, the heat induced by the SPIONs is also dependent on their size, shape, crystallinity, intrinsic magnetic properties, and the magnitude (H)/frequency (f) of the applied AMF. For instance, Néel and Brownian relaxations are dominant for the smaller and larger size SPIONs, respectively. Apart from susceptibility loss, other heating mechanisms such as hysteresis loss/viscous heating (stirring) are negligible for the SPIONs, since they occur mainly in single/multi-domain magnetic particles (Bedanta and Kleemann 2009; Kozissnik et al. 2013; Laurent et al. 2011; Suto et al. 2009).

Moreover, the generated heat (i.e., heating efficacy) by the SPIONs is quantitatively determined in terms of specific absorption rate (SAR) which is expressed by the following relation (Kandasamy and Maity 2015; Lahiri et al. 2016).

$$SAR\left(\mathrm{W}/\mathrm{g}_{Fe}\right)=\frac{\rho_{\mathrm{samp}}\times C_{\mathrm{samp}}}{m_{Fe}}\frac{\Delta T}{\Delta t}$$

where ρ_{samp} and C_{samp} are, respectively, the solvent density and specific heat capacity. Moreover, m_{Fe} and $\Delta T/\Delta t$ are the weight fraction of the magnetic element (i.e., Fe) in the sample and initial slope of the time-dependent temperature curve, respectively. However, some researchers prefer to calculate the heating efficiency in terms of intrinsic loss power (ILP) via normalization of SAR by taking into account the AMF (frequency (f) and amplitude (H)) for better comparison of the reported results from different research laboratories. The ILP is calculated by the following equation (Lahiri et al. 2016).

$$ILP\left(nH\mathrm{m}^2/\mathrm{kg}\right)=\frac{SAR}{H^2 f}$$

12.4.1.2 SPIONs as Heating Agents

Initially in 1957, Gilchrist et al. have used the iron oxides nanoparticles (i.e., Fe_2O_3 nanoparticles with size of 20–100 nm) as heating agents for the lymph node treatment of dogs via MFH. This is performed by primarily injecting these nanoparticles into the lymphatic channels of dogs, and subsequently exposing the affected area to AMF for induction heating to destroy tumor cells (Gilchrist et al. 1957).

However lately, different research groups have performed various experimental investigations on the heat induction process of the SPIONs having various sizes, shapes, distributions, and organic/inorganic surface coatings under calorimetric/in vitro conditions to evaluate their potentiality for further in vivo applications or clinical trials. For instance, recently Maity et al. have synthesized triethylene glycol (TREG)/triethanolamine (TEA) coated magnetite (Fe_3O_4 with superparamagnetic character) nanoclusters and have performed the calorimetric studies under the applied AMF with $H = 89$ kA/m and $f = 240$ kHz, where these nanoclusters have shown SAR values in the range of 135–500 W/g. Moreover, 74% inhibition in the proliferation of MCF-7 cancerous cells is also attained (via MFH through these nanoclusters) when treated at a therapeutic temperature of 45 °C for 1 h (Maity et al. 2011b). Similarly, Gkanas has also demonstrated that TREG/decanethiol/polyethylene glycol (PEG-800)-coated SPIONs have induced considerable inhibition in the proliferation of three different cancer cell lines (DA3, MCF-7 and HeLa cancer) while applying the AMF with $H = 26.48$ kA/m and $f = 765$ kHz (Gkanas 2013).

Likewise, few in vivo MFH experiments in different cancerous animal models are also recently performed (by using SPIONs). For example, Rabias et al. (2010) have infused 150 μL of maghemite nanoparticles (10–12 nm) into the glioma tumors inside the rats, where MFH therapy is done for damaging the cancerous tissue by inducing a therapeutic temperature (for 20 min) under the applied AMF with $H = 11$ kA/m and $f = 150$ kHz (Rabias et al. 2010). In another MFH study, Ohno et al. have reported significant survival period for glioma-bearing rats while treated with carboxymethyl cellulose-coated SPIONs (for 10 nm) by applying the AMF with $H = 30.3$ kA/m and $f = 88.9$ kHz (Ohno et al. 2002). In similar fashion, arginyl glycylaspartic acid (RGD) peptide-conjugated and poly(maleic anhydride-alt-1-octadecene)-coated Fe_3O_4 nanoparticles have resulted in significant reduction in viability of liver cancers after MFH treatment on application of AMF with $H = 14$ kA/m and $f = 606$ kHz (Arriortua et al. 2016).

In addition, researchers have executed few clinical trials (at different phases (I/II/III)) by using MFH therapy for cancer treatment but with moderate achievements. However, very recently, researchers are majorly involved in (1) reducing the side effects associated with this MFH therapy and also (2) achieving high therapeutic efficacy by improving the inherent properties of SPIONs or by combining SPION-based MFH therapy with other therapies (Silva et al. 2011; Kitture et al. 2012; Thanh 2018).

12.4.2 Magnetic Resonance Imaging

Magnetic resonance imaging (MRI) is a diagnosis technique in radiology that uses the principles of radio waves, magnetism, and computer technology to produce anatomical images of human body. An MRI used with a nanoparticle-based contrast agent (for instance, SPIONs) is an effective way to detect the cancerous tumors located deep inside the body (Grover et al. 2015; Jo et al. 2016).

12.4.2.1 Relaxation Process

In general, the following process occurs in MRI: (1) initial alignment of nuclei (containing an odd number of protons and/or neutrons) in the direction of applied magnetic field; (2) then rotation of the aligned nuclei (also known as excitation) by applying a pulse of radio waves; and (3) finally emission of a radio signal by the nuclei while returning (also known as relaxation/realignment) to their equilibrium state, which is recorded by a radio-frequency (R.F.) detector (as shown in Fig. 12.8) and subsequently processed via a computer for images. In human MRI, the water molecules (especially hydrogen/proton nuclei) from different tissues are utilized to image the entire body in slices. Herein, the nuclei tend to excite and relax/realign (while applying the magnetic field/radio waves) at a specific time period, which is called as relaxation time that can be either T1 or T2 relaxation time ascribed, respec-

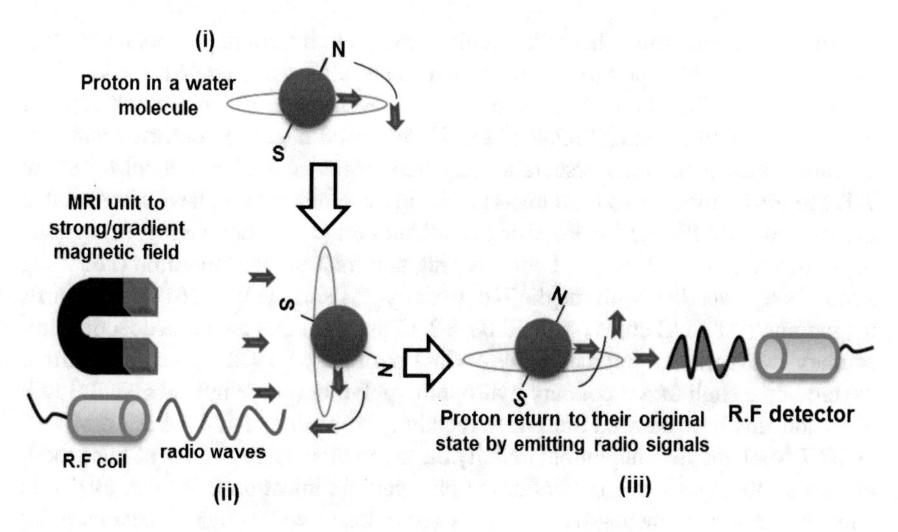

Fig. 12.8 Schematic representation of basic principle behind magnetic resonance imaging (MRI) based on hydrogen nuclei/proton excitation and relaxation

tively, to longitudinal (spin–lattice) or transverse (spin–spin) relaxations (Jacques et al. 2010; Stephen et al. 2012). Moreover, these relaxations differ diversely for various tissues which gives positive or negative contrast with corresponding relaxivity values of r1(1/T1) or r2(1/T2) (Nitz and Reimer 1999; Yin et al. 2004; Pooley 2005; McCarthy 2011).

12.4.2.2 SPIONs as Contrast Agents

Initially, the most widely used MRI contrast agents in clinical application are based on gadolinium (Gd(III)) complexes (used as positive contrast agent) including Magnevist® (Gadopentetic acid, Gd-DTPA—one of the oldest contrast agent approved by FDA in 1988), Multihance® (Gadobenate disodium, Gd-BOPTA), OptiMARK® (Gadoversetamide, Gd-DTPA-BMEA), and Omniscan® (Gadodiamide, Gd-DTPA-BMA). Nonetheless, various research investigations have raised concerns about the potential health risks (including nephrotoxicity) of these gadolinium-based complexes. Therefore, it is essential to find an alternative for gadolinium-based contrast agents for MRI imaging. Then recently, SPIONs are newly introduced as negative contrast agents for MRI applications since these nanoparticles tend to relax in transverse directions and they are more biocompatible/biodegradable and safe as compared to gadolinium complexes (Maity et al. 2010b; Tan et al. 2011; Corot and Warlin 2013; Chopra et al. 2016). Further, some of the SPION-based negative contrast agents are approved by FDA and in clinical trials also (Corot and Warlin 2013). In addition, very recently it has also been proved that the SPIONs having extremely small particle size (<5 nm) could potentially act as a positive contrast agent (Shen et al. 2017).

However, many researchers are widely involved in improving the relaxivity values of the SPIONs for providing better contrast and also to overcome the side effects associated with the usage of surfactants and concentration (Shen et al. 2017). For example, dextran-coated SPIONs (12 ± 2 nm) based negative contrast agents are prepared with high transverse relaxivity (r2) value of 90.5 ± 0.8 mM^{-1} s^{-1} at 7 T. Moreover, the in vivo MRI experiments in nude mice model have revealed that the as-prepared SPIONs are promising candidates for (1) tumor imaging, (2) specific organ imaging, and (3) whole body imaging at minimum concentration (i.e., 3 mg Fe/kg body weight) with negligible toxicity (Mishra et al. 2016). Similarly, maghemite (γ-Fe_2O_3) cores (with 13.08 ± 2.33 nm size) that are embedded inside a primary hydrophobic polymer (poly(4-vinyl pyridine), P4VP) matrix, and further covered by a shell of a secondary hydrophilic polymer (polyethylene glycol, PEG) are made, and have showed transverse relaxivity (r2) value of 104.7 ± 9.7 mM^{-1} s^{-1} at 4.7 T MRI and also negligible toxicity during in vivo study in liver of mice models (up to 30 days)—observed after the nanoparticle injection (Ali et al. 2017). In another study, nanometer-sized SPIONs (complexed with amylose nanoparticles that are cationized with spermine, ASP-SPIONs) labeled with transgenic green fluorescent protein (GFP)-mesenchymal stem cells (MSCs) were prepared for in vivo MRI tracking. Moreover, these SPIONs have exhibited a high transverse relaxivity (r2) value of 296.2 mM^{-1} s^{-1} in comparison to some commercially available SPION-based contrast agents such as Sinerem (~65 mM^{-1} s^{-1}), Endorem (120 mM^{-1} s^{-1}), and Resovist (180–202 mM^{-1} s^{-1}) (Lin et al. 2017). Recently, Wei et al. have developed zwitterion-coated ultra-small SPIONs—composed of ~3-nm-sized magnetic core and ~1-nm-sized hydrophilic shell—for use as positive contrast agents, which have displayed longitudinal (*r1*) relaxivity value of 5.2 mM^{-1} s^{-1} (at 1.5 T) that is slightly higher than the value of commercially available gadolinium-based positive contrast agent (i.e., 4.8 mM^{-1} s^{-1} at 1.5 T) (Wei et al. 2017). Despite the abovementioned improvements, SPION-based contrast agents are required to be crucially optimized in terms of better physicochemical (particle size/shape), water dispersibility, and magnetic (including relaxivities) properties for better clinical acceptance.

12.4.3 Magnetic Drug Delivery

Targeting techniques which are involved in the delivery of the chemotherapeutic drugs by using the SPIONs are classified as (1) passive targeting, (2) active targeting, (3) magnetic targeting, and (4) combined targeting (i.e., active + magnetic targeting). Passive targeting is based on the targeting of the drugs after their conjugation/encapsulation with SPIONs to the tumor sites through leaky vasculatures based on the enhanced permeability and retention (EPR) effect (Danhier et al. 2010; Kim et al. 2013; Xu and Sun 2013; Shin et al. 2015; Revia and Zhang 2016). However, passive targeting may be inadequate to achieve efficient targeting in the tumor sites, since it is solely relied on the EPR effect (Rosenblum et al. 2018). Therefore, active targeting of drugs is preferred over passive targeting to increase efficiency of their delivery. The active targeting is achieved by the attachment/conjugation of targeting moieties (i.e., antibodies) and drugs onto the SPIONs for effective tumor antigen targeting based on ligand-receptor interactions (Hairston 1996; Prijic and Sersa 2011; Yu et al. 2012). Drug delivery is also performed by utilizing the SPIONs as a magnetic targeting agent to effectively deliver the chemotherapeutic drugs at the site of tumors for cancer treatment with/without the influence of an externally applied static magnetic field (Polyak and Friedman 2009; Laurent et al. 2014; Burgess et al. 2016; Zhu et al. 2017; Ansari et al. 2018). Moreover, MRI via SPIONs in real-time scenario can also be performed for simultaneous navigation, localization, and the release of the drugs at tumor sites to track their bioavailability (Kokura et al. 2016). Moreover, combined targeting involves active and magnetic targeting techniques, and a typical combined drug delivery system may consist of SPIONs, anticancer/chemotherapeutic drugs, targeting moieties (i.e., antibodies), and/or carriers (i.e., polymeric micelles/nanoparticles for encapsulation of SPIONs along with drugs) (Veiseh et al. 2011; Thomsen et al. 2015).

In a recent study, a multifunctional hybrid biocompatible nanoplatform (having a size of ~100 nm) has been prepared by constituting poly(lactic-co-glycolic acid) (PLGA) nanoparticles stabilized with chitosan and poly(vinyl alcohol) (PVA) and co-loaded with SPIONs and anticancer drug cisplatin (Ibarra et al. 2018). Then, they have confirmed that the as-prepared nanoplatform has been successfully internalized into both HeLa and MDA-MB-231 cells (determined through the cellular uptake studies) via passive targeting. In another recent study, antiCD44 antibodies and gemcitabine-conjugated multifunctional iron oxide nanoparticles were primarily developed, which have demonstrated the targeting (i.e., active targeting) of nanoparticles to different CD44-positive cancer cell lines using a CD44-negative non-tumorigenic cell line as a control to verify the specificity by ultrastructural characterization and downregulation of CD44 expression (Aires et al. 2016). Then, they have also showed the selective drug delivery potential of the nanoparticles by the killing of CD44-positive cancer cells using a CD44-negative non-tumorigenic cell line as a control. Herein, (1) CD44 is a lymphocyte homing receptor that has been overexpressed usually in a large variety of cancer cells, cancer stem cells (CSCs) and circulating tumor cells (CTCs), which is actively targeted via antiCD44 antibodies; and (2) gemcitabine is a chemotherapeutic drug—currently used for pancreatic cancer treatment in clinical scenarios.

Very recently, a 250-nm-sized nanocarrier system is prepared by constituting paclitaxel (a chemotherapeutic drug)/SPIONs co-loaded PEG-ylated PLGA-based nanoparticles (Ganipineni et al. 2018). An ex vivo bio-distribution study have showed an enhanced accumulation of the SPIONs in the brain of glioblastoma (GBM) bearing orthotopic U87MG mice with magnetic targeting. In addition, they have observed that the blood–brain barrier is disrupted in the GBM area via magnetic resonance imaging (MRI) studies, which confirms the entry of the SPIONs. Moreover, the magnetic targeting treatment based on these nanocarrier system have significantly prolonged the median survival time of GBM bearing mice models as compared with the passive targeting and control treatments, when tested for in vivo antitumor efficacy. In another investigation, multifunctional nanoparticles are pre-

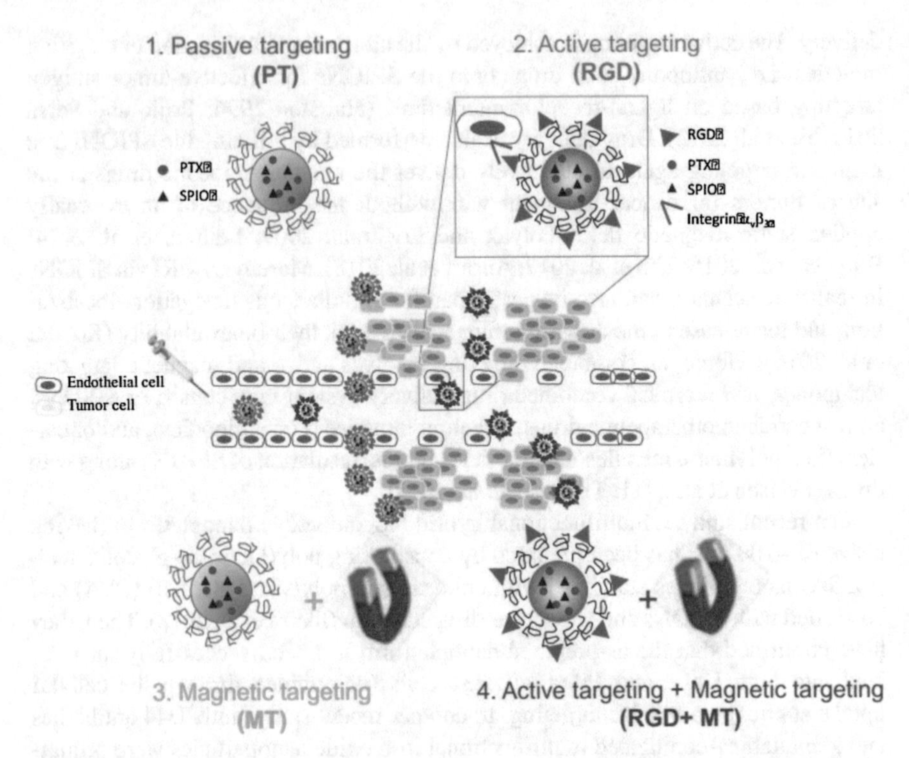

Fig. 12.9 Schematic representation of the different tumor targeting strategies. (1) *passive targeting* (PT) via the enhanced permeability and retention (EPR) effect, (2) *active targeting* of $\alpha_v\beta_3$ integrin via RGD grafting (RGD), (3) *magnetic targeting* (MT) via a magnet of 1.1 T placed on the tumor, and (4) the combination of the magnetic targeting and the active targeting of $\alpha_v\beta_3$ integrin (RGD + MT). (Adapted from Schleich et al. 2014)

pared by using PLGA-based nanoparticles loaded with paclitaxel and SPIONs and then compared through different in vivo targeting strategies: (1) passive targeting (PT) via the EPR effect, (2) active targeting of $\alpha_v\beta_3$ integrin via arginine–glycine–aspartic acid (RGD) grafting, (3) magnetic targeting (MT) via a magnet placed on the tumor, and (4) the combination of the active targeting of $\alpha_v\beta_3$ integrin and the magnetic targeting (i.e., RGD + MT)—as shown in Fig. 12.9 (Schleich et al. 2014). They have demonstrated that—as compared to non-targeted (i.e., PT) or single-targeted nanoparticles (i.e., RGD or MT)—the combination of both active and magnetic targeted strategies (i.e., RGD and MT) which have drastically enhanced (1) the nanoparticle accumulation into the tumor tissue with an eightfold increase as compared to passive targeting (1.12% and 0.135% of the injected dose, respectively), (2) contrast in MRI, and (3) anticancer efficacy with a median survival time of 22 days as compared to 13 days for the passive targeting. Finally, they have concluded that the double targeting of nanoparticles to tumors by different mechanisms could be a promising translational approach for the management of therapeutic treatment or personalized therapy.

12.5 Combined Cancer Theranostic Applications

The use of SPIONs is not only limited in their intrinsic cancer theranostic applications (as discussed above), but also for combined cancer theranostic applications to improve the effectiveness of cancer diagnosis and therapy. The combined cancer theranostics applications include: (1) MRI as adjuvant to fluorescence imaging, (2) thermo-chemotherapy (i.e., a combination of MFH therapy and chemotherapy), (3) thermo-radiotherapy (a combination of MFH therapy and radiotherapy), and (4) thermo-immunotherapy (a combination of MFH therapy and immunotherapy) (Zhang et al. 2013; Hervault and Thanh 2014; Mclaughlin et al. 2015; Grifantini et al. 2018), which are discussed below.

12.5.1 MRI as Adjuvant to Fluorescence Imaging

MRI is a significant diagnosis tool used for the detection and planning of cancer treatment. However, a single imaging technique is not adequate enough to examine multiple aspects of cancers due to the limitations in detection sensitivity, resolution, and specificity. In cancer treatment, accurate discrimination between the cancerous tissue from healthy tissue should be done during diagnosis to avoid severe damages to normal/surrounding tissues and effective treatment (Stephen et al. 2012; Mclaughlin et al. 2015). This could be achieved by utilizing a multimodal diagnostic approach (such as MRI as an adjuvant to fluorescence imaging) to enhance the spatial resolution and detection sensitivity in cellular imaging, which might be done by developing a multifunctional nanoparticle system by conjugating/co-encapsulating the SPIONs (for MRI) along with other fluorescence agents (for fluorescence imaging) such as fluorescence dyes (fluorophores, for example, Alexa Fluor 647/750) and quantum dots (QDs, for example—carbon QDs) (Josephson et al. 2002; Fass 2008; Maxwell et al. 2008; Kosaka et al. 2009; Ito et al. 2014; Wu et al. 2016).

Recently, multifunctional Fe_3O_4-gold (Au) hybrid nanoparticles are prepared by simultaneous conjugation of two fluorescent dyes or alternatively the combination of a drug and a dye, selectively binding to Fe_3O_4 and Au, which are as follows: (1) Fe_3O_4-Au nanoparticles functionalized with covalently attached Sulfo-Cyanine5 NHS ester derivative (Cy5) fluorescent dye via thiol-Au bonds for nanoparticle tracking; and (2) an anticancer drug doxorubicin (DOX) or Nile Red dye is loaded into the polymeric shell at the Fe_3O_4 surface (Efremova et al. 2018). The in vitro studies have revealed (1) high accumulation of fluorescent (Cy5) labeled Fe_3O_4-Au hybrids on 4T1 tumors (via passive targeting), (2) successful therapeutic payload (i.e., Dox) delivery and release to the tumors via Fe_3O_4-Au hybrids labeled with Cy5 and co-loaded with Nile Red dye, and (3) high diagnostic potential of Fe_3O_4-Au hybrids in 4T1 cells via MRI based on the enhanced relaxivity (r_2) values as compared to commercial T2 contrast agents. Similarly, near-infrared (NIR) two-photon fluorescence emitting Fe_3O_4 nanostructures are prepared through trimesic acid/citrate-mediated reaction process for live cell multimodal imaging (Liao et al. 2013). In another study, SPIONs and IR-780 dye are co-encapsulated inside stearic acid-modified polyethylenimine to form a dual-modality contrast agent, which have effectively labeled and tracked the stem cells through MRI and near-infrared fluorescence (NIRF) imaging, respectively (Liu et al. 2016).

12.5.2 *Thermo-chemotherapy*

Thermo-chemotherapy that constitutes MFH therapy and chemotherapy is an effective dual/combined therapy used for cancer treatment, which is performed via SPION-based nano-carriers (SNCs, i.e., SPIONs are co-encapsulated along with chemotherapeutic drugs (CHDs) inside polymeric carrier/vehicles) (Hervault and Thanh 2014). Herein, MFH therapy is utilized to enhance the sensitivity of cancer cells toward CHDs, and this phenomenon is known as "chemo-sensitization" that occurs due to (1) increase in cell membrane permeability, and (2) reduction of the interstitial fluid pressure due to heating from MFH therapy (via SPIONs). Consequently, the uptake of the CHDs by the cancerous cells might significantly increase. Thus, thermo-chemotherapy is more effective to treat cancer than MFH alone. In addition, MRI could also be performed for simultaneous tumor imaging while performing thermo-chemotherapy due to the presence of SPIONs (contrast agent) (Rao et al. 2010; Li et al. 2018). Figure 12.10 represents the possible mechanism involved in cancer treatment via thermo-chemotherapy by using CHDs and SPIONs co-loaded SNCs.

In a recent investigation, the as-prepared 150-nm-sized magnetoliposomes, consisting of magnetite nanoparticle cores and the anticancer drug gemcitabine encapsulated by a phospholipid bilayer, have displayed 70% drug release and better elevated temperature—i.e., from 32 to 52 °C in 5 min—in time-dependent temperature curves, when exposed to AMF with $H = 30$ kA/m and $f = 356$ kHz (Ferreira et al. 2016). In another investigation, 370-nm-sized thermosensitive nanocomposites are prepared by having SPIONs/5-fluorouracil (anticancer drug) in the core and poly(*N*-isopropylacrylamide) (PNIPAM) polymer/silica (SiO_2) in the shell. These nanocomposites have displayed good thermal efficacy (rise to therapeutic temperature of 45 °C in 3.7 min) and faster drug release at relatively lower magnetic field (Shen et al. 2016). Moreover very recently, 64-nm-sized core-shell nanoparticles are prepared with tightly clustered Fe_3O_4 nanoparticles (having 17 nm size) in the core and doxorubicin (Dox—anticancer drug)-containing polymer in the shell (Hayashi et al. 2017). These nanoparticles have displayed a higher therapeutic efficacy against intraperitoneal tumors (located in BALB/c-nu/nu mice) via thermo-chemotherapy (i.e., MFH and chemotherapy) as compared to only thermotherapy (i.e., MFH) or chemotherapy alone. Herein, good heating efficacy (i.e., SAR values of 194–353 W/g) and faster drug release (49% in 20 min) under the influence of an

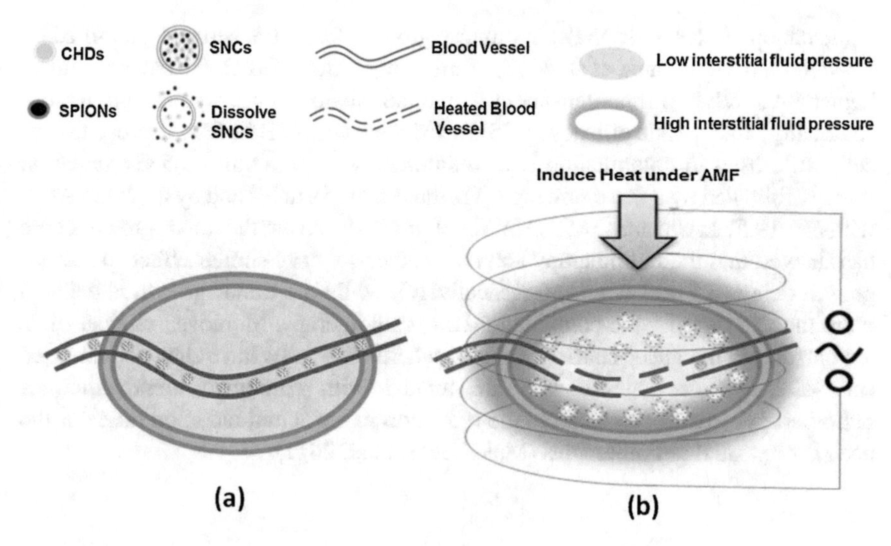

Fig. 12.10 Schematic representation of mechanism of the thermo-chemotherapy, (**a**) chemotherapy alone, and (**b**) thermo-chemotherapy (combined MFH and chemotherapy)

externally applied AMF (with $H = 8$ kA/m and $f = 217$ kHz) have resulted in the overall improvement of cancer therapeutic efficiency.

12.5.3 Thermo-radiotherapy

Thermo-radiotherapy that comprises MFH-based thermotherapy and radiation therapy/radiotherapy is another effective dual/combined therapy employed in cancer treatments, which is implemented via utilization of SPIONs (for MFH) and application of high-energy X-rays/gamma radiations (for radiotherapy). Generally in radiotherapy, high energy radiations destroy the cancer cells by damaging their genetic material (such as deoxyribonucleic acid (DNA)) either via direct interaction with cellular DNA, or interaction of free radicals (generated due to ionization/excitation of the water component) with DNA. Consequently, the shrinkage of tumor happens due to the death of cancerous cells through apoptosis, necrosis, mitotic catastrophe, or inhibition of the cell proliferation (Baskar et al. 2014; Desouky et al. 2015). However, the cancer cells may recover via (1) repairing their DNA damage, and (2) proliferation from their residual cells, after radiotherapy. Thus, the combined therapy might be preferred for the treatment of cancer to resolve this challenge. MFH can be effectively used as an adjuvant therapy for radiation therapy also, which might enhance the radio-sensitization effect in cancer cells or used to destroy the remaining cancer cells left after radiation therapy (Datta et al. 2015; Kaur et al. 2011). Besides, MRI can be combined with thermo-radiotherapy for effective tumor imaging also. For instance, in a study, the accumulation of gadolinium-doped iron oxide nanoparticles (GdIONPs) in tumor region is clearly tracked and quantified by T2-weighted MRI (Jiang et al. 2017). Furthermore, these GdIONPs have displayed higher SAR value in time-dependent temperature/calorimetric studies and demonstrated high therapeutic efficacy in C57BL/6J mice with (TRAMP)-C1 prostate cancer, while used in combination with radiotherapy by utilizing a 25 Gy radiation therapy followed by a thermotherapy (via these nanoparticles and by applying AMF with $H = 19.57$ kA/m and $f = 52$ kHz) for 30 min. Moreover, the tumor growth curve has showed that the radiotherapy and thermotherapy have similar effects on tumor growth delay (2.5 and 4.5 days, respectively), while the tumor growth is delayed more than 10 days by the combined/thermo-radiotherapy. Moreover, clinical trials are performed in cancer-bearing patients, where the results have showed increased survival rates for patients who are treated with combined therapy/thermo-radiotherapy—mainly ascribed to the reduction in the actual radiation doses or the need for repeated radiotherapies (Maier-Hauff et al. 2011).

12.5.4 Thermo-immunotherapy

Now-a-days, immunotherapy has gained significant interest in cancer treatments, which is based on the killing of cancerous cells via artificial activation of immune system in human body. Usually, the cancer cells have antigens on their surface that act as a marker to reorganize and bind with the antibodies of the immune system, which may help in proliferation inhibition or killing of cancer cells (Farkona et al. 2016; Evans et al. 2018). The three prominent mechanisms that are involved in the cell death via immunotherapy include: (1) antibody-dependent cell-mediated cytotoxicity (ADCC) against tumor cells, which is triggered by the interaction of the fragment crystallizable (Fc) portion of the monoclonal antibodies (mAb) with the Fc receptors on the effector cells like natural killer cells, macrophages, and dendritic cells, (2) antibody-dependent cellular phagocytosis (ADCP), which is described as the target tumor cell elimination by the innate network of phagocytic cells, primarily neutrophils, monocytes, and macrophages, and (3) complement-dependent cytotoxicity, which is the result of the Fc region of an antigen–immunoglobulin complex triggering a cascade of more than 30 proteins that culminates in the formation of the membrane-attack complex, an amalgam of subunits that functions to perforate the phospholipid bilayer of the target cell and induce lysis (Jena 1997; Kohrt et al. 2015; Wang 2015). However, the delivery of antibodies alone to antigen-presenting cancer cells is insufficient to induce improved immunity due to their rapid degradation during the systemic administration. SPIONs can be used to improve the delivery of immunotherapeutic agents at the targeted tumor sites for enhanced cancer treatment via immunotherapy. For instance, the therapeutic efficacy of immunotherapy in FAT1-positive colorectal cancer is improved by magnetically targeting of the murine monoclonal antibodies (mAb198.3) conjugated SPIONs toward the cancer cells (Grifantini et al. 2018). Moreover, the activation of immune system against cancer can be done through heat stimulation via MFH by using the SPIONs (Yanase et al. 1998; Toraya-Brown and Fiering 2014; Yagawa et al. 2017; Lin et al. 2018; Park et al. 2018).

12.6 Conclusions and Future Perspectives

In summary, nowadays new advancements are achieved in the synthesis of high-quality SPIONs using different synthesis routes in comparison to other types of magnetic nanoparticles, and also in the formation/stabilization of the ferrofluids (by using the SPIONs as major building blocks). Moreover, based on the abovementioned investigations, it can be clearly evident that the SPIONs have attained great importance in intrinsic cancer theranostics applications (such as MFH, MRI and magnetic drug delivery) and also in combined cancer theranostics applications (including MRI as an adjuvant to fluorescence imaging, thermo-chemotherapy, thermo-radiotherapy and thermo-immunotherapy). But, the overall utilization of the SPIONs is mainly restricted to in vitro levels. Nevertheless, very few high impacting efforts are being made to involve the SPIONs as an effective part of in vivo or clinical theranostics. For example, the global researchers have made worldwide consortiums that mainly focus to generate new treatment concepts (combined targeting radio-sensitization and thermotherapy via MFH), strengthen the existing synergies between technical advances, and inspect biocompatible coatings for magnetic nanoparticles (including SPIONs) in order to achieve breakthroughs in clinical cancer treatments. In addition, companies like "MagForce" (who developed ferrofluid and patented) have obtained FDA approval to conduct clinical studies by using "NanoTherm" for focal tumor ablation in prostate cancer with immediate risks. However, more fruitful attempts are needed so that, in near future the SPION-based theranostics will be used in full potential for treating the cancers effectively.

References

Abadeer NS, Murphy CJ. Recent progress in cancer thermal therapy using gold nanoparticles. J Phys Chem C. 2016;120(9):4691–716.

Ahn T, Kim JH, Yang HM, Lee JW, Kim JD. Formation pathways of magnetite nanoparticles by coprecipitation method. J Phys Chem C. 2012;116(10):6069–76.

Aires A, Ocampo SM, Simões BM, Josefa Rodríguez M, Cadenas JF, Couleaud P, Cortajarena AL. Multifunctionalized iron oxide nanoparticles for selective drug delivery to CD44-positive cancer cells. Nanotechnology. 2016;27(6):065103.

Ali A, Zafar H, Zia M, ul Haq I, Phull AR, Ali JS, Hussain A. Synthesis, characterization, applications, and challenges of iron oxide nanoparticles. Nanotechnol Sci Appl. 2016;9:49–67.

Ali LMA, Marzola P, Nicolato E, Fiorini S, De M. Polymer-coated superparamagnetic iron oxide nanoparticles as T2 contrast agent for MRI and their uptake in liver. Future Sci OA. 2017;5:FSO235.

Ansari MO, Ahmad MF, Shadab GGHA, Siddique HR. Superparamagnetic iron oxide nanoparticles based cancer theranostics: a double edge sword to fight against cancer. J Drug Deliv Sci Technol. 2018;45:177–83.

Araki EH, Ehlers J, Hepp K, Tvergaard JSV, Potier-ferry M. Colloidal magnetic fluids: basics, development and application of ferrofluid, Lecture notes in physics. Berlin: Springer Science & Business Media; 2009.

Arriortua OK, Garaio E, de la Parte BH, Insausti M, Lezama L, Plazaola F, et al. Antitumor magnetic hyperthermia induced by RGD-functionalized Fe3O4 nanoparticles, in an experimental model of colorectal liver metastases. Beilstein J Nanotechnol. 2016;7(1):1532–42.

Baskar R, Dai J, Wenlong N, Yeo R, Yeoh K-W. Biological response of cancer cells to radiation treatment. Front Mol Biosci. 2014;1:1–9.

Bastow TJ, Trinchi A. NMR analysis of ferromagnets: Fe oxides. Solid State Nucl Magn Reson. 2009;35(1):25–31.

Bedanta S, Kleemann W. Supermagnetism. J Phys D Appl Phys. 2009;42(1):013001.

Boutonnet M, Lögdberg S, Elm Svensson E. Recent developments in the application of nanoparticles prepared from w/o microemulsions in heterogeneous catalysis. Curr Opin Colloid Interface Sci. 2008;13(4):270–86.

Burgess A, Shah K, Hough O, Hynynen K. Next generation superparamagnetic iron oxide nanoparticles for cancer theranostics. Drug Discov Today. 2016;15(5):477–91.

Chopra M, Kandasamy G, Maity D. Multifunctional magnetic nanoparticles—a promising approach for cancer treatment. J Nanomed Res. 2016;4(1):3–4.

Cornell RM, Schwertmann U. Introduction to iron oxides. In: The iron oxides: structure, properties, reactions, occurences and uses. Weinheim: Wiley; 2004. p. 1–7. https://doi.org/10.1002/3527602097. ISBN: 9783527302741 (Print); 9783527602094 (Online).

Corot C, Warlin D. Superparamagnetic iron oxide nanoparticles for MRI: contrast media pharmaceutical company R&D perspective. Wiley Interdiscip Rev Nanomed Nanobiotechnol. 2013;5(5):411–22.

Danhier F, Feron O, Préat V. To exploit the tumor microenvironment: passive and active tumor targeting of nanocarriers for anti-cancer drug delivery. J Control Release. 2010;148(2):135–46.

Datta NR, Ordóñez SG, Gaipl US, Paulides MM, Crezee H, Gellermann J, et al. Local hyperthermia combined with radiotherapy and-/or chemotherapy: recent advances and promises for the future. Cancer Treat Rev. 2015;41(9):742–53.

Demirer GS, Okur AC, Kizilel S. Synthesis and design of biologically inspired biocompatible iron oxide nanoparticles for biomedical applications. J Mater Chem B. 2015;3(40):7831–49.

Desouky O, Ding N, Zhou G. Targeted and non-targeted effects of ionizing radiation. J Radiat Res Appl Sci. 2015;8(2):247–54.

Efremova MV, Naumenko VA, Spasova M, Garanina AS, Abakumov MA, Blokhina AD, et al. Magnetite-gold nanohybrids as ideal all-in-one platforms for theranostics. Sci Rep. 2018;8(1):1–19.

Evans ER, Bugga P, Asthana V, Drezek R. Metallic nanoparticles for cancer immunotherapy. Mater Today. 2018;21(6):673–85.

Farkona S, Diamandis EP, Blasutig IM. Cancer immunotherapy: the beginning of the end of cancer? BMC Med. 2016;14(1):1–18.

Fass L. Imaging and cancer: a review. Mol Oncol. 2008;2:115–52.

Ferreira RV, Martins TM, Goes AM, Fabris JD, Cavalcante LCD, Outon LEF, Domingues RZ. Thermosensitive gemcitabine-magnetoliposomes for combined hyperthermia and chemotherapy. Nanotechnology. 2016;27(8):085105.

Fu C, Ravindra NM. Magnetic iron oxide nanoparticles: synthesis and applications. Bioinspir Biomim Nanobiomater. 2012;1(4):229–44.

Ganipineni LP, Ucakar B, Joudiou N, Bianco J, Danhier P, Zhao M, et al. Magnetic targeting of paclitaxel-loaded poly(lactic-co-glycolic acid)-based nanoparticles for the treatment of glioblastoma. Int J Nanomed. 2018;13:4509–21.

Gilchrist RK, Medal R, Shorey WD, Hanselman RC, Parrott JC, Taylor CB. Selective inductive heating of lymph nodes. Ann Surg. 1957;146(4):596–606.

Gkanas EI. In vitro magnetic hyperthermia response of iron oxide MNP's incorporated in DA3, MCF-7 and HeLa cancer cell lines. Cent Eur J Chem. 2013;11(7):1042–54.

Grifantini R, Taranta M, Gherardini L, Naldi I, Parri M, Grandi A, Cinti C. Magnetically driven drug delivery systems improving targeted immunotherapy for colon-rectal cancer. J Control Release. 2018;280:76–86.

Grover VPB, Tognarelli JM, Crossey MME, Cox IJ, Taylor-Robinson SD, McPhail MJW. Magnetic resonance imaging: principles and techniques: lessons for clinicians. J Clin Exp Hepatol. 2015;5(3):246–55.

Gupta KM, Gupta N. Advanced electrical and electronics materials: processes and applications. Hoboken, NJ: Wiley; 2015.

Hairston RJ. The management of cytomegalovirus-associated retinal detachments. J Int Assoc Phys AIDS Care. 1996;2(5):31–4.

Hayashi K, Sato Y, Sakamoto W, Yogo T. Theranostic nanoparticles for MRI-guided thermochemotherapy: "tight" clustering of magnetic nanoparticles boosts relaxivity and heat-generation power. ACS Biomater Sci Eng. 2017;3(1):95–105.

Hervault A, Thanh NTK. Magnetic nanoparticle-based therapeutic agents for thermo-chemotherapy treatment of cancer. Nanoscale. 2014;6(20):11553–73.

Ibarra J, Encinas D, Blanco M, Barbosa S, Taboada P, Juárez J, Valdez MA. Co-encapsulation of magnetic nanoparticles and cisplatin within biocompatible polymers as multifunctional nanoplatforms: synthesis, characterization, and in vitro assays. Mater Res Express. 2018;5(1):015023.

Issa B, Obaidat IM, Albiss BA, Haik Y. Magnetic nanoparticles: surface effects and properties related to biomedicine applications. Int J Mol Sci. 2013;14(11):21266–305.

Ito A, Ito Y, Matsushima S, Tsuchida D, Ogasawara M, Hasegawa J, et al. New whole-body multimodality imaging of gastric cancer peritoneal metastasis combining fluorescence imaging with ICG-labeled antibody and MRI in mice. Gastric Cancer. 2014;17(3):497–507.

Jacques V, Dumas S, Sun W-C, Troughton J, Greenfield MT, Caravan P. High relaxivity MRI contrast agents part 2: optimization of inner- and second-sphere relaxivity. Investig Radiol. 2010;45(10):613–24.

Jena BP. Atomic force microscope: providing new insights on the structure and function of living cells. Cell Biol Int. 1997;21(11):683–4.

Jiang PS, Tsai HY, Drake P, Wang FN, Chiang CS. Gadolinium-doped iron oxide nanoparticles induced magnetic field hyperthermia combined with radiotherapy increases tumour response by vascular disruption and improved oxygenation. Int J Hyperth. 2017;33:1–9.

Jo SD, Ku SH, Won YY, Kim SH, Kwon IC. Targeted nanotheranostics for future personalized medicine: recent progress in cancer therapy. Theranostics. 2016;6:1362–77.

Josephson L, Kircher MF, Mahmood U, Tang Y, Weissleder R. Near-infrared fluorescent nanoparticles as combined MR/optical imaging probes. Bioconjug Chem. 2002;13(3):554–60.

Kalikmanov VI. Ferrofluids. In: Statistical physics of fluids Texts and Monographs in Physics. https://doi.org/10.1007/978-3-662-04536-7. Springer, Berlin, Heidelberg 2001. pp. 223–238.

Kandasamy G, Maity D. Recent advances in superparamagnetic iron oxide nanoparticles (SPIONs) for in vitro and in vivo cancer nanotheranostics. Int J Pharm. 2015;496(2):191–218.

Kandasamy G, Sudame A, Bhati P, Chakrabarty A, Maity D. Systematic investigations on heating effects of carboxyl-amine functionalized superparamagnetic iron oxide nanoparticles (SPIONs) based ferrofluids for in vitro cancer hyperthermia therapy. J Mol Liq. 2018a;256:224–37.

Kandasamy G, Sudame A, Luthra T, Saini K, Maity D. Functionalized hydrophilic superparamagnetic iron oxide nanoparticles for magnetic fluid hyperthermia application in liver cancer treatment. ACS Omega. 2018b;3(4):3991–4005.

Kandasamy G, Khan S, Giri J, Bose S, Veerapu NS, Maity D. One-pot synthesis of hydrophilic flower-shaped iron oxide nanoclusters (IONCs) based ferrofluids for magnetic fluid hyperthermia applications. J Mol Liq. 2019a;275:699–712.

Kandasamy G, Soni S, Sushmita K, Veerapu NS, Bose S, Maity D. One-step synthesis of hydrophilic functionalized and cytocompatible superparamagnetic iron oxide nanoparticles (SPIONs) based aqueous ferrofluids for biomedical applications. J Mol Liq. 2019b;274:653–63.

Kaur P, Hurwitz MD, Krishnan S, Asea A. Combined hyperthermia and radiotherapy for the treatment of cancer. Cancers. 2011;3(4):3799–823.

Kim DH, Vitol EA, Liu J, Balasubramanian S, Gosztola DJ, Cohen EE, et al. Stimuli-responsive magnetic nanomicelles as multifunctional heat and cargo delivery vehicles. Langmuir. 2013;29(24):7425–32.

Kitture R, Ghosh S, Kulkarni P, Liu XL, Maity D, Patil SI, et al. Fe3O4-citrate-curcumin: promising conjugates for superoxide scavenging, tumor suppression and cancer hyperthermia. J Appl Phys. 2012;111(6):064702.

Kohrt H, Rajasekaran N, Chester C, Yonezawa A, Zhao X. Enhancement of antibody-dependent cell mediated cytotoxicity: a new era in cancer treatment. Immunotargets Ther. 2015;4:91.

Kokura S, Yoshikawa T, Ohnishi T. Hyperthermic oncology from bench to bedside. Singapore: Springer; 2016.

Kosaka N, Ogawa M, Choyke PPL, Kobayashi H. Clinical implications of near-infrared fluorescence imaging in cancer. Future Oncol. 2009;5(9):1501–11.

Kozissnik B, Bohorquez AC, Dobson J, Rinaldi C. Magnetic fluid hyperthermia: advances, challenges, and opportunity. Int J Hyperth. 2013;29(8):706–14.

Kudr J, Haddad Y, Richtera L, Heger Z, Cernak M, Adam V, Zitka O. Magnetic nanoparticles: from design and synthesis to real world applications. Nanomaterials. 2017;7(9):243.

Kumar CSSR, Leuschner C. Nanoparticles for cancer drug delivery. In: Nanofabrication towards biomedical applications: techniques, tools, applications, and impact. Weinheim: Wiley-VCH; 2005.

Lahiri BB, Muthukumaran T, Philip J. Magnetic hyperthermia in phosphate coated iron oxide nanofluids. J Magn Magn Mater. 2016;407:101–13.

Laurent S, Forge D, Port M, Roch A, Robic C, Vander Elst L, Muller RN. Magnetic iron oxide nanoparticles: synthesis, stabilization, vectorization, physicochemical characterizations and biological applications. Chem Rev. 2008;108(6):2064–110.

Laurent S, Dutz S, Häfeli UO, Mahmoudi M. Magnetic fluid hyperthermia: focus on superparamagnetic iron oxide nanoparticles. Adv Colloid Interf Sci. 2011;166(1–2):8–23.

Laurent S, Saei AA, Behzadi S, Panahifar A, Mahmoudi M. Superparamagnetic iron oxide nanoparticles for delivery of therapeutic agents: opportunities and challenges. Expert Opin Drug Deliv. 2014;11(9):1449–70.

Li W, Hinton CH, Lee SS, Wu J, Fortner JD. Surface engineering superparamagnetic nanoparticles for aqueous applications: design and characterization of tailored organic bilayers. Environ Sci Nano. 2015;3:1–20.

Li X, Wei J, Aifantis KE, Fan Y, Feng Q, Cui FZ, Watari F. Current investigations into magnetic nanoparticles for biomedical applications. J Biomed Mater Res A. 2016;104:1285–96.

Li Q, Kartikowati CW, Horie S, Ogi T, Iwaki T, Okuyama K. Correlation between particle size/domain structure and magnetic properties of highly crystalline Fe3O4 nanoparticles. Sci Rep. 2017;7(1):1–4.

Li M, Bu W, Ren J, Li J, Deng L, Gao M, et al. Enhanced synergism of thermo-chemotherapy for liver cancer with magnetothermally responsive nanocarriers. Theranostics. 2018;8(3):693–709.

Liao MY, Wu CH, Lai PS, Yu J, Lin HP, Liu TM, Huang CC. Surface state mediated NIR two-photon fluorescence of iron oxides for nonlinear optical microscopy. Adv Funct Mater. 2013;23:2044–51.

Lin BL, Zhang JZ, Lu LJ, Mao JJ, Cao MH, Mao XH, Shen J. Superparamagnetic iron oxide nanoparticles-complexed cationic amylose for in vivo magnetic resonance imaging tracking of transplanted stem cells in stroke. Nanomaterials. 2017;7(5):107.

Lin FC, Hsu CH, Lin YY. Nano-therapeutic cancer immunotherapy using hyperthermia-induced heat shock proteins: insights from mathematical modeling. Int J Nanomed. 2018;13:3529–39.

Liu S, Jia B, Qiao R, Yang Z, Yu Z, Liu Z, et al. A novel type of dual-modality molecular probe for MR and nuclear imaging of tumor: preparation, characterization and in vivo application. Mol Pharm. 2009;6(4):1074–82.

Liu H, Tan Y, Xie L, Yang L, Zhao J, Bai J, et al. Self-assembled dual-modality contrast agents for non-invasive stem cell tracking via near-infrared fluorescence and magnetic resonance imaging. J Colloid Interface Sci. 2016;478:217–26.

Mahmoudi M, Sant S, Wang B, Laurent S, Sen T. Superparamagnetic iron oxide nanoparticles (SPIONs): development, surface modification and applications in chemotherapy. Adv Drug Deliv Rev. 2011;63(1–2):24–46.

Maier-Hauff K, Ulrich F, Nestler D, Niehoff H, Wust P, Thiesen B, et al. Efficacy and safety of intratumoral thermotherapy using magnetic iron-oxide nanoparticles combined with external beam radiotherapy on patients with recurrent glioblastoma multiforme. J Neurooncol. 2011;103(2):317–24.

Maity D, Agrawal DC. Synthesis of iron oxide nanoparticles under oxidizing environment and their stabilization in aqueous and non-aqueous media. J Magn Magn Mater. 2007;308(1):46–55.

Maity D, Ding J, Xue JM. Synthesis of magnetite nanoparticles by thermal decomposition: time, temperature, surfactant and solvent effects. Funct Mater Lett. 2008;01(3):189–93.

Maity D, Ding J, Xue JM. One-pot synthesis of hydrophilic and hydrophobic ferrofluid. Int J Nanosci. 2009;8(1–2):65–9.

Maity D, Chandrasekharan P, Si-Shen F, Xue JM, Ding J. Polyol-based synthesis of hydrophilic magnetite nanoparticles. J Appl Phys. 2010a;107(9):09B310.

Maity D, Chandrasekharan P, Yang CT, Chuang KH, Shuter B, Xue JM, Feng SS. Facile synthesis of water-stable magnetite nanoparticles for clinical MRI and magnetic hyperthermia applications. Nanomedicine. 2010b;5(10):1571–84.

Maity D, Chandrasekharan P, Pradhan P, Chuang K-H, Xue J-M, Feng S-S, Ding J. Novel synthesis of superparamagnetic magnetite nanoclusters for biomedical applications. J Mater Chem. 2011a;21(38):14717.

Maity D, Pradhan P, Chandrasekharan P, Kale SN, Shuter B, Bahadur D, Ding J. Synthesis of hydrophilic superparamagnetic magnetite nanoparticles via thermal decomposition of Fe(acac)3 in 80 Vol% TREG + 20 Vol% TREM. J Nanosci Nanotechnol. 2011b;11(3):2730–4.

Maxwell DJ, Bonde J, Hess DA, Hohm SA, Lahey R, Zhou P, et al. Fluorophore-conjugated iron oxide nanoparticle labeling and analysis of engrafting human hematopoietic stem cells. Stem Cells. 2008;26(2):517–24.

McCarthy MJ. Introduction to magnetic resonance imaging (MRI). In: Magnetic resonance imaging in foods; 2011. pp. 1–29.

Mclaughlin R, Hylton N, Imaging B. MRI in breast cancer therapy monitoring. NMR Biomed. 2015;24(6):712–20.

Merbach A, Helm L, Tóth É. The chemistry of contrast agents in medical magnetic resonance imaging. 2nd ed. Oxford: Wiley-Blackwell; 2013.

Mishra SK, Kumar BSH, Khushu S, Tripathi RP, Gangenahalli G. Increased transverse relaxivity in ultrasmall superparamagnetic iron oxide nanoparticles used as MRI contrast agent for biomedical imaging. Contrast Media Mol Imaging. 2016;11(5):350–61.

Mojica Pisciotti ML, Lima E, Vasquez Mansilla M, Tognoli VE, Troiani HE, Pasa AA, Zysler RD. In vitro and in vivo experiments with iron oxide nanoparticles functionalized with dextran or polyethylene glycol for medical applications: magnetic targeting. J Biomed Mater Res B Appl Biomater. 2014;102(4):860–8.

Morel A, Nikitenko SI, Gionnet K, Wattiaux A, Lai-kee-him J, Labrugere C, Faculte A. Sonochemical approach to the synthesis of Fe_3O_4@SiO_2 core–shell nanoparticles with tunable properties. ACS Nano. 2008;2(5):847–56.

Moskowitz BM. Hitchhiker's guide to magnetism. Environmental magnetism workshop, vol. 279(1), 1991. p. 48.

Mutin PH, Vioux A. Nonhydrolytic processing of oxide-based materials: simple routes to control homogeneity, morphology, and nanostructure. Chem Mater. 2009;21(4):582–96.

Nitz WR, Reimer P. Contrast mechanisms in MR imaging. Eur Radiol. 1999;9(6):1032–46.

Odenbach S. Ferrofluids—magnetically controllable fluids and their applications. Berlin: Springer; 2002.

Odenbach S. Ferrofluids—magnetically controlled suspensions. Colloids Surf A Physicochem Eng Asp. 2003;217(1–3):171–8.

Ohno T, Wakabayashi T, Takemura A, Yoshida J, Ito A, Shinkai M, Kobayashi T. Effective solitary hyperthermia treatment of malignant glioma using stick type CMC-magnetite. In vivo study. J Neurooncol. 2002;56(3):233–9.

Okoli C, Boutonnet M, Mariey L, Jaras S, Rajarao G. Application of magnetic iron oxide nanoparticles prepared from microemulsions for protein purification. J Chem Technol Biotechnol. 2011;86(11):1386–93.

Ortega RA, Giorgio TD. A mathematical model of superparamagnetic iron oxide nanoparticle magnetic behavior to guide the design of novel nanomaterials. J Nanopart Res. 2012;14(12):1282.

Park W, Heo Y-J, Han DK. New opportunities for nanoparticles in cancer immunotherapy. Biomater Res. 2018;22(1):24.

Piñeiro Y, Vargas Z, Rivas J, Lõpez-Quintela MA. Iron oxide based nanoparticles for magnetic hyperthermia strategies in biological applications. Eur J Inorg Chem. 2015;2015(27):4495–509.

Pinkas J, Reichlova V, Zboril R, Moravec Z, Bezdicka P, Matejkova J. Sonochemical synthesis of amorphous nanoscopic iron(III) oxide from Fe(acac)3. Ultrason Sonochem. 2008;15(3):257–64.

Polyak B, Friedman G. Magnetic targeting for site-specific drug delivery: applications and clinical potential. Expert Opin Drug Deliv. 2009;6(1):53–70.

Pooley R. Fundamental Physics of MR Imaging. Radiographics. 2005;25(4):1087–99.

Prashant C, Dipak M, Yang CT, Chuang KH, Jun D, Feng SS. Superparamagnetic iron oxide—loaded poly (lactic acid)-d-alpha-tocopherol polyethylene glycol 1000 succinate copolymer nanoparticles as MRI contrast agent. Biomaterials. 2010;31(21):5588–97.

Prijic S, Sersa G. Magnetic nanoparticles as targeted delivery systems in oncology. Radiol Oncol. 2011;45(1):1–16.

Qiao R, Yang C, Gao M. Superparamagnetic iron oxide nanoparticles: from preparations to in vivo MRI applications. J Mater Chem. 2009;19(35):6274.

Rabias I, Tsitrouli D, Karakosta E, Kehagias T, Diamantopoulos G, Fardis M, Papavassiliou G. Rapid magnetic heating treatment by highly charged maghemite nanoparticles on Wistar rats exocranial glioma tumors at microliter volume. Biomicrofluidics. 2010;4(2):024111.

Raj K, Boulton RJ. Ferrofluids—properties and applications. Mater Des. 1987;8(4):233–6.

Raj K, Moskowitz B, Casciari R. Advances in ferrofluid technology. J Magn Magn Mater. 1995;149(1–2):174–80.

Rao W, Deng ZS, Liu J. A review of hyperthermia combined with radiotherapy/chemotherapy on malignant tumors. Crit Rev Biomed Eng. 2010;38(1):101–16.

Revia RA, Zhang M. Magnetite nanoparticles for cancer diagnosis, treatment, and treatment monitoring: recent advances. Mater Today. 2016;19(3):157–68.

Rosenblum D, Joshi N, Tao W, Karp JM, Peer D. Progress and challenges towards targeted delivery of cancer therapeutics. Nat Commun. 2018;9(1):1410.

Rosensweig RE. Ferrohydrodynamics. Mineola: Dover Publications; 1997.

Scherer C, Neto AMF. Ferrofluids: properties and applications. Braz J Phys. 2005;35(3):718–27.

Schleich N, Po C, Jacobs D, Ucakar B, Gallez B, Danhier F, Préat V. Comparison of active, passive and magnetic targeting to tumors of multifunctional paclitaxel/SPIO-loaded nanoparticles for tumor imaging and therapy. J Control Release. 2014;194:82–91.

Shen B, Ma Y, Yu S, Ji C. Smart multifunctional magnetic nanoparticle-based drug delivery system for cancer thermo-chemotherapy and intracellular imaging. ACS Appl Mater Interfaces. 2016;8(37):24502–8.

Shen Z, Wu A, Chen X. Iron oxide nanoparticle based contrast agents for magnetic resonance imaging. Mol Pharm. 2017;14(5):1352–64.

Shin TH, Choi Y, Kim S, Cheon J. Recent advances in magnetic nanoparticle-based multi-modal imaging. Chem Soc Rev. 2015;44(14):4501–16.

Silva AC, Oliveira TR, Mamani JB, Malheiros SMF, Malavolta L, Pavon LF, Gamarra LF. Application of hyperthermia induced by superparamagnetic iron oxide nanoparticles in glioma treatment. Int J Nanomed. 2011;6:591–603.

Spaldin N. Magnetic materials-fundamentals and applications. Cambridge: Cambridge University Press; 2003.

Stephen ZR, Kievit FM, Zhang M. Magnetite nanoparticles for medical MR imaging. Mater Today. 2012;14(11):330–8.

Suto M, Hirota Y, Mamiya H, Fujita A, Kasuya R, Tohji K, Jeyadevan B. Heat dissipation mechanism of magnetite nanoparticles in magnetic fluid hyperthermia. J Magn Magn Mater. 2009;321(10):1493–6.

Tan YF, Chandrasekharan P, Maity D, Yong CX, Chuang KH, Zhao Y, Feng SS. Multimodal tumor imaging by iron oxides and quantum dots formulated in poly (lactic acid)-d-alpha-tocopheryl polyethylene glycol 1000 succinate nanoparticles. Biomaterials. 2011;32(11):2969–78.

Tartaj P, Del M, Morales P, Veintemillas-Verdaguer S, González-Carr T, Serna CJ. The preparation of magnetic nanoparticles for applications in biomedicine. J Phys D Appl Phys. 2003;36(36):182–97.

Thanh NTK. Magnetic nanoparticles: from fabrication to clinical applications. Boca Raton, FL: CRC Press; 2012.

Thanh NTK. Clinical applications of magnetic nanoparticles, vol. 91. Boca Raton, FL: CRC Press; 2018.

Thomsen LB, Thomsen MS, Moos T. Targeted drug delivery to the brain using magnetic nanoparticles. Ther Deliv. 2015;6(10):1145–55.

Toraya-Brown S, Fiering S. Local tumour hyperthermia as immunotherapy for metastatic cancer. Int J Hyperth. 2014;30(8):531–9.

Trohidou K, editor. Magnetic nanoparticle assemblies. Singapore: Pan Stanford Publishing; 2014.

Turcheniuk K, Tarasevych AV, Kukhar VP, Boukherroub R, Szunerits S. Recent advances in surface chemistry strategies for the fabrication of functional iron oxide based magnetic nanoparticles. Nanoscale. 2013;5(22):10729.

Veiseh O, Gunn J, Zhang M. Design and fabrication of magnetic nanoparticles for targeted drug delivery and imaging. Adv Drug Deliv Rev. 2011;62(3):284–304.

Wang W. NK cell-mediated antibody-dependent cellular cytotoxicity in cancer immunotherapy. Front Immunol. 2015;6(11):683–4.

Wang YJ, Xuan S, Port M, Idee J. Recent advances in superparamagnetic iron oxide nanoparticles for cellular imaging and targeted therapy research. Curr Pharm Des. 2013;19:6575–93.

Wei H, Bruns OT, Kaul MG, Hansen EC, Barch M, Wiśniowska A, et al. Exceedingly small iron oxide nanoparticles as positive MRI contrast agents. Proc Natl Acad Sci. 2017;114(9):2325–30.

Wu W, He Q, Jiang C. Magnetic iron oxide nanoparticles: synthesis and surface functionalization strategies. Nanoscale Res Lett. 2008;3(11):397–415.

Wu W, Wu Z, Yu T, Jiang C, Kim W-S. Recent progress on magnetic iron oxide nanoparticles: synthesis, surface functional strategies and biomedical applications. Sci Technol Adv Mater. 2015;16(2):023501.

Wu F, Su H, Zhu X, Wang K, Zhang Z, Wong WK. Near-infrared emissive lanthanide hybridized carbon quantum dots for bioimaging applications. J Mater Chem B. 2016;4(38):6366–72.

Xu C, Sun S. New forms of superparamagnetic nanoparticles for biomedical applications. Adv Drug Deliv Rev. 2013;65:732–43.

Yagawa Y, Tanigawa K, Kobayashi Y, Yamamoto M. Cancer immunity and therapy using hyperthermia with immunotherapy, radiotherapy, chemotherapy, and surgery. J Cancer Metastasis Treat. 2017;3(10):218.

Yanase M, Shinkai M, Honda H, Wakabayashi T, Yoshida J, Kobayashi T. Antitumor immunity induction by intracellular hyperthermia using magnetite cationic liposomes. Jpn J Cancer Res. 1998;89(7):775–82.

Yin X, Zhang J, Wang X. Sequential injection analysis system for the determination of arsenic by hydride generation atomic absorption spectrometry. Fenxi Huaxue. 2004;32(10):1365–7.

Yoffe S, Leshuk T, Everett P, Gu F. Superparamagnetic iron oxide nanoparticles (SPIONs): synthesis and surface modification techniques for use with MRI and other biomedical applications. Curr Pharm Des. 2013;19:493–509.

Yu MK, Park J, Jon S. Targeting strategies for multifunctional nanoparticles in cancer imaging and therapy. Theranostics. 2012;2(1):3–44.

Zhang Y, Zhang B, Liu F, Luo J, Bai J. In vivo tomographic imaging with fluorescence and MRI using tumor-targeted dual-labeled nanoparticles. Int J Nanomed. 2013;9(1):33–41.

Zhu L, Zhou Z, Mao H, Yang L. Magnetic nanoparticles for precision oncology: theranostic magnetic iron oxide nanoparticles for image-guided and targeted cancer therapy. Nanomedicine. 2017;12(1):73–87.

Chapter 13
Theranostic Applications of Nanobiotechnology in Cancer

Rabia Javed, Muhammad Arslan Ahmad, and Qiang Ao

Abstract Cancer is a devastating disease and leading cause of death worldwide. Although many furtherance has been made regarding conventional diagnosis and treatment of cancerous tumors, still many gaps remain to be filled in overcoming this menace. Many benchmarks have been achieved regarding clinical translation with special emphasis on implementation of nanotheranostics in healthcare sector. This has led to mitigation in nanoparticles toxicity and patient's resistance to drugs. Nanotheranostics has remarkable applications in patient stratification, treatment of cancer stem cells (CSCs) and drug-resistant cancer. Novel strategies including synthesis and engineering of different nanoparticles have been developed that fulfill the purpose of early diagnosis and specified drug delivery to target site. Multifunctional nanoformulations have also been made to perform diagnosis, targeting, and treatment simultaneously. This chapter describes the biology of cancer and concept of nanobiotechnology and nanomedicine for various biomedical applications, having particular emphasis on cancer remedy via nanotheranostics. The advances in nanotheranostics have been discussed in detail and the ways to overcome respective challenges to this technology have been narrated.

Keywords Cancer · Nanoparticles · Targeted drug delivery · Nanotheranostics · Nano-oncology

R. Javed (✉) · Q. Ao
Department of Tissue Engineering, China Medical University, Shenyang, China

M. A. Ahmad
Department of Tissue Engineering, China Medical University, Shenyang, China

Key Lab of Eco-restoration of Regional Contaminated Environment, Shenyang University, Ministry of Education, Shenyang, China

M. Rai, B. Jamil (eds.), *Nanotheranostics*,
https://doi.org/10.1007/978-3-030-29768-8_13

13.1 Introduction

Theranostics is a term described in which the diagnostics and therapeutics are catered for broad range of applications in the field of medicine. A single procedure has to be designed in which both detection and treatment modalities are combined to get more cost-effective clinical healthcare. The most prominent advanced technology that shows coherence in this regard is nanobiotechnology. Such nanotheranostic agents/nanoformulations are developed having the capability to undergo targeted drug delivery with desirable specificity and sensitivity. The inherent properties of nanoparticles enhance the theranostic applications for the cure of broad spectrum of diseases. The nano size and large surface area of nanoparticles make them able to bind with multiple ligands and stabilizers. Moreover, the selective accumulation of nanoparticles on target site and bloodstream for longer periods and mitigation of their side effects, for instance toxicity, make them appropriate candidates for nanotheranostics (Wang et al. 2012).

Cancer is a complex ailment controlled by multiple factors, so it is hard to get rid of this disease. The conventional chemotherapeutic treatment of cancer has not been proved successful due to non-targeted drug delivery; however, nanobiotechnology offers many advantages for enhancement of solubility of nanoparticles and their specificity of action in delivery of drugs. Nanobiotechnology can be adequately tailored for nanotheranostic applications. Delivery of anticancer agents to the cancer cells occurs selectively by enhanced permeability and retention (EPR) effect in case of passive targeting. This effect is further escalated in active targeting via ligand-receptor interaction on the surface of cancer cells (Chowdhury et al. 2016; Ramzy et al. 2017).

The aim of this chapter is to elucidate the concept of theranostics and illustrate how nanobiotechnology plays a pivotal role in the design of nanotheranostics for various biomedical applications. It also describes the biology of deadly cancer disease and properties of nanoparticles used in cancer diagnosis and therapy. The mechanism of smart drug delivery to cancer cells and basics of cancer nanotheranostics have been explained diagrammatically. Lastly, the challenges faced by the nano-oncology have been addressed and future prospects of nanotheranostics of cancer have been appraised.

13.2 Cancer

The abnormal growth of cells is termed as cancer. In normal conditions, the old cells die and newly formed cells take their place. In contrast, the cancer cells do not die and divide uncontrollably and form proliferations called tumors.

Tumors are of three types, i.e.,

- Benign (non-cancerous)
- Malignant (cancerous)
- Precancerous

Benign tumors cannot spread to other body parts and do not grow back after removal from body. Malignant tumors have the ability to spread everywhere in the body by means of bloodstream and lymphatic system. The treatment of malignant tumors is difficult as compared to benign tumors because the former can uncontrollably metastasize. The new tumors form and are named metastases. Precancerous tumors are immature abnormal cells having the ability to become cancerous with the passage of time by the transfer of genetic changes. The normal cells get mutated into hyperplasia (abnormal cells that look normal under microscope) which then form dysplasia (lethal condition in which appearance of cells is abnormal under microscope) and finally a cancer tumor is formed (Meyenfeldt 2005).

Cancer stem cells (CSCs) are the cancer cells that have the ability to differentiate into new progeny. Like normal stem cells, CSCs possess the property of self-renewal and differentiation. They are responsible for excessive growth and development of tumors. These are the key drivers of tumor progression. These cells are responsible for heterogenous phenotypes and are resistant to conventional therapies, hence play a pivotal role in tumor recurrence. There are reports about the cross talks between CSCs leading to progressive metastasis angiogenesis (Ayob and Ramasamy 2018).

Cancer cells are categorized on the basis of their origin, appearance, and genetic alterations. However, sometimes the origin of cancer is not determined because at the start these cells are differentiated and they get undifferentiated after progression. Various cancer types have been shown in Fig. 13.1. The shape of various types of cancer is different from one another. Cancer cannot occur by a single mutation, it occurs by the series of mutations. Different types of genetic alterations are caused in different types of cancerous tumors. Cancer is derived by means of genetic alterations affecting proto-oncogenes, tumor suppressor genes, and DNA repair genes. Proto-oncogenes get converted into oncogenes by unwanted cell divisions. Tumor suppressor genes also get altered causing cancer. Moreover, DNA repair genes get mutated and produce cancerous cells (Hassanpour and Dehghani 2017).

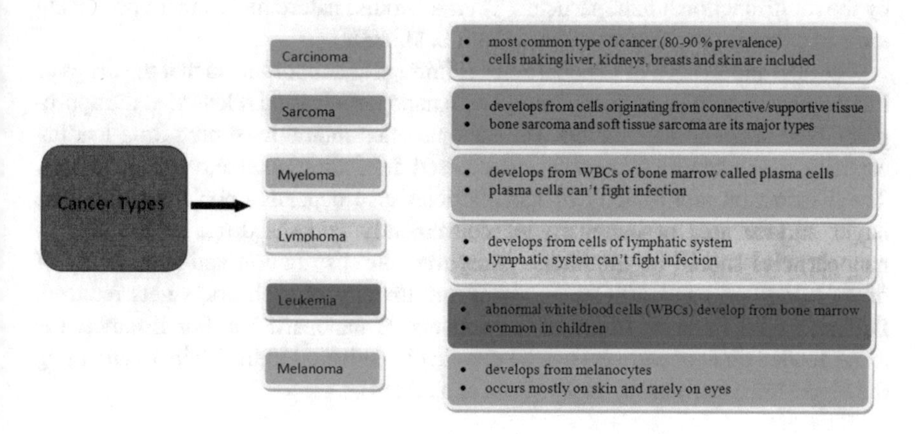

Fig. 13.1 Different types of cancer

Cancer is the leading cause of deaths worldwide and the number of cancer cases is expected to escalate to 23.6 million by 2030. Thus, control and management of cancer by health professionals, researchers, and policy makers is essential. Cancer is one of the most common genetic diseases that is caused by inheritance from ancestors as well as by exposure to toxic environmental conditions. Still, it is preventable by improvement of diet, regular exercise, and protection from alcohol and carcinogens (Sue-Kyung et al. 2000). Many human lives can be saved by taking these preventable measures. For instance, taking vitamin D and calcium supplements reduces the risks of cancer (Lappe et al. 2007).

13.3 Nanobiotechnology

Nanotechnology involve nanomaterials having 1–100 nm size and are capable of manipulating physical and chemical features of substances at molecular level. Biotechnology is the subdiscipline of biology that uses techniques of biological sciences for manipulation or engineering of molecular, cellular, and genetic substances. It is sometimes called genetic engineering of biological molecules forming novel products for industries. The combination of nanotechnology and biotechnology is known as nanobiotechnology. It is cutting-edge technology that removes boundaries between chemistry, physics, and biology. It can also be called multidisciplinary field of study (Bharali et al. 2011).

Nanobiotechnology is doing wonders in the field of medicine since last two decades. It is crux technology in disease diagnosis, therapy, and targeted drug delivery. It improves the way of detection and treatment of different human diseases and prevents them from occurring in future. Targeted drug delivery is carried out in which the nanoparticles are engineered in such a way that the delivery of drugs occurs only to diseased cells while neighboring healthy cells are protected. Additionally, the targeted delivery, therapy and imaging can occur simultaneously by the multifunctional nanoparticles. Their all-round nature makes them promising tools in encountering cancer disease (Park et al. 2009).

The size and surface of nanoparticles is immensely important so that the drugs or imaging agents can be perfectly bound with nanoparticles and released at the appropriate time without degradation. The nanoparticles should have high drug loading capacity which can be increased or decreased depending on size of nanoparticles. The distribution and toxicity of nanoparticles also depends upon their size. The larger surface area of nanoparticles conveniently releases drug, so the size of nanoparticles should be minimum. Moreover, the distribution and availability of drugs loaded on nanoparticles increases and toxicity of such drugs gets reduced. Different routes can be followed for delivery of nanoparticles. For instance, the drugs loaded on nanoparticles can be effectively delivered to the brain, overcoming the blood-brain barrier/meninges (Patel et al. 2014).

13.4 Nanomedicine in Diagnostics and Therapeutics

The application of nanobiotechnology to the field of medicine is called nanomedicine. The quality of life of patients can be improved by learning the pathophysiology of diseases using nanotechnology. The long-term effectivity of nanomedicine formulations in the diagnostic and therapeutic intercession has been depicted (Rizzo et al. 2013).

The diagnostic methods currently available are able to detect the particular disease when the disease symptoms appear and patient starts suffering from illness. However, the technology should be able to diagnose the disease way before the appearance of symptoms so that it could be treated earlier. Polymerase chain reaction (PCR) based tests are recently devised technologies. Likewise, nanobiotechnology offers a promising approach which is more sensitive and efficient (Savaliya et al. 2015).

Usually, the diseases are detected by means of antibodies attached with fluorescent dyes. However, the specificity of dyes is much less, so the use of nanotechnology is a dire need to overcome this discrepancy. Quantum dots (QDs) are supposed to be useful in this regard which are very helpful in intracellular imaging. Moreover, if the genetic compositions of biological specimens are to be detected, gold (Au) nanoparticles bound with DNA would be promising. Nanosystems also help in isolation and detection of sparse cells (for example, cancer cells) from the bloodstream (Qasim et al. 2014).

The conventional drugs for different diseases are very expensive because the conventional techniques require many reagents and solutions. To overcome this, biopharmaceuticals are currently developing at a very fast rate. Nano-drug formulations are designed for efficient and economic drug delivery to targeted cells. The drugs that can't be orally administered can be given to the patients in the form of nanoformulations. The nano-drugs can be quickly delivered to the locations where standard drugs cannot reach. These drugs are triggered by laser or infrared light or by specific internal molecules present for controlled release of drugs to their destined location. It is the property of nanoparticles to protect drugs from degradation or denaturation on getting exposed to lower pH or other extreme conditions. The half-life of drugs also increases by adhesion of drugs to the cell membranes (Zhang et al. 2008).

Few other applications of nanobiotechnology in therapeutics have been described below:

13.4.1 Gene Therapy

The transportation of viral vectors is problematic, so nanoparticle-based nonviral vectors have been devised for successful delivery of plasmid DNA. Many trials are ongoing to prepare nanometer-scale constructs as gene carriers (Liu and Zhang 2011).

13.4.2 Engineering of Biomolecules

It was once a very tough task, but with the advent of nanobiotechnology, it has become easier by performing biochemical reactions on solid substrates rather than liquid solutions. This technology has applications in drug discovery and functional genomics and it holds a great promise for future generations regarding biosensors, biocatalysts, etc. (Nagamune 2017).

13.4.3 Cardiovascular (CDV) Diseases

The immune signals of CDV diseases can be sensed by quantum dots and other nanosystems which are otherwise difficult to be monitored. Different nanomachines for the treatment of CDV diseases can also be devised (Jiang et al. 2017a; Chandarana et al. 2018).

13.4.4 Dentistry

Dental healthcare can be facilitated by nanorobots and other such nanomaterials made by using nanodental techniques. Natural tooth will be maintained by repairing the tooth tissues in this way (Abiodun-Solanke et al. 2014; AlKahtani 2018).

13.4.5 Orthopedics

Nanobiotechnology plays an important role in orthopedic surgery by the deposition of nanomaterials on implants. This enhances the efficiency of orthopedic implants by enhancing nanomaterials and bones interactions (Garimella and Eltorai 2017; Smith et al. 2018).

13.5 Role of Nanomedicine in Cancer

Nanobiotechnology offers a great potential in combating cancer. Nanoparticles for targeted drug delivery to cancer cells cause excessive drug accumulation on the sites of tumor development, while preventing the closely located normal cells from drug exposure. This causes cytotoxicity only in cancerous cells and improves the pharmacokinetic profiles. The nanodrugs remain stable during and after its delivery to desired cells. These drugs are able to invade inside the tumor cells through enhanced permeability and retention (EPR) effect. However, the biomarkers of EPR effect should be identified that will lead to the development of personalized nanomedicine for cancer nanotherapeutics (Shi et al. 2017).

Nano-oncology is the term derived for applications of nanobiotechnology in cancer. It involves early diagnosis via magnetic resonance imaging (MRI), therapeutics via nanocarrier for targeted drug delivery, and cancer prevention via nanochemoprevention. Its future implications involve the improvement in molecular individualized medicine (Mark et al. 2013). Hitherto, cancer nanomedicine reveals difficulty in targeted drug delivery because of uptake of nanodrug by the healthy cells along with the unhealthy ones. Nanodrugs for cancer are also inefficient in penetrating tumor cells, i.e., intracellular delivery. Therefore, the field of immune-oncology could be combined with nano-oncology so that the host's own immune system could be stimulated to recognize and eradicate tumors by means of nanomedicines. In this way, cancer patients are much benefitted (Jiang et al. 2017b).

Cancer stem cells (CSCs) have the ability to escape conventional therapies. Therefore, the isolation of CSCs should be done that may help in understanding the mechanism of metastasis and signaling pathways. Different cell surface markers have been discovered such as CD24, CD44, and CD133. Similarly, various regulatory pathways have also been reported like Wnt/β-catenin and Hedgehog pathways. These discoveries pave the way for novel anticancer therapies using nanoparticles (Kievit and Zhang 2011; Yu et al. 2012; Sisay et al. 2014).

13.6 Major Nanoparticles Used in Diagnosis and Therapy of Cancer

Different types of nanoparticles have their strengths and weaknesses and their selection to be used as theranostic agents is based on the application. The commonly used nanoparticles for diagnosis and therapy of cancer disease have been shown in Fig. 13.2 and are listed below.

13.6.1 Quantum Dots (QDs)

QDs are inorganic semiconductor nanoparticles having applications in medicine and biology. QDs are biocompatible and biologically inert materials of low toxicity. These are used as nanosensors and nanocarriers (Granada-Ramirez et al. 2018). QDs are even more stable than fluorescent proteins at fluctuating temperature and pH conditions, hence used for prolonged fluorescent monitoring and image-guided therapy. This unique property of QDs, i.e., their ability to visualize on the targeted tissue, makes them desirable semiconductor molecules to be used in cancer nanotheranostics. QDs can be multifunctionalized for cancer targeting and imaging, and delivery of drug (Luk and Zhang 2014; Sisay et al. 2014).

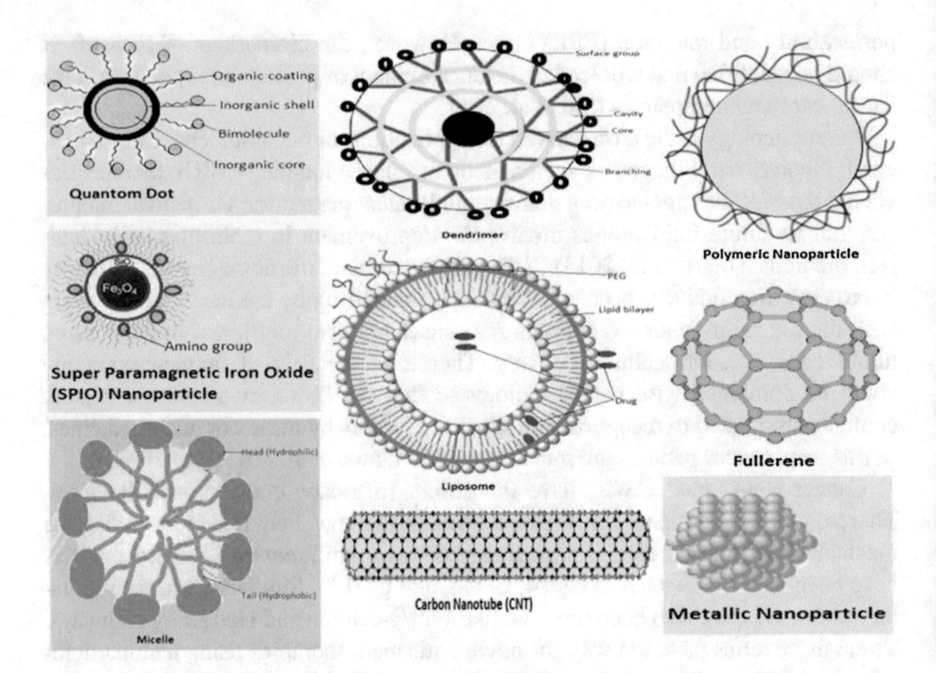

Fig. 13.2 Major nanoparticles used in cancer nanotheranostics

Huang et al. (2017) developed a nanotheranostic system in which they coated the QDs with a polymer that was loaded with paclitaxel drug and lipoic acid. The ester linkage was formed between polymer and drug. This system was proved favorable for in vitro diagnosis and therapy of cancer cell lines. AbdElhamid et al. (2018) developed highly fluorescent QDs that were layer by layer coupled with gelatin/chondroitin that showed anti-breast cancer efficacy and non-immunogenicity. Imaging and drug delivery were performed simultaneously in this study.

13.6.2 Liposomes

The first nanoformulation to be approved by the Food and Drug Administration (FDA, USA) was a liposomal formulation. Liposomes are organic nanovectors used for targeted delivery of drugs and imaging agents. Personalized therapeutic medicines are made by the combination of nanovectors with targeting moieties. These vesicles having lipid bilayers are biocompatible, biodegradable, and non-immunogenic that make them potent candidates for encapsulation of broad range of drugs. Polyethylene glycol (PEG) is attached to liposomal surface due to which the liposomes do not get recognized by immune system, thereby stay for a longer time period in the body (Muthu and Feng 2012). Liposomes have the ability to work as multifunctional nanocarriers by attaching multiple functional groups on the surface of liposomes and they enhance targeted delivery by accumulating for longer period on targeted sites (Deshpande et al. 2013). Liposomes are used for diagnosis of cancer by magnetic resonance imaging (MRI), positron emission tomography (PET), and single photon emission computed tomography (SPECT) (Sisay et al. 2014).

13.6.3 Micelles

Micelles are polymeric spherical nanoparticles having hydrophobic interior that can hold drugs that are otherwise insoluble in aqueous solution. They have tremendous ability to deliver cargo to the site of action and least cytotoxicity. They are highly stable but stimuli-sensitive and undergo controlled drug release. These nanoparticles have the capability to reduce intracellular drug resistance in cancer patients (Tyrrell et al. 2010; Biswas et al. 2016).

13.6.4 Dendrimers

Dendrimers are organic molecules with monodisperse morphology that can generate different polydisperse molecules. They possess low cytotoxicity and are chemically stable. Dendrimers undergo controlled and targeted delivery of drugs against various diseases including cancer (Abbasi et al. 2014). These polymeric nanoparticles are multifunctional, versatile, and biocompatible. They are upgradable, i.e., their surfaces can be modified with targeting ligands. They can carry both the therapeutic and imaging moiety for simultaneous imaging and therapy (Sisay et al. 2014).

13.6.5 Carbon Nanotubes (CNTs)

CNTs are organic allotropes of carbon having cylindrical shape. These can be single-walled or multi-walled in structure. These are very light weight and have a structure on which drug could easily be loaded. They can easily get conjugated to diagnostic and therapeutic agents like drugs, genes, DNA, RNA, etc. These are widely used in optical imaging and drug delivery (He et al. 2013).

13.6.6 Super Paramagnetic Iron Oxide (SPIO) Nanoparticles

SPIO are inorganic nanoparticles having unique physicochemical and biological properties. SPIO nanoparticles lack the ability to get solubilized in water. They can be aggregated intracellularly which can be prevented by adding hydrophilic polymers to their surface. The water solubility of these nanoparticles can also be enhanced in this way (Sisay et al. 2014). These nanoparticles are capable of magnetic targeting and magnetic resonance imaging (MRI) due to their excellent magnetic properties. These are biocompatible and stable, thereby undergo specified drug delivery (Wang et al. 2013; Kandasamy and Maity 2015).

Zou et al. (2010) reported the study in which SPIO nanoparticles were capped with PEG polymer, and then antibody and fluorescent dye were conjugated with PEGylated polymer. Anticancer drug was loaded on this molecule. Later on, cancer targeting and imaging was conducted in colon cancer cell line.

13.6.7 Metallic Nanoparticles

Au nanoparticles can be found as nanorods, nanocages, and nanoshells having excellent thermal and optical properties. These nanoparticles can be conveniently prepared, have facile structure and function, and can cause thermal ablation of cancer cells to be targeted (Sisay et al. 2014). These nanoparticles are naturally inorganic and are used in targeted drug delivery, sensing, and imaging (Yeh et al. 2012). AgO nanoparticles are inorganic in nature and have excellent potential to be used as anticancer agents. Besides, they are capable to be used as antimicrobial, antioxidant, anti-inflammatory, and anti-angiogenic agents (Zhang et al. 2016).

13.6.8 Fullerenes

Fullerenes are organic molecules, also called buckyballs. They consist of carbon-cages that are very stable structures used for cure of various diseases (Mansoori et al. 2007). Metallofullerenes have potential applications in diagnostics because these are used as contrast-enhancing agent in magnetic resonance imaging (MRI). Fullerenes are also used in therapeutics and have potential applications in radiotherapy, photodynamic therapy, and chemotherapy. Photoacoustic imaging is performed by carboxyl and polyhydroxylated fullerenes (Chen et al. 2012).

13.6.9 Polymers

Natural polymers are biodegradable, biocompatible, non-immunogenic, and non-toxic. These are broadly classified as polysaccharides and proteins. Polysaccharides include dextran, starch, chitosan, and alginate, while examples of proteinaceous natural polymers are albumin and gelatin. All of them are fabricated into nanoparticles for nanodrug delivery and its controlled release to the specific site (Anwunobi and Emeje 2011). Synthetic polymers include polyethylene glycol (PEG), polyvinylpyrrolidone (PVP), polystyrene, etc. These are similar to natural polymers with respect to their role in drug delivery and medicine (Gardel 2013). Zhao et al. (2017) used peptide aptamer targeted polymers for imaging and delivery of doxorubicin (DOX) in cancer cells.

13.7 Cancer Prevention via Nanomedicine

Removal of carcinogens can result in cancer prevention because these chemicals result in chromosomal aberrations. Conventional cancer preventive treatments do not have potential to remove such oxidizing agents from the environment. However, nanomedicine can play a crucial role in this scenario. Nanoparticles attached with UV-scattering and UV-absorbing substances could eliminate carcinogens effectively (Zare-Zardini et al. 2015). Food and Drug Administration (FDA) has approved various conventional drugs for destruction of precancerous cells. These drugs reduce the risk of cancer. Application of nanotechnology for cancer prevention will increase specificity and sensitivity of drug delivery and will also evoke immune response in precancer patients. However, the nano-based cancer prevention requires heavy funds and resources. Moreover, the nanoformulations for cancer prevention should not be toxic or least toxic to human body. A legislation needs to be passed for fighting the cancer by nanobiotechnology (Menter et al. 2014).

13.8 Cancer Diagnostics/Imaging via Nanomedicine

Various conventional diagnosis techniques involve fluorescence in situ hybridization (FISH), fine-needle aspiration cytology (FNAC), immunohistochemistry, and biopsy. Hence, only in vitro diagnosis is possible with traditional methods. Biosensors are one of the applications of nanobiotechnology in cancer diagnostics. QDs are used for early diagnosis of cancer markers such as Her2, and real-time monitoring of release of drug can take place via in vivo biosensors. CNTs may also be used for detecting cancer markers by atomic force microscopy (AFM). Another common example of nanosensors used for diagnosing cancer tumors are cadmium selenide (CdSe) nanoparticles (Zare-Zardini et al. 2015).

13.9 Cancer Therapeutics via Nanomedicine

The current methods for drug delivery in case of cancer are oral, intravenous, intra-arterial, or intra-muscular. The conventional drug delivery system has lot of serious problems, one of the major discrepancies being the development of toxicity to non-cancerous tissues due to non-targeted delivery. Using nanobiotechnology, the issue of bioavailability of drugs can be sorted out using encapsulated nanoparticles in which the surfaces of nanoparticles are fabricated with different polymeric agents or proteins, etc. Thus, the effectivity and efficiency of targeted drug delivery is enhanced. Even small doses can be delivered in this way by which toxicity is reduced to minimum (Zare-Zardini et al. 2015).

Small interfering RNAs (siRNAs) are synthesized by the cleavage of RNA molecules. These are 20–25 nucleotides long and function in silencing the gene transcription by forming RNA-protein complex called RNA-induced silencing complex (RISC). The technique of RNA silencing has also applications in nanotherapeutics. Nanoparticles can be used as carriers for delivery of siRNAs to gastrointestinal tract (GI) of diseased individuals, hence making them free of oral or cervical cancer (Menter et al. 2014).

13.9.1 Mechanism of Targeted Drug Delivery

The mechanism of targeted delivery of drugs is based on the principle of enhanced permeability and retention (EPR) effect of nanoparticles. Professor Hiroshi Maeda first proposed this phenomenon. The different strategies of drug targeting are:

1. Passive targeting
2. Active targeting/smart drug delivery
3. Physical targeting/next generation photothermal and magnetic hybrid Nanoparticles-based targeting

When the drug is distributed to other body organs besides tumor through blood circulation and extravasation, such type of targeting is known as passive targeting. It relies on EPR effect. On the other hand, the drug distribution to only tumor is termed active targeting/"smart" drug delivery shown in Fig. 13.3. It involves the interaction between drugs and targeted cells through ligand-receptor bonding. The drug/drug carrier is guided to reach the specific target by doing polyethylene glycol (PEG) capping of nanoparticles which increases the time of blood circulation, thereby enhancing EPR effect. The recent targeting strategy is dependent on the heat released by the photothermal and magnetic nanoparticles for controlled drug release (Bae and Park 2011; Yu et al. 2016; Unsoy and Gunduz 2018).

Targeted drug delivery reduces the possible side effects due to accumulation of nanodrug in the bloodstream for longer time. It maximizes the pharmacokinetic efficiency and therapeutic effects by efficiently clearing toxic substances from the body (Fahmy et al. 2005; Vasir and Labhasetwar 2005). Wang et al. (2019) used mesenchymal stem cells for targeted delivery of anticancer drug-encapsulated nanoparticles to the lungs. Since this strategy was effective, it produced incomparable inhibition of lung cancer in different animals (rabbit and monkey) even in low dosage. Recently, an effective protocol for treatment of metastatic ovarian cancer was developed using her2-targeted theranostic nanoparticles (Satpathy et al. 2019).

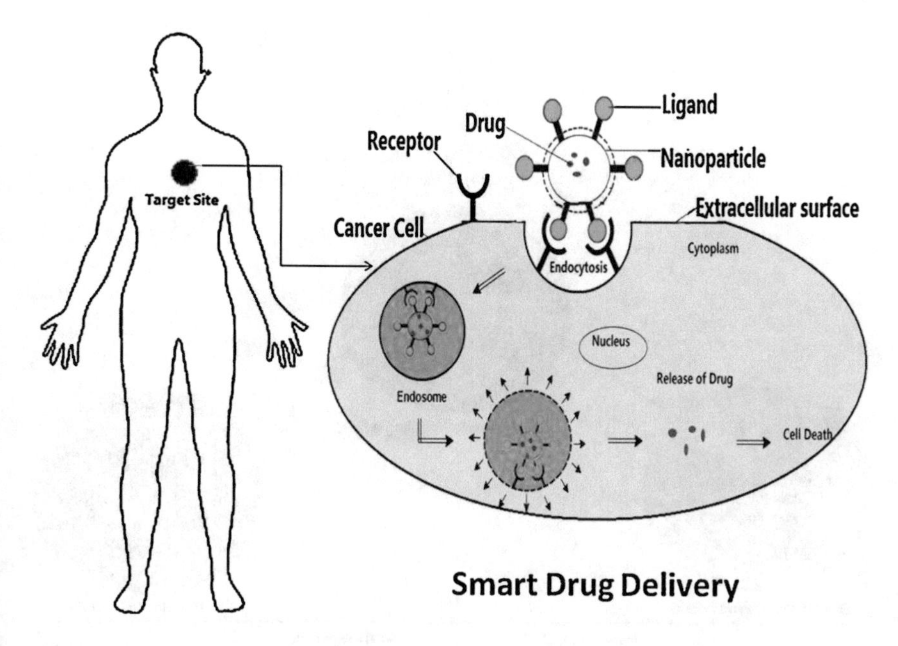

Fig. 13.3 Mechanism of smart drug delivery system

13.10 Cancer Detection/Monitoring via Nanomedicine

Different pathways involved in progression of cancer are quite complicated, so there is difficulty in study. Positron emission tomography (PET) and computed tomography (CT) scan have been used for monitoring and detection of cancer so far. Usage of nanomaterials for early detection of cancer holds a great promise. Recently, QDs, magnetic nanoparticles, and dendrimers have been tried for optical imaging and magnetic resonance imaging (MRI). Ultrasonography is also used for this purpose but the recognition of neo-vascularization in case of cancer requires more sophisticated and sensitive instrumentation. Ultrasonography contrast media (UCM) can identify early-stage cancer and is a necessary solution to this problem. Nevertheless, the recent advancements in this domain will cause early detection of cancer and its pattern of occurrence which indicates future strategy for an easy, noninvasive detection in the lymphatic system of patients (Hartman et al. 2008).

13.11 Cancer Nanotheranostics

The amalgamation of nanodiagnostics and nanotherapeutics is termed "nanotheranostics." Figure 13.4 explains the concept of nanotheranostics of cancer in detail. The plausible applications of nanoparticles in which the individuals are diagnosed, treated, and monitored simultaneously in real-time and in noninvasive tis-

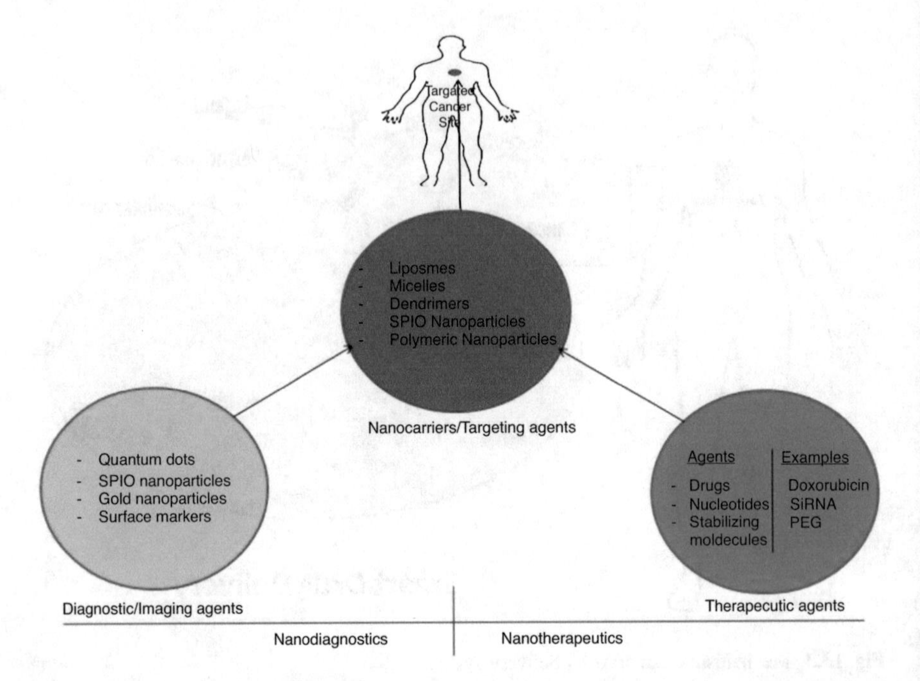

Fig. 13.4 Phenomenon of nanotheranostics of cancer

sues are the most promising tools for winning battle against cancer using a single nanoformulation. The patient tumor-nanodrug interaction plays a significant role in scale-up fabrication of nanoparticles for cancer nanotheranostics. The combined diagnosis and therapy by attaching the corresponding theranostic agents to nanoparticles allows researchers to get images of mechanism of action of nanoparticles along with treatment ultimately leading to damage of cancer cells. In this way, different physicochemical properties of targeting, imaging, and therapeutic agents can be harnessed to discover novel cancer nanotheranostic methods of drug delivery. In spite of all efforts and deliberate attempts to optimize cancer nanotheranostics, it has been facing exigent barriers to clinical translation (Melancon et al. 2012; Chen et al. 2017; Silva et al. 2019).

13.12 Challenges in Nano-Oncology

The major challenge in the field of cancer study is early diagnosis. Almost all types of cancer (99%) are epithelial in origin, so the anatomy of superficial cells is of utmost importance (Frank 2007). Therefore, three-dimensional (3-D) images of microanatomy should be obtained. Nanobiotechnology can resolve this issue by inexpensive and convenient nanodiagnostic methods. However, there are few challenges due to some clinical failures of nanomedicine, particularly nano-oncology. The list of hurdles faced by nano-oncology are given below.

13.12.1 Toxicity of Nanoparticles (NPs)

There are innumerable reports about nanostructures declared to be toxic for humans. The vulnerable body parts such as respiratory and gastrointestinal tract (GI) may develop toxicity of these substances. Nanoparticles are attached with peptides so that after functioning they can easily get degraded by proteases and excrete out from the body in the form of urine. If these NPs are not excreted, they may cause significant distress to human body. For instance, they might be absorbed by the normal body tissues. Their long-term presence in blood might result in blood clotting (Zare-Zardini et al. 2015).

13.12.2 In Vivo Studies

Lots of nanotheranostic studies performed to date are in vitro. However, biological system is very complex and nano-drugs need to be tested for their efficiency in in vivo environment. A framework should be established for nanomedicine development and translation where the animal models and patients who respond well to nanomedicine need to be selected (Sisay et al. 2014).

13.12.3 Multifunctional Nanoparticles

"Smart" multifunctional nanoparticles should be largely prepared for simultaneous diagnosis, targeting, and therapy of cancer tumors (Riggio et al. 2011).

13.12.4 Targeting Efficacy of Nanovectors

The nanocarriers should be able to cross the physiological barriers such as blood-brain barrier and multidrug resistance in order to reach at their target site (Riggio et al. 2011).

13.12.5 Phagocytosis

The reticuloendothelial system that undergoes excessive phagocytosis by nanoformulations needs to be minimized. The nanomedicine design must be optimized to fulfill this purpose (Kuncic 2015).

13.12.6 Immuno-oncology

The immune-oncology should utilize new techniques in order to avoid previous mistakes. Fundamental challenges facing immune-oncology should be minimized. Molecularly targeted therapies must be amalgamated with immunotherapies to obtain next generation nanomedicines (Kuncic 2015).

13.12.7 Interdisciplinary Collaboration

Previously, clinical and basic science researchers have not been brought to a single platform where they could get an opportunity to explore current opportunities in nano-oncology. Scientists from such areas of research should be brought together in order to discover something novel in nano-oncology that would prove beneficial for translational healthcare (Kuncic 2015).

13.13 Conclusions and Future Perspectives

Cancer is a lethal disease that is prevalent all over the world. The biology of cancer has largely been understood by advanced scientific practices. Conventional cancer diagnostic and therapeutic procedures have failed to eradicate cancerous cells effectively from the body. From the past few years, nanotherapeutics is rapidly emerging to solve many issues related to conventional chemotherapeutics. Moreover, nanotheranostics has been budding into a plausible combined solution for all issues related to cancer imaging and treatment. Nanomedicine development has introduced advanced nanosystems for theranostics. It has huge potential to reduce multidrug resistance and enhance selectivity and solubility of nanoparticles. It is evident that the efficacy of nanoparticles depends on their size, surface charge, surface area, and multipotency. Nanoparticles for effective drug delivery need to be stable, easy to fabricate, biocompatible, biodegradable, non-immunogenic, and nontoxic and possess a capability to release the loaded drug to only targeted site with minimum side effects. Tumor heterogeneity is a major challenge in the development of nanomedicines for targeted delivery in the remedy of cancer. Nevertheless, many such nanoformulations are in clinical trial phase and there is need for more preclinical models so that their targeting effectiveness could be tested. Multiscale modeling and computational models should be used for designing of nanoparticles in order to escalate more advanced targeted drug delivery. Furthermore, the cancer nanotherapeutics should be customized for individual patients leading to advances in personalized medicine.

References

Abbasi E, Aval SF, Akbarzadeh A, Milani M, Nasrabadi HT, Joo SW, Hanifehpour Y, Nejati-Koshki K, Pashaei-Asl R. Dendrimers: synthesis, applications, and properties. Nanoscale Res Lett. 2014;9(1):247.

AbdElhamid AS, Helmy MW, Ebrahim SM, Bahey-El-Din M, Zayed DG, Zein El Dein EA, El-Gizawy SA, Elzoghby AO. Layer-by-layer gelatin/chondroitin quantum dots-based nanotheranostics: combined rapamycin/celecoxib delivery and cancer imaging. Nanomedicine. 2018;13(10). https://doi.org/10.2217/nnm-2018-0028.

Abiodun-Solanke IMF, Ajayi DM, Arigbede AO. Nanotechnology and its application in dentistry. Ann Med Health Sci Res. 2014;4(3):S171–7.

AlKahtani RN. The implications and applications of nanotechnology in dentistry. A review. Saudi Dent J. 2018;30(2):107–16.

Anwunobi AP, Emeje MO. Recent applications of natural polymers in nanodrug delivery. J Nanomed Nanotechnol. 2011;S4:002.

Ayob AZ, Ramasamy TS. Cancer stem cells as key drivers of tumor progression. J Biomed Sci. 2018;25(1):20.

Bae YH, Park K. Targeted drug delivery to tumors: myths, reality and possibility. J Control Release. 2011;153(3):198–205.

Bharali DJ, Siddiqui IA, Adhami VM, Chamcheu JC, Aldahmash AM, Mukhtar H, Mousa SA. Nanoparticle delivery of natural products in the prevention and treatment of cancers: current status and future prospects. Cancers. 2011;3:4024–45.

Biswas S, Kumari P, Lakhani PM, Ghosh B. Recent advances in polymeric micelles for anti-cancer drug delivery. Eur J Pharm Sci. 2016;83:184–202.

Chandarana M, Curtis A, Hoskins C. The use of nanotechnology in cardiovascular disease. Appl Nanosci. 2018;8(7):1607–19.

Chen Z, Ma L, Liu Y, Chen C. Applications of functionalized fullerenes in tumor theranostics. Theranostics. 2012;2(3):238–50.

Chen H, Zhang W, Zhu G, Xie J, Chen X. Rethinking cancer nanotheranostics. Nat Rev Mater. 2017;2. https://doi.org/10.1038/natrevmats.2017.2.

Chowdhury MR, Schumann C, Bhakta-Guha D, Guha G. Cancer nanotheranostics: strategies, promises and impediments. Biomed Pharmacother. 2016;84:291–304.

Deshpande PP, Biswas S, Torchilin VP. Current trends in the use of liposomes for tumor targeting. Nanomedicine. 2013;8(9):1509–28. https://doi.org/10.2217/nnm.13.118.

Fahmy TM, Fong PM, Goyal A, Saltzman WM. Targeted for drug delivery. Mater Today. 2005;8(8):18–26.

Frank SA. Chapter 12, Stem cells: tissue renewal. In: Dynamics of cancer: incidence, inheritance, and evolution. Princeton, NJ: Princeton University Press; 2007. https://www.ncbi.nlm.nih.gov/books/NBK1566/.

Gardel ML. Synthetic polymers with biological rigidity. Nature. 2013;493:618–9.

Garimella R, Eltorai AEM. Nanotechnology in orthopedics. J Orthop. 2017;22(1):30–3.

Granada-Ramirez DA, Arias-Ceron JS, Rodriguez-Fragoso P, Vazquez-Hernandez F, Luna-Arias JP, Herrera-Perez JL, Mendoza-Alvarez JG. Quantum dots for biomedical applications. In: Narayan R, editor. Nanobiomaterials. Sawston: Woodhead Publishing; 2018. p. 411–36. https://doi.org/10.1016/B978-0-08-100716-7.00016-7.

Hartman KB, Wilson LJ, Rosenblum MG. Detecting and treating cancer with nanotechnology. Mol Diagn Ther. 2008;12(1):1–14.

Hassanpour SH, Dehghani M. Review of cancer from perspective of molecular. J Cancer Res Pract. 2017;4(4):127–9.

He H, Pham-Huy LA, Dramou P, Xiao D, Zuo P, Pham-Huy C. Carbon nanotubes: applications in pharmacy and medicine. BioMed Res Int. 2013;2013:578290.

Huang H-K, Yan J, Liu P, Zhao B-Y, Cao Y, Zhang X-F. A novel cancer nanotheranostics system based on quantum dots encapsulated by a polymer-prodrug with controlled release behavior. Aust J Chem. 2017;70(12):1302–11.

Jiang W, Rutherford D, Vuong T, Liu H. Nanomaterials for treating cardiovascular diseases: a review. Bioact Mater. 2017a;2(4):185–98.

Jiang W, von Roemeling CA, Chen Y, Qie Y, Liu X, Chen J, Kim YS. Designing nanomedicine for immuno-oncology. Nat Biomed Eng. 2017b;1(2). https://doi.org/10.1038/s41551-017-0029.

Kandasamy G, Maity D. Recent advances in superparamagnetic iron oxide nanoparticles (SPIONs) for *in vitro* and *in vivo* cancer nanotheranostics. Int J Pharm. 2015;496(2):191–218.

Kievit FM, Zhang M. Cancer nanotheranostics: improving imaging and therapy by targeted delivery across biological barriers. Adv Mater. 2011;20:1–31.

Kuncic Z. Cancer nanomedicine: challenges and opportunities. Med J Aust. 2015;203(5):204–5.

Lappe JM, Travers-Gustafson D, Davies KM, Recker RR, Heaney RP. Vitamin D and calcium supplementation reduces cancer risk: results of a randomized trial. Am J Clin Nutr. 2007;85(6):1586–91.

Liu C, Zhang N. Nanoparticles in gene therapy principles, prospects, and challenges. Prog Mol Biol Transl Sci. 2011;104:509–62.

Luk BT, Zhang L. Current advances in polymer-based nanotheranostics for cancer treatment and diagnosis. ACS Appl Mater Interfaces. 2014;6(24):21859–73.

Mansoori GA, Mohazzabi P, McCormack P, Jabbari S. Nanotechnology in cancer prevention, detection and treatment: bright future lies ahead. World Rev Sci Technol Sustain Develop. 2007;4:226–57.

Mark Y, Xuchuan J, Liuen L, Jia-Lin Y. Application of nanobiotechnology in cancer: creation of nano-oncology and revolution in cancer research and practice. World J Cancer Res. 2013;1(1):24–36.

Melancon MP, Stafford RJ, Li C. Challenges to effective cancer nanotheranostics. J Control Release. 2012;164(2):177–82.

Menter DG, Patterson SL, Logsdon CD, Kopetz S, Sood AK, Hawk ET. Convergence of nanotechnology and cancer prevention: are we there yet? Cancer Prev Res. 2014;7(10):973–92. https://doi.org/10.1158/1940-6207.

Meyenfeldt MV. Cancer-associated malnutrition: an introduction. Eur J Oncol Nurs. 2005;9(2):S35–8.

Muthu MS, Feng S-S. Theranostic liposomes for cancer diagnosis and treatment: current development and pre-clinical success. Expert Opin Drug Deliv. 2012;10(2):151–5.

Nagamune T. Biomolecular engineering for nanobio/bionanotechnology. Nano Converg. 2017;4(1):9.

Park K, Lee S, Kang E, Kim K, Choi K, Kwon IC. New generation of multifunctional nanoparticles for cancer imaging and therapy. Adv Funct Mater. 2009;19(10):1553–66.

Patel SP, Patel PB, Parekh BB. Application of nanotechnology in cancer prevention, early detection and treatment. J Cancer Res Therap. 2014;10(3):479–86.

Qasim M, Lim D-J, Park H, Na D. Nanotechnology for diagnosis and treatment of infectious diseases. J Nanosci Nanotechnol. 2014;14(10):7374–87.

Ramzy L, Nasr M, Metwally AA, Awad GAS. Cancer nanotheranostics: a review of the conjugated ligands for overexpressed receptors. Eur J Pharm Sci. 2017;104:273–92.

Riggio C, Pagni E, Raffa V, Cuschieri A. Nano-oncology: clinical application for cancer therapy and future perspectives. J Nanomater. 2011;2011:17.

Rizzo LY, Theek B, Storm G, Kiessling F, Lammers T. Recent progress in nanomedicine: therapeutic, diagnostic and theranostic applications. Curr Opin Biotechnol. 2013;24(6):1159–66.

Satpathy M, Wang L, Zielinski RJ, Qian W, Wang YA, Mohs AM, Kairdolf BA, Ji X, Capala J, Lipowska M, Nie S, Mao H, Yang L. Targeted drug delivery and image-guided therapy of heterogeneous ovarian cancer using Her2-targeted theranostic nanoparticles. Theranostics. 2019;9(3):778–95.

Savaliya R, Shah D, Singh R, Kumar A, Shankar R, Dhawan A, Singh S. Nanotechnology in disease diagnostic techniques. Curr Drug Metab. 2015;16:645–61.

Shi J, Kantoff PW, Wooster R, Farokhzad OC. Cancer nanomedicine: progress, challenges and opportunities. Nat Rev Cancer. 2017;17:20–37.
Silva CO, Pinho JO, Lopes JM, Almeida AJ, Gaspar MM, Reis C. Current trends in cancer nanotheranostics: metallic, polymeric, and lipid-based systems. Pharmaceutics. 2019;11(1):1–40.
Sisay B, Abrha S, Yilma Z, Assen A, Molla F, Tadese E, Wondimu A, Gebre-Samuel N, Pattnaik G. Cancer nanotheranostics: a new paradigm of simultaneous diagnosis and therapy. J Drug Deliv Therap. 2014;4(5):79–86.
Smith WR, Hudson PW, Ponce BA, Manoharan SRR. Nanotechnology in orthopedics: a clinically oriented review. BMC Musculoskelet Disord. 2018;19(1):67.
Sue-Kyung P, Keung-Young Y, Seung-Joon L, Sook-Un K, Se-Hyun A, Dong-Young N, Kuk-Jin C, Paul ST, Ari H, Daehee K. Alcohol consumption, glutathione S-transferase M1 and T1 genetic polymorphisms and breast cancer risk. Pharmacogenetics. 2000;10(4):301–9.
Tyrrell ZL, Shen Y, Radosz M. Fabrication of micellar nanoparticles for drug delivery through the self-assembly of block copolymers. Prog Polym Sci. 2010;35(9):1128–43.
Unsoy G, Gunduz U. Smart drug delivery systems in cancer therapy. Curr Drug Targets. 2018;19(3):202–12.
Vasir JK, Labhasetwar V. Targeted drug delivery in cancer therapy. Technol Cancer Res Treat. 2005;4(4):363–74.
Wang L-S, Chuang M-C, Ho JA. Nanotheranostics—a review of recent publications. Int J Nanomed. 2012;7:4679–95.
Wang Y-XJ, Xuan S, Port M, Idee J-M. Recent advances in superparamagnetic iron oxide nanoparticles for cellular imaging and targeted therapy research. Curr Pharm Des. 2013;19(37):6575–93.
Wang X, Chen H, Zeng X, Guo W, Jin Y, Wang S, Tian R, Han Y, Guo L, Han J, Wu Y, Mei L. Efficient lung cancer-targeted drug delivery via a nanoparticle/MSC system. Acta Pharm Sin B. 2019;9(1):167–76. https://doi.org/10.1016/j.apsb.2018.08.006.
Yeh Y-C, Creran B, Rotello V-M. Gold nanoparticles: preparation, properties, and applications in bionanotechnology. Nanoscale. 2012;4(6):1871–80.
Yu Z, Pestell TG, Lisanti MP, Pestell RG. Cancer stem cells. Int J Biochem Cell Biol. 2012;44(12):2144–51.
Yu X, Trase I, Ren M, Duval K, Guo X, Chen Z. Design of nanoparticle-based carriers for targeted drug delivery. J Nanomater. 2016;3:1–15. https://doi.org/10.1155/2016/1087250.
Zare-Zardini H, Amiri A, Shanbedi M, Taheri-Kafrani A, Sadri Z, Ghanizadeh F, Neamatzadeh H, Sheikhpour R, Boroujeni FK, Dehshiri RM, Hashemi A, Aminorroaya MM, Dehgahnzadeh MR, Shahriari SH. Nanotechnology and pediatric cancer: prevention, diagnosis and treatment. Iranian J Pediat Hematol Oncol. 2015;5(4):233–48.
Zhang L, Gu FX, Chan JM, Wang AZ, Langer RS, Farokhzad OC. Nanoparticles in medicine: therapeutic applications and developments. Clin Pharmacol Ther. 2008;83(5):761–9.
Zhang X-F, Liu Z-G, Shen W, Gurunathan S. Silver nanoparticles: synthesis, characterization, properties, applications, and therapeutic approaches. Int J Mol Sci. 2016;17(9):1534.
Zhao Y, Houston ZH, Simpson JD, Chen L, Fletcher NL, Fuchs AV, Blakey I, Thurecht KJ. Using peptide aptamer targeted polymers as a model nanomedicine for investigating drug distribution in cancer nanotheranostics. Mol Pharm. 2017;14(10):3539–49.
Zou P, Yu Y, Wang YA, Zhong Y, Welton A, Galban C, Wang S, Sun D. Superparamagnetic iron oxide nanotheranostics for targeted cancer cell imaging and pH-dependent intracellular drug release. Mol Pharm. 2010;7(6):1974–84.

Chapter 14
Magnetic/Superparamagnetic Hyperthermia as an Effective Noninvasive Alternative Method for Therapy of Malignant Tumors

Costica Caizer

Abstract Superparamagnetic hyperthermia (SPMHT) is noninvasive, nontoxic, and with increased efficiency in destroying malignant tumors compared with magnetic hyperthermia (MHT), and conventional chemo- and radiotherapy (RT) currently used in medical clinics in this issue. Nowadays SPMHT appears as the most promising alternative method in future therapy of cancer. In this chapter, SPMHT/MHT with bioencapsulated/biofunctionalized ferrimagnetic nanoparticles, best suited for this therapy, and the recent significant results obtained in vitro and in vivo on different animal models and for different types of cancers with high incidence among the people, with the greatest potential for application in clinical trials, will be presented. Moreover, the new concept of nanotheranostic as a result of advanced nanobiotechnology for increasing the efficiency in cancer therapy to 100% and nontoxicity on the heath tissues also will be presented.

Keywords Superparamagnetic hyperthermia · Nanotheranostic · In vitro · In vivo · Cancer therapy

Nomenclature

2-DG	2-Deoxyglucose
5-FU	5-Fluorouracil (anticancer drug)
ADR	Drug-resistive cancer cells
AMF	Alternating magnetic field
ATA	Aminoterephthalic acid
Bio-FiMNPs	Biocompatible ferrimagnetic nanoparticles
Bio-MNPs	Biocompatible magnetic nanoparticles
Bio-SPMNPs	Biocompatible superparamagnetic nanoparticles

C. Caizer (✉)
Department of Physics, West University of Timisoara, Timisoara, Romania
e-mail: costica.caizer@e-uvt.ro

M. Rai, B. Jamil (eds.), *Nanotheranostics*,
https://doi.org/10.1007/978-3-030-29768-8_14

CDs	Cyclodextrins
CMC	Carboxymethyl cellulose
CT	Computed tomography
Cy7	Cyanine7 (lipophilic fluorescent dye)
DOX	Doxorubicin (anticancer drug)
FiM	Ferrimagnetic
FiMNPs	Ferrimagnetic nanoparticles
FIMO	Ferromagnetic iron-manganese oxide
FITC	Fluorescent nanoparticles for imaging
FMI	Fluorescence molecular imaging
HAP	Hydroxyapatite
HER	Herceptin
HPMC	Hydroxyl-propyl methyl cellulose
IONPs	Iron oxide nanoparticles
IR	Infrared
Ls	Liposome
MagTSLs	Thermo-sensitive magnetoliposomes
MCL	Magnetic cationic liposomes
MFHT	Magnetic fluid hyperthermia
mHAP	Magnetic hydroxyapatite
MHT	Magnetic hyperthermia
MNCs	Micellar magnetic nanoclusters
MNPs	Magnetic nanoparticles
MRI	Magnetic resonance imaging
MTB	Magnetic tactic bacteria
MTT	MTT assay
MTX	Methotrexate
NFs	Nano-flowers
NIR	Near-infrared
NMHT	Nano-magnetic hyperthermia
NPs	Nanoparticles
NPTT	Nano-photothermal therapy
OA	Oleic acid
PAI	Photoacoustic imaging
PBS	Phosphate buffer solution
PDT	Photodynamic therapy
PEG	Polyethylene glycol
PES	Poly(3,4-ethylenedioxythiophene):poly(4-styrenesulfonate)
PET	Positron emission tomography
PM	Polymeric micelle
PSA	Prostate-specific antigen
PVA	Polyvinyl alcohol
ROS	Reactive oxygen specie
RT	Radiotherapy
SAR	Specific absorption rate

SERS	Surface-enhanced Raman scattering
SPECT	Single-photon emission computed tomography
SPIONs	Superparamagnetic iron oxide nanoparticles
SPM	Superparamagnetic relaxation
SPMHT	Superparamagnetic hyperthermia
SPMNPs	Superparamagnetic nanoparticles
TA	Terephthalic acid
US	Ultrasound

14.1 Introduction

In principle, magnetic hyperthermia (MHT) or superparamagnetic hyperthermia (SPMHT) represents an increase in temperature to 42.5–43 °C inside tumor, where biocompatible magnetic nanoparticles (BioMNPs) are locally stored (deposited there by various techniques), after applying on the tumor an external alternating magnetic field (AMF) with amplitude (H) and frequency (f) suitable for this therapy (most often: H = 1–50 kA/m, f = 100–500 kHz) (Fig. 14.1) (Ito et al. 2005). The temperature developed in the tumor by superparamagnetic relaxation (SPM) in SPMHT or hysteresis in MHT, or sometimes both, is sufficient to destroy tumor cells by apoptosis and/or necrosis sometimes at temperatures higher than 43 °C. This technique is noninvasive and apparently nontoxic, within the permissible biological limit: $H \times f < 5 \times 10^9$ A/m Hz (Hergt and Dutz 2007).

Recent data shows a high increase in interest and studies on magnetic- and superparamagnetic hyperthermia (MHT/SPMHT), or the combination of this therapy

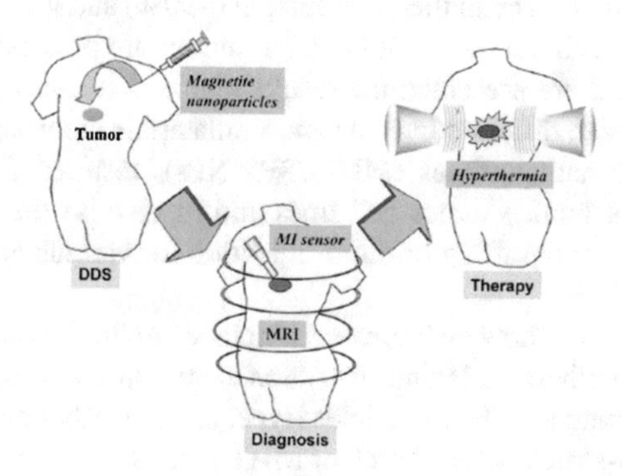

Fig. 14.1 Schematic illustration of the therapeutic strategy using magnetic particles. Functionalized magnetic nanoparticles accumulate in the tumor tissues via the drug delivery system (DDS). Magnetic nanoparticles can be used as a tool for cancer diagnosis by magnetic resonance imaging (MRI) or for magnetoimpedance (MI) sensor. Hyperthermia can then be induced by AMF exposure. Thus, magnetic nanoparticles can be used for cancer therapy at the same time as diagnosis. (Reprinted from Ito et al. 2005, with permission from Elsevier, © (2005) The Society for Biotechnology, Japan)

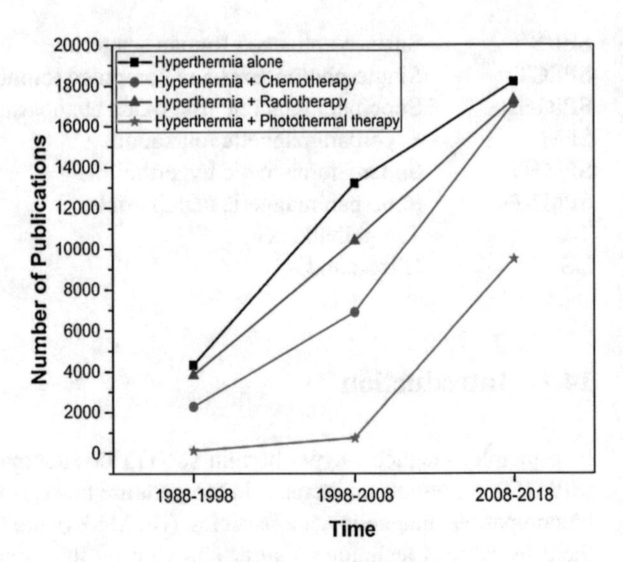

Fig. 14.2 Emergence of magnetic hyperthermia-mediated cancer therapy over the years. Data collected from Google Scholar (10 March 2018). (Reprinted with permission from Gupta and Sharma 2019, © (2019) American Chemical Society)

with other techniques used in cancer therapy, due to the benefits of this alternative noninvasive therapy compared to conventional methods (chemo- and radiotherapy) currently used in cancer therapy, with a high degree of toxicity on the human body, and sometimes even ineffective in cancer therapy in more advanced stages (Fig. 14.2) (Gupta and Sharma 2019).

State of the art in alternative cancer therapy using SPMHT/MHT is presented in Sect. 14.2.1, and recent developments and innovations using nanobiotechnology for new biofunctionalized and multifunctional hybrid nanobiomaterials (surfactation, encapsulation, bioconjugation, biofunctionalization), which contain magnetic nanoparticles (MNPs) such as thermal vectors and nanocarriers, targeting tumor cells with increased efficacy in their destruction (~90%) and with as low as possible toxicity on healthy cells in the vicinity of the tumor, are presented in Sect. 14.2.2. Also, in Sect. 14.2.2 are presented the results obtained in cancer therapy by using SPMHT or MHT with different hybrid nanobiomaterials incorporating magnetic or superparamagnetic nanoparticles (MNPs/SPMNPs), resulted from their in vitro testing on different human tumor cell lines and in vivo on the laboratory animal model (mouse) by xenografting or tumor injection for various human cancers with high risk degree.

The results obtained show an increased efficiency in the destruction of tumors by SPMHT/MHT hyperthermia, but not 100%, as there is also a certain dose of toxicity depending on the magnetic nanoparticles (type, size, distribution, magnetic behavior, etc.), nano-bio-structures, SPMHT or MHT parameters, doses, exposure times, etc. Reduction of toxicity on health cells to zero and increase to 100% of SPMHT/MHT efficiency in the destruction of cancer cells is a target to be achieved in the near future, through current researches in the field. In pursuit of this goal, it is intended to achieve a practically integrative therapy, nanotheranostics, which allows both tumor therapy and diagnosis in real time using nanobiotechnology and various current imaging techniques used in medical clinics. The nanotheranostic procedure also allows increased tumor therapy efficiency by guiding targeted biocompatible magnetic nanoparticles to tumor cells, as well as their proper cellular internalization and location with precision. These issues are presented in Sect. 14.3 of this chapter.

The current trend in cancer therapy is to use in nanotheranostics the current imaging techniques with high resolution and depth in biological tissues, such as magnetic resonance imaging (MRI), computed tomography (CT), and photoacoustic imaging (FAI) together with combined therapies simultaneously, dual- or even triple-therapy (e.g., magnetic thermal therapy (MTT) with chemo- or radiotherapy (RT) (DUAL-Therapy) or MTT with photodynamic therapy (PDT) and chemotherapy (TRIPLE-Therapy), so that efficacy of the "all in one" cancer therapy soon leads to 100% and reduced to zero toxicity.

This chapter presents some important issues regarding superparamagnetic hyperthermia/magnetic hyperthermia (SPMHT/MHT) as a noninvasive alternative method in cancer therapy, and with very low toxicity compared to conventional chemo- and radiotherapy used today in cancer therapy.

14.2 Magnetic/Superparamagnetic Hyperthermia as Noninvasive Alternative Method in Cancer Therapy

14.2.1 State of the Art

Nowadays major efforts are made around the world in finding alternative, noninvasive, and feasible magnetic hyperthermia therapy (MHT) in order to destroy malignant tumors without toxicity on living organism (Pankhurst et al. 2003; Matsuoka et al. 2004; Ito et al. 2005; Johannsen et al. 2005, 2007; Sunderland et al. 2006; Hergt et al. 2006; Purushotham and Ramanujan 2010; Alphandéry et al. 2013; Sivakumar et al. 2013; Qu et al. 2014; Lee et al. 2013; Di Corato et al. 2015; Liu et al. 2016a, b; Almaki et al. 2016; Thorat et al. 2017; Caizer 2017; Kandasamy et al. 2018; Yan et al. 2018; Guoa et al. 2018; Zhou et al. 2018). As is well known, nowadays the commonly used methods in this area include surgery, chemotherapy, and radiation therapy, conventional methods which do not solve entirely the issue, and in advanced cases they are inefficient. As a result, the modern society is more affected by this disease (Jansen et al. 2001; Mancuso et al. 2009; Denoyer et al. 2010; Li et al. 2011; Yamanaka et al. 2011). An alternative method must be found as soon as possible, that leads, in nearest future, to the effectiveness of the cancer therapy.

In this area, starting from 2002–2003 the possible use of ferrimagnetic (FiM) nanoparticles (NPs) as a support in producing superparamagnetic hyperthermia (SPMHT) (temperature ~43 °C) has been proposed. SPMHT is obtained as a result of the superparamagnetic (SPM) relaxation in ferrimagnetic nanoparticles (FiMNPs) (Rosensweig 2002), found inside the tumors, under the action of an external alternating magnetic field having a frequency in the range of hundreds of kHz, which leads to their irreversible thermal destruction (Safarik and Safarikova 2002; Tartaj et al. 2003; Ito et al. 2003a, b). From then until now the problem advanced and the research results obtained so far (Shinkai and Ito 2004; Ito et al. 2004; Matsuoka et al. 2004; Hilger et al. 2005; Tanaka et al. 2005; Gazeau et al. 2008; Habib et al. 2008; Sivakumar et al. 2013; Di Corato et al. 2015; Liu et al. 2016a, b; Almaki et al. 2016; Yan et al. 2018; Zhou et al. 2018; Caizer 2012, 2014; Caizer et al. 2013) are very promising, and show that SPMHT might be the future method in cancer therapy, noninvasive and with very low toxicity or nontoxicity on the health tissues.

NPs most used in magnetic hyperthermia (MHT) are those of iron oxide nanoparticles (IONPs), the base being Fe_3O_4 (magnetite) and γ-Fe_2O_3 (maghemite) (Ito et al. 2004; Gazeau et al. 2008), which give the best results in MHT and have reduced toxicity on the living body.

There are MHT studies also made for other ferrimagnetic NPs (FiMNPs), such as the Mg ferrite, Mn or their combinations Mg–Mn, Mn–Zn or even the Co ferrite ($CoFe_2O_4$) (Kumar and Mohammad 2011; Pradhan et al. 2007; Hejase et al. 2012; Saldívar-Ramírez et al. 2014), but with poorer results. Cobalt ferrite ($CoFe_2O_4$) NPs presents a high interest as a SPMHT material and has not been systematically studied. It has almost the same saturation magnetization as Fe_3O_4 but with a magneto-crystalline anisotropy one order of magnitude larger.

However, finding the most suitable NPs to achieve maximum efficiency of the SPMHT remains an open question.

In order to use magnetic nanoparticles (MNPs) in the SPMHT, they must be made biocompatible with the biological tissues in which it will be introduced. There are different ways for doing this: covering NPs with a biocompatible organic layer (Safarik and Safarikova 2002), bioencapsulation of the NPs in biological membranes, such as liposomes (Ls) (Shinkai and Ito 2004; Tanaka et al. 2005), or biofunctionalizing of the NPs with specific molecular biostructures (Zhu et al. 2009) depending on the intended purpose.

On the other hand, in most of the applications NPs have a hybrid nature of the surface, consisting of organic functional molecules bonded to the magnetic nanoparticles, and this plays a crucial role in defining their properties. A key challenge for NPs applications is the development of controlled surface reaction, strategy that delivers a versatile and stable interface between NPs and the biological environment and allows the incorporation of functional groups (Gupta and Gupta 2005; Selvan et al. 2010).

However, in the last years, the researches in magnetic hyperthermia (MHT) which gave good results, focused on liposomes (Ls) for encapsulating NPs (Shinkai and Ito 2004; Ito et al. 2004; Tanaka et al. 2005; Gazeau et al. 2008) due to their very good biocompatibility, biodegradability, and ability to transport safely the NPs release, and therefore showing the feasibility of MHT method in cancer therapy. Ls offer an advantage for NPs delivery because they allow water-soluble and -insoluble NPs to be encapsulated together. Liposomal NPs-delivery systems have a large size range, the ability to guard encapsulated entities and change surface chemistry, extended plasma life-time and highly efficient targeting capabilities, using different surface coatings. Therefore, NPs delivery nanosystem using NPs encapsulated in Ls is considered as an effective strategy for passive tumor targeting, which can increase NPs circulation in blood but also reduce side effects in living organism.

Caizer et al. (2017) reported other possible magnetic bionanostructures, such as salicylic acid-coated magnetite nanoparticles, which could be used successfully in SPMHT/MHT for cancer therapy. The authors further reported other aspects of improving/optimizing the SPMHT/MHT method for cancer therapy.

However, for the effective application of SPMHT/MHT in cancer therapy, there still remained some problems, both fundamental and applied, which needs to be well known before proceeding to a more advanced research on animal model, and finally to humans (in clinical trials). Mostly these problems are:

1. Finding the best magnetic nanoparticles for this type of experiment (material, size (mean diameter, shape, distribution), magnetic behavior in external field (ferro- or ferrimagnetic (FM or FiM)), superparamagnetic (SPM)), observed in terms of toxicity and effectiveness of superparamagnetic hyperthermia effect on the cells.
2. The better biocompatibility (over 95%) (with lower toxicity) of magnetic NPs with the biological tissues (where they will be introduced), which need to take into account finding the organic compounds suited to the desired aim (NPs encapsulation/surfactation/functionalization).
3. The magnetic NPs biofunctionalization, which has biochemical/biophysical affinity to tumor cells (a totally open problem); solving it might lead to curing cancer even in advanced stages (metastasis) and even discovering cancer (tumor cells) in early stages.
4. The practical achievement of SPMHT, which is more suited for obtaining a maximum specific absorption rate (SAR) using the superparamagnetic relaxation phenomena in ferrimagnetic NPs in an external alternating magnetic field of hundreds of kHz, targeted to a specific type of cancer.
5. Finding appropriate and effective protocols (types of nanoparticles, appropriate biofunctionalization, dosages/concentrations appropriate to the type of cells, cellular viability, SPMHT effect efficiency (the amplitude and frequency of the alternating magnetic field, inductor geometry and time of field exposure)), starting from in vitro studies to those in vivo, and finally to applying in future to humans (clinical trials).

The worldwide researches in the field are currently working on each of the above directions (Ito et al. 2005; Qiang et al. 2006; Baker et al. 2006; Zeng et al. 2007; Hergt and Dutz 2007; Fortin et al. 2008; Ondeck et al. 2009; Mehdaoui et al. 2010; Alphandéry et al. 2013; Di Corato et al. 2015; Liu et al. 2016a, b; Almaki et al. 2016; Thorat et al. 2017; Yan et al. 2018; Zhou et al. 2018), neither of them being in an advanced stage.

Taking this into consideration, I have proposed (Caizer 2013, 2014) further research of the above issues by using iron-cobalt ferrite nanoparticles with superparamagnetic behavior (Fe–Co ferrite NPs) such as: $Co_{\delta}Fe_{3-\delta}O_4$, with magnetic structure of $Fe^{3+}[Co_{\delta}^{2+}Fe_{(3-\delta)}^{2+}, Fe^{3+}]O_4^{2-}$ (Fe^{3+} is magnetic ion in sublattice A, and [...] contains the Co^{2+} and Fe^{2+} magnetic ions in sublattice B) (Smit and Wijin 1961), for Co^{2+} ions concentration $\delta = 0$–1. Using this innovative route, the Fe^{2+} ions (which have 4 μ_B, where μ_B is the Bohr magneton) from the octahedral magnetic network will be substituted by magnetic ions Co^{2+} (having 3 μ_B). Therefore, this replacement makes the magnetic anisotropy of the ferrite NPs to change in a very wide range of values (up to ~10^2 as order of magnitude), although the saturation magnetization changes only slightly (decreases by up to ~1.14). This strategy that I proposed will allow us to find the most suitable NPs to be used in SPMHT, that will have a maximum efficiency and also give the possibility to be used in intracellular therapy that is more efficient in SPMHT than the extracellular therapy.

Moreover, this approach involves the development of new ferrimagnetic bionanostructures, namely: Fe–Co ferrite NPs bioconjugated with cyclodextrins (CDs) (Caizer 2014), in order to increase magnetic NPs delivery and to be more efficient in cancer treatment by using SPMHT, and also lack of toxicity.

The motivation of the choice of these ferrimagnetic nanoparticles is their remarkable magnetic properties in the nanometers—tens of nanometers field, comparable with those of Fe_3O_4 and γ-Fe_2O_3 NPs, which make them very suitable in SPMHT, besides their reduced toxicity on cells.

Besides the material (saturation magnetization, magnetic susceptibility), the NPs (mean) diameter is a critical parameter in obtaining SPMHT (Caizer 2017). In addition to this, the magnetic anisotropy of the nanoparticles is a metric very important in the superparamagnetic relaxation, with direct effect on the SPMHT effectiveness. This was another reason for choosing the NPs type, the anisotropy of $Co_\delta Fe_{3-\delta}O_4$ nanoparticles being more different (Valenzuela 1994), depending on the concentration δ of Co^{2+} ions. Thus, the magnetocrystalline anisotropy constant of the Fe–Co ferrite NPs can be modified in a very wide range of values, increasing from ~10^3 to ~10^5 J/m^3 (with approx. two orders of magnitude) when the concentration of Co^{2+} ions changes from 0 to 1. This aspect is very important for superparamagnetic hyperthermia because by changing the magnetic anisotropy the optimum value of the NPs diameter of Fe–Co ferrite (increased anisotropy leads to diameter reduction) can be found, which can give the maximum effect in SPMHT (the highest SAR value) taking into consideration the superparamagnetic behavior (SPM) of NPs in the external magnetic field (the necessary condition for obtaining SPMHT). These two aspects are the key in obtaining maximum efficiency in SPMHT.

For values of NPs diameter larger or smaller than the optimum diameter, SAR decreases rapidly, tending toward zero, making the SPMHT totally inefficient, because of the rapid decrease or even disappearance of the hyperthermia effect. Therefore, the diameter of NPs in SPMHT is a critical parameter and its optimal value must be accurately found in order to achieve the maximum efficiency in SPMHT (a maximum SAR with a maximum thermal effect at around 42.5–43 °C).

Also, reducing the size of Fe–Co ferrite NPs by increasing the magnetic anisotropy allows the introduction of additional benefits, such as substantially reducing the toxicity and increasing the half-life time in the blood.

In order to obtain superparamagnetic hyperthermia (SPMHT), which was recently proved (Pavel et al. 2008; Pavel and Stancu 2009) to be significantly more efficient than the magnetic hyperthermia (MHT), obtained, for example, as a result of magnetic loses by hysteresis (in high fields and, in the same time, large sizes of the nanoparticles (generally >20–25 nm)), I will take into consideration for study the Fe–Co ferrite NPs (Caizer and Tura 2006; Caizer et al. 2010a, b; Caizer 2012, 2013, 2017) with the (mean) diameter in the range 5–10 nm, and low fields in the range 1–20 kA/m. I think that these small NPs would be very suitable for the intracellular therapy (destruction from the inside of tumor cells more efficiently through SPMHT, by entering small NPs within them).

From the point of view of medical application the major goals in designing bionanocarriers as a NPs delivery system is the control of the particle size, surface properties, and release of active bioagents. Other issues regarding the biocompatibility and biofunctionalization of the magnetic nanoparticles, and practical achievement of SPMHT/MHT in vitro and in vivo, are presented in the next sections.

14.2.2 Recent Developments and Innovations in the Area of Biocompatible Magnetic Nanoparticles Using Advanced Nanobiotechnology: In Vitro and In Vivo Magnetic/Superparamagnetic Hyperthermia

In this section, I present some significant results obtained in vitro and in vivo regarding MHT/SPMHT with biocompatible magnetic nanoparticles (BioMNPs) as magnetic induction thermal vectors with high potential in alternative and noninvasive cancer therapy, and which also show the viability of this method. Other current results in this matter can be found by consulting other references in the field (Kikumori et al. 2009; Kobayashi 2011; Lee et al. 2011; Hilger 2013; Goya et al. 2013; Kossatz et al. 2014; Yi et al. 2014; Lima-Tenório et al. 2015; Wang et al. 2015a; Lee et al. 2013; Nedyalkova et al. 2017; Mahmoudi et al. 2018; Engelmann et al. 2018).

Thus, an important result was reported in 2018 by Kandasamy and his colleagues, which was obtained in vitro experiment on MCF-7 cancer cells using SPMHT with superparamagnetic iron oxide nanoparticles (SPIONs) of 9 nm mean diameter, biofunctionalized with dual-surfactants TA-ATA (terephthalic acid-aminoterephthalic acid) (Fig. 14.3i). After application of SPMHT (MFHT) the rise in nanofluid temperature with SPIONs during the 30 min for concentrations of Fe ions in suspension from 0.5 to 8 mg/mL and frequency (*f*) of the alternating magnetic field (AMF) in the 100–1000 kHz range (under Hergt's biological limit (Hergt and Dutz 2007) are shown in Fig. 14.3ii–iv. The 42 °C temperature required in MHT is reached faster with increasing the frequency of the magnetic field as well as increasing the Fe concentration in the suspension. However, it is shown that SAR in this case is maximal at the concentration of 2 mg/mL, this decreasing to lower or higher concentrations. Using the ferrofluid with the concentrations of 0.5 and 1 mg/mL in SPMHT made at 751.5 kHz frequency of the external magnetic field has been

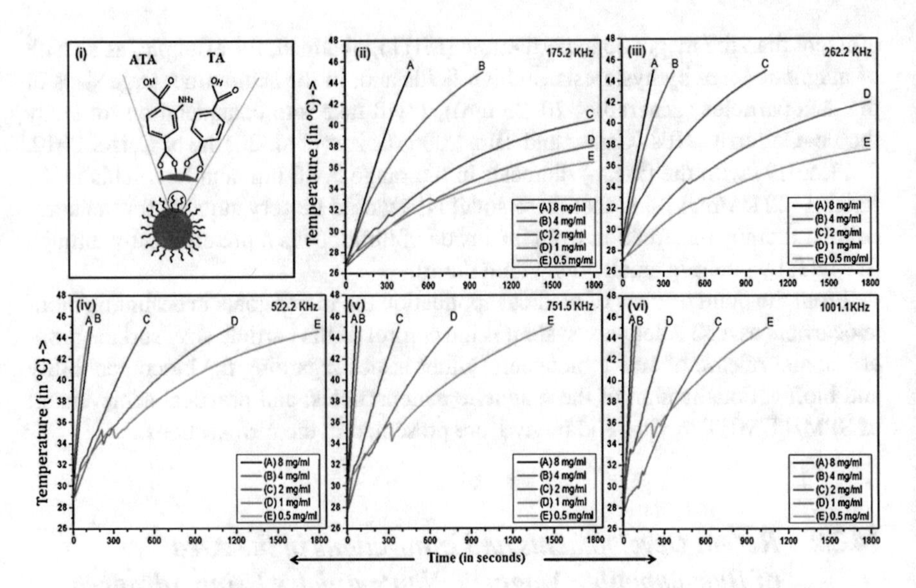

Fig. 14.3 (**i**) TA-ATA molecules attached on the surface of the SPION. Time-dependent temperature rise of 1 mL aqueous suspension of F6 with concentrations of (A) 8, (B) 4, (C) 2, (D) 1, and (E) 0.5 mg/mL, on exposure to AMFs at frequencies of (**ii**) 175.2 kHz, (**iii**) 262.2 kHz, (**iv**) 522.2 kHz, (**v**) 751.5 kHz, and (**vi**) 1001.1 kHz, respectively. (Reprinted from Kandasamy et al. 2018, © (2018) Elsevier Ltd., with permission from Elsevier)

achieved the destruction of the cancer cells of MCF-7 over 90% (Fig. 14.4), and even more at slightly higher temperatures.

This result shows the viability of the SPMHT method in alternative cancer therapy and their potential for future application in preclinical and clinical trials.

In order to increase efficiency of SPMHT in destroying cancer cells was used other types of nanoparticles than those known of iron oxides (most notably Fe_3O_4 (magnetite) and γ-Fe_2O_3 (maghemite)), such as ferrimagnetic nanoparticles of $MnFe_2O_4$, $ZnFe_2O_4$, and their combination (Mn–Zn)Fe_2O_4 ($Mn_xZn_{(1-x)}Fe_2O_4$ with $x = 0$–1) with an average diameter of 8 nm in the form of nanoclusters (Qu et al. 2014). These have been made biocompatible using core-shell nanobiotechnology by encapsulating nanoclusters into amphiphilic block copolymer (mPEG) for the development of hydrophilic micellar magnetic nanoclusters (MNCs) (Fig. 14.5). These nanobiostructures, $Mn_xZn_{(1-x)}Fe_2O_4$/MNCs, show a very good stability and excellent biocompatibility, of approx. 100% (cytotoxicity close to 0%) for $MnFe_2O_4$/MNCs and $Mn_{0.6}Zn_{0.4}Fe_2O_4$/MNCs micellars. Also these nanobiostructures have a high SAR, with over 90% efficiency in cellular apoptosis of MCF-7 (drug-sensitive cancer cells) and MCF-7/ADR (drug-resistive cancer cells) in a relatively short time (Fig. 14.6b), only during 10–15 min after applying SPMHT (MFHT) (Fig. 14.6a) and for a reduced frequency of magnetic field, around 100 kHz, to reach the optimal temperature of 43 °C.

Recently, a new magnetic nanobiomaterial was developed with highly efficient MHT ablation of tumors. This magnetic nanobiomaterial, noted with HPMC/Fe_3O_4,

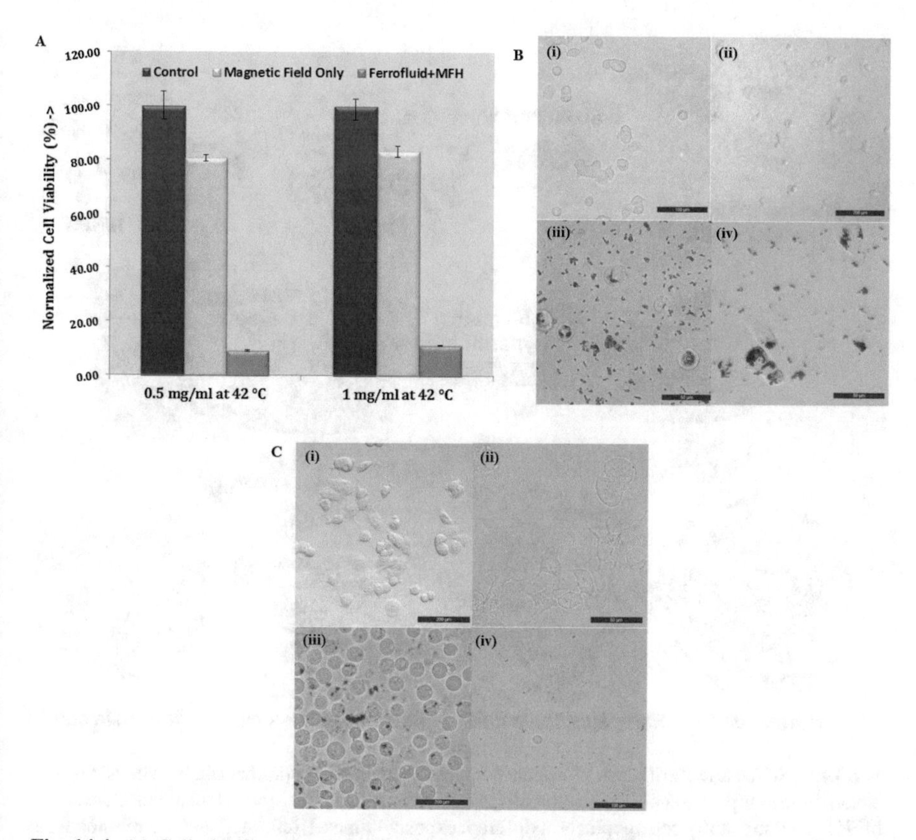

Fig. 14.4 (**a**) Cell viability plot shows the cytotoxic effect on MCF-7 breast cancer cells treated with MFH (~42 °C) by using TA-ATA-coated SPION-based ferrofluids (i.e., F6) at 0.5 and 1 mg/mL concentration on exposure to AMF (751.5 kHz) as compared to control and the cells (without F6) treated with magnetic field only. (**b**, **c**) Comparison of optical microscope images of the MCF-7 breast cancer cells, (i) control (after washing), (ii) cells (without F6) treated with magnetic field only (after washing), (iii, iv) cells treated with MFHT by using F6 at 0.5 and 1 mg/mL concentrations (before and after washing), respectively. (Reprinted from Kandasamy et al. 2018, © (2018) Elsevier Ltd., with permission from Elsevier)

consists of hydroxyl-propyl methyl cellulose (HPMC), polyvinyl alcohol (PVA), and Fe_3O_4 nanoparticles of 10–50 nm (Wang et al. 2017). The obtained HPMC/Fe_3O_4 was mixed with PVA solution for practical use. The HPMC is nontoxic, known as hypromellose (Muralidhar et al. 2012), being a semi-synthetic hydrophilic polymer used in dermatology (Lee et al. 2014), oral medicine (Li et al. 2003; Siepmann and Peppas 2012), and ophthalmic lubricant (Muralidhar et al. 2012). The PVA is also a nontoxic, biocompatible, and biodegradable polymer (Huang and Yang 2008; Parhi et al. 2015) used in various areas of biomedical field (Paradossi et al. 2003). This new nanobiomaterial is thermally contractible, injectable, biodegradable, and with very good biocompatibility, tested both in vitro and in vivo.

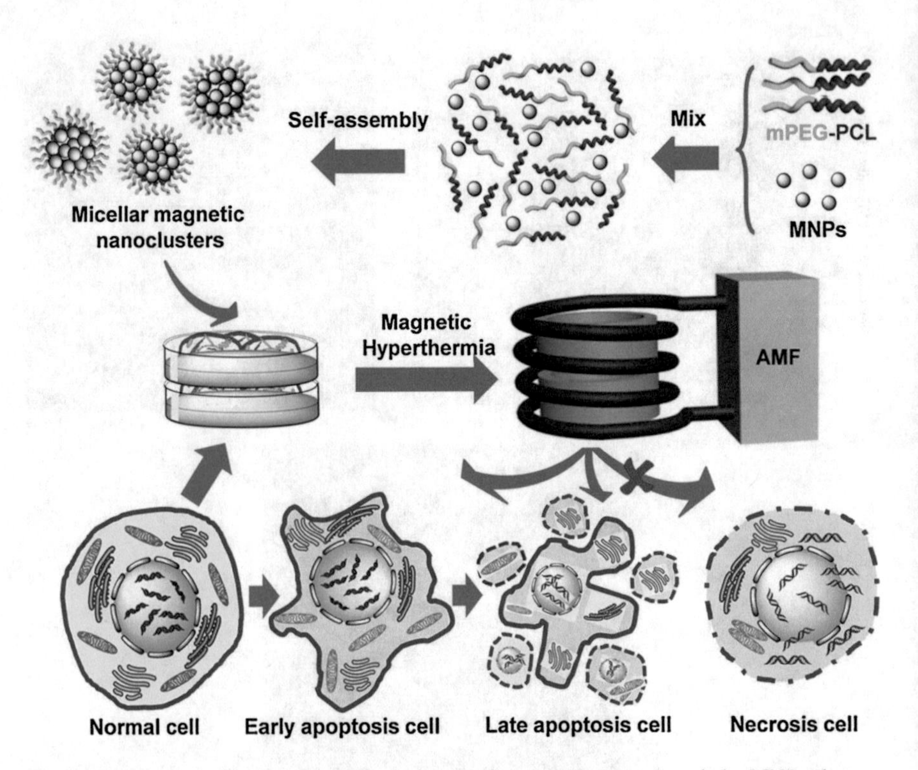

Fig. 14.5 Schematic illustration of effective apoptotic magnetic hyperthermia by MNPs clusters, including the preparation of the micellar $Mn_xZn_{1-x}Fe_2O_4$ nanoclusters and cell death mechanism by MFHT, predominantly cell apoptosis at different exposure times. (Reprinted with permission from Qu et al. 2014, © (2014) American Chemical Society)

The results obtained in vivo on tumor-bearing mice after intratumoral injection of HPMC/Fe_3O_4 and exposure to an AMF by MHT (frequency: 626 kHz and current output: 28.6 A; coil diameter: 3 cm) show that tumors can be fully ablated using 0.06 mL of HPMC/Fe_3O_4 with 60% Fe_3O_4 concentration after 180 s of induction heating by MHT, compared to tumor control (without HPMC/Fe_3O_4 + AMF) and tumor only with HPMC/Fe_3O_4 (no AMF applied) which develop continuously during the 14 days of experiment (Fig. 14.7).

For the targeting of metastatic breast cancer cells a new SPIONs-PEG-HER complex (SPIONs: superparamagnetic iron oxide nanoparticles, PEG: polyethylene glycol, HER: herceptin) was synthesized using core-shell nanobiotechnology (Almaki et al. 2016). SPIONs of γ-Fe_2O_3, having an average diameter of approx. 9 nm, were first salinized using 3-aminopropyltrimethoxysilane (Si) to allow covalent PEG binding (SPIONs-PEG) and to improve the bioavailability of SPIONs. Afterward, the SPIONs-PEG nanobiostructure was bioconjugated with Herceptin (HER) forming the final SPIONs-PEG-HER biocomplex having a mean hydrodynamic diameter of ~18 nm. This biocomplex allows the targeting of surface-specific receptors of HER2+ metastatic breast cancer cells (Fig. 14.8).

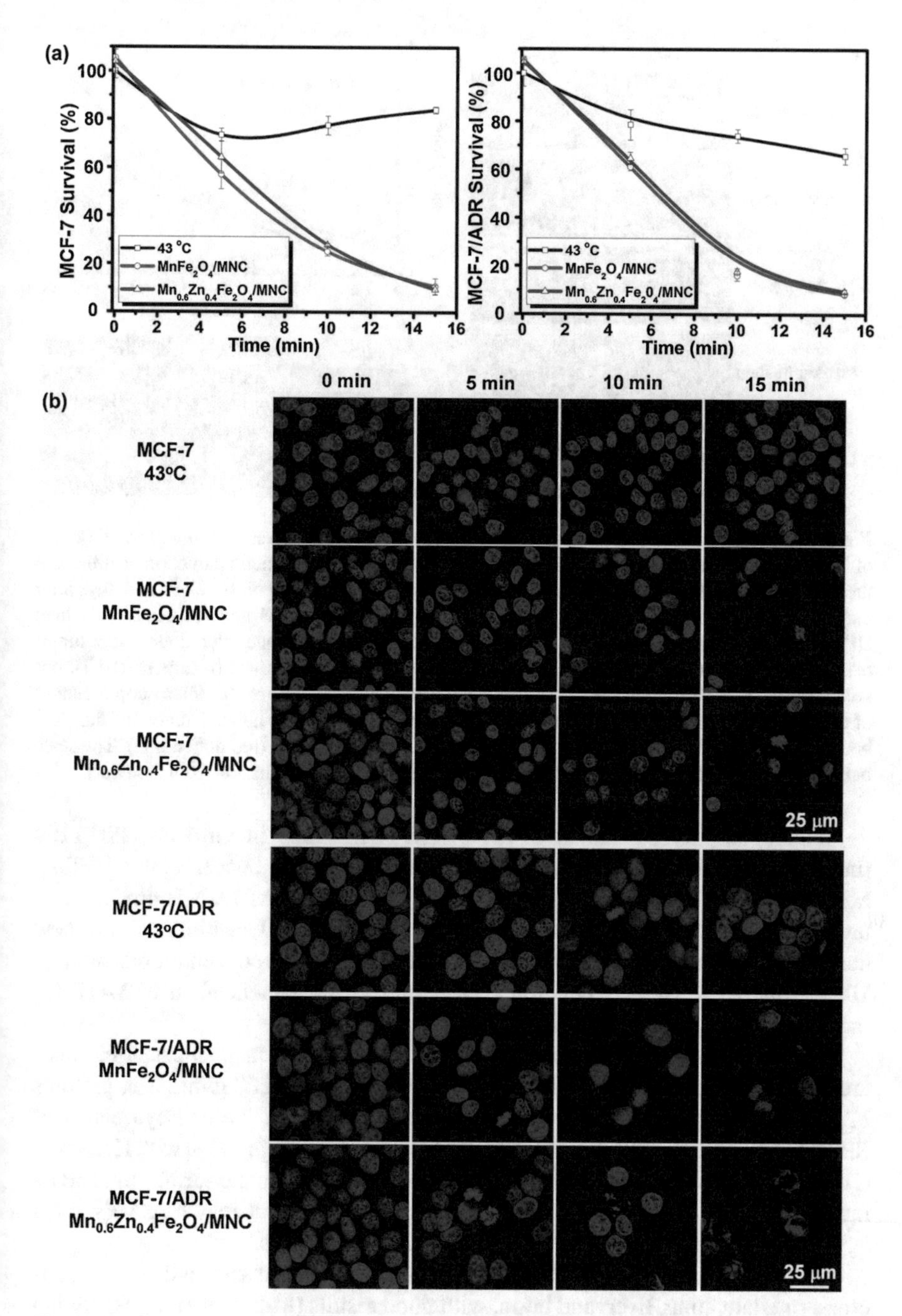

Fig. 14.6 Cytotoxicity of MFH and mild hyperthermia on MCF-7 and MCF-7/ADR with different exposure times: (**a**) MTT result of MCF-7 and MCF-7/ADR, and (**b**) fluorescence imaging of MCF-7 and MCF-7/ADR. (Reprinted with permission from Qu et al. 2014, © (2014) American Chemical Society)

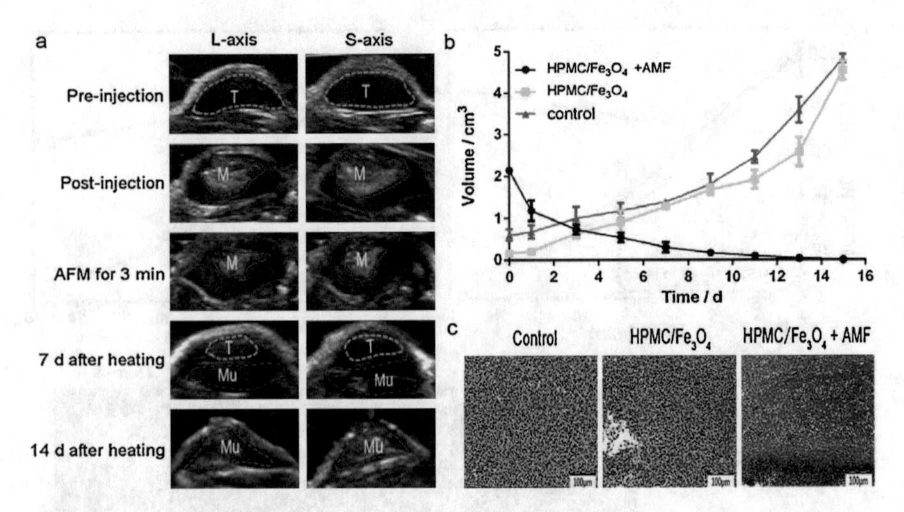

Fig. 14.7 Magnetic hyperthermia ablation efficiency in vivo. (**a**) Ultrasound images of 0.06 mL of HPMC/Fe_3O_4 injected inside a tumor before and after 180 s of magnetic induction heating, and the area changes of the HPMC/Fe_3O_4 material in the tumor at the time points of 7 and 14 days after magnetic heating (L-axis: long axis; S-axis: short axis; T: tumor; M and red continuous line: HPMC/Fe_3O_4 implant; Mu and pink dotted line: leg muscle of the mice; blue dotted line: tumor tissue after ablation; red arrow: the pit was caused by the disappearance of tumor). (**b**) Tumor volume-time curves showing changes in the treated and untreated groups. (**c**) Microscopic images of the tumor tissue (H&E staining) showing coagulation necrosis of the ablated tumor and the clear boundary between the coagulation necrosis and the liver tumor tissue (red dotted line). The scale bar is 100 mm. (Reprinted from Wang et al. 2017, © (2017) with permission from Elsevier)

Biocompatibility studies of SPIONs-PEG-HER made in vitro on HSF-1184 (human skin fibroblast cells), SK-BR-3 (human breast cancer cells, HER+), MDA-MB-231 (human breast cancer cells, HER−), and MDA-MB-468 (human breast cancer cells, HER−) have shown a very good cell viability of this new nanobiocompound, this being over 85–90% even for high Fe concentrations of up to 1000 g/mL (Fig. 14.9). High iron concentrations are beneficial in SPMHT for increasing the thermal efficiency of this method.

The target effect on tumor cells and the internalization of magnetic nanoparticles are shown in Fig. 14.10 for a Fe concentration of 100 and 200 g/mL. The pictures b2 and b3 in Fig. 14.10 show increased internalization in the cell cytoplasm of SPIONs-PEG-HER in SK-BR-3 tumor cells as a result of conjugation with Herceptin (HER), which is a targeting agent for HER2/neu receptors on the surface of HER2+ metastatic breast cancer cells. The SPIONs-PEG-HER agent may be a very good agent for the target SPMHT of metastatic breast cancer cells.

In vivo MHT has been studied for treating a variety of cancers: breast, pancreas, prostate, lung, liver, and brain, with good results (Moroz et al. 2002; Jordan et al. 2006; Hu et al. 2011; Wang et al. 2012; Kossatz et al. 2015). Using magnetic cationic liposomes (MCL) by encapsulating SPIONs, Matsuoka et al. (2004) applied MHT treatment on an osteosarcoma hamster by injecting MCL directly into the osteosarcoma (Laurent et al. 2011). Applying an alternating magnetic field

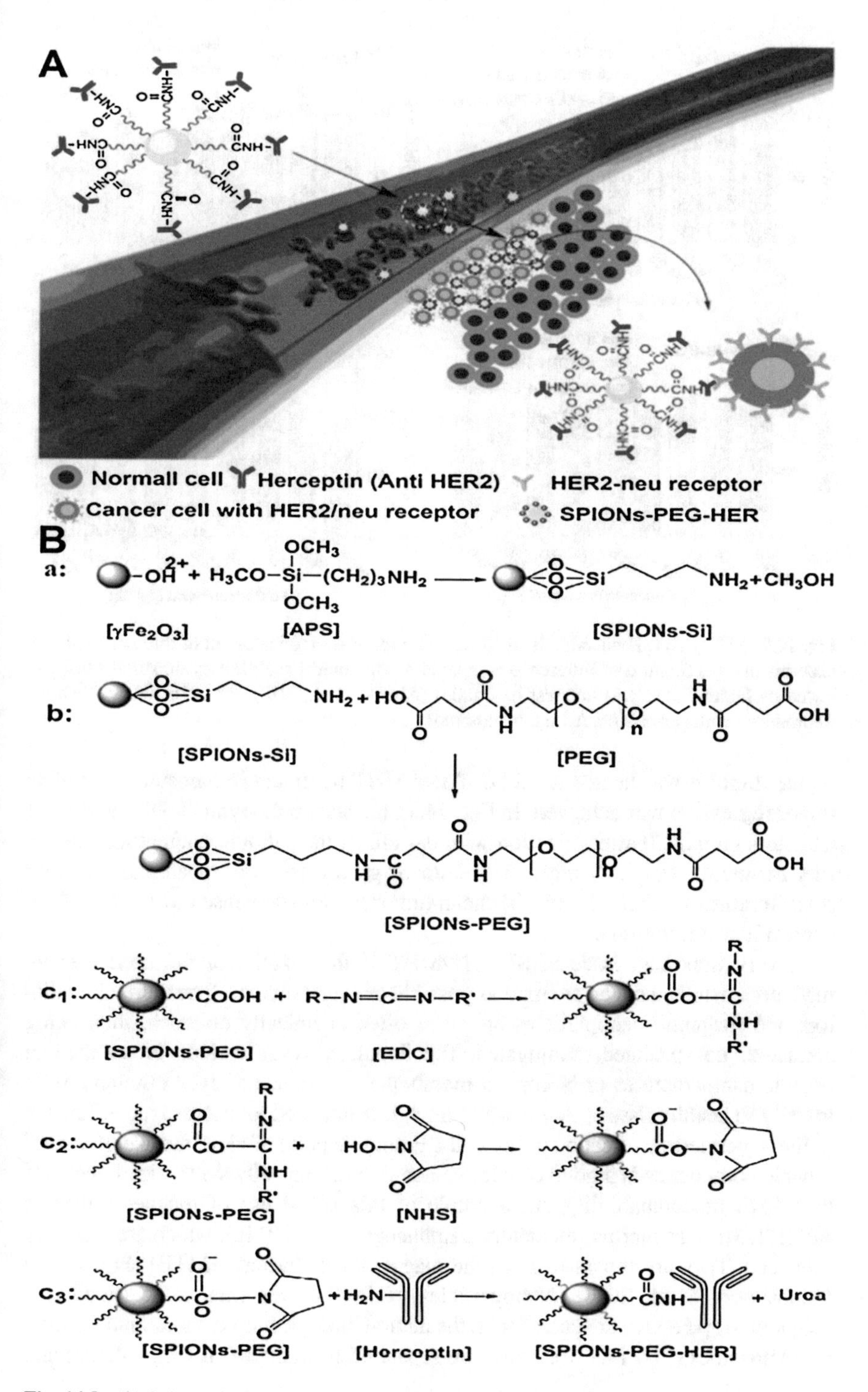

Fig. 14.8 (**A**) Schematic presentation of SPIONs–PEG–HER binding to the HER2/neu receptors on the SK-BR-3 cancer cells (Almaki et al. 2016). (© 2016 IOP Publishing. Reproduced with permission. All rights reserved) (**B**) Synthesis route of silanized (a), PEGylated (b) and HER conjugated (c) ferrofluids

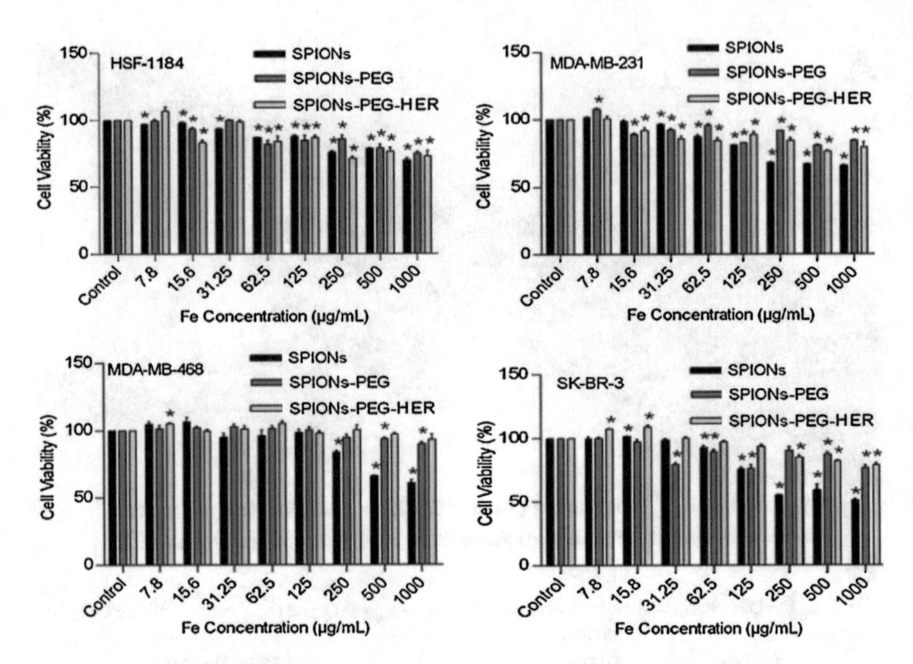

Fig. 14.9 MTT assay. Each value is the mean ± s.e.m. of six replicates out of three independent experiments. (∗) Significant difference compared to the control analyzed by unpaired *t*-test followed by Holm–Sidak post hoc test ($p < 0.05$) (Almaki et al. 2016). (© 2016 IOP Publishing. Reproduced with permission. All rights reserved)

on the tumor it was heated to 42 °C. After MHT treatment in hamster a complete tumor regression was achieved. In Fig. 14.11 are shown the typical photographs of hamsters on day 20 after injection with 0.4 mL of the cationic magnetoliposomes (net magnetite weight 3 mg): (a) treatment group; (b) control animals without AMF treatment. After 12 days the mean tumor volume decreased to 1/1000 of the control hamster volume.

A very important issue in SPMHT/MHT is the toxicity of the used thermal mediators, which must be as small as possible or even nontoxic. The thermal mediators and magnetic nanoparticles are most often chemically obtained, then being surfacted, encapsulated, conjugated, functionalized with various biocompatible organic nanostructures or biological membranes, to reduce or even eliminate their toxicity to healthy tissues. At the same time, it is intended to increase the efficiency of the hyperthermic effect on tumor cells. From this point of view, the use of natural organic compounds is a good choice replacing the chemically synthesized ones, for their high biocompatibility. An example of this is the use of magnetosomes in SPMHT/MHT as thermal mediators (Alphandéry et al. 2011a), which are naturally produced. They are separated from the magnetotactic bacteria (MTB) (Fig. 14.12) (Alphandéry et al. 2013). Nanoparticles of iron oxides (magnetite (Fe_3O_4) or maghemite (γ-Fe_2O_3)) are coated with the natural biological lipid membrane forming magnetosomes (Fig. 14.12b). These are separated from the MTB (Fig. 14.12a) and

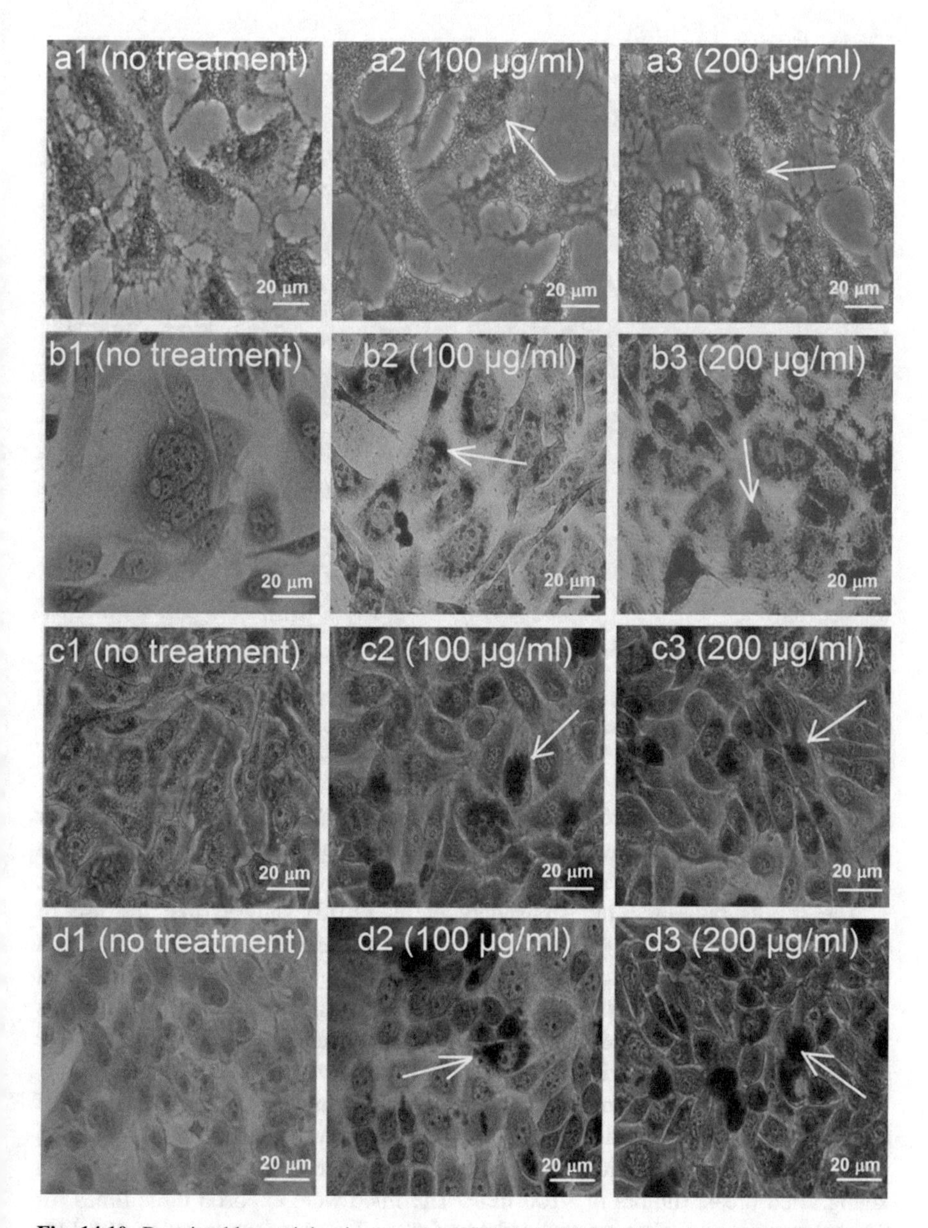

Fig. 14.10 Prussian blue staining images: (**a**) HSF-1184, (**b**) SK-BR-3, (**c**) MDA-MB-468, and (**d**) MDA-MB-231 under light microscope (×320) (Almaki et al. 2016).

can be used in the form of magnetosome chains (Fig. 14.12b) or distinct magnetosomes (Fig. 14.12c). Up to the concentrations of 1.3 mg/mL of magnetosomes in suspension no toxicity was observed in mouse fibroblast cells incubated in the presence of bacterial magnetomes. A study done on mice and rats showed that the

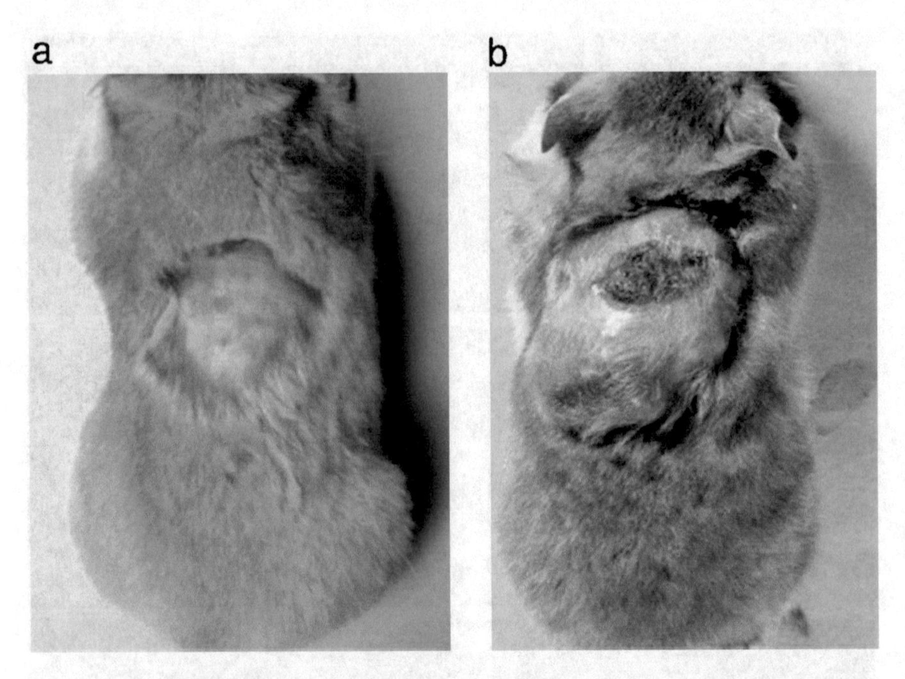

Fig. 14.11 Typical photographs of hamsters on day 20 after the MCL injection. These photographs show one hamster of the treatment (**a**) group and one of the control group (**b**) (Matsuoka et al. 2004). (CCL © 2004 Matsuoka et al.; licensee BioMed Central Ltd.)

dose of magnetosomes injected intravenously leads to the survival of mice even for 360 mg/kg, compared to the standard SPIONs chemical that resulted in death for a low dose of injected nanoparticles 135 mg/kg (Liu et al. 2012; Alphandéry et al. 2013).

The results presented by Alphandéry et al. (2013) showed that the use of magnetosome chains (Fig. 14.12b) is more effective in the treatment of tumors through MHT (Fig. 14.13) than the use of separated bacterial magnetosomes (Fig. 14.12c) that leads to aggregates of MNPs with effect on toxicity in vivo. The magnetosomes arrangement in the chain prevents their aggression and leads to a high rate of cellular internalization and uniform internal heating (Alphandéry et al. 2011b, 2012).

MHT therapy was performed on breast tumors xenografted under the skin of mice (Fig. 14.13) (Alphandéry et al. 2011b) using each suspension (Fig. 14.12a–c) with 10 mg/mL of iron oxide and injected 100 μL each time into the center of the xenografted breast tumors of ~100 mm^3. The mice were exposed three times for 20 min to an alternating magnetic field of 40 mT and a frequency of 198 kHz (Fig. 14.13a, b). Following the injection of MTB suspensions and magnetosomes without magnetic field application the tumor growth could not be stopped. However, after applying the magnetic field (MHT therapy) tumors in several mice have disappeared (Fig. 14.13c).

However, although magnetosomes are more magnetically stable, a major drawback of using MTB's separated magnetosomes is that iron oxide magnetic nanoparticles are larger than those of SPIONs, heating being produced more by hysteresis

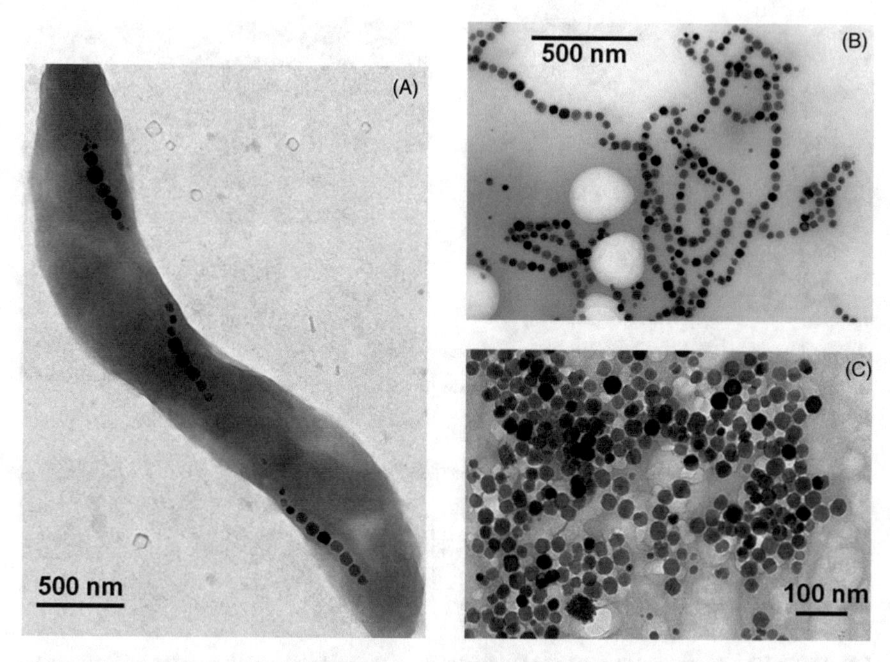

Fig. 14.12 Transmission electron microscopy images of whole MTB (**a**), chains of magnetosomes isolated from MTB (**b**), and individual magnetosomes detached from the magnetosome chains by heat and SDS treatment (**c**) (Alphandéry et al. 2013). (© 2013 Informa UK Ltd. Francis & Taylor)

rather than magnetic relaxation, which leads to a lower SAR in the same magnetic field (Rosensweig 2002; Pavel et al. 2008; Pavel and Stancu 2009).

Hou et al. (2009) developed a new nanobiomaterial for magnetic hyperthermia of cancer: mHAP powder (magnetic hydroxyapatite nanoparticles), consisting of magnetite (m) nanoparticles with hydroxyapatite (HAP) (with molar ratio of Fe:Ca of 1:1), mixed with phosphate buffer solution (PBS), that is injected subcutaneously around the tumor. MHT efficacy using mHAP was tested for 15 days on a mouse model (balb/c mouse) inoculating 5×10^6 murine colorectal cancer cells (CT-26 cell line). A suspension of 0.16 g/mL m-HAP was used in the solution for which the temperature of 45–46 °C was reached within 20 min after the application of the external alternating magnetic field. The mice used were divided into six study groups:

1. A lot consisting of group 1, group 2, and group 3 that were not exposed to the magnetic field
2. Another lot consisting of group 4, group 5, and group 6 exposed to the magnetic field

Groups 1 and 4 were control groups (not injected with the colloidal suspension), groups 2 and 5 are mice injected only with HAP (hydroxyapatite), and groups 3 and 6 are mice injected with mHAP (magnetic hydroxyapatite suspension nanoparticles) solution. MHT therapies were applied in several sessions of

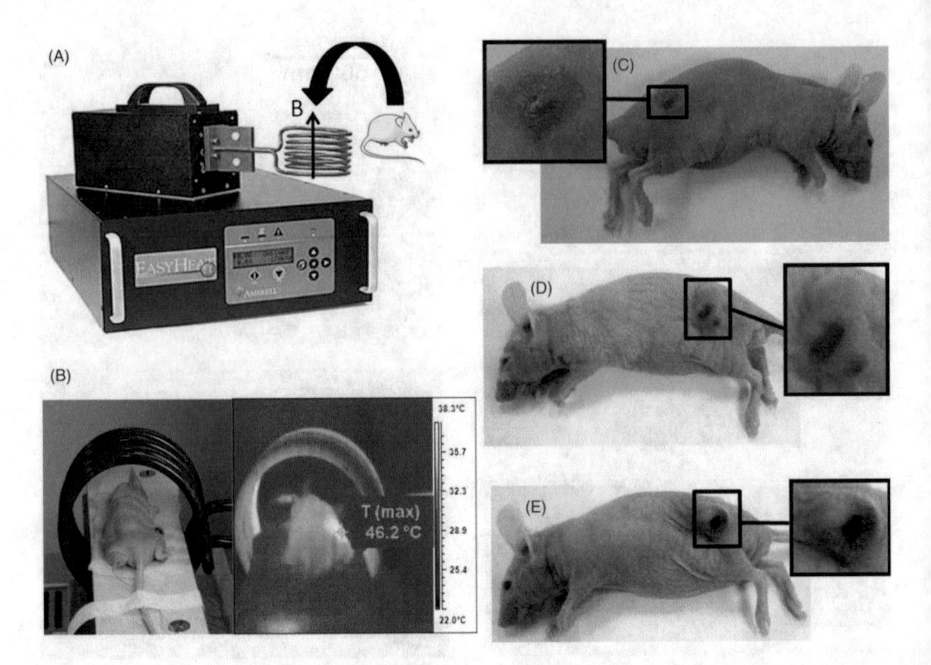

Fig. 14.13 The setup used to carry out the treatment of the mice by positioning the mice inside the copper coil and by applying an alternating magnetic field (**a**). The measurement of the temperature during the treatment (**b**). Photographs of the mice treated with a suspension containing chains of magnetosomes (**c**), individual magnetosomes (**d**), or superparamagnetic iron oxide nanoparticles (SPIONs) (**e**) (Alphandéry et al. 2013). (© 2013 Informa UK Ltd. Francis & Taylor)

20 min: every day for the first 3 days, then the sessions were repeated on days 5, 7, 9, 11, 13, and 15.

After treatment with MHT (after applying the magnetic field) a dramatic reduction of tumor volume was found without local recurrences, only in the case of group 6 (Fig. 14.14II(c)). In group 3, which were injected with mHAP and the magnetic field was not applied, the tumor increased continuously until day 14 (Fig. 14.14I(c)), not being stopped by mHAP nanoparticles or HAP. Likewise, at the magnetic field: the tumor developed in control mice (group 4) and mice injected only with HAP (no magnetic nanoparticles) (group 5).

In Fig. 14.15 is showed the increase of tumor (case I) and the decrease of tumor size (case II), respectively, during 15 days of study. It is clear that only in the case of group 6, which were injected with mHAP and subjected to the alternating magnetic field, tumor volume decreases to zero. This result shows the MHT efficacy in destroying tumor cells without significant side effects (Hou et al. 2009).

Another case is the one in which MHT was studied for lung cancer using a slightly higher temperature than the usual one in hyperthermia, respectively 46 °C, and in a shorter time of 30 min, which induces tumor cell death by apoptosis and partial necrosis (Hu et al. 2012). In the in vivo MHT experiment, superparamagnetic

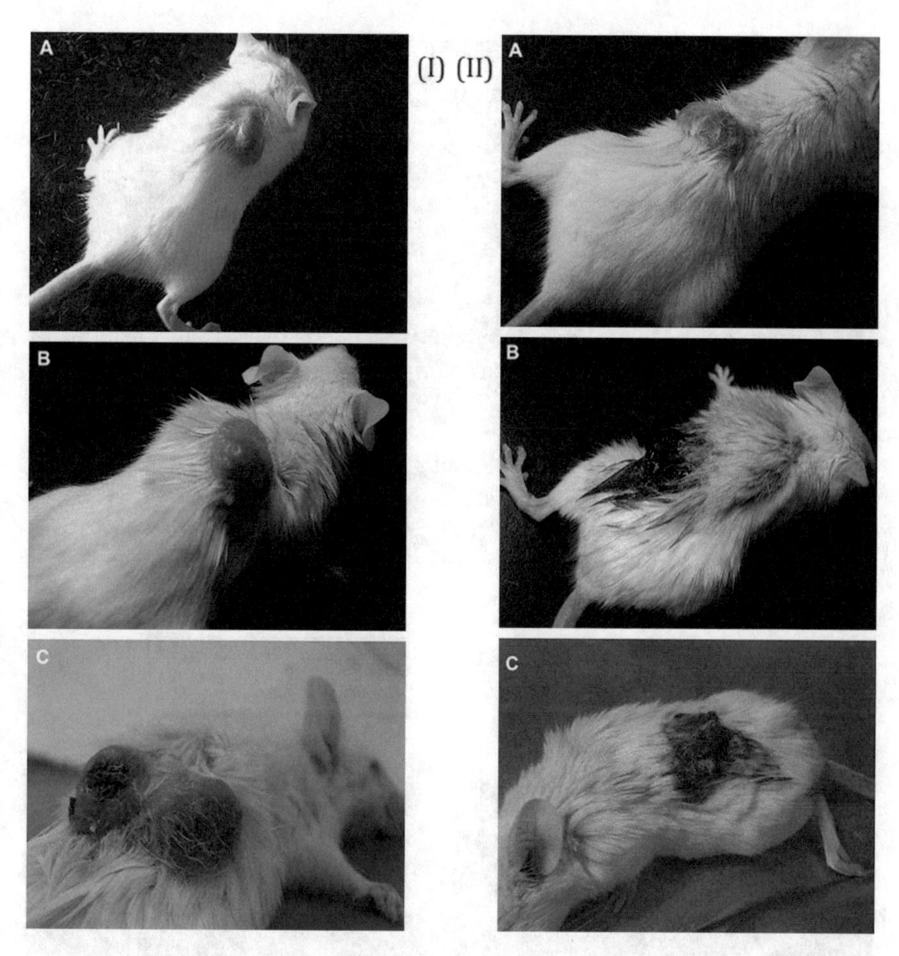

Fig. 14.14 (**I**) The clinical photographs of the tumor in Group 3 (mHAP without magnetic field). The tumor on day 1 (a), day 5 (b), and day 14 (c). (**II**) The clinical photographs of the tumor in Group 6 (mHAP with magnetic field). The tumor on day 1 (a), day 5 (b), and day 14 (c). (Reprinted from Hou et al. 2009, © 2009 Elsevier Ltd., with permission from Elsevier)

iron oxide nanoparticles (SPIONs) of 10 nm (dispersed in water) with 10 mg/mL injected into the tumor, and a 150 kHz alternating magnetic field were used. In this experiment, it was shown that at this temperature, tumor growth in mice was significantly inhibited ($p < 0.05$) from mouse control, and in several mice the tumors regressed completely after 14 days of MHT treatment (Fig. 14.16).

The results presented above and others show that SPMHT is a viable alternative method in cancer therapy, with very low toxicity or even no toxicity compared to chemo- and radiotherapy, with real application prospects in human trials.

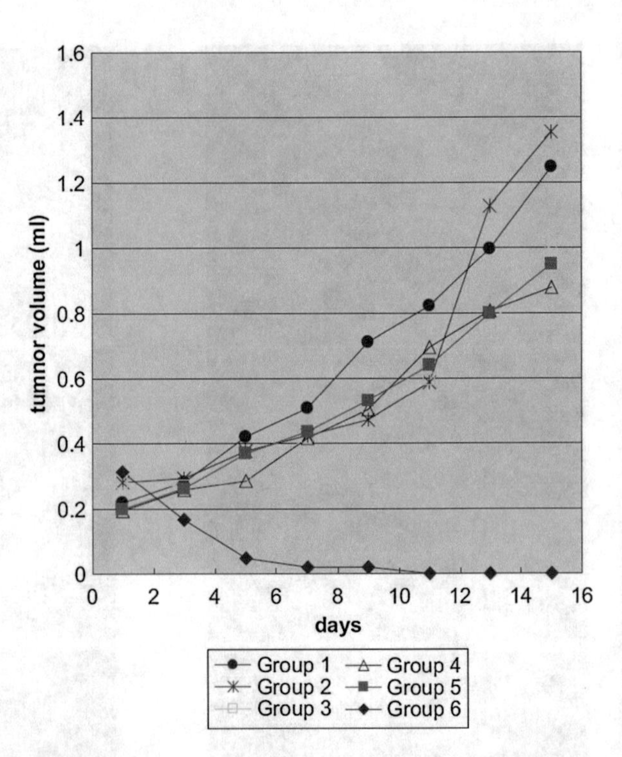

Fig. 14.15 The tumor size of different groups in the experimental period of 15 days. Among the six groups, only the tumors in Group 6 (mHAP with magnetic field) shrinked significantly. The tumors in Group 1 (control group without magnetic field) grew faster than any other groups; except on days 13 and 15, the size of tumor in Group 2 (HAP without magnetic field) was larger than Group 1. (Reprinted from Hou et al. 2009, © 2009 Elsevier Ltd., with permission from Elsevier)

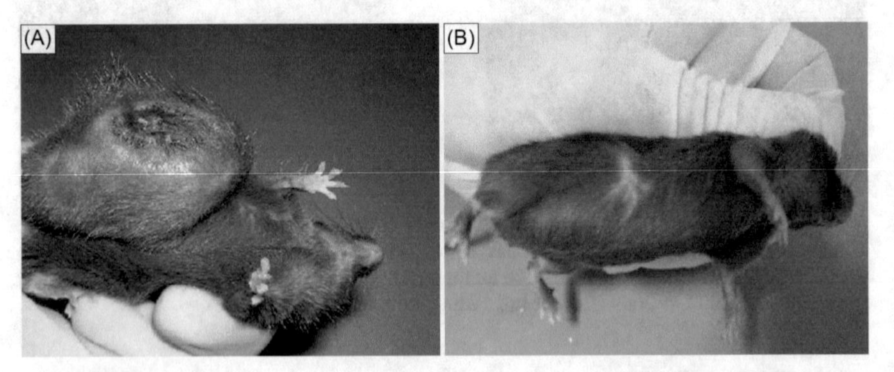

Fig. 14.16 Mice photographed on the 28th day after hyperthermia. (**a**) A mouse from the control group (A). (**b**) A mouse from experimental group (D) (Hu et al. 2012). (With permission from John Wiley & Sons (2012), CCL, © 2011 Tianjin Lung Cancer Institute and Blackwell Publishing Asia Pty. Ltd.)

14.3 Nanotheranostics in Magnetic/Superparamagnetic Hyperthermia for Cancer Therapy

Nanotheranostic (Xue et al. 2018) is an integrated technique of advanced nanobiotechnology that allows simultaneous diagnosis and effective therapy at nano level through a single nanoformulation (Kim et al. 2013; Chen 2017; Kievit and Zhang 2011) with real-time visualization of ADMET (absorption, distributions,

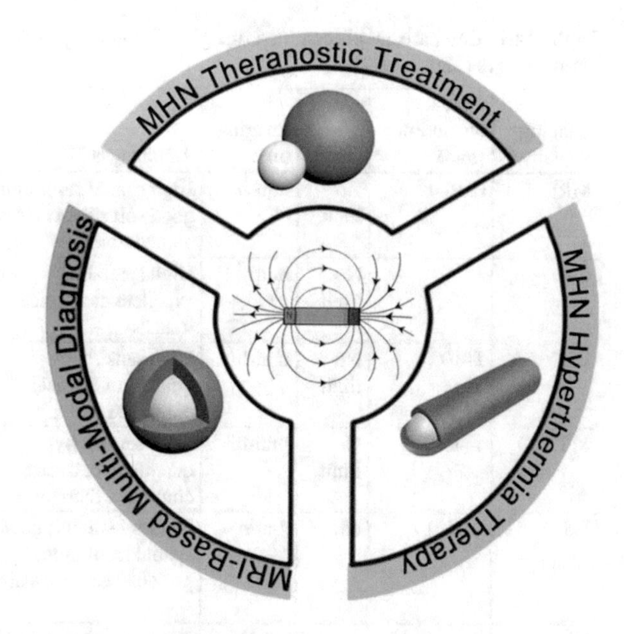

Fig. 14.17 The illustration of the multifunction of MHNs for diagnosis, therapy, and theranostic treatment (Tian et al. 2019). (© (2019) IOP Publishing. Reproduced with permission. All rights reserved)

metabolism, elimination, and toxicity) (Hodgson 2001), and synergistic, therapeutic, precision, and even personalized nanomedicine (Kim et al. 2013). This concept has also been applied recently to magnetic nanoparticles as hybrid nanobiomaterials (Tian et al. 2018) used in magnetic/superparamagnetic hyperthermia therapy (MHT/SPMHT) (Yang et al. 2013; Lee et al. 2013; Lima-Tenório et al. 2015; Datta et al. 2016; Iatridi et al. 2016; Liu et al. 2016a, b; Tian et al. 2019) by simultaneous application of MHT/SPMHT together with others, noninvasive, e.g., PDT (photodynamic therapy), or conventional therapies, chemotherapy and radiotherapy, and/or imaging methods used to guide therapy and/or diagnosis (see below).

In Fig. 14.17, the concept of theranostic treatment has been illustrated (Tian et al. 2019) by the simultaneous integration of therapy and diagnosis in the treatment of cancer in the case of magnetic hyperthermia as therapy and the use of imaging methods such as magnetic resonance imaging (MRI) and/or computed tomography (CT), to diagnose and visualize the therapeutic effect or even guide the therapy.

The best treatments with the fewest side effects are obtained by optimizing the nanobiomaterial used, biofunctionalized to be stable in time and in the biological environment, effective in therapy and safe with the combination of the functions of multimodal imaging and various treatments (the integration of cancer diagnosis and treatment) (Tian et al. 2019). In addition to MRI and CT in clinical imaging, other techniques are also used, depending on their resolution, the depth of penetration into biological tissue, and other practical parameters. In Table 14.1 the common imaging techniques are presented (Jokerst and Gambhir 2011; Chen et al. 2013; Naumova et al. 2014; Shin et al. 2015; Liu et al. 2018) that can be used in single-, bi-, and multimodal imaging: MRI, CT, PET (positron emission tomography), SPECT (single-photon emission computed tomography), US (ultrasound), and PA (photoacoustic imaging).

Table 14.1 Comparison of common imaging technology performance and advantages and disadvantages (Tian et al. 2019)

Imaging modality	Spatial resolution (mm)	Depth	Imaging time	Advantages	Disadvantages
MRI	0.01–1	No limit	min-h	High spatial resolution; good soft tissue contrast; quantitative	Expensive; long imaging time; low sensitivity
CT	1	No limit	min	High spatial resolution; excellent anatomical images	Radioactive; poor contrast of soft tissue
PET	1–10	No limit	min-h	High sensitivity; quantitative; tracer; combined therapy	Radioactive; low spatial resolution
SPECT	1–15	No limit	min-h	High sensitivity; quantitative; tracer; combined therapy	Radioactive; low spatial resolution; long imaging time
US	0.05–1	cm	s-min	High sensitivity; good spatial resolution; portable; economical	Depending on the operator's technical level; lacking available probe
PA	0.01–1	cm	s-min	Provide high contrast and resolution tissue imaging and combined with the advantages of US	Limited detecting depth; limited in bone and air organization

In this area, recently it has been proposed to use synergistic therapeutic photo-magnetic hyperthermia by combining nano-magnetic hyperthermia (NMHT) and nano-photothermal therapy (NPTT) methods to increase together their effectiveness on thermal tumor cell destruction (Espinosa et al. 2015; Wang et al. 2015b). Using MHT alone has the disadvantage of limiting the increase of nanoparticle concentration in tumor to increase the hyperthermic effect. Similarly, increasing the laser radiation dose in the NPTT, or the dependence of its intensity on the penetration into tissue, limits the applications in clinical trials (Durymanov et al. 2015; Datta et al. 2016). However, an important aspect of reducing the effectiveness of the photo-magnetic therapy is that magnetic particles are generally injected directly intratumoral and not administered intravenously that raises toxicity problems for the body or the action of macrophages (which tends to eliminate magnetic nanoparticles). In addition, there is lack of nanoparticles homogeneity in the entire tumor volume. In order to increase the effectiveness of cancer therapy was used accurately guided nanoparticles to cancer cells, then obtaining the hyperthermal effect simultaneously with tumor evaluation using current imaging methods: magnetic resonance imaging (MRI), photoacoustic imaging (PAI), or fluorescence molecular imaging (FMI) (Kircher et al. 2012; Liu et al. 2016a, b). Imaging information obtained allows the optimal guidance of tumor hyperthermia and diagnosis at the same time. This procedure is known as "Nanotheranostic" in cancer therapy, which allows simultaneous tumor therapy and diagnosis. In the area of MHT and in the perspective of its application in clinical trials, this idea of nanotheranostic advances further, having in mind other possible therapeutic combinations in the field of nano-formulation of magnetic nanoparticles, as nanocarrier vectors (magnetic, photo, thermal, drugs, photo luminescent, magnetic sensitive, etc.) to cancer cells.

Yan et al. (2018) proposed a complex nanobiomaterial (multifunctional theranostic nanoplatform) as multishell nanostructure: MNPs@PES-Cy7/2-DG with magnetic nanoparticles (MNPs), poly(3,4-ethylenedioxythiophene):poly(4-styrenesulfonate) (PES), Cyanine7 (Cy7), and 2-deoxyglucose (2-DG)-polyethylene glycol (PEG) (Fig. 14.18I), under the "all-in-one" concept of nanotheranostic, with intravenous injection of the multifunctional suspension (not in the solid tumor). Dual therapy is obtained in this case by combined hyperthermia experiments, using photo-magnetic hyperthermia. The technique and effects recorded in this case by the nanotheranostic (dual therapy and imaging (diagnosis)) are shown in Fig. 14.18II.

Recently, biocompatible nanoparticles of Fe_3O_4 having a diameter of ~10 nm coated with *poly-L-lysine* have been proposed to combine magnetic hyperthermia (MHT) with magnetic resonance imaging (MRI) in order to unify the therapeutic and diagnostic approach (Kubovcikova et al. 2019). These bio-nanoparticles were tested for MHT at a frequency of 190 kHz and magnetic field amplitude of ~8 kA/m, the results leading to a suitable SAR for hyperthermia. At the same time, there was a significant effect of bio-nanoparticles on transversal relaxation time T2 in MRI, demonstrating possible future application in targeted synergistic cancer treatment.

Another new nanobiomaterial was recently developed and tested in vitro and in vivo by combining magnetic hyperthermia (MHT) with photodynamic therapy (PDT) and a photosensitizer (Di Corato et al. 2015) as a smart nanoplatform for therapeutic and diagnostic methods in tumor therapy (nanotheranostic). Iron oxide nanoparticles were high loaded in the aqueous core of hybrid liposomes (6 fg of iron/liposome) having lipid bilayer supplied with a photosensitizer which generates oxygen (high-toxicity reactive oxygen species (ROS)) under laser excitation (near-infrared (NIR)). Heat is produced by coupling PDT to MHT.

In vitro (on ovarian cancer cells) using PDT-MHT therapy as a new synergistic approach shows complete cancer cell death, and in vivo model (on tumor-bearing mice), a total solid-tumor ablation (Fig. 14.19), overcome current limitations of a single thermotherapy. Using high-resolution MRI (imaging technique), the distribution of injected magnetoliposomes was monitored in vivo.

Thorat et al. (2017) reported a new chemo-thermo complex agent for synergic cancer therapy as nanotheranostic was prepared and then in vitro and in vivo studied to reduce drug resistance in cancer. It consists of superparamagnetic nanoparticles (SPMNPs) of $La_{0.7}Sr_{0.3}MnO_3$ biofunctionalized with oleic acid (OA)-polyethylene glycol (PEG) polymeric micelle (PM) structure, and high loaded (~60%) with anti-cancer drug doxorubicin DOX (Fig. 14.20). SPMNPs are used for heat in MHT therapy guided by magnetic resonance imaging (MRI). Thus, this SPMNPs polymeric micelle can be used for effective multimodal cancer theranostics by combined chemotherapy with magnetic hyperthermia in external alternating magnetic field. In vitro results show the cancer cell death rate of ~90%.

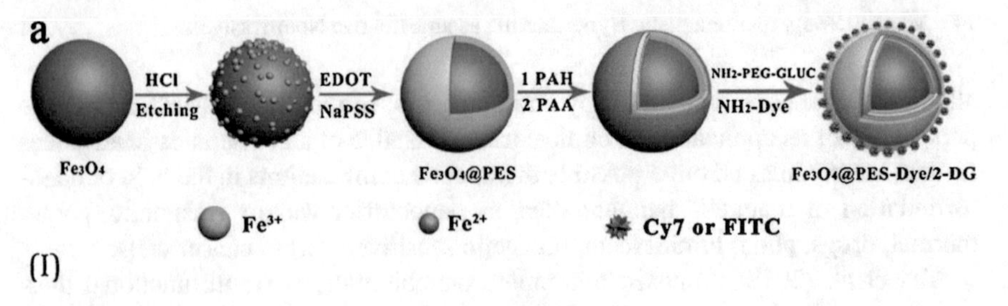

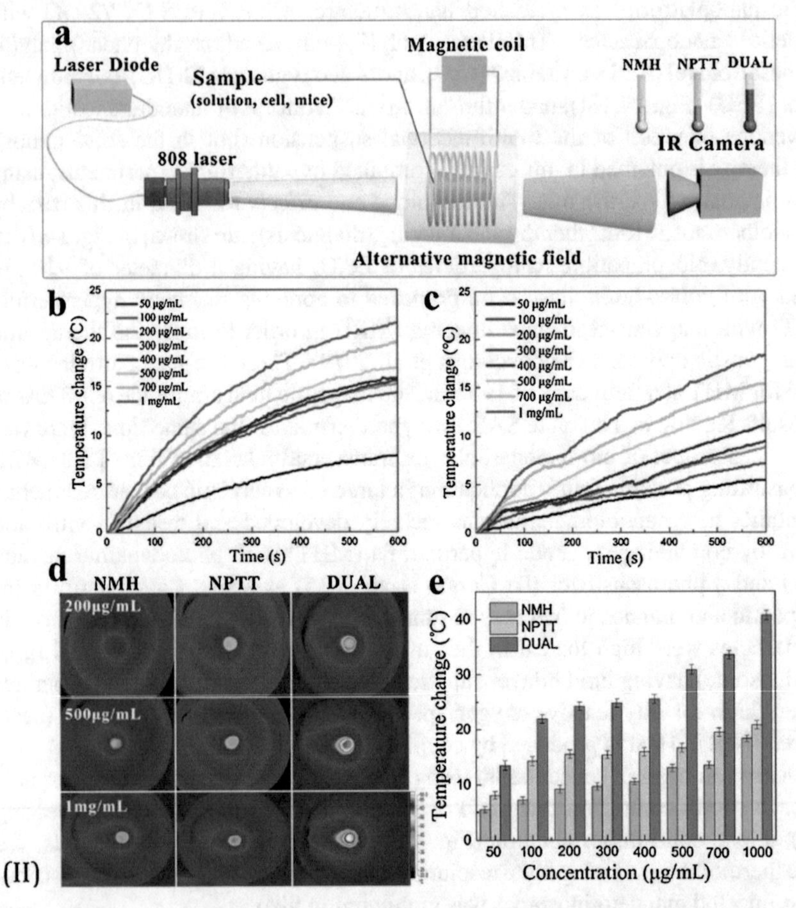

Fig. 14.18 (**I**) Synthesis and characterization of MNP@PES-Cy7/2-DG. (a) A schematic diagram of the fabrication process; (**II**) Photo-magnetic hyperthermia in aqueous suspension. (a) Scheme of the experimental device for combined hyperthermia experiments, consisting of a magnetic coil in which the sample is placed so that it can be stimulated by the near-infrared (NIR) laser (808 nm). The temperature increase was recorded using an infrared thermal imaging (IR) camera located at the end of the coil cavity. (b) Heating curves of MNP@PES-Cy7/2-DG solutions at various concentrations under 808 nm laser irradiation at a power density of 0.75 W/cm^2. (c) Heating curves of suspensions of nanocomposites at various concentrations in an alternating magnetic field (200 kHz, 38 kA/m). (d) Panel of thermal images acquired by the IR camera on samples at three concentrations. (e) Average temperature increase recorded for nanocomposites under the three heating protocols. Data were given as mean ± SD (n = 3) (Yan et al. 2018). (Reproduced with permission of John Wiley & Sons (2018) conveyed through Copyright Clearance Center. © 2018 Wiley-VCH Verlag GmbH & Co. KGaA)

Folate-conjugated FITC/5-FU/CMC MNPs for magnetic hyperthermia (MHT), chemotherapy, targeted cancer cells, and cellular imaging has been proposed by Sivakumar et al. (2013) as a new multifunctional nanovector and theranostic system. This multifunctional nanovector consists of Fe_3O_4 magnetic nanoparticles (MNPs) (10 nm average diameter) encapsulated in carboxymethyl cellulose (CMC) in a rather spherical form (100–150 nm mean size) (CMC MNPs), loaded with 5-fluorouracil (5-FU) anticancer drugs and folate, and green fluorescent nanoparticles (FITC-labeled) for imaging studies (diagnostic imaging) (Fig. 14.21). The folate used as the targeting moiety target over expressed folate receptors in cancer cells (MCF7, G1). The nanovector suspension has a superparamagnetic behavior (SPIONs) in the external magnetic field, making it very suitable for superparamagnetic hyperthermia therapy (SPMHT).

Through simultaneous application of hyperthermia and drug, the in vitro experimental results obtained on human breast cancer cell lines (MCF7-cells) and glial cell line (G1-cells) at 24 h after applying magnetic hyperthermia at a concentration of 4 mg/mL showed 94–96% in the destruction of tumor cells (Table 14.2). The internalization of magnetic nanoparticles (MNPs) by MCF7 and G1 cells was studied using cellular imaging and flow cytometry. The results obtained demonstrate both the efficacy of hyperthermia therapy combined with drug cancer therapy and the viability of these multifunctional nanovectors in cancer therapy.

A schematic representation of a current dual-treatment with bi-modal imaging monitorized (theranostic) is shown in Fig. 14.22 (Guoa et al. 2018). There, methotrexate (MTX)-modified thermo-sensitive magnetoliposomes (MTX-MagTSLs) is used. Magnetic nanoparticles and lipophilic fluorescent dye Cy5.5 for dual imaging have been encapsulated into bilayer of liposomes and has provided an appropriate laser irradiation (near-infrared (808 nm)) region to release doxorubicin (DOX) under alternating magnetic field (AMF) (DUAL-mode treatment). The results obtained in vitro and in vivo show that MTX-MagTSLs possessed an excellent targeting ability toward HeLa cells and HeLa tumor-bearing mice, and these multifunctional liposomes can be used for treatments and diagnoses in precise cancer therapy.

Another innovative theranostic agent is FIMO-NFs (ferromagnetic iron-manganese oxide-nano-flowers) consisting of uniform ferromagnetic wüstite $Fe_{0.6}Mn_{0.4}O$ (FIMO) nanoflowers (NFs), with T1–T2 dual-mode magnetic resonance imaging (MRI) (Liu et al. 2016a, b). FIMO-NFs have the diameters of ~100 nm and the Fe/Mn ratio of 1.5. After administration of $Fe_{0.6}Mn_{0.4}O$ nanoflowers, in vivo MRI on the mouse glioma model shows clear delineation in both T1- and T2-weighted MR images. In vitro and in vivo experiments show that novel FIMO-NFs induce MCF-7 breast cancer cell apoptosis and complete tumor regression (Fig. 14.23) without significant toxicity. This FIMO-NFs agent can be used both in diagnostic applications and in therapy by magnetic hyperthermia (magnetic theranostic platform).

A multifunctional Fe_3O_4/Au cluster/shell nanocomposite for surface-enhanced Raman scattering (SERS)-assisted theranostic strategy was proposed by Han et al. (2016). This nanocomposite was used for detection of free prostate-specific antigen (free-PSA), magnetic resonance imaging (MRI), and magnetic hyperthermia.

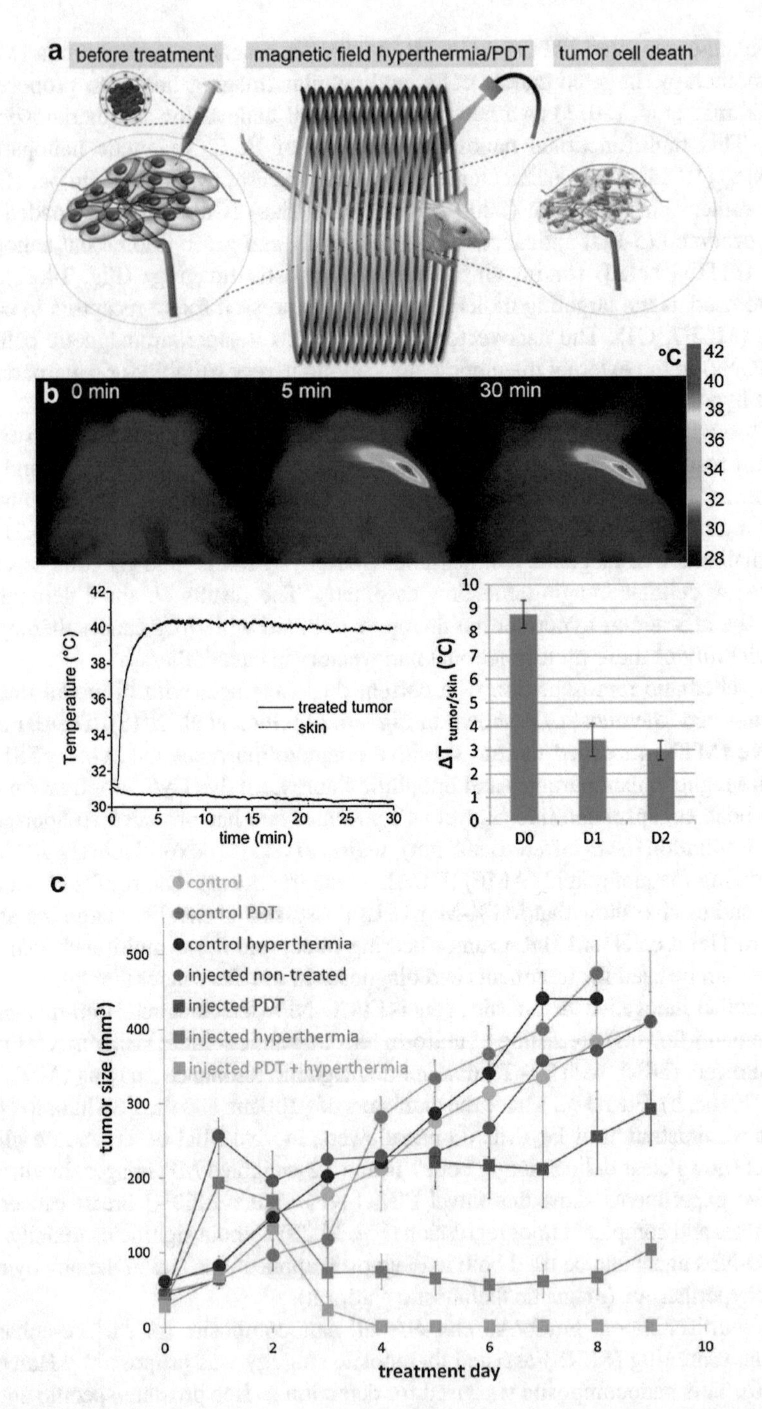

Fig. 14.19 Treatment efficacy on tumor-bearing mice. (**a**) Therapeutic strategy sketch: liposomes were injected intratumorally, and mice were subsequently subjected to combined treatment with magnetic hyperthermia and laser irradiation. (**b**) Increase in local temperature during magnetic hyperthermia treatment was monitored with an infrared thermocamera. The maximum temperature

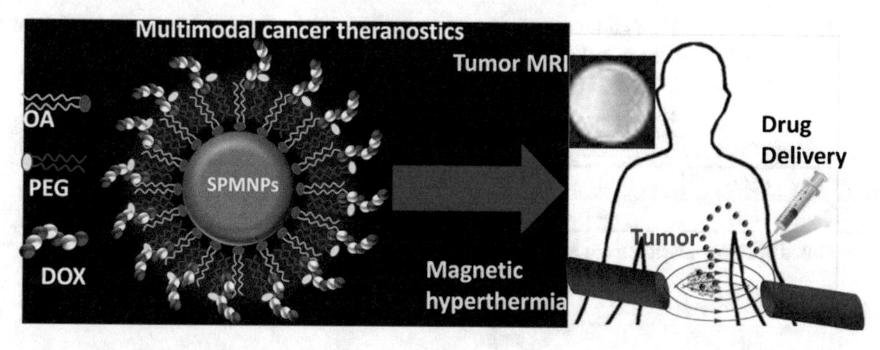

Fig. 14.20 Graphical abstract in Thorat et al. (2017). (Reprinted with permission from Thorat et al. 2017, © (2016) American Chemical Society)

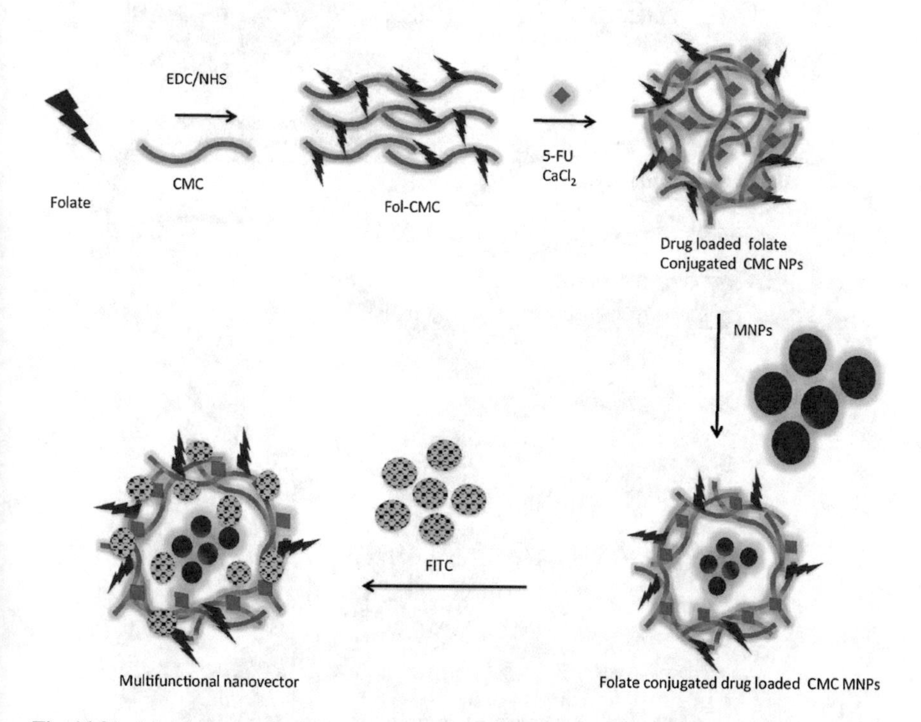

Fig. 14.21 Schematic representation of the synthesis of the multifunctional nanovector illustrating a folate-targeted CMC nanoparticle incorporating MNPs, anticancer drug 5-FU, and imaging moiety FITC. (Reprinted with permission from Sivakumar et al. 2013, © (2013) American Chemical Society)

Fig. 14.19 (continued) was reached about 5 min after field application and maintained for the entire treatment cycle (30 min). Because of cell rearrangement, nanoparticle dilution, and aggregation, the heating efficacy decayed after one cycle of treatment, assessing a local temperature increase of a few degrees during magnetic hyperthermia. It is noteworthy to highlight that the camera measures the surface temperature of the skin, so the temperature within the tumor is expected to be higher. (**c**) Tumor growth curves of different control groups and treatment groups. D0 on the graph corresponds to the day of liposome injection, which also corresponds to the first day of treatment. (Reprinted with permission from Di Corato et al. 2015, © (2015) American Chemical Society)

Table 14.2 Percentage of cell viability after magnetic hyperthermia with plain CMC MNPs and 5-FU-loaded CMC MNPs at different time intervals in cancer cells

Cell lines	Control	Hyperthermia with CMC MNPs			Hyperthermia with CMC MNPs + drug		
		Immediately after MHT	12 h	24 h	Immediately after MHT	12 h	24 h
MCF7	100	41	32	23	25	19	6
G1	100	43	30	21	23	15	4

Reprinted with permission from Sivakumar et al. 2013, © 2013 American Chemical Society

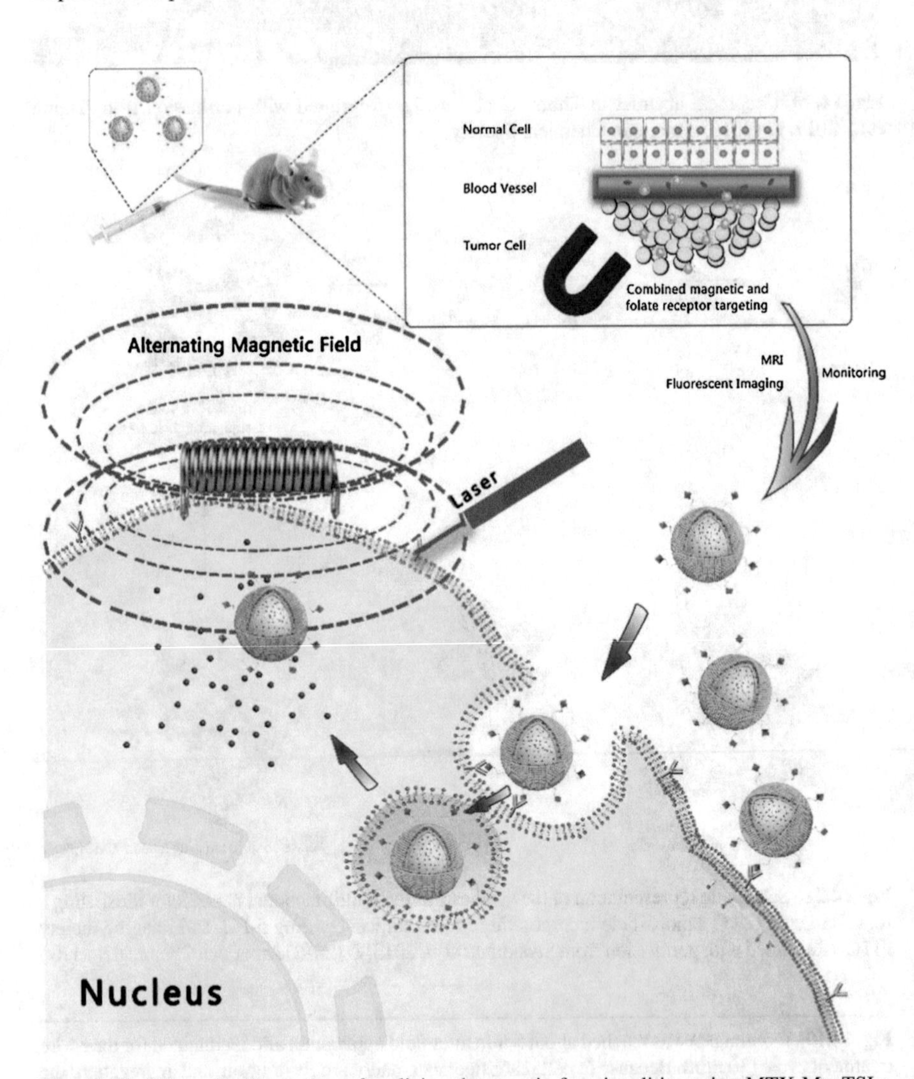

Fig. 14.22 Schematic illustration of realizing theranostic functionalities using MTX-MagTSLs. The multifunctional MTX-MagTSLs can be targeted to tumor site under constant magnetic field (CMF) and folate receptor targeting, followed by triggering DOX release under light/magnetic hyperthermia simulation synchronously to achieve the effect of drug treatment. The process can be monitored by fluorescence and MR imaging. (Reprinted from Guoa et al. 2018, © (2017) Published by Elsevier B.V., with permission from Elsevier)

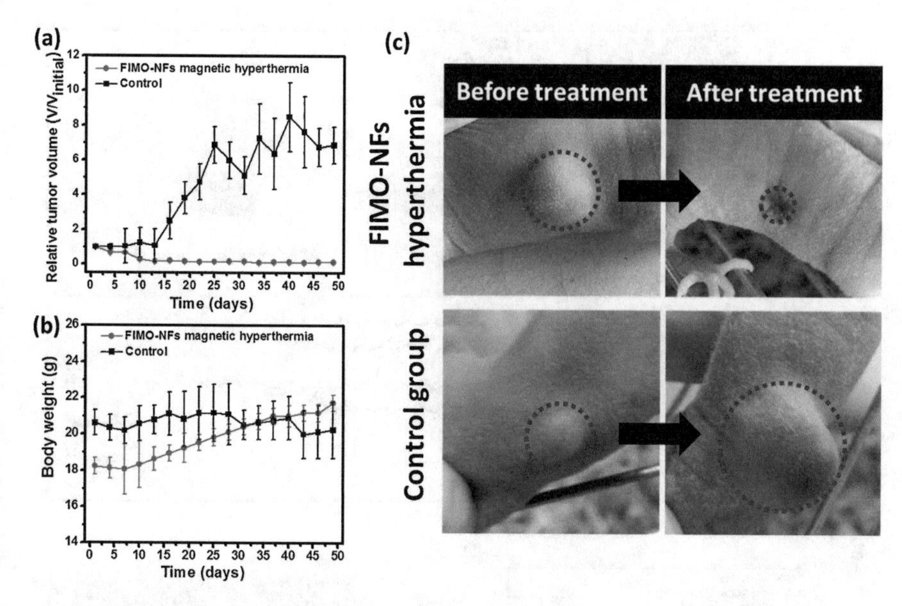

Fig. 14.23 (**a**) Tumor growth curves of FIMO-NFs magnetic hyperthermia group of tumor-bearing mice after treatment and control group. The tumor volumes were normalized to their initial sizes. Error bars represent the standard deviations of three mice per group. (**b**) Body weight of mice in two groups. The error bars represent the standard deviations (three mice per group). (**c**) Photographs at day 50 of representative mice from two groups: FIMO-NFs magnetic hyperthermia treatment group and control group (Liu et al. 2016a, b). (Reproduced with permission of John Wiley & Sons (2016) conveyed through Copyright Clearance Center. © 2016 Wiley-VCH Verlag GmbH & Co. KGaA)

In 2018, a new theranostic complex nanobiomaterial was developed by Zhou et al. being made up of c(RGDyK) peptide conjugated PEGylated Fe@Fe_3O_4 magnetic nanoparticles (RGD-PEG-MNPs) for photoacoustic (PA)-enabled self-guidance in tumor-targeting magnetic hyperthermia therapy (MHT) in vivo (U87MG glioblastoma xenograft mouse model) (Fig. 14.24I). The hydrodynamic diameter of RGD-PEG-MNPs of hybrid nanoparticles is 33.7 nm.

After the intravenous (i.v.) administration of RGD-PEG-MNPs suspension in an aqueous solution, with a dosage of 40 mg/kg body weight, results obtained in vivo show an effective magnetic hyperthermia of tumor using the guidance of PA, and excellent targeting property of RGD-PEG-MNPs. Tumor size was monitored daily over 16 days to evaluate the magnetic hyperthermia efficiency on tumor size. In vivo results are shown in Fig. 14.24II. It has been demonstrated that using RGD-PEG-MNPs in MHT (272 G, 242.5 kHz) for 40 min after 6 and 12 h i.v., guided by PA, after the mice were treated for ten times, the tumor growth in the mice group under magnetic hyperthermia therapy was inhibited.

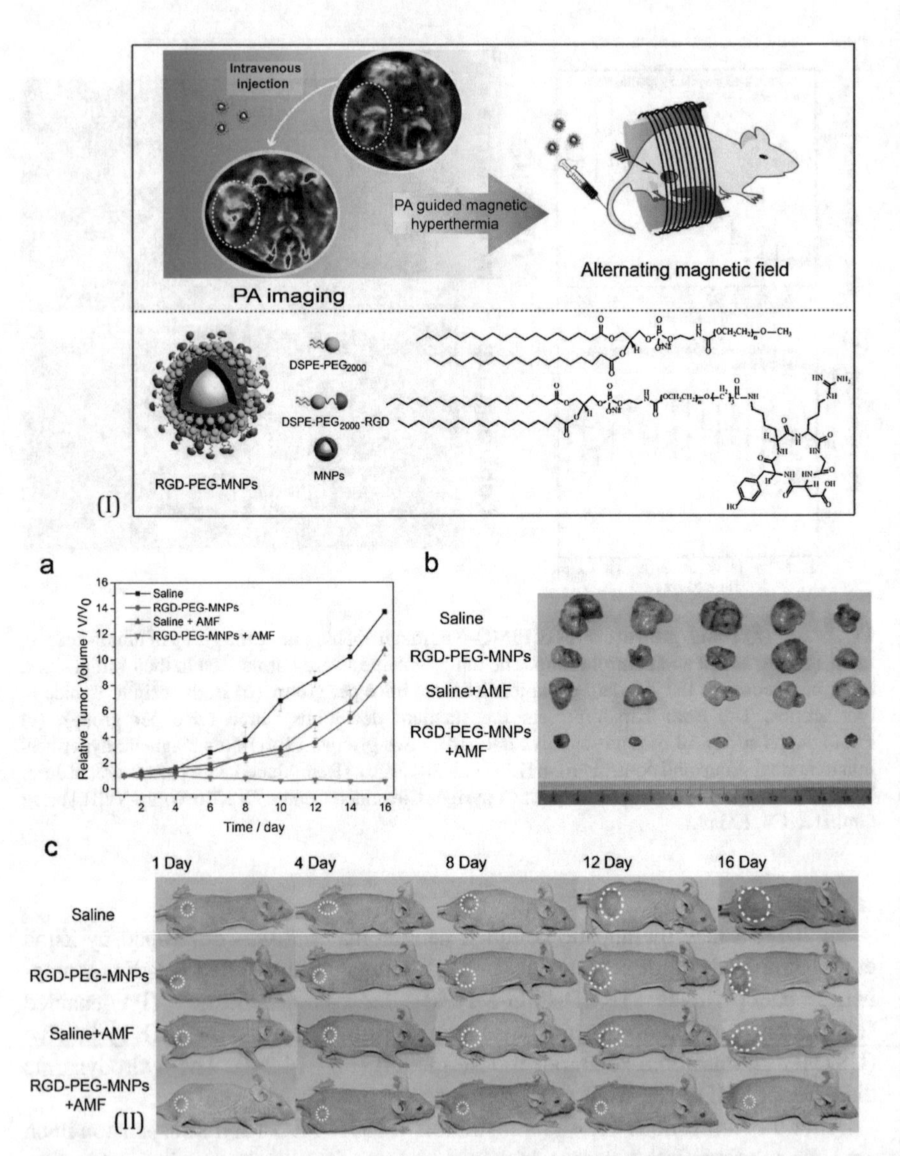

Fig. 14.24 (**I**) The schematic synthetic illustration of RGD-PEG-MNPs and PA-guided magnetic hyperthermia in vivo; (**II**) Magnetic hyperthermia ablation efficiency in a glioblastoma xenograft model. (a) The changes of the relative tumor volume after various treatments. (b) Photographs of tumors harvested on 16th day posttreatments from mice. (c) Photographs on the 1st, 4th, 8th, 12th, and 16th day of mice after the magnetic hyperthermia. Saline group: mice were intravenously injected with saline only; RGD-PEG-MNPs group: mice were intravenously injected with RGD-PEG-MNPs, respectively; AMF group: mice were intravenously injected with saline and exposed under AMF (272 G and 242.5 kHz); RGD-PEG-MNPs plus group (magnetic hyperthermia therapy group): mice were intravenously injected with RGD-PEG-MNPs and exposed under AMF (272 G and 242.5 kHz). Error bars were based on the standard deviations of five parallel samples. (Reproduced from Zhou et al. 2018 with permission of John Wiley & Sons (2018) conveyed through Copyright Clearance Center. © 2018 Wiley-VCH Verlag GmbH & Co. KGaA)

14.4 Conclusions

Recent results obtained in vitro and in vivo in alternative therapy of malignant tumors using noninvasive SPMHT/MHT, and presented in this chapter, currently show an effectiveness of this therapy up to ~90% in the thermal destruction of tumors by apoptosis, and partially necrosis, depending on the technique and nanobiomaterial used as a result of advanced nanobiotechnology. The success of this therapy greatly depends on both the magnetic nanoparticles used as nanocarriers and thermal mediators as a result of superparamagnetic relaxation, but also on the biological nanostructures used to encapsulate magnetic nanoparticles with specific biocompatibility properties and affinity for tumor cells, targeting intracellular therapy.

At the same time, the results presented in SPMHT/MHT, as noninvasive and low-toxicity alternative therapy, show that it is a viable therapy in cancer treatment that can be successfully applied in the near future in human clinical trials.

However, in order to increase the SPMHT/MHT efficiency to ~100%, and reduce the toxicity to zero, the two key factors should be considered: (1) the type of magnetic nanoparticles used, and (2) the biofunctional nanobiostructure most suitable for such a therapy.

In addition, combining SPMHT/MHT with other modern techniques used in cancer therapy, such as photodynamic therapy, simultaneously or alternating with other techniques, such as using conventional methods, chemotherapy and radiotherapy, and using advanced imaging techniques (MRI, CT, FAI, etc.) can increase the effectiveness in cancer therapy to 100%, while reducing the toxicity of healthy tissues and increasing safety in therapy.

Certainly, the future will require the use of alternative, noninvasive and lowest- or nontoxicity therapies, such as SPMHT, and the reduction of conventional high-toxicity chemotherapy and radiotherapy as much as possible. The researchers hope that this goal will be achieved in the future by advanced nanobiotechnology, nanotheranostics, and new results from optimization of the SPMHT using new target hybrid nanobiomaterials.

References

Almaki JH, Nasiri R, Idris A, Majid FAA, Salouti M, Wong TS, Dabagh S, Marvibaigi M, Amini N. Synthesis, characterization and in vitro evaluation of exquisite targeting SPIONs–PEG–HER in HER2+ human breast cancer cells. Nanotechnology. 2016;27:105601, 13pp.

Alphandéry E, Faure S, Raison L, Duguet E, Howse PA, Bazylinski DA. Heat production by bacterial magnetosomes exposed to an oscillating magnetic field. J Phys Chem C. 2011a;115:18–22.

Alphandéry E, Faure S, Seksek O, Guyot F, Chebbi I. Chains of magnetosomes extracted from AMB 1 magnetotactic bacteria for application in alternative magnetic field cancer therapy. ACS Nano. 2011b;5:6279–96.

Alphandéry E, Guyot F, Chebbi I. Preparation of chains of magnetosomes, isolated from Magnetospirillum magneticum AMB-1 magnetotactic bacteria, yielding efficient treatment of tumors using magnetic hyperthermia. Int J Pharm. 2012;434:444–52.

Alphandéry E, Chebbi I, Guyot F, Durand-Dubief M. Use of bacterial magnetosomes in the magnetic hyperthermia treatment of tumours: a review. Int J Hyperthermia. 2013;29:801–9.

Baker I, Zeng Q, Li W, Sullivan CR. Heat deposition in iron oxide and iron nanoparticles for localized hyperthermia. J Appl Phys. 2006;99:08H106–08H106-3.

Caizer C. Magnetic anisotropy of $Co_\delta Fe_{3-\delta}O_4$ nanoparticles for applications in magnetic hyperthermia. In: The 19th international conference on magnetism (ICM 2012), 8–13 Jul, Busan, Korea, 2012.

Caizer C. SPMHT with biocompatible SPIONs for destroy the cancer cells. In: The 8th international conference on fine particle magnetism (ICFPM-2013), 24–27 Jun, Perpignan, France, 2013.

Caizer C. Computational study on superparamagnetic hyperthermia with biocompatible SPIONs to destroy the cancer cells. J Phys Conf Ser. 2014;521:012015–4.

Caizer C. Magnetic hyperthermia using magnetic metal/oxide nanoparticles with potential in cancer therapy, Ch. 10. In: Rai M, Shegokar R, editors. Metal nanoparticles in pharma. Cham: Springer; 2017.

Caizer C, Tura V. Magnetic relaxation/stability of Co ferrite nanoparticles embedded in amorphous silica particles. J Magn Magn Mater. 2006;301:513–20.

Caizer C, Hadaruga N, Hadaruga D, Tanasie G, Vlazan P. The Co ferrite nanoparticles/liposomes: magnetic bionanocomposites for applications in malignant tumors therapy. In: The 7th international conference on inorganic materials, 12–14 Sept, Biarritz, France, 2010a.

Caizer C, Stancu A, Postolache P, Dumitru I, Bodale I, Vlazan P. The magnetic properties of the $Co_\delta Fe_{(3-\delta)}O_4$ surfacted nanoparticles with potential applications in cancer therapy. In: The 7th international conference on fine particle magnetism (ICFPM-2010), 21–24 Jun, Uppsala, Sweden, 2010b.

Caizer C, Soica C, Dehelean C, Radu A, Caizer IS. Study on toxicity of the superparamagnetic nanoparticles on the cells in order to use them in cancer therapy. In: The 8th international conference on fine particle magnetism, 24–27 Jun, Perpignan, France, 2013.

Caizer C, Buteica A, Mindrila I. Biocompatible magnetic oxide nanoparticles with metal ions coated with organic shell as potential therapeutic agents in cancer, Ch. 11. In: Rai M, Shegokar R, editors. Metal nanoparticles in pharma. Cham: Springer; 2017.

Chen F, Ellison PA, Lewis CM, Hong H, Zhang Y, Shi S, Hernandez R, Meyerand ME, Barnhart TE, Cai W. Chelator-free synthesis of a dual-modality PET/MRI agent. Angew Chem. 2013;52:13319–23.

Chen H, Zhang W, Zhu G, Xie J, Chen X. Rethinking cancer nanotheranostics. Nat Rev Mater. 2017;2:17024.

Datta NR, Krishnan S, Speiser DE, Neufeld E, Kuster N, Bodis S, Hofmann H. Magnetic nanoparticle-induced hyperthermia with appropriate payloads: Paul Ehrlich's "magic (nano) bullet" for cancer theranostics? Cancer Treat Rev. 2016;50:217–27.

Denoyer D, Greguric I, Roselt P, Neels OC, Aide N, Taylor SR, Katsifis A, Dorow DS, Hicks RJ. High-contrast PET of melanoma using (18)F-MEL050, a selective probe for melanin with predominantly renal clearance. J Nucl Med. 2010;51:441–7.

Di Corato R, Béalle G, Kolosnjaj-Tabi J, Espinosa A, Clément O, Silva AK, Ménager C, Wilhelm C. Combining magnetic hyperthermia and photodynamic therapy for tumor ablation with photoresponsive magnetic liposomes. ACS Nano. 2015;9:2904–16.

Durymanov MO, Rosenkranz AA, Sobolev AS. Current approaches for improving intratumoral accumulation and distribution of nanomedicines. Theranostics. 2015;5:1007–20.

Engelmann U, Roeth A, Eberbeck D, Buhl E, Neumann U, Schmitz-Rode T, Slabu I. Combining bulk temperature and nanoheating enables advanced magnetic fluid hyperthermia efficacy on pancreatic tumor cells. Sci Rep. 2018;8:13210.

Espinosa A, Bugnet M, Radtke G, Neveu S, Botton GA, Wilhelm C, Abou-Hassan A. Can magneto-plasmonic nanohybrids efficiently combine photothermia with magnetic hyperthermia? Nanoscale. 2015;7:18872–7.

Fortin JP, Gazeau F, Wilhelm C. Intracellular heating of living cells through Néel relaxation of magnetic nanoparticles. Eur Biophys J. 2008;37:223–8.

Gazeau F, Lévy M, Wilhelm C. Optimizing magnetic nanoparticle design for nanothermotherapy. Nanomedicine. 2008;3:831–44.

Goya G, Asín L, Ibarra R. Cell death induced by AC magnetic fields and magnetic nanoparticles: current state and perspectives. Int J Hyperthermia. 2013;29:810–8.

Guoa Y, Zhang Y, Ma J, Li Q, Li Y, Zhou X, Zhao D, Song H, Chen Q, Zhu X. Light/magnetic hyperthermia triggered drug released from multi-functional thermo-sensitive magnetoliposomes for precise cancer synergetic theranostics. J Control Release. 2018;272:145–58.

Gupta AK, Gupta M. Synthesis and surface engineering of iron oxide nanoparticles for biomedical applications. Biomaterials. 2005;26:3995–4021.

Gupta R, Sharma D. Evolution of magnetic hyperthermia for glioblastoma multiforme therapy. ACS Chem Neurosci. 2019;10:1157–72.

Habib AH, Ondeck CL, Chaudhary P, Bockstaller MR, McHenry ME. Evaluation of iron-cobalt/ferrite core-shell nanoparticles for cancer thermotherapy. J Appl Phys. 2008;103:07A307-1–3.

Han Y, Lei S, Lu J, He Y, Chen Z, Ren L, Zhoua X. Potential use of SERS-assisted theranostic strategy based on Fe3O4/Au cluster/shell nanocomposites for bio-detection, MRI, and magnetic hyperthermia. Mater Sci Eng C. 2016;64:199–207.

Hejase H, Hayek S, Qadri S, Haik Y. MnZnFe nanoparticles for self-controlled magnetic hyperthermia. J Magn Magn Mater. 2012;324:3620–8.

Hergt R, Dutz S. Magnetic particle hyperthermia—biophysical limitations of a visionary tumour therapy. J Magn Magn Mater. 2007;311:187–92.

Hergt R, Dutz S, Muller R, Zeisberger M. Magnetic particle hyperthermia: nanoparticle magnetism and materials development for cancer therapy. J Phys Condens Matter. 2006;18:S2919–34.

Hilger I. *In vivo* applications of magnetic nanoparticle hyperthermia. Int J Hyperthermia. 2013;29:828–34.

Hilger I, Hergt R, Kaiser WA. Towards breast cancer treatment by magnetic heating. J Magn Magn Mater. 2005;293:314–9.

Hodgson J. ADMET-turning chemicals into drugs. Nat Biotechnol. 2001;19:722–6.

Hou CH, Hou SM, Hsueh YS, Lin J, Wu HC, Lin FH. The in vivo performance of biomagnetic hydroxyapatite nanoparticles in cancer hyperthermia therapy. Biomaterials. 2009;30:3956–60.

Hu R, Ma S, Li H, Ke X, Wang G, Wei D, Wang W. Effect of magnetic fluid hyperthermia on lung cancer nodules in a murine model. Oncol Lett. 2011;2:1161–4.

Hu R, Zhang X, Liu X, Xu B, Yang H, Xia Q, Li L, Chen C, Tang J. Higher temperature improves the efficacy of magnetic fluid hyperthermia for Lewis lung cancer in a mouse model. Thorac Cancer. 2012;3:34–9.

Huang MH, Yang MC. Evaluation of glucan/poly(vinyl alcohol) blend wound dressing using rat models. Int J Pharm. 2008;346:38e46.

Iatridi Z, Vamvakidis K, Tsougos I, Vassiou K, Dendrinou-Samara C, Bokias G. Multifunctional polymeric platform of magnetic ferrite colloidal superparticles for luminescence, imaging, and hyperthermia applications. ACS Appl Mater Interfaces. 2016;8:35059–70.

Ito A, Matsuoka F, Honda H, Kobayashi T. Heat shock protein 70 gene therapy combined with hyperthermia using magnetic nanoparticles. Cancer Gene Ther. 2003a;10:918–25.

Ito A, Tanaka K, Honda H, Abe S, Yamaguchi H, Kobayaschi T. Complete regression of mouse mammary carcinoma with a size greater than 15 mm by frequent repeated hyperthermia using magnetite nanoparticles. J Biosci Bioeng. 2003b;96:364–9.

Ito A, Kuga Y, Honda H, Kikkawa H, Horiuchi A, Watanabe Y, Kobayashi T. Magnetite nanoparticle-loaded anti-HER2 immunoliposomes for combination of antibody therapy with hyperthermia. Cancer Lett. 2004;212:167–75.

Ito A, Shinkai M, Honda H, Kobayashi T. Medical application of functionalized magnetic nanoparticles. J Biosci Bioeng. 2005;100:1–11.

Jansen AP, Verwiebe EG, Dreckschmidt NE, Wheeler DL, Oberley TD, Verma AK. Protein kinase C-epsilon transgenic mice: a unique model for metastatic squamous cell carcinoma. Cancer Res. 2001;61:808–12.

Johannsen M, Thiesen B, Jordan A, Taymoorian K, Gneveckow U, Waldofner N. Magnetic fluid hyperthermia (MFH) reduces *prostate* cancer growth in the orthotopic Dunning R3327 rat model. Prostate. 2005;64:283–92.

Johannsen M, Gneveckow U, Thiesen B, Taymoorian K, Cho CH, Waldofner N, Scholz R, Jordan A, Loening SA, Wust P. Thermotherapy of prostate cancer using magnetic nanoparticles: feasibility, imaging, and three-dimensional temperature distribution. Eur Urol. 2007;52:1653–62.

Jokerst JV, Gambhir SS. Molecular imaging with theranostic nanoparticles. Acc Chem Res. 2011;44:1050–60.

Jordan A, Scholz R, Maier-Hauff K, van Landeghem FK, Waldoefner N, Teichgraeber U, Pinkernelle J, Bruhn H, Neumann F, Thiesen B, von Deimling A, Felix R. The effect of thermotherapy using magnetic nanoparticles on rat malignant glioma. J Neurooncol. 2006;78:7–14.

Kandasamy G, Sudame A, Bhati P, Chakrabarty A, Maity D. Systematic investigations on heating effects of carboxyl-amine functionalized superparamagnetic iron oxide nanoparticles (SPIONs) based ferrofluids for in vitro cancer hyperthermia therapy. J Mol Liq. 2018;256:224–37.

Kievit FM, Zhang M. Cancer nanotheranostics: improving imaging and therapy by targeted delivery across biological barriers. Adv Mater. 2011;23:H217–47.

Kikumori T, Kobayashi T, Sawaki M, Imai T. Anti-cancer effect of hyperthermia on breast cancer by magnetite nanoparticle-loaded anti-HER2 immunoliposomes. Breast Cancer Res Treat. 2009;113:435–41.

Kim TH, Lee S, Chen X. Nanotheranostics for personalized medicine. Expert Rev Mol Diagn. 2013;13:257–69.

Kircher MF, Zerda A, Jokerst JV, Zavaleta CL, Kempen PJ, Mittra E, Pitter K, Huang R, Campos C, Habte F, Sinclair R, Brennan CW, Mellinghoff IK, Holland EC, Gambhir SS. A brain tumor molecular imaging strategy using a new triple-modality MRI-photoacoustic-Raman nanoparticle. Nat Med. 2012;18:829–34.

Kobayashi T. Cancer hyperthermia using magnetic nanoparticles. Biotechnol J. 2011;6:1342–7.

Kossatz S, Ludwig R, Dähring H, Ettelt V, Rimkus G, Marciello M, Salas G, Patel V, Teran FJ, Hilger I. High therapeutic efficiency of magnetic hyperthermia in xenograft models achieved with moderate temperature dosages in the tumor area. Pharm Res. 2014;31:3274–88.

Kossatz S, Grandke J, Couleaud P, Latorre A, Aires A, Crosbie-Staunton K, Ludwig R, Dähring H, Ettelt V, Lazaro-Carrillo A, Calero M, Sader M, Courty J, Volkov Y, Prina-Mello A, Villanueva A, Somoza Á, Cortajarena AL, Miranda R, Hilger I. Efficient treatment of breast cancer xenografts with multifunctionalized iron oxide nanoparticles combining magnetic hyperthermia and anti-cancer drug delivery. Breast Cancer Res. 2015;17:66.

Kubovcikova M, Koneracka M, Strbak O, Molcana M, Zavisova V, Antal I, Khmara I, Lucanska D, Tomco L, Barathova M, Zatovicov M, Dobrota D, Pastorekova S, Kopcansky P. Poly-L-lysine designed magnetic nanoparticles for combined hyperthermia, magnetic resonance imaging and cancer cell detection. J Magn Magn Mater. 2019;475:316–26.

Kumar CS, Mohammad F. Magnetic nanomaterials for hyperthermia-based therapy and controlled drug delivery. Adv Drug Deliv Rev. 2011;63:789–808.

Laurent S, Dutz S, Häfeli U, Mahmoudi M. Magnetic fluid hyperthermia: focus on superparamagnetic iron oxide nanoparticles. Adv Colloid Interface Sci. 2011;166:8–23.

Lee JH, Jang JT, Choi JS, Moon SH, Noh SH, Kim JW, Kim JG, Kim IS, Park KI, Cheon J. Exchange-coupled magnetic nanoparticles for efficient heat induction. Nat Nanotechnol. 2011; 6:418–22.

Lee JH, Chen KJ, Noh SH, Garcia MA, Wang H, Lin WY, Jeong H, Kong BJ, Stout DB, Cheon J, Tseng HR. On-demand drug release system for in vivo cancer treatment through self-assembled magnetic nanoparticles. Angew Chem Int Ed Engl. 2013;52:4384–4388.

Lee YB, Song EJ, Kim SS, Kim JW, Yu DS. Safety and efficacy of a novel injectable filler in the treatment of nasolabial folds: polymethylmethacrylate and cross-linked dextran in hydroxypropyl methylcellulose. J Cosmet Laser Ther. 2014;16:185e190.

Li S, Lin S, Daggy BP, Mirchandani HL, Chien YW. Effect of HPMC and carbopol on the release and floating properties of gastric floating drug delivery system using factorial design. Int J Pharm. 2003;253:13e22.

Li C, Chi S, Xie J. Hedgehog signaling in skin cancers. Cell Signal. 2011;23:1235–43.

Lima-Tenório MK, Edgardo A, Pineda G, Ahmad NM, Fessi H, Elaissari A. Magnetic nanoparticles: in vivo cancer diagnosis and therapy. Int J Pharm. 2015;493:313–27.

Liu RT, Liu J, Tong JQ, Tang T, Kong WC, Wang XW, Li Y, Tang JT. Heating effect and biocompatibility of bacterial magnetosomes as potential materials used in magnetic fluid hyperthermia. Prog Nat Sci Mater Int. 2012;22:31–9.

Liu XL, Ng CT, Chandrasekharan P, Yang HT, Zhao LY, Peng E, Lv YB, Xiao W, Fang J, Yi JB, Zhang H, Chuang CH, Bay BH, Ding J, Fan HM. Synthesis of ferromagnetic Fe0.6Mn0.4O nanoflowers as a new class of magnetic theranostic platform for *in vivo* T1-T2 Dual-Mode magnetic resonance imaging and magnetic hyperthermia therapy. Adv Healthcare Mater. 2016a;5:2092–104.

Liu Y, Kang N, Lv J, Zhou Z, Zhao Q, Ma L, Chen Z, Ren L, Nie L. Deep photoacoustic/luminescence/magnetic resonance multimodal imaging in living subjects using high-efficiency upconversion nanocomposites. Adv Mater. 2016b;28:6411–9.

Liu J, Chen Y, Wang G, Lv Q, Yang Y, Wang J, Zhang P, Liu J, Xie Y, Zhang L, Xie M. Ultrasound molecular imaging of acute cardiac transplantation rejection using nanobubbles targeted to T lymphocytes. Biomaterials. 2018;162:200–7.

Mahmoudi K, Alexandros Bouras A, Dominique Bozec D, Robert Ivkov R, Constantinos HC. Magnetic hyperthermia therapy for the treatment of glioblastoma: a review of the therapy's history, efficacy and application in humans. Int J Hyperthermia. 2018;34:1316–28.

Mancuso M, Gallo D, Leonardi S, Pierdomenico M, Pasquali E, De Stefano I, Rebessi S, Tanori M, Scambia G, Di Majo V, Covelli V, Pazzaglia S, Saran A. Modulation of basal and squamous cell carcinoma by endogenous estrogen in mouse models of skin cancer. Carcinogenesis. 2009;30:340–7.

Matsuoka F, Shinkai M, Honda H, Kubo T, Sugita T, Kobayashi T. Hyperthermia using magnetite cationic liposomes for hamster osteosarcoma. Biomagn Res Technol. 2004;2(3):1–6.

Mehdaoui B, Meffre A, Lacroix LM, Carrey J, Lachaize S, Gougeon M, Respaud M, Chaudret B. Large specific absorption rates in the magnetic hyperthermia properties of metallic iron nanocubes. J Magn Magn Mater. 2010;322:L49–52.

Moroz P, Jones SK, Gray BN. Tumor response to arterial embolization hyperthermia and direct injection hyperthermia in a rabbit liver tumor model. J Surg Oncol. 2002;80:149–56.

Muralidhar R, Swamy GS, Vijayalakshmi P. Completion rates of anterior and posterior continuous curvilinear capsulorrhexis in pediatric cataract surgery for surgery performed by trainee surgeons with the use of a low-cost viscoelastic. Indian J Ophthalmol. 2012;60:144–6.

Naumova AV, Modo M, Moore A, Murry CE, Frank JA. Clinical imaging in regenerative medicine. Nat Biotechnol. 2014;32:804–18.

Nedyalkova M, Donkova B, Romanova J, Tzvetkov G, Madurga S, Simeonov V. Iron oxide nanoparticles—in vivo/in vitro biomedical applications and in silico studies. Adv Colloid Interface Sci. 2017;249:192–212.

Ondeck CL, Habib AH, Ohodnicki P, Miller K, Sawyer CA. Theory of magnetic fluid heating with an alternating magnetic field with temperature dependent materials properties for self-regulated heating. J Appl Phys. 2009;105:07B324-1–3.

Pankhurst QA, Connolly J, Jones SK, Dobson J. Applications of magnetic nano-particles in biomedicine. J Phys D Appl Phys. 2003;36:R167–81.

Paradossi G, Cavalieri F, Chiessi E, Spagnoli C, Cowman MK. Poly (vinyl alcohol) as versatile biomaterial for potential biomedical applications. J Mater Sci Mater Med. 2003;14:687e691.

Parhi R, Suresh P, Patnaik S. Formulation optimization of PVA/HPMC cryogel of Diltiazem HCl using 3-level factorial design and evaluation for ex vivo permeation. J Pharm Investig. 2015;45:319–27.

Pavel M, Stancu A. Study of the optimum injection sites for a multiple metastases region in cancer therapy by using MFH. IEEE Trans Magn. 2009;45:4825–8.

Pavel M, Gradinariu G, Stancu A. Study of the optimum dose of ferromagnetic nanoparticles suitable for cancer therapy using MFH. IEEE Trans Magn. 2008;44:3205–8.

Pradhan P, Giri J, Samanta G, Sarma HD, Mishra KP, Bellare J, Banerjee R, Bahadur D. Comparative evaluation of heating ability and biocompatibility of different ferrite-based magnetic fluids for hyperthermia application. J Biomed Mater Res B Appl Biomater. 2007;81B:12–22.

Purushotham S, Ramanujan RV. Modeling the performance of magnetic nanoparticles in multimodal cancer therapy. J Appl Phys. 2010;107:114701-1–9.

Qiang Y, Antony J, Sharma A, Nutting J, Sikes D, Meyer D. Iron/iron oxide core-shell nanoclusters for biomedical applications. J Nanopart Res. 2006;8:489–96.

Qu Y, Li J, Ren J, Leng J, Lin C, Shi D. Enhanced magnetic fluid hyperthermia by micellar magnetic nanoclusters composed of $Mn_xZn_{1-x}Fe_2O_4$ nanoparticles for induced tumor cell apoptosis. ACS Appl Mater Interfaces. 2014;6:16867–79.

Rosensweig RE. Heating magnetic fluid with alternating magnetic field. J Magn Magn Mater. 2002;252:370–4.

Safarik I, Safarikova M. Magnetic nanoparticles and biosciences. Monatch Chem. 2002;133:737–59.

Saldívar-Ramírez MM, Sánchez-Torres CG, Cortés-Hernández DA, Escobedo-Bocardo JC, Almanza-Robles JM, Larson A, Reséndiz-Hernández PJ, Acuña-Gutiérrez IO. Study on the efficiency of nanosized magnetite and mixed ferrites in magnetic hyperthermia. J Mater Sci Mater Med. 2014;25:2229–36.

Selvan ST, Tan TTY, Yi DK, Jana NR. Functional and multifunctional nanoparticles for bioimaging and biosensing. Langmuir. 2010;26:11631–41.

Shin TH, Choi Y, Kim S, Cheon J. Recent advances in magnetic nanoparticle-based multi-modal imaging. Chem Soc Rev. 2015;44:4501–16.

Shinkai M, Ito A. Functional magnetic particles for medical application. Adv Biochem Eng/Biotechnol. 2004;91:191–220.

Siepmann J, Peppas NA. Modeling of drug release from delivery systems based on hydroxypropyl methylcellulose (HPMC). Adv Drug Deliv Rev. 2012;64:163e174.

Sivakumar B, Aswathy RG, Nagaoka Y, Suzuki M, Fukuda T, Yoshida Y, Maekawa T, Sakthikumar DN. Multifunctional carboxymethyl cellulose-based magnetic nanovector as a theragnostic system for folate receptor targeted chemotherapy, imaging, and hyperthermia against cancer. Langmuir. 2013;29:3453–66.

Smit J, Wijin HPJ. Les ferrites. Paris: Bibl Tech Philips; 1961.

Sunderland CJ, Steiert M, Talmadge JE, Derfus AM, Barry SE. Targeted nanoparticles for detecting and treating cancer. Drug Develop Res. 2006;67:70–93.

Tanaka K, Ito A, Kobayashi T, Kawamura T, Shimada S, Matsumoto K, Saida T, Honda H. Intratumoral injection of immature dendritic cells enhances antitumor effect of hyperthermia using magnetic nanoparticles. Int J Cancer. 2005;116:624–33.

Tartaj P, Veintemillas-Verdaguer S, Serna CJ. The preparation of magnetic nanoparticles for applications in biomedicine. J Phys D Appl Phys. 2003;36:R182.

Thorat N, Bohara R, Noor MR, Dhamecha D, Soulimane T, Tofail S. Effective cancer theranostics with polymer encapsulated superparamagnetic nanoparticles: combined effects of magnetic hyperthermia and controlled drug release. ACS Biomater Sci Eng. 2017;3:1332–40.

Tian X, Zhang L, Yang M, Bai L, Dai Y, Yu Z, Pan Y. Functional magnetic hybrid nanomaterials for biomedical diagnosis and treatment. Wiley Interdiscip Rev Nanomed Nanobiotechnol. 2018;10:e1476.

Tian X, Liu S, Zhu J, Qian Z, Bai L, Pan Y. Biofunctional magnetic hybrid nanomaterials for theranostic applications. Nanotechnology. 2019;30:032002, 10pp.

Valenzuela R. Magnetic ceramics. Cambridge: Cambridge University Press; 1994. p. 137–42.

Wang L, Dong J, Ouyang W, Wang X, Tang J. Anticancer effect and feasibility study of hyperthermia treatment of pancreatic cancer using magnetic nanoparticles. Oncol Rep. 2012;27:719–26.

Wang J, Zhou Z, Wang L, Wei J, Yang H, Yang S, Zhao J. $CoFe_2O_4$@$MnFe_2O_4$/polypyrrole nanocomposites for *in vitro* photothermal/magnetothermal combined therapy. RSC Adv. 2015a;5:7349–55.

Wang P, Xie X, Wang J, Shi Y, Shen N, Huang X. Ultra-small superparamagnetic iron oxide mediated magnetic hyperthermia in treatment of neck lymph node metastasis in rabbit pyriform sinus VX2 carcinoma. Tumor Biol. 2015b;36:8035–40.

Wang F, Yang Y, Ling Y, Liu J, Cai X, Zhou X, Tang X, Liang B, Chen Y, Chen H, Chen D, Li C, Wang Z, Hu B, Zheng Y. Injectable and thermally contractible hydroxypropyl methyl cellulose/ Fe3O4 for magnetic hyperthermia ablation of tumors. Biomaterials. 2017;128:84e93.

Xue X, Huang Y, Bo R, Jia B, Wu H, Yuan Y, Wang Z, Ma Z, Jing D, Xu X, Yu W, Lin TY, Li Y. Trojan Horse nanotheranostics with dual transformability and multifunctionality for highly effective cancer treatment. Nat Commun. 2018;9:3653.

Yamanaka K, Nakahara T, Yamauchi T, Kita A, Takeuchi M. Antitumor activity of YM155, a selective small-molecule survivin suppressant, alone and in combination with docetaxel in human malignant melanoma models. Clin Cancer Res. 2011;17:5423–31.

Yan H, Shang W, Sun X, Zhao L, Wang J, Xiong Z, Yuan J, Zhang R, Huang Q, Wang K, Li B, Tian J, Kang F, Feng SS. "All-in-One" nanoparticles for trimodality imaging-guided intracellular photo-magnetic hyperthermia therapy under intravenous administration. Adv Funct Mater. 2018;28:1705710-1–12.

Yang HW, Hua MY, Hwang TL, Lin KJ, Huang CY, Tsai RY, Ma CC, Hsu PH, Wey SP, Hsu PW, Chen PY, Huang YC, Lu YJ, Yen TC, Feng LY, Lin CW, Liu HL, Wei KC. Non-invasive synergistic treatment of brain tumors by targeted chemotherapeutic delivery and amplified focused ultrasound-hyperthermia using magnetic nanographene oxide. Adv Mater. 2013;25:3605–11.

Yi G, Gu B, Chen L. The safety and efficacy of magnetic nano-iron hyperthermia therapy on rat brain glioma. Tumor Biol. 2014;35:2445–9.

Zeng Q, Baker I, Loudis JA, Liao Y, Hoopes PJ, Weaver JB. Fe/Fe oxide nanocomposite particles with large specific absorption rate for hyperthermia. Appl Phys Lett. 2007;90:233112.

Zhou P, Zhao H, Wang Q, Zhou Z, Wang J, Deng G, Wang X, Liu Q, Yang H, Yang S. Photoacoustic-enabled self-guidance in magnetic hyperthermia Fe@Fe3O4 nanoparticles for theranostics *in vivo*. Adv Healthcare Mater. 2018;7:e1701201.

Zhu L, Ma J, Jia N, Zhao Y, Shen H. Chitosan-coated magnetic nanoparticles as carriers of 5-fluorouracil: preparation, characterization and cytotoxicity studies. Colloids Surf B Biointerface. 2009;68:1–6.

Chapter 15
Emerging Role of Aminolevulinic Acid and Gold Nanoparticles Combination in Theranostic Applications

Lilia Coronato Courrol, Karina de Oliveira Gonçalves, and Daniel Perez Vieira

Abstract Major advancements in theranostic agents for cancer and inflammatory processes, such as atherosclerosis, have been reported in recent years. The theranostic agents can be used for both diagnosis and treatment. In this chapter, we show that cancer and atheroma plaques exhibit accumulation of protoporphyrin IX (PpIX), which is transferred to the blood and feces. PpIX may therefore be a biomarker for atherosclerosis and cancer, enabling minimally invasive and inexpensive diagnosis. PpIX is the immediate precursor in the heme biosynthesis. Tumor cells tend to retain more PpIX owing to cellular energy metabolism. Analysis of these spectroscopic properties of PpIX enables monitoring of its concentrations in tissues and biological fluids. Additionally, increases in PpIX fluorescence are proportional to tumor progression; thus, this tool can be used to detect and stage tumors. Exogenous administration of aminolevulinic acid (ALA) enhances endogenous PpIX production and allows its use in photodynamic diagnosis and as a photosensitizer/sonosensitizer for photodynamic/sonodynamic therapies. However, ALA cannot easily penetrate target cells. Accordingly, we propose the use of gold nanoparticles produced with ALA or its methyl ester to solve this problem.

Keywords Fluorescence · Porphyrin · Gold nanoparticle · Atherosclerosis · Prostate cancer

L. C. Courrol (✉) · K. de Oliveira Gonçalves
Laboratory of Applied Biomedical Optics, Physics Department, Federal University of São Paulo, Diadema, São Paulo, Brazil

D. P. Vieira
Radiobiology Laboratory, Nuclear and Research Institute, IPEN/CNEN-SP, São Paulo, Brazil

M. Rai, B. Jamil (eds.), *Nanotheranostics*,
https://doi.org/10.1007/978-3-030-29768-8_15

15.1 Introduction

According to the World Health Organization (WHO), chronic noncommunicable diseases (NCDs) (cardiovascular diseases, chronic respiratory, cancer, diabetes, between others) account for about 71% of all deaths worldwide, estimating 41 million deaths annually. Atherosclerotic cardiovascular disease and cancer are the leader cause of these deaths (Sum et al. 2018). Their etiologies are complexes and involves non-modifiable risk factors (genetics, sex, and age) and modifiable (smoking, physical inactivity, unhealthy eating habits, and excessive alcohol consumption). The socio-economic costs associated with these diseases have influences in the economies of the countries.

The first description of an artery with cholesterol plaques, responsible for the atherosclerosis, was done around 1790 by Edward Jenner (Linden et al. 2014). Atherosclerosis is the main cause of most cardiovascular diseases. Atherosclerosis is an inflammatory disease and not a degenerative process (Ross 1999). Understanding this inflammatory phenomenon and of the type of complication that causes contributed enormously to new therapeutic solutions that greatly improved the evolution of the disease. Atherogenesis and neoplasia have similarities, such as oxidative stress, cellular damage that results in angiogenesis (Tapia-Vieyra et al. 2017), and accumulation of protoporphyrin IX (PpIX), supporting the use of similar approaches for diagnosis and therapy (Ross et al. 2001).

Application of nanoparticles to diagnose and treat atherosclerosis and cancer has shown growing potential. In this chapter, we describe the use of metallic nanoparticles produced with aminolevulinic acid (ALA) or its methylated form (MALA) as nanotheranostic agents. These nanomaterials can act as PpIX precursors, accumulating in rapidly growing tissues, such as cancerous and atherosclerotic tissues. PpIX is a fluorescent molecule and due its specific accumulation in tumor or atherosclerotic plaques can be used to distinguish tumor from normal tissues in fluorescence-guided surgery. PpIX is an ideal photosensitizer and sonosensitizer of photodynamic and sonodynamic therapies (PDT and SDT), respectively. We also describe the properties of PpIX, the mechanisms of PpIX accumulation in tumor tissues, transference of PpIX to the blood and feces, and the use of these properties in theranostic applications. Details of nanoparticles synthesis, characterization, and cytotoxicity will also be presented.

15.2 Fluorescence

Fluorescence spectroscopy has been used extensively for the analysis of biological and industrial compounds (Boens et al. 2007). This approach is known to be highly sensitive and have a low detection limit. Indeed, compared with ultraviolet (UV)/ visible spectrophotometry, fluorescence exhibits detection limits three orders of magnitude smaller, i.e., in the range of ng/mL, owing to the low background signal of fluorescent measurements. Moreover, fluorescence spectroscopy is highly selective, with fluorescent molecules having characteristic wavelengths of excitation and/or emission, and shows a wide linear response range. Finally, the instruments required for this approach are simple and low cost compared with other analytical methods.

The characterization of biological tissues for cancer diagnosis by fluorescence was reported for the first time by Policard in 1924, in a study of the autofluorescence of tumors after illumination with UV light. Moreover, Xu et al. (1988) observed intense luminescence in the red range (630 nm) in samples from patients with cancer when excited in 450 nm. The observed fluorescence at 630 nm was correlated with the presence of porphyrinic compounds. Masilamani et al. (2004) studied the autofluorescence of blood components from healthy individuals and patients with cancer of different etiologies and reported that the blood of patients with gastric cancer, breast cancer, and Hodgkin's lymphoma showed increased emission bands at around 630 nm owing to the increased presence of endogenous porphyrin, depending on the tumor stage.

15.3 PpIX

PpIX is a compound formed by four pyrrolic rings linked by methylene bridges (Fig. 15.1).

The absorption spectrum of PpIX (Fig. 15.2) presents five bands: the Soret band (~400 nm) and four bands known as Q Bands (450 and 750 nm) (Courrol et al. 2007b). PpIX fluorescence is observed at ~635 and 705 nm. Analyzing changes in the spectroscopic properties of PpIX, such as emission intensity, differences in the

Fig. 15.1 Protoporphyrin IX structure (Wikipedia)

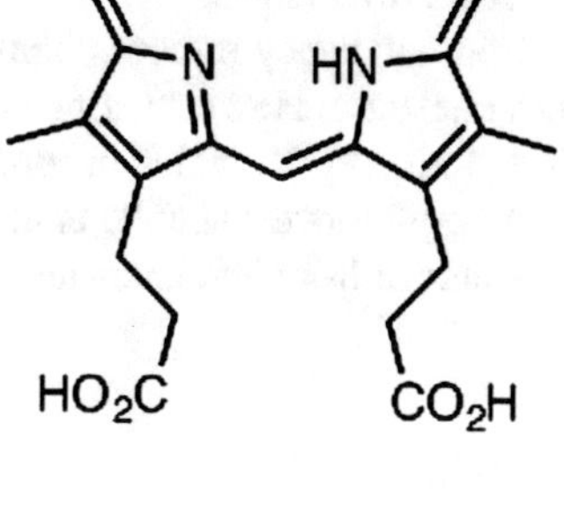

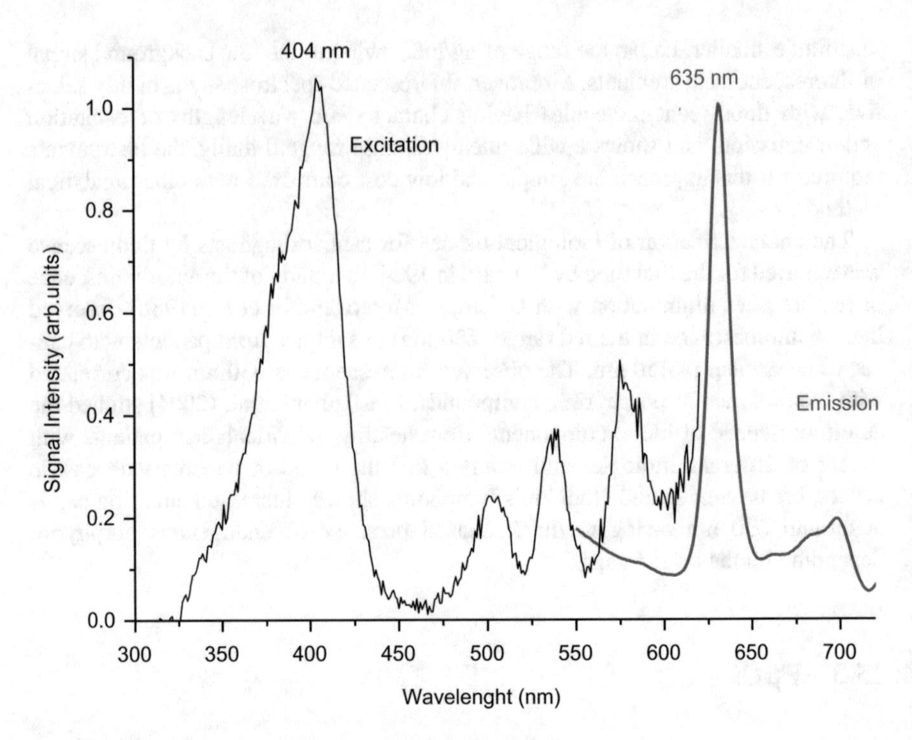

Fig. 15.2 The absorption and emission and spectra of PpIX

proportions of emission bands, and the presence of bands at distinct wavelengths, enables monitoring of the concentration of PpIX in tissues and biological fluids and facilitates disease diagnosis.

PpIX complexed with ionic iron forms heme, a constituent of hemoproteins, which play critical roles in oxygen transport, cellular oxidation and reduction, electron transport, and drug metabolism (Sachar et al. 2016). Synthesis of heme occurs primarily in the mitochondria and partly in the cytosol. Several enzymes are involved in the formation of heme, as illustrated in Fig. 15.3 (Taketani et al. 2007). The pathway is initiated by the synthesis of ALA from the amino acid glycine and succinyl-CoA. Heme is generated by the insertion of ferrous iron into the tetrapyrrole macrocycle of PpIX, a reaction catalyzed by ferrochelatase, which resides in the mitochondrial matrix.

Iron deficiency prevents heme formation because ferrous iron is a substrate for ferrochelatase. However, zinc can substitute for iron, and an increase in the formation of zinc PpIX, which accumulates in reticulocytes, has been observed in circulating erythrocytes (Labbe et al. 1999). An increase in erythrocyte zinc PpIX is an indicator of iron deficiency and the development of anemia.

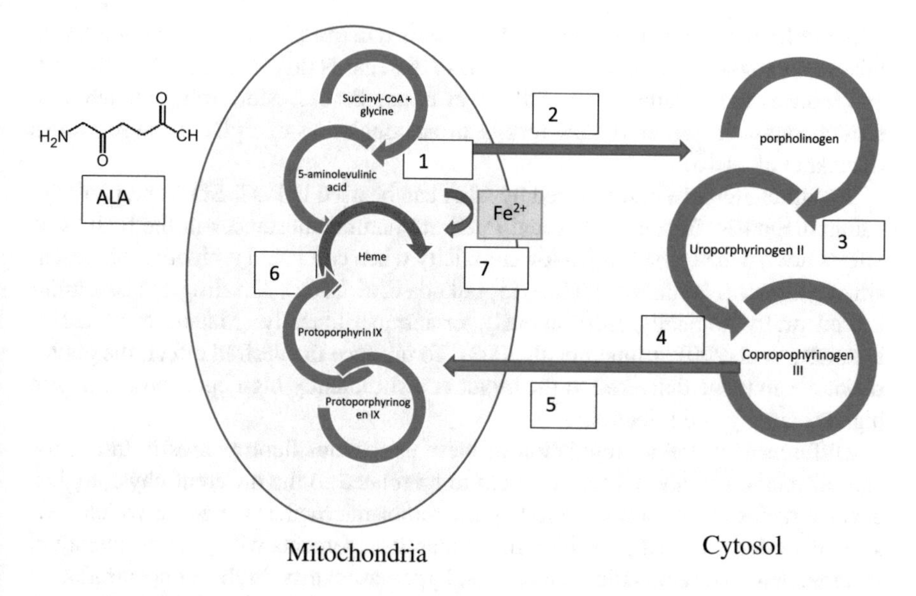

Fig. 15.3 Heme biosynthesis. Schematic illustrating the interaction of the heme biosynthesis pathway with exogenous 5-aminolevulinic acid to give intracellular Protoporphyrin IX. The process begins in the mitochondria. Many of the intermediate steps are cytoplasmic. Catalyzing enzymes: (1) delta-aminolevulinic acid synthase (ALA synthase). (2) δ-Aminolevulinic acid dehydratase (ALA dehydratase). (3) Uroporphyrinogen I synthase and uroporphyrinogen III cosynthase. (4) Uroporphyrinogen decarboxylase. (5) Coproporphyrinogen III oxidase. (6) Protoporphyrinogen IX oxidase. (7) Ferrochelatase

15.4 ALA

Fast-growing tissues, such as cancerous lesions and atheromatous plaques, produce increased numbers of monocytes, which in turn lead to PpIX accumulation and enhancement of the fluorescence of this fluorophore (Zhu et al. 2018).

In 1960, Ghadially and Neish (1960) suggested the use of ALA as a precursor to induce PpIX fluorescence in tumors. Today, ALA is one of the most selective photosensitizers for cancer treatment because accumulation of PpIX is higher in malignant cells than in normal tissues (Sunar et al. 2013; Gibbs et al. 2006). ALA is not a fluorophore by itself, but a natural precursor of PpIX in the pathway of heme biosynthesis (Fig. 15.3), and exogenous administration of ALA induces the accumulation of PpIX selectively in tumor cells.

Typically, heme synthesis regulates PpIX synthesis through "feedback" control and inhibits excess endogenous ALA formation through a negative feedback control mechanism. The presence of free heme inhibits the synthesis of ALA. When all heme synthesis occurs, there is a negative "feedback" and an increase in the synthesis of ALA. Under these conditions, the ALA concentration increases, leading to enhanced heme synthesis. Under normal conditions, heme demand controls the heme synthesis rate and consequently the PpIX synthesis rate. If the conversion rate of the PpIX in heme is lower than the synthesis rate of PpIX, PpIX accumulates. In this way, exogenous administration of ALA disregards this "feedback" mechanism and induces the accumulation of PpIX in tumor tissues. Moreover, deficiency of ferrochelatase in tumor tissues results in accumulation of PpIX in these tissues (Sachar et al. 2016).

Accumulation of PpIX induced by ALA can be used in PDT, SDT, and photodynamic diagnosis. Because ALA and PpIX are natural substances in the body, side effects are reduced, resulting in lower toxicity when compared with other photosensitizing agents (Namikawa et al. 2015; Fukuda et al. 2006). The drug can be administered orally, topically, intravenously, or intraperitoneally (Mateus et al. 2014; Donnelly et al. 2005; Lippert et al. 2003). To produce the desired effect, the photosensitizer must be delivered to the target at a sufficiently high concentration, with high specificity and selectivity.

Differences in the accumulation of these exogenous fluorophores in tumor tissues in relation to normal tissues seem to be related to the different physiological characteristics of the tumor, including the tumor microenvironment in which cells show disordered growth, resulting in cellular arrangements with greater interstitial volume; deficient lymphatic drainage and hypervascularity; higher concentration of low-density lipoprotein receptors than in the normal tissue (porphyrin derivatives exhibit high affinity for low-density lipoproteins); and lower pH than the healthy tissue (Padera et al. 2016). In this acidified environment, hydrophilic photosensitizing agents become more hydrophobic (lipophilic) through protonation, thus facilitating entry into the cells (Courrol et al. 2007a).

The efficiency of fluorescence in the diagnosis of tumor tissues is dependent on the synthesis capacity of PpIX in the target tissue. Different tissues can accumulate different amounts of PpIX. Gibbs et al. (2006) studied the production capacity of PpIX in eight different cell types, including breast cancer, prostate cancer (DU145), and brain cancer cells, and the amount of PpIX varied more than tenfold between the different cell types after administration of ALA. Additionally, Chakrabarti et al. (1998) studied the production of PpIX in vitro in benign prostatic cells (TP-2) and two more types of prostate cancer cells (LNCaP and PC-3) and showed that these cells produced different amounts of PpIX. Benign cells always showed the lowest PpIX levels, independent of the incubation time with ALA.

Many studies have been conducted using ALA as a photosensitizing agent, mainly for use in PDT. However, studies for the photodynamic diagnosis of cancer have also been reported in vitro and in vivo (Zaak et al. 2004; Fukuda et al. 2006).

15.5 PDT and SDT

PDT is a light-activated photochemical reaction that is used for the selective destruction of tissues or cells. Three components are required to perform PDT: a photosensitizing agent, a source of light, and oxygen (Huang et al. 2015). These three components together cause apoptosis of tumor cells or microorganisms via the actions of reactive oxygen species promoted by energy transfer or electron transfer from the photosensitizer in the excited state to oxygen molecules present in the medium.

Studies of PDT were first carried out by Raab and Tappeiner in 1900 (Ackroyd et al. 2001). These researchers observed the death of the Protozoan *Paramecium caudatum* after exposure to sunlight and in the presence of acridine dye. The presence of light modified the effects of the dye, demonstrating photosensitization. Tappeiner expanded on these discoveries, performing other experiments, and demonstrated the need for the presence of oxygen for the reaction to occur.

In 1911, Haussmann et al. reported the use of hematoporphyrins in combination with light to kill tumor cells (Daniell and Hill 1991). Lipson and Baldes (1960) used PDT to synthesize the first photosensitizers for clinical use; photofrin I, also derived from hematoporphyrin, was found to have applications in tumors, where it emitted fluorescence (Lipson and Baldes 1960). Dougherty (1987) purified this compound, producing photofrin II, which was the first drug approved by the US Food and Drug Administration for the treatment of cancers. Subsequently, several studies and clinical tests have been conducted.

The clinical use of these photosensitizing agents is limited by the disadvantages they present, such as low selectivity, prolonged action time, low absorption, and difficulties in developing appropriate formulations for drug use in vivo (Abrahamse and Hamblin 2016). Therefore, the development of new photosensitizing agents with minimal or no side effects is essential for the success of therapy and the photodynamic diagnosis of cancer.

Kennedy and Pottier (1992) revolutionized this technique when they proposed the use of ALA, which is the precursor of PpIX. The development of topical photosensitizing using ALA, which does not induce prolonged generalized photosensitivity, led to a breakthrough in the popularity of PDT in dermatology (Thunshelle et al. 2016).

Although PDT has been developed for applications in cancer therapy, such as in esophageal and lung cancers, other applications have also been reported, such as the treatment of mycoses and macular degeneration of the retina, removal of warts in the larynx, and destruction of bacterial infestations resistant to traditional antibiotic-based treatments (Peng et al. 2001; Namikawa et al. 2015; Wakui et al. 2010; Feuerstein et al. 2011).

Peng et al. (2011) performed PDT with rabbits consuming a hypercaloric diet after administration of ALA intravenously. Notably, at 2 h after administration, PpIX fluorescence increased 12-fold compared with that in arteries without atheroma. Thus, PpIX-induced ALA could be detected and reflected the macrophage contents of the plaque. Because ALA and PpIX are natural substances in the body, the side effects are minimal and thus cause lower toxicity when compared with other photosensitizing agents (Georges et al. 2019).

Changing the energy source to excite sensitizers from light to low-frequency ultrasound led to the development of SDT (Costley et al. 2015). SDT uses ultrasound, oxygen, and sonosensitizers for the treatment of tumors. The possible mechanisms of SDT therapy include the generation of reactive oxygen species radicals derived from sonosensitizer agents, which cause peroxidation of the membrane lipid chain via peroxyl radicals and/or alkoxyls, and the physical destabilization of the cell membrane by sonosensitizers (Chen et al. 2014).

Ultrasounds are mechanical waves with periodic vibrations with frequencies equal to or greater than 20 kHz. This radiation type has good tissue penetration capacity without energy attenuation, as occurs with visible light in PDT (Foster et al. 2000). Ultrasound has been used for diagnostic imaging of soft tissues and therapeutic applications related to the thermal effects caused by ultrasound absorption. Hyperthermia caused by ultrasound application occurs when the tissue temperature increases to 40–45 °C and may have roles in inflammation relief. The increase in temperature to ~60 or 85 °C during exposure ultrasound for a few seconds generates thermal ablation, which has been used for noninvasive surgery, such as necrosis of solid tumors, sealing of blood vessels, and correction of arrhythmias (Rosenthal et al. 2004).

Ultrasonic waves cause a physical phenomenon known as cavitation, which results in the formation of bubbles or cavities (steam or gas bubbles) in a liquid through a reduction in the total pressure. Antitumor drugs can be encapsulated in microspheres and transported to the target through circulation, and ultrasound can be used to induce collapse and release the drug. The collapse of cavitation bubbles allows the release of the drug and the permeabilization of the cells (Lawrie et al. 2000).

Focusing the ultrasound on a defined region and choosing compounds with tumor affinity can enhance drug cytotoxicity with ultrasound with minimal damage to neighbor healthy tissue. SDT is based on the synergistic effects of ultrasound and chemical compounds or sonosensitizers. After ultrasonic irradiation, some drugs can create reactive oxygen species, such as superoxide radicals and singlet oxygen (Yumita et al. 2012).

SDT applications were demonstrated for the first time by Umemura et al. (1989). In vitro and in vivo data have demonstrated the effectiveness of SDT in the treatment of cancer (McEwan et al. 2016; Zhu et al. 2010; Li et al. 2015; Ju et al. 2016; Mehrad and Farhoudi 2016; Wang et al. 2017; Trendowski 2014; Umemura et al. 1996; Wood and Sehgal 2015; McHale et al. 2016). Methods to improve tumor oxygenation during SDT have benefits since oxygen essentially fuels the generation of ROS. In 2002, Arakawa et al. (Arakawa et al. 2002) reported that SDT inhibits neointimal hyperplasia in a rabbit model of iliac artery stent, suggesting that SDT could also be beneficial in cardiovascular disease and cancer. SDT displays good safety without obvious side effects. In a pilot clinical trial recruiting patients suffering atherosclerotic peripheral artery disease, combination of SDT with atorvastatin efficiently reduced the progression of atherosclerotic plaques within 4 weeks, and its efficacy was able to last for at least 40 weeks. Recently, Wang et al. (2017) observed rapid inhibition of atherosclerotic plaque progression by SDT.

Porphyrins, which have traditionally been used as sensitizing agents in PDT, have also been evaluated in ultrasound-induced reactions because ultrasonic energy can cause electron excitation of porphyrins and initiate a photochemical process resulting in the formation of singlet oxygen (Sun et al. 2018). In vitro studies have indicated that ROS may play important roles in ultrasound-induced cell death in the presence of PpIX. Moreover, an in vivo study by Li et al. (2015) showed that ALA plus SDT induced macrophage apoptosis and enhanced apoptotic cell clearance, resulting in reduced macrophage build-up within plaques. In addition, these authors found that an ultrasound intensity of 1.5 W/cm^2 exerted the greatest ability to reduce macrophage amounts in plaques by 71%.

15.6 ALA and Nanoparticles

The biggest disadvantage of the ALA is its limited ability to penetrate into target cells (Risaliti et al. 2018) owing to the hydrophilic nature of ALA. Various ALA modifications and different carriers have been reported; however, the most successful ALA derivatives are its esters, i.e., MALA and the hexyl ester hexyl aminolevulinate (Wachowska et al. 2011). The advantages of ALA derivatives include increased lipophilicity and higher membrane and skin permeability. In ALA-induced PpIX formation, ALA esters show varying effects depending on esterase activity, which varies within tissues and cell lines (Korbelik and Dougherty 1999).

Another way to increase the effectiveness of ALA is to attach ALA or ALA esters to gold nanoparticles (AuNPs). On the nanometric scale, gold shows special characteristics owing to surface plasmon resonance (SPR) and the resonant oscillation of electrons of the conduction band of the metal, which gives rise to a sharp, intense absorption band in the visible range (~530 nm) depending on the shape, size, and agglomeration of particles (Navarro and Lerouge 2017). AuNPs have been used as biomarkers in the diagnosis of heart and brain diseases, cancers, and infectious agents (Smolsky et al. 2017; Vio et al. 2017; Hasanzadeh et al. 2017). Differences in the size, shape, and surface properties of AuNPs can be manipulated for specific therapeutic purposes, depending on the nature of preclinical or clinical applications (Wozniak et al. 2017). Recent data indicate that AuNPs covered with polyethylene glycol show pharmacokinetic properties suitable for systemic injection and exhibit excellent biodistribution in vivo (Zheng et al. 2012). Additionally, AuNPs can also be used in PDT and SDT, yielding surface drugs that may be photosensitizers or sonosensitizers, in addition to promoting on-site heating.

Several authors have described the applications of ALA-conjugated AuNPs for therapy of cancer (Oo et al. 2008; Mohammadi et al. 2013, 2017; Benito et al. 2013; Thunshelle et al. 2016; Wu et al. 2017; Xu et al. 2016; Zhang et al. 2015). Notably, it is possible to synthesize AuNPs directly from ALA or MALA, without any other toxic chemicals, using a photoreduction method (Karina et al. 2015; Goncalves et al. 2015, 2018).

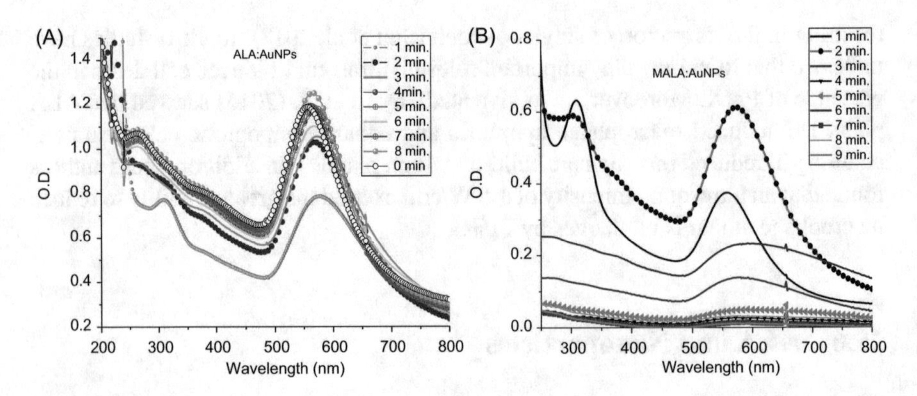

Fig. 15.4 UV-vis spectra of nanoparticle solutions (**a**) ALA:AuNPs, (**b**) MALA:AuNPs

15.7 Synthesis and Characterization of ALA AuNPs

ALA AuNPs can be produced by photoreduction with a white light from xenon (Xe) lamp irradiation. Before Xe illumination, an absorption band at approximately 310 nm, characteristic of ALA, is observed. During Xe light exposure, the solution color changes from colorless to purple, and after illumination ends, an absorption peak at approximately 540 nm appears due to the SPR effect, indicating the formation of AuNPs (Hou and Cronin 2013). Light acts as a catalyst for metal reduction (oxidation/photoinduced reduction). At the same time, an increase in the temperature of the solution promotes agitation of the particles.

Figure 15.4 shows the formation of AuNPs with ALA and MALA. Notably, for the solution with ALA, illumination times from 1 to 9 min lead to the formation of nanoparticles; longer illumination times are associated with higher absorbance intensities (i.e., number of nanoparticles present in the solution). With MALA, irradiation times of 1, 2, and 3 min promote the formation of nanoparticles, whereas longer illumination times (4–9 min) promote particle agglomeration. ALA-conjugated AuNPs and MALA-conjugated AuNPs can be obtained after illumination at a pH of approximately 3.14 owing to the release of H^+ by the oxidation of ALA/MALA.

The results obtained adding polyethylene glycol (100 mg) to the solution containing $HauCl_4$ and ALA or MALA and illumination for 5 min before and after changing the pH to ~7.0 are shown in Fig. 15.5. The pH adjustment promotes displacement to the left and narrowing of the absorption band of AuNPs with ALA and with MALA, indicating an improvement in the synthesis in relation to size and polydispersity. For MALA, pH adjustment promotes the formation of nanoparticles in solutions irradiated between 5 and 9 min, and the presence of bands in the region of 700 nm indicates the formation of no spherical particles, such as prims, in the solution irradiated for 2 min (Table 15.1).

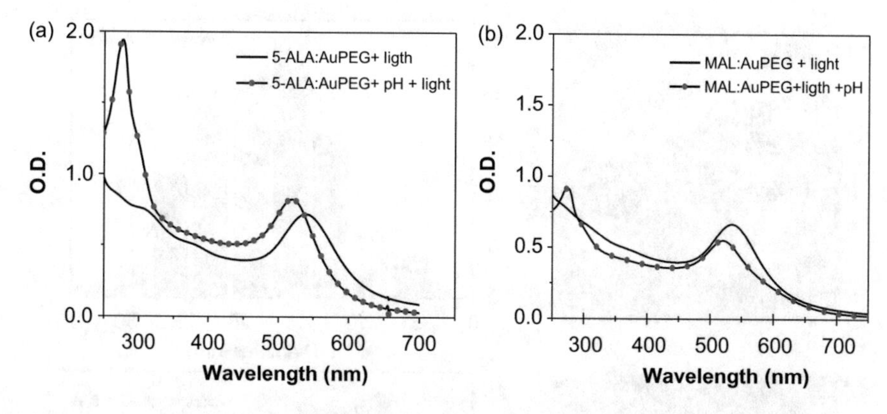

Fig. 15.5 Absorption spectra of nanoparticle solutions (5 min) with PEG and pH changed from ~3.1 to 7.2: (**a**) ALA:AuNPs and (**b**) MALA:AuNPs

Table 15.1 Concentration of reagents, time of exposure to light for solutions

Solution	Reagents	Exposure time (light)
ALA:AuNPs	15.0 mg $HAuCl_4$	0–9 min (10 mL)
	45.0 mg ALA	
	100 mL solution	
MALA:AuNPs	15.0 mg $HAuCl_4$	1–10 min (10 mL)
	45.0 mg of MALA	
	100 mL solution	

15.7.1 Physicochemical Characterization of Nanoparticles

15.7.1.1 Morphological Analysis of Nanoparticles

Morphological evaluation of the samples was performed using transmission electron microscopy images (Fig. 15.6). The nanoparticles obtained from synthesis with ALA and MALA were spherical, with sizes of around 18 nm, and showed good homogeneity of sizes.

15.7.1.2 Zeta Potential

The physicochemical parameters of samples irradiated for 2 min are presented in Table 15.2. The measurements were performed at 25 °C using samples diluted in Milli-Q water, and the results are presented as the means of ten measurements. The polydispersity index, which provides information on the homogeneity of the size distribution, was low (<0.7) for all the dispersions obtained, indicating the formation of monodisperse systems. The zeta potential values indicated moderate stability.

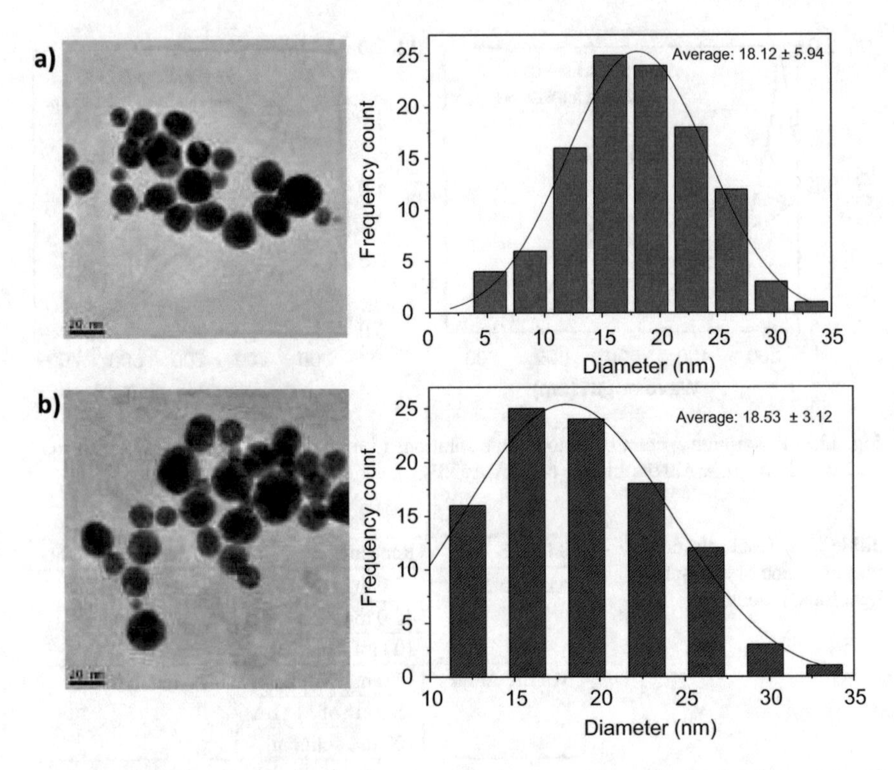

Fig. 15.6 MET images of samples irradiated by 2 min and size distribution histogram: (**a**) ALA:AuNPs and (**b**) MALA:AuNPs

Table 15.2 Sizes, zeta potential, polydispersity index, and hydrodynamic diameter of gold nanoparticles with ALA and MALA

	Samples size (nm)	Zeta potential (mV)	IP	Hydrodynamic diameter (nm)
ALA:AuNPs	18.12 ± 5.94	−23.1 ± 1.00	0.483	56.84
MALA:AuNPs	18.53 ± 3.12	−21.9 ± 0.98	0.437	51.16

15.7.1.3 Infrared Spectroscopy

The presence of functional groups in nanoparticle solutions was determined by vibrational spectroscopy. The spectra for ALA, MALA, ALA:Au, and MALA:Au are shown in Fig. 15.7.

The spectra for ALA and MALA provided evidence of the main functional groups, i.e., the amine group (NH_2) at 3500 cm^{-1} and the carboxyl group (–COOH) at 1710 cm^{-1}. These two groups are important for bonding with metals through electrostatic interactions. For ALA:AuNPs and MALA:AuNPs, the first region (3600–2700 cm^{-1}) is associated with vibrations of axial deformation in the hydrogen atoms linked to carbon, oxygen, and nitrogen (C–H, O–H, and N–H). These spectra also contain a band for NH_2 representing the free amines (~3440 cm^{-1}), for

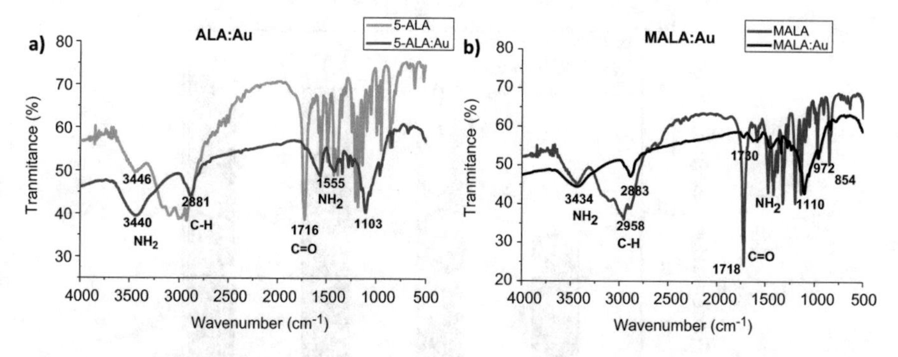

Fig. 15.7 FTIR spectra of: (**a**) ALA and ALA:AuNPs, (**b**) MALA and MALA:AuNPs and their respective functional groups

ALA, MALA, ALA:Au, and MALA:Au. In the region of 2300–1900 cm^{-1}, the axial deformation vibrations of double bonds and angular deformations of N–H and –NH are present. Additionally, the C=O band of the carboxyl (~1716 cm^{-1}) present in ALA and MALA decreased in the spectra for the nanoparticles, suggesting that the bonds with the nanoparticles were made using this functional group.

15.7.2 Liquid Biopsy Using ALA and ALA-Conjugated AuNPs

In a study by Courrol et al. (2007b), PpIX extracted from the blood of nude mice inoculated with DU145 prostate cancer cells revealed that PpIX fluorescence intensity was higher in animals with prostate cancer tumors in relation to that in animals without tumors and that the PpIX concentration was correlated with tumor size. Although from week 1 after induction of the tumor it was possible to confirm changes in the intensity of blood emission in comparison with that in the control group (animals inoculated with physiological serum), a significant difference in PpIX signal was observed only after 3 weeks of tumor growth.

In order to improve the sensitivity of this approach, ALA was used as a pro-drug capable of inducing the selective production of PpIX in the tumoral region. The blood PpIX fluorescence area obtained from animals inoculated with DU145 prostate cancer cells or with phosphate-buffered saline as a control and treated with ALA, as a function of days of tumor growth, is shown in Fig. 15.8. The difference in the signal between tumor-bearing and control animals was statistically significant after 1 week, demonstrating that ALA could be used in the photodynamic diagnosis of prostate cancer to detect the first malignant alterations that occur in the tissue.

ALA selectively targets the tumor region through differences in pH and is then metabolized to PpIX. Excess PpIX in the tissue is transferred to the blood, as demonstrated by experiments with culture cells treated with 60 mg/kg or 100 mg/mL ALA, which produced PpIX and then transferred this product to the blood (de Oliveira Silva et al. 2011). Moreover, as shown in Fig. 15.9, PpIX has also been detected in human feces (Gotardelo et al. 2018).

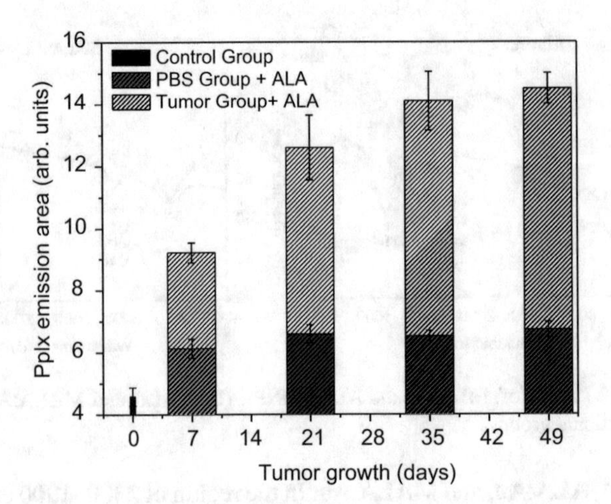

Fig. 15.8 Integrated area of the PpIX emission spectrum, extracted from the blood of the animals according to the days of tumor growth, in days. Day 0 corresponds to the control group (animals before surgical intervention) and days 7, 21, 35, and 49 correspond to the animals that received, or PBS or animals inoculated with DU145 cells. Except for control animals, all others received the doses of ALA orally

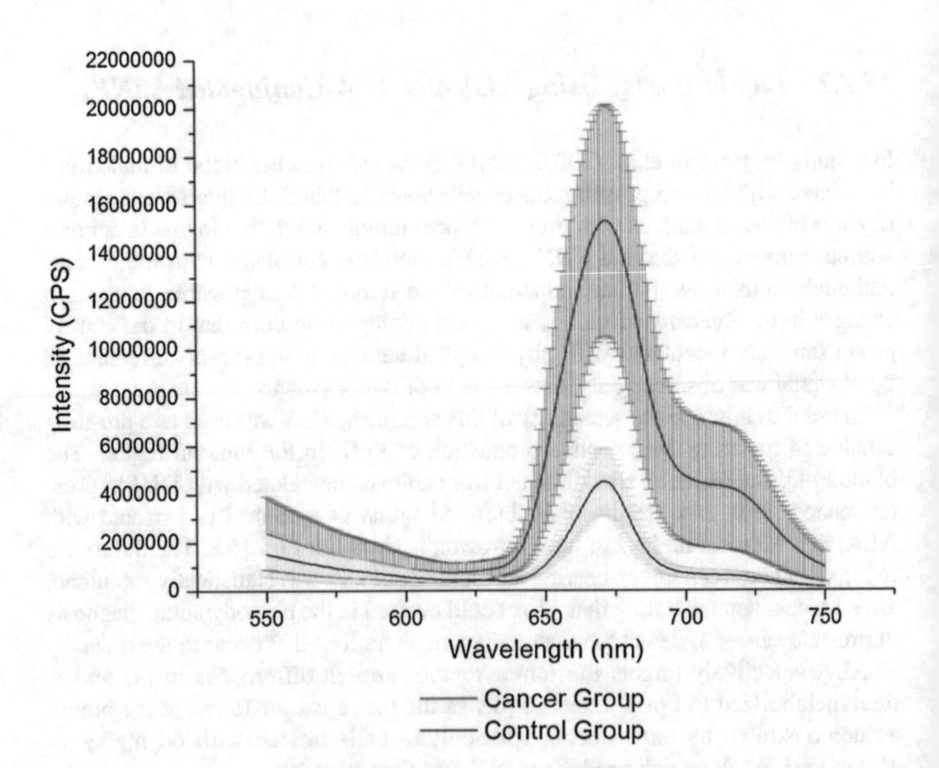

Fig. 15.9 Analysis of porphyrins extracted from the human feces of control and cancer group

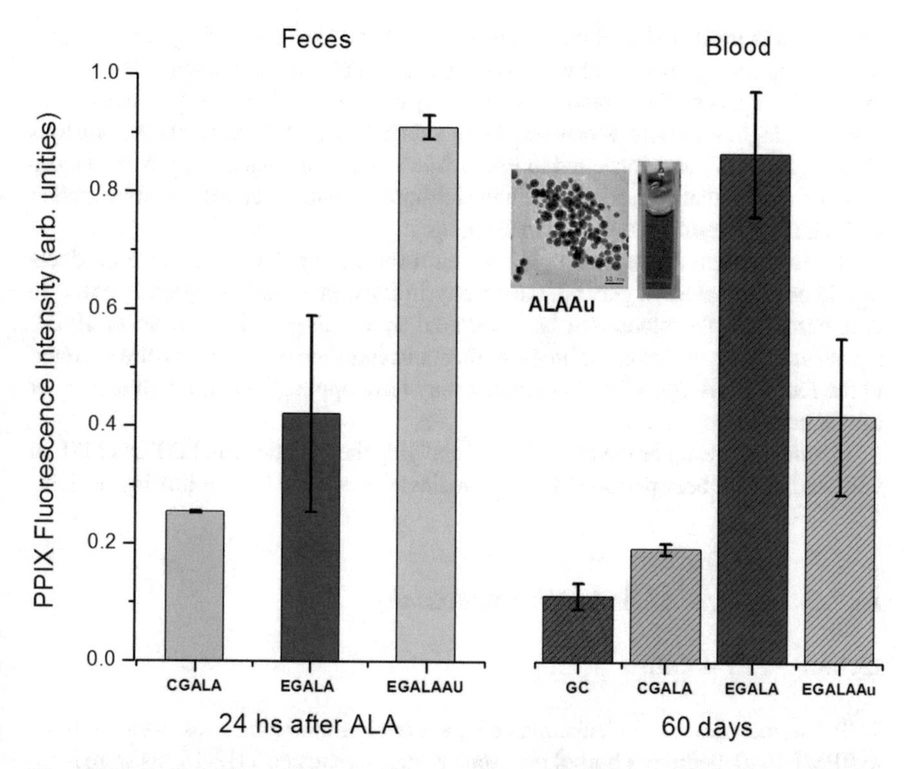

Fig. 15.10 (Left) Differences observed in the coproporphyrin fluorescence in feces collected 33 days after diet procedure beginning from the control and experimental groups before and after 24 h of ALA and ALA:AuNPs administration. (Right) Differences observed in the PpIX fluorescence in blood collected from the control (CG—normal diet) and experimental groups (EG—hypercholesteremic diet) before and after 24 h of ALA and ALA:AuNPs administration (60 days after diet procedure beginning)

Nascimento da Silva et al. (2014) concluded that PpIX is a potential marker of atherosclerosis, supporting the development of a minimally invasive and inexpensive diagnostic method based on detection of this target. In their study, New Zealand rabbits received a normal diet or a diet with 1% cholesterol, and variations in the quantity of PpIX extracted from tissues, such as blood and feces, with acetone were detected according to atherosclerosis staging.

Karina et al. (2015) administered ALA-conjugated AuNPs functionalized with polyethylene glycol to rabbits and showed that porphyrin release into the blood and feces was increased after ALA:AuNP administration (Fig. 15.10). For this experiment, blood samples were extracted from animals at day 0 and then 34 and 60 days after consumption of a specific diet, and fecal samples were collected after 33 days. The ALA:AuNps were prepared at a concentration of ~57.74 mg/kg ALA. ALA and ALA:AuNPs were orally administered, and blood was collected before and 4 h after ALA and ALA:AuNP administration. The results suggested that ALA was incorporated by the AuNPs, its structure was preserved, and rapid conversion into endogenous porphyrins occurred, overloading the synthetic pathway and leading to PpIX accumulation. This finding indicated that this method could aid in the early diagnosis and treatment of atherosclerosis with high sensitivity. Moreover, the functionalized AuNPs reached atheromatous plaques, and ALA was converted to PpIX. Selective accumulation of PpIX in plaques provided contrast between control animals and those with atherosclerosis. Administration of ALA did not result in any adverse reactions.

Because excess heme and PpIX are harmful, accumulation of these products should be eliminated. In general, the porphyrin fraction of the hemoglobin molecule is released into the blood and later secreted by the liver in bile (Welcker 1945). However, in atherosclerotic animals, higher concentrations of porphyrin are excreted in the feces. Thus, ALA/MALA:AuNPs may have applications in the diagnosis of cancer and atherosclerosis.

In order to investigate the possibility of applying these particles in PDT and SDT, in vitro studies have been performed. These studies are described in the following sections.

15.7.3 ALA/MALA:AuNP Cytotoxicity

15.7.3.1 THP-1 Cells

THP-1 human monocytic leukemia cells, a model of atherosclerosis, were cultured in RPMI-1640 medium. Phorbol myristate acetate-pretreated THP-1 cells were incubated with ALA/MALA:AuNPs for 24 h, and the cytotoxicity of ALA/MALA:AuNPs was evaluated using MTS assays. As shown in Fig. 15.11, both ALA and MALA did not produce cell death. For macrophages, the results showed that higher concentrations of ALA:AuNPs induced dramatic cytotoxicity, whereas lower concentrations showed only moderate effects. Similar results were observed for MALA:AuNPs. Macrophages are cells of phagocytic lineage, which remove dead cells, tissues, and particles foreign to the host by incorporation and digestion (Jiang et al. 2017; Virmani et al. 2006) and may have the tendency to incorporate nanoparticles easily.

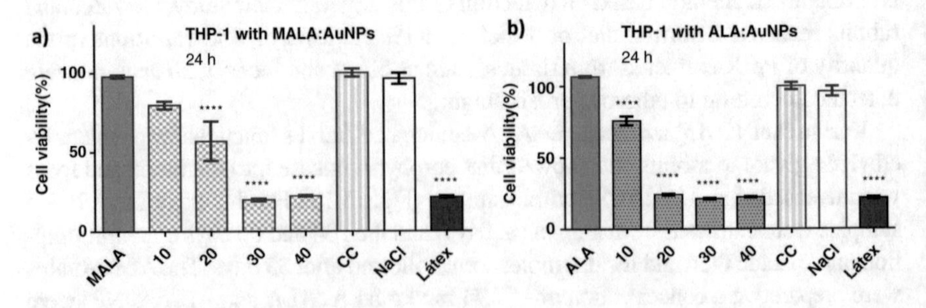

Fig. 15.11 Cell viability test (THP-1) for gold nanoparticles with ALA and MALA: (**a**) THP-1 macrophages with ALA:AuNPs and (**b**) with MALA:AuNPs, incubated for 24 h. The data were compared using the ANOVA test followed by the Dunnett's test with *P* values: ****<0.0001

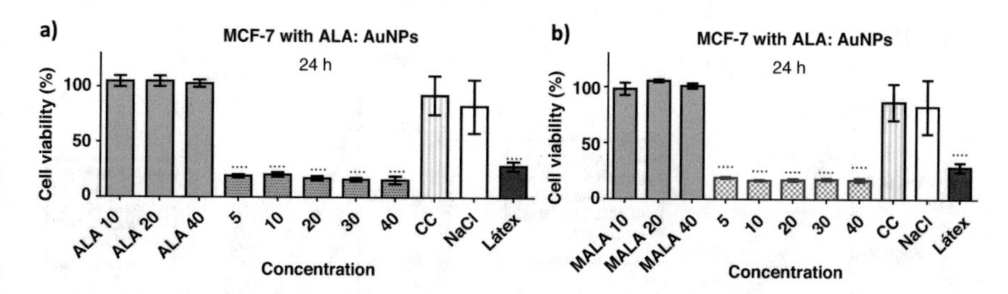

Fig. 15.12 Cell viability test (MCF-7) with gold nanoparticles: (**a**) with ALA incubated for 24 h. The data were compared using the ANOVA test followed by the Turkey test with *P*-value <0.0001 and (**b**) with MALA incubated for 24 h. The data were compared using the ANOVA test followed by the Dunnett's test with *P*-value <0.0001

15.7.3.2 MCF-7 Cells

In tests with breast cancer cells, we analyzed various concentrations of ALA and MALA to examine their cytotoxicities (Fig. 15.12). Notably, ALA and MALA alone showed no toxicity in MCF-7 cells; however, with the addition of AuNPs, cell death was increased in a concentration-dependent manner.

MCF-7 cells were more susceptible to death than THP-1 macrophages. Epithelial lineages have no way to combat oxidative stress (Hecht et al. 2016), and nanoparticles tend to accumulate in various cell types, with special affinity for macrophages and reticuloendothelial cells throughout the body. In tissues, they may accumulate in lymph nodes, bone marrow, spleen, adrenal glands, liver, and kidneys (Bailly et al. 2019). Different cell types may have different nanoparticle input mechanisms and death may occur through a series of events, which depend on the damage caused in them. Two types of cell death are commonly distinguished: accidental cell death or necrosis and apoptotic, programmed cell death.

During the process of death by necrosis, cells may have undergone some physical or chemical trauma, such as oxygen deprivation or extreme temperature, which causes the cell to absorb and accumulate a large amount of water. This in turn causes an increase in internal volume and consequently leads to the disruption and dissolution of organelles and their digestion by cell enzymes. Subsequently, the plasma membrane also breaks, and the cellular content is extravasated, triggering a process of inflammation, which recruits cells of the immune system, such as macrophages; these macrophages then phagocytose the remains of the cell. Apoptosis is an organized process, in which the contents of the cell are processed and compacted in small packets of membrane to "garbage collection" by cells of the immune system. This process is different from that of necrosis (death by injury), in which the cell contents are expelled, resulting in inflammation (Rock 2008).

Cell death caused by nanoparticles is expected to be related to apoptosis because the nanoparticles induce the production of ROS (Valko 2006; Li 2017).

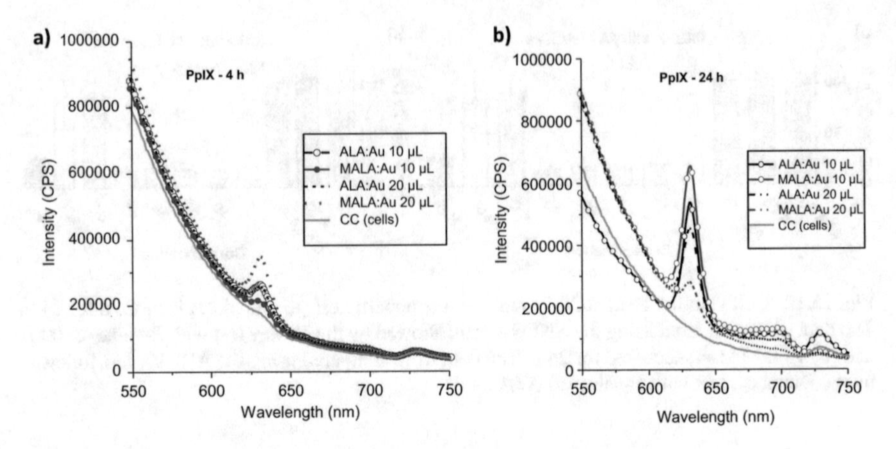

Fig. 15.13 THP-1 cells PpIX fluorescence: (**a**) 4 h of incubation with ALA/MALA:AuNPs (10 and 20 μL) and (**b**) 24 h of incubation (10 and 20 μL)

15.7.4 Extraction of PpIX

Possible alterations in the fluorescence of PpIX produced after the cells were incubated with nanoparticles were analyzed. The cells were cultivated, differentiated into macrophages, and incubated with nanoparticle solutions for 4 or 24 h. After incubation, PpIX was extracted using acetone. Supernatants were then analyzed in a fluorimeter under excitation at 400 nm, yielding the spectra presented in Fig. 15.13. The results showed that after 4 h of incubation, the nanoparticles already entered the cells, and part of their ALA/MALA was converted into PpIX. Concentrations of 10 and 20 μL were chosen because they showed lower toxicity in the cells. Increased fluorescence intensity, indicative of the production of PpIX, was observed after 4 h of incubation with MALA:AuNPs.

15.7.5 Effects of PDT on THP-1 Cells

THP-1 cells were incubated for 24 h with ALA/MALA or ALA/MALA:AuNPs and subjected to PDT with an LED at 590 nm for 2 min (~100 mW); this wavelength can excite AuNPs, promoting photothermal effects and PpIX production. As shown in Fig. 15.14, cells exposed to ALA:AuNPs or MALA:AuNPs were toxic and caused cell death.

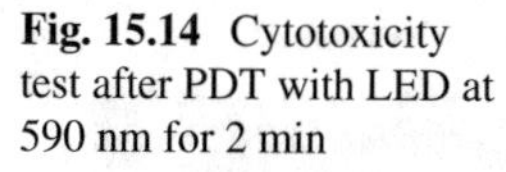

Fig. 15.14 Cytotoxicity test after PDT with LED at 590 nm for 2 min

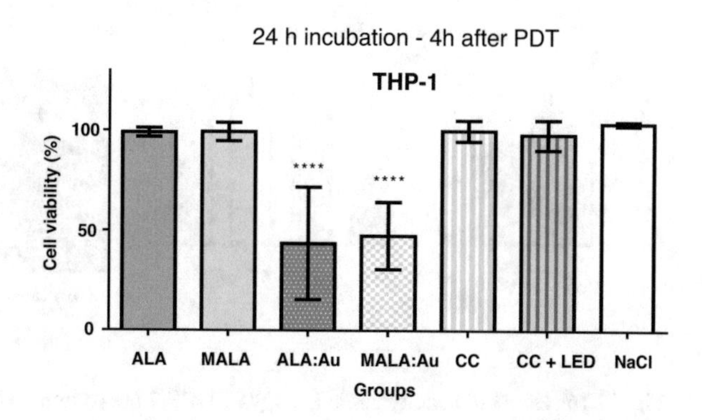

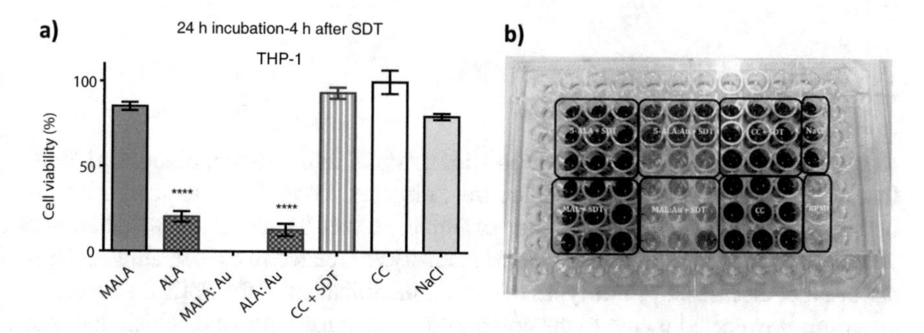

Fig. 15.15 Effect of sonodynamic therapy on macrophages: (**a**) 4 h of incubation and 24 post-therapy and (**b**) plaque; red—high cell viability; yellow—low cell viability

15.7.6 Effects of SDT

15.7.6.1 THP-1 Cells

Goncalves et al. (2018) showed that ultrasound combined with MALA:AuNPs exhibited impressive results in in vitro studies. SDT with MALA:AuNPs after 2 min of ultrasound exposure (1 MHz and 1 W/cm^2) culminated in total macrophage reduction (Fig. 15.15). Thus, SDT combined with MALA:AuNPs may have promising potential for atherosclerosis treatment.

15.7.6.2 MCF-7 Cells

In breast cancer cells, the effects of ALA/MALA:AuNPs plus SDT were evaluated (Fig. 15.16). Incubation of MCF-7 cells with ALA/MALA:AuNps for 24 h and ultrasound irradiation (1 MHz and 1 W/cm^2) for 2 min demonstrated that SDT was efficacious. Notably, NPs showed efficacy in both THP-1 and MCF-7 cells, and MALA:AuNPs produced greater cell death.

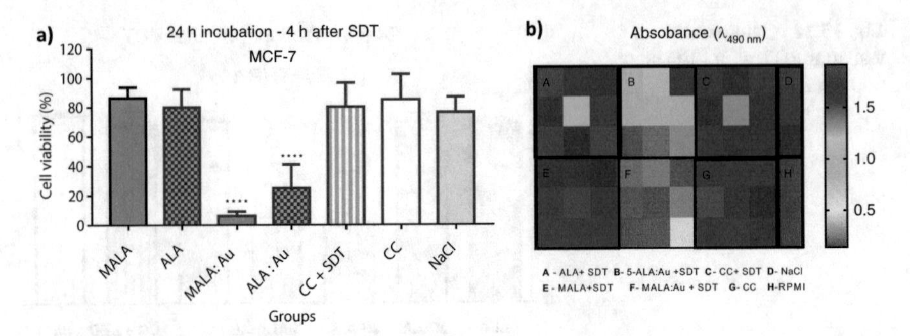

Fig. 15.16 Effect of sonodynamic therapy in MCF-7 breast tumor cells (5 μL). Cell viability test: (**a**) 24 h of incubation and 24 post-therapy and (**b**) graphic representation of cell death in the groups

15.8 Conclusions

In this chapter, we discussed the properties of PpIX in the tissue, blood, and feces that can be exploited to diagnose and treat diseases. Poor ability to screen for illnesses is also another problem. For example, it will be necessary to determine whether increases in PpIX fluorescence intensity are related to cancer, atherosclerosis, or other diseases. In some types of cancer, modification of the PpIX fluorescence spectrum is expected owing to the complexation of PpIX with zinc, which does not occur in the case of atherosclerosis. PpIX fluorescence can have applications as a diagnostic tool. Moreover, the use of AuNPs may permit diagnosis by contrast methods, such as computed tomography, magnetic resonance tomography, positron emission tomography, or fluorescence angioscopy, enabling recognition of disease. Additionally, ALA/MALA:AuNps show high stability, which may affect drug administration, although ALA is typically prepared fresh, and its pH is adjusted as needed.

From the results of this analysis, we concluded that ALA/MALA:AuNP-mediated PDT and SDT could be safe and effective treatment modalities for cancer and atherosclerosis. To evaluate the effectiveness of these therapies, further clinical trials must be carried out using larger sample sizes with long-term follow-up.

References

Abrahamse H, Hamblin MR. New photosensitizers for photodynamic therapy. Biochem J. 2016;473:347–64. https://doi.org/10.1042/BJ20150942.

Ackroyd R, Kelty C, Brown N, Reed M. The history of photodetection and photodynamic therapy. Photochem Photobiol. 2001;74:656–69.

Arakawa K, Isoda K, Ito T, Nakajima K, Shibuya T, Ohsuzu F. Fluorescence analysis of biochemical constituents identifies atherosclerotic plaque with a thin fibrous cap. Arterioscl Thromb Vasc Biol. 2002;22:1002–7.

Bailly AL, Correard F, Popov A, Tselikov G, Chaspoul F, Appay R, Al-Kattan A, Kabashin AV, Braguer D, Esteve MA. In vivo evaluation of safety, biodistribution and pharmacokinetics of laser-synthesized gold nanoparticles. Scientific Reports 2019;9.

Benito M, Martin V, Blanco MD, Teijon JM, Gomez C. Cooperative effect of 5-aminolevulinic acid and gold nanoparticles for photodynamic therapy of cancer. J Pharm Sci. 2013;102:2760–9. https://doi.org/10.1002/jps.23621.

Boens N, Qin W, Basaric N, Hofkens J, Ameloot M, Pouget J, Lefevre J-P, Valeur B, Gratton E, vandeVen M, Silva ND Jr, Engelborghs Y, Willaert K, Sillen A, Rumbles G, Phillips D, Visser AJWG, van Hoek A, Lakowicz JR, Malak H, Gryczynski I, Szabo AG, Krajcarski DT, Tamai N, Miura A. Fluorescence lifetime standards for time and frequency domain fluorescence spectroscopy. Anal Chem. 2007;79:2137–49. https://doi.org/10.1021/ac062160k.

Chakrabarti P, Orihuela E, Egger N, Neal DE, Gangula R, Adesokun A, Motamedi M. Delta-aminolevulinic acid-mediated photosensitization of prostate cell lines: implication for photodynamic therapy of prostate cancer. Prostate. 1998;36:211–8.

Chen HJ, Zhou XB, Gao Y, Zheng BY, Tang FX, Huang JD. Recent progress in development of new sonosensitizers for sonodynamic cancer therapy. Drug Discov Today. 2014;19:502–9. https://doi.org/10.1016/j.drudis.2014.01.010.

Costley D, Mc Ewan C, Fowley C, McHale AP, Atchison J, Nomikou N, Callan JF. Treating cancer with sonodynamic therapy: a review. Int J Hyperth. 2015;31:107–17. https://doi.org/10.3109/02656736.2014.992484.

Courrol LC, de Oliveira Silva FR, Bellini MH, Mansano RD, Schor N, Vieira Junior ND, Kessel D. Blood porphyrin luminescence and tumor growth correlation - art. no. 64270Y. Optical Methods for Tumor Treatment and Detection: Mechanisms and Techniques in Photodynamic Therapy XVI, 2007a. PROCEEDINGS OF THE SOCIETY OF PHOTO-OPTICAL INSTRUMENTATION ENGINEERS (SPIE) Volume: 6427 Páginas:Y4270–Y4270.

Courrol LC, de Oliveira Silva FR, Coutinho EL, Piccoli MF, Mansano RD, Vieira Junior ND, Schor N, Bellini MH. Study of blood porphyrin spectral profile for diagnosis of tumor progression. J Fluoresc. 2007b;17:289–92. https://doi.org/10.1007/s10895-007-0171-7.

Daniell MD, Hill JS. A history of photodynamic therapy. Aust N Z J Surg. 1991;61:340–8.

de Oliveira Silva FR, Bellini MH, Nabeshima CT, Schor N, Vieira ND Jr, Courrol LC. Enhancement of blood porphyrin emission intensity with aminolevulinic acid administration: a new concept for photodynamic diagnosis of early prostate cancer. Photodiagn Photodyn Ther. 2011;8:7–13. https://doi.org/10.1016/j.pdpdt.2010.12.006.

Donnelly RF, McCarron PA, Woolfson AD. Drug delivery of aminolevulinic acid from topical formulations intended for photodynamic therapy. Photochem Photobiol. 2005;81:750–67. https://doi.org/10.1562/2004-08-23-IR-283.

Dougherty TJ. PHOTOSENSITIZERS - THERAPY AND DETECTION OF MALIGNANT-TUMORS. Photochemistry and Photobiology 1987;45:879–89.

Feuerstein T, Berkovitch-Luria G, Nudelman A, Rephaeli A, Malik Z. Modulating ALA-PDT efficacy of mutlidrug resistant MCF-7 breast cancer cells using ALA prodrug. Photochem Photobiol Sci. 2011;10:1926–33. https://doi.org/10.1039/c1pp05205e.

Foster FS, Pavlin CJ, Harasiewicz KA, Christopher DA, Turnbull DH. Advances in ultrasound biomicroscopy. Ultrasound Med Biol. 2000;26:1–27.

Fukuda H, Casas A, Batlle A. Use of ALA and ALA derivatives for optimizing ALA-based photodynamic therapy: a review of our experience. J Environ Pathol Toxicol Oncol. 2006;25:127–43.

Georges JF, Valeri A, Wang H, Brooking A, Kakareka M, Cho SS, Al-Atrache Z, Bamimore M, Osman H, Mach J, Yu S, Li C, Appelt D, Lee JYK, Nakaji P, Brill K, Yocom S. Delta-Aminolevulinic Acid-Mediated Photodiagnoses in Surgical Oncology: A Historical Review of Clinical Trials. Frontiers in Surgery 2019;6.

Ghadially FN, Neish WJP. Porphyrin fluorescence of experimentally produced squamous cell carcinoma. Nature. 1960;188:1124.

Gibbs SL, Chen B, O'Hara JA, Hoopes PJ, Hasan T, Pogue BW. Protoporphyrin IX level correlates with number of mitochondria, but increase in production correlates with tumor cell size. Photochem Photobiol. 2006;82:1334–41. https://doi.org/10.1562/2006-03-11-RA-843.

Goncalves KD, Cordeiro TD, Silva FRD, Samad RE, Vieira ND, Courrol LC. In: Kurachi C, et al., editors. Biophotonics South America. Bellingham, WA: SPIE Press; 2015.

Goncalves KD, Vieira DP, Courrol LC. Study of THP-1 macrophage viability after sonodynamic therapy using methyl ester of 5-aminolevulinic acid gold nanoparticles. Ultrasound Med Biol. 2018;44:2009–17. https://doi.org/10.1016/j.ultrasmedbio.2018.05.012.

Gotardelo DR, Courrol LC, Bellini MH, Silva FRD, Soares CRJ. Porphyrins are increased in the faeces of patients with prostate cancer: a case-control study. BMC Cancer. 2018;18:1090. https://doi.org/10.1186/s12885-018-5030-1.

Hasanzadeh M, Shadjou N, de la Guardia M. Early stage screening of breast cancer using electrochemical biomarker detection. TrAC Trends Anal Chem. 2017;91:67–76. https://doi.org/10.1016/j.trac.2017.04.006.

Hecht F, Cazarin JM, Lima CE, Faria CC, Leitao AAD, Ferreira ACF, Carvalho DP, Fortunato RS. Redox homeostasis of breast cancer lineages contributes to differential cell death response to exogenous hydrogen peroxide. Life Sciences 2016;158:7–13.

Hou W, Cronin SB. A review of surface plasmon resonance-enhanced photocatalysis. Adv Funct Mater. 2013;23:1612–9. https://doi.org/10.1002/adfm.201202148.

Huang Z, Hsu Y-C, Li L-B, Wang L-W, Song X-D, Yow CMN, Lei X, Musani AI, Luo R-C, Day BJ. Photodynamic therapy of cancer—challenges of multidrug resistance. J Innov Opt Health Sci. 2015;8:1530002. https://doi.org/10.1142/S1793545815300025.

Jiang LQ, Wang TY, Webster TJ, Duan HJ, Qiu JY, Zhao ZM, Yin XX, Zheng CL. Intracellular disposition of chitosan nanoparticles in macrophages: intracellular uptake, exocytosis, and intercellular transport. International Journal of Nanomedicine 2017;12:6383–98.

Ju DH, Yamaguchi F, Zhan GZ, Higuchi T, Asakura T, Morita A, Orimo H, Hu SS. Hyperthermotherapy enhances antitumor effect of 5-aminolevulinic acid-mediated sonodynamic therapy with activation of caspase-dependent apoptotic pathway in human glioma. Tumor Biol. 2016;37:10415–26. https://doi.org/10.1007/s13277-016-4931-3.

Karina OG, Monica NS, Leticia BS, Lilia CC. Green synthesis of gold nanoparticles with aminolevulinic acid of: a novel theranostic agent for atherosclerosis. BBA Clin. 2015;3:S13. https://doi.org/10.1016/j.bbacli.2015.05.038.

Kennedy JC, Pottier RH. ENDOGENOUS PROTOPORPHYRIN-IX, A CLINICALLY USEFUL PHOTOSENSITIZER FOR PHOTODYNAMIC THERAPY. Journal of Photochemistry and Photobiology B-Biology 1992; 14:275–92.

Korbelik M, Dougherty GJ. Photodynamic therapy-mediated immune response against subcutaneous mouse tumors. Cancer Res. 1999;59:1941–6.

Labbe RF, Vreman HJ, Stevenson DK. Zinc protoporphyrin: a metabolite with a mission. Clin Chem. 1999;45:2060–72.

Lawrie A, Brisken AF, Francis SE, Cumberland DC, Crossman DC, Newman CM. Microbubble-enhanced ultrasound for vascular gene delivery. Gene Ther. 2000;7:2023–7. https://doi.org/10.1038/sj.gt.3301339.

Li ZT, Sun X, Guo SY, Wang LP, Wang TY, Peng CH, Wang W, Tian Z, Zhao RB, Cao WW, Tian Y. Rapid stabilisation of atherosclerotic plaque with 5-aminolevulinic acid-mediated sonodynamic therapy. Thrombosis and Haemostasis 2015;114:793–803.

Li Y, Zhao JW, You WL, Cheng DH, Ni WH. Gold nanorod@iron oxide core-shell heterostructures: synthesis, characterization, and photocatalytic performance. Nanoscale 2017;9:3925–33.

Linden F, Domschke G, Erbel C, Akhavanpoor M, Katus HA, Gleissner CA. Inflammatory therapeutic targets in coronary atherosclerosis—from molecular biology to clinical application. Front Physiol. 2014;5:455.

Lippert BM, Grosse U, Klein M, Kuelkens C, Klahr N, Brossmann P, Teymoortash A, Ney M, Doss MO, Werner JA. Excretion measurement of porphyrins and their precursors after topical administration of 5-aminolaevulinic acid for fluorescence endoscopy in head and neck cancer. Res Commun Mol Pathol Pharmacol. 2003;113:75–85.

Lipson RL, Baldes EJ. The photodynamic properties of a particular hematoporphyrin derivative. Arch Dermatol. 1960;82:508–16.

Masilamani V, Al-Zhrani K, Al-Salhi M, Al-Diab A, Al-Ageily M. Cancer diagnosis by autofluorescence of blood components. J Lumin. 2004;109:143–54. https://doi.org/10.1016/j.jlumin.2004.02.001.

Mateus JE, Valdivieso W, Hernandez IP, Martinez F, Paez E, Escobar P. Cell accumulation and antileishmanial effect of exogenous and endogenous protoporphyrin IX after photodynamic treatment. Biomedica. 2014;34:589–97. https://doi.org/10.1590/S0120-41572014000400012.

McEwan C, Nesbitt H, Nicholas D, Kavanagh ON, McKenna K, Loan P, Jack IG, McHale AP, Callan JF. Comparing the efficacy of photodynamic and sonodynamic therapy in non-melanoma and melanoma skin cancer. Bioorg Med Chem. 2016;24:3023–8. https://doi.org/10.1016/j.bmc.2016.05.015.

McHale AP, Callan JF, Nomikou N, Fowley C, Callan B. Sonodynamic therapy: concept, mechanism and application to cancer treatment therapeutic. Ultrasound. 2016;880:429–50. https://doi.org/10.1007/978-3-319-22536-4_22.

Mehrad H, Farhoudi M. Investigation of protoporphyrin IX-mediated sonodynamic therapy on intermediate stage atherosclerosis using a new computerized B-mode ultrasound analyzing method. Atherosclerosis. 2016;252:E192.

Mohammadi Z, Sazgarnia A, Rajabi O, Soudmand S, Esmaily H, Sadeghi HR. An in vitro study on the photosensitivity of 5-aminolevulinic acid conjugated gold nanoparticles. Photodiagn Photodyn Ther. 2013;10:382–8. https://doi.org/10.1016/j.pdpdt.2013.03.010.

Mohammadi Z, Sazgarnia A, Rajabi O, Toosi MS. Comparative study of X-ray treatment and photodynamic therapy by using 5-aminolevulinic acid conjugated gold nanoparticles in a melanoma cell line. Artif Cells Nanomed Biotechnol. 2017;45:467–73. https://doi.org/10.3109/21691401.2016.1167697.

Namikawa T, Yatabe T, Inoue K, Shuin T, Hanazaki K. Clinical applications of 5-aminolevulinic acid-mediated fluorescence for gastric cancer. World J Gastroenterol. 2015;21:8769–75. https://doi.org/10.3748/wjg.v21.i29.8769.

Nascimento da Silva M, Sicchieri LB, Rodrigues de Oliveira Silva F, Andrade MF, Courrol LC. Liquid biopsy of atherosclerosis using protoporphyrin IX as a biomarker. Analyst. 2014;139:1383–8. https://doi.org/10.1039/c3an01945d.

Navarro JRG, Lerouge F. From gold nanoparticles to luminescent nano-objects: experimental aspects for better gold-chromophore interactions. Nanophotonics. 2017;6:71–92. https://doi.org/10.1515/nanoph-2015-0143.

Oo MKK, Yang X, Wang H, Du H. 5-Aminolevulinic acid conjugated gold nanoparticles for cancer treatment. Nanomedicine (Lond). 2008;3:777–86. https://doi.org/10.2217/17435889.3.6.777.

Padera TP, Meijer EFJ, Munn LL. The lymphatic system in disease processes and cancer progression. Annu Rev Biomed Eng. 2016;18:125–58. https://doi.org/10.1146/annurev-bioeng-112315-031200.

Peng Q, Warloe T, Moan J, Godal A, Apricena F, Giercksky KE, Nesland JM. Antitumor effect of 5-aminolevulinic acid-mediated photodynamic therapy can be enhanced by the use of a low dose of photofrin in human tumor xenografts. Cancer Res. 2001;61:5824–32.

Peng CH, Li YS, Liang HJ, Cheng JL, Li QS, Sun X, Li ZT, Wang FP, Guo YY, Tian Z, Yang LM, Tian Y, Zhang ZG, Cao WW. Detection and photodynamic therapy of inflamed atherosclerotic plaques in the carotid artery of rabbits. J Photochem Photobiol B Biol. 2011;102:26–31. https://doi.org/10.1016/j.jphotobiol.2010.09.001.

Risaliti L, Piazzini V, Di Marzo MG, Brunetti L, Cecchi R, Lencioni P, Bilia AR, Bergonzi MC. Topical formulations of delta-aminolevulinic acid for the treatment of actinic keratosis: characterization and efficacy evaluation. Eur J Pharm Sci. 2018;115:345–51. https://doi.org/10.1016/j.ejps.2018.01.045.

Rock KL, Kono H. The inflammatory response to cell death. Annual Review of Pathology-Mechanisms of Disease 2008;3:99–126.

Rosenthal I, Sostaric JZ, Riesz P. Sonodynamic therapy—a review of the synergistic effects of drugs and ultrasound. Ultrason Sonochem. 2004;11:349–63. https://doi.org/10.1016/j.ultsonch.2004.03.004.

Ross R. Mechanisms of disease—atherosclerosis—an inflammatory disease. N Engl J Med. 1999;340:115–26.

Ross JS, Stagliano NE, Donovan MJ, Breitbart RE, Ginsburg GS. Atherosclerosis and cancer: common molecular pathways of disease development and progression. Ann N Y Acad Sci. 2001;947:271–93.

Sachar M, Anderson KE, Ma XC. Protoporphyrin IX: the Good, the Bad, and the Ugly. J Pharmacol Exp Ther. 2016;356:267–75. https://doi.org/10.1124/jpet.115.228130.

Smolsky J, Kaur S, Hayashi C, Batra SK, Krasnoslobodtsev AV. Surface-enhanced raman scattering-based immunoassay technologies for detection of disease biomarkers. Biosensors (Basel). 2017;7:7. https://doi.org/10.3390/bios7010007.

Sum G, Hone T, Atun R, Millett C, Suhrcke M, Mahal A, Koh GCH, Lee JT. Multimorbidity and out-of-pocket expenditure on medicines: a systematic review. BMJ Glob Health. 2018;3:e000505. https://doi.org/10.1136/bmjgh-2017-000505.

Sun SJ, Xu YX, Fu P, Chen M, Sun SH, Zhao RR, Wang JR, Liang XL, Wang SM. Ultrasound-targeted photodynamic and gene dual therapy for effectively inhibiting triple negative breast cancer by cationic porphyrin lipid microbubbles loaded with HIF1 alpha-siRNA. Nanoscale. 2018;10:19945–56. https://doi.org/10.1039/c8nr03074j.

Sunar U, Rohrbach DJ, Morgan J, Zeitouni N, Henderson BW. Quantification of PpIX concentration in basal cell carcinoma and squamous cell carcinoma models using spatial frequency domain imaging. Biomed Opt Express. 2013;4:531–7. https://doi.org/10.1364/BOE.4.000531.

Taketani S, Ishigaki M, Mizutani A, Uebayashi M, Numata M, Ohgari Y, Kitajima S. Heme synthase (ferrochelatase) catalyzes the removal of iron from heme and demetalation of metalloporphyrins. Biochemistry. 2007;46:15054–61. https://doi.org/10.1021/bi701460x.

Tapia-Vieyra JV, Delgado-Coello B, Mas-Oliva J. Atherosclerosis and cancer; a resemblance with far-reaching implications. Arch Med Res. 2017;48:12–26. https://doi.org/10.1016/j.arcmed.2017.03.005.

Thunshelle C, Yin R, Chen QQ, Hamblin MR. Current advances in 5-aminolevulinic acid mediated photodynamic therapy. Curr Dermatol Rep. 2016;5:179–90. https://doi.org/10.1007/s13671-016-0154-5.

Trendowski M. The promise of sonodynamic therapy. Cancer Metastasis Rev. 2014;33:143–60. https://doi.org/10.1007/s10555-013-9461-5.

Umemura S, Kawabata K, Sasaki K, Yumita N, Umemura K, Nishigaki R. Recent advances in sonodynamic approach to cancer therapy. Ultrason Sonochem. 1996;3:S187–91. https://doi.org/10.20892/j.issn.2095-3941.2016.0068.

Valko M, Rhodes CJ, Moncol J, Izakovic M, Mazur M. Free radicals, metals and antioxidants in oxidative stress-induced cancer. Chemico-Biological Interactions 2006;160:1–40.

Virmani R, Burke AP, Farb A, Kolodgie FD. Pathology of the vulnerable plaque. Journal of the American College of Cardiology 2006;47:C13–C18

Vio V, Marchant MJ, Araya E, Kogan MJ. Metal nanoparticles for the treatment and diagnosis of neurodegenerative brain diseases. Curr Pharm Des. 2017;23:1916–26. https://doi.org/10.2174/1381612823666170105152948.

Wachowska M, Muchowicz A, Firczuk M, Gabrysiak M, Winiarska M, Wanczyk M, Bojarczuk K, Golab J. Aminolevulinic acid (ALA) as a prodrug in photodynamic therapy of cancer. Molecules. 2011;16:4140–64. https://doi.org/10.3390/molecules16054140.

Wakui M, Yokoyama Y, Wang H, Shigeto T, Futagami M, Mizunuma H. Efficacy of a methyl ester of 5-aminolevulinic acid in photodynamic therapy for ovarian cancers. J Cancer Res Clin Oncol. 2010;136:1143–50. https://doi.org/10.1007/s00432-010-0761-7.

Wang Y, Wang W, Xu HB, Sun Y, Sun J, Jiang YX, Yao JT, Tian Y. Non-lethal sonodynamic therapy inhibits atherosclerotic plaque progression in ApoE(−/−) mice and attenuates ox-LDL-mediated macrophage impairment by inducing heme oxygenase-1. Cell Physiol Biochem. 2017;41:2432–46. https://doi.org/10.1159/000475913.

Welcker ML. The porphyrins. N Engl J Med. 1945;232:11–9.

Wood AKW, Sehgal CM. A review of low-intensity ultrasound for cancer therapy. Ultrasound Med Biol. 2015;41:905–28. https://doi.org/10.1016/j.ultrasmedbio.2014.11.019.

Wozniak A, Malankowska A, Nowaczyk G, Grzeskowiak BF, Tusnio K, Slomski R, Zaleska-Medynska A, Jurga S. Size and shape-dependent cytotoxicity profile of gold nanoparticles for biomedical applications. J Mater Sci Mater Med. 2017;28:11. https://doi.org/10.1007/s10856-017-5902-y.

Wu JN, Han HJ, Jin Q, Li ZH, Li H, Ji J. Design and proof of programmed 5-aminolevulinic acid prodrug nanocarriers for targeted photodynamic cancer therapy. ACS Appl Mater Interfaces. 2017;9:14596–605. https://doi.org/10.1021/acsami.6b15853.

Xu XR, Meng JW, Hou SG, Ma HP, Wang DS. The characteristic fluorescence of the serum of cancer-patients. J Lumin. 1988;40–41:219–20. https://doi.org/10.1016/0022-2313(88)90163-9.

Xu H, Yao CP, Wang J, Chang ZN, Zhang ZX. Enhanced 5-aminolevulinic acid-gold nanoparticle conjugate-based photodynamic therapy using pulse laser. Laser Phys Lett. 2016;13:025602.

Yumita N, Nishigaki R, Umemura K, Umemura S. HEMATOPORPHYRIN AS A SENSITIZER OF CELL-DAMAGING EFFECT OF ULTRASOUND. Japanese Journal of Cancer Research 1989;80:219–22.

Yumita N, Iwase Y, Nishi K, Komatsu H, Takeda K, Onodera K, Fukai T, Ikeda T, Umemura S, Okudaira K, Momose Y. Involvement of reactive oxygen species in sonodynamically induced apoptosis using a novel porphyrin derivative. Theranostics. 2012;2:880–8. https://doi.org/10.7150/thno.3899.

Zaak D, Sroka R, Stocker S, Bise K, Lein M, Hoppner M, Frimberger D, Schneede P, Reich O, Kriegmair M, Knuchel R, Baumgartner R, Hofstetter A. Photodynamic therapy of prostate cancer by means of 5-aminolevulinic acid-induced protoporphyrin IX—in vivo experiments on the dunning rat tumor model. Urol Int. 2004;72:196–202. https://doi.org/10.1159/000077114.

Zhang ZX, Wang SJ, Xu H, Wang B, Yao CP. Role of 5-aminolevulinic acid-conjugated gold nanoparticles for photodynamic therapy of cancer. J Biomed Opt. 2015;20:51043. https://doi.org/10.1117/1.JBO.20.5.051043.

Zheng SY, Li XL, Zhang YB, Xie Q, Wong YS, Zheng WJ, Chen TF. PEG-nanolized ultrasmall selenium nanoparticles overcome drug resistance in hepatocellular carcinoma HepG2 cells through induction of mitochondria dysfunction. Int J Nanomedicine. 2012;7:3939–49. https://doi.org/10.2147/IJN.S30940.

Zhu B, Liu QH, Wang YA, Wang XB, Wang P, Zhang LN, Su S. Comparison of accumulation, subcellular location, and sonodynamic cytotoxicity between hematoporphyrin and protoporphyrin IX in L1210 cells. Chemotherapy. 2010;56:403–10. https://doi.org/10.1159/000317743.

Zhu ZX, Scalfi-Happ C, Ryabova A, Grafe S, Wiehe A, Peter RU, Loschenov V, Steiner R, Wittig R. Photodynamic activity of Temoporfin nanoparticles induces a shift to the M1-like phenotype in M2-polarized macrophages. J Photochem Photobiol B Biol. 2018;185:215–22. https://doi.org/10.1016/j.jphotobiol.2018.06.015.

Chapter 16
Gold Nanorods as Theranostic Nanoparticles for Cancer Therapy

Maria Mendes, Antonella Barone, João Sousa, Alberto Pais, and Carla Vitorino

Abstract Inadequate therapies and clinical methods for overcoming multidrug-resistant cancer constitute the major barrier for cancer treatment. Also, early detection of this disease is fundamental and new nanotechnologies emerge with clear relevance. Considering their distinctive chemical and physical properties, plasmonic nanoparticles have been proposed and are regarded as promising carriers for cancer treatment. Gold nanoparticles (AuNPs) are the most studied plasmonic nanoparticles because of their special optical and electronic properties. Depending on size and shape, AuNPs are able to perform, simultaneously, several therapeutic functions, including photothermal therapy (PTT), photodynamic therapy (PDT), and imaging. The synergistic effect between PTT/PDT and chemotherapeutic drugs, to cooperatively suppress cancer cells, has also been studied, wherein rod-shaped AuNPs has been pointed out as suitable theranostic NP. This demonstrates their ability to integrate multiple functions in a single system. However, their performance is highly dependent on several experimental parameters including size, aspect ratio, surface modification, and morphology. All these parameters strongly affect both the physical and biological processes involved. This review focuses on AuNRs properties, their multiple applications, and the trends for the integration of

M. Mendes
Faculty of Pharmacy, University of Coimbra, Coimbra, Portugal

Centre for Neurosciences and Cell Biology (CNC), University of Coimbra, Coimbra, Portugal

A. Barone
Department of Health Sciences, University "Magna Græcia" of Catanzaro, Catanzaro, Italy

J. Sousa · C. Vitorino (✉)
Faculty of Pharmacy, University of Coimbra, Coimbra, Portugal

Coimbra Chemistry Centre, Department of Chemistry, University of Coimbra, Coimbra, Portugal
e-mail: csvitorino@ff.uc.pt

A. Pais
Coimbra Chemistry Centre, Department of Chemistry, University of Coimbra, Coimbra, Portugal

M. Rai, B. Jamil (eds.), *Nanotheranostics*,
https://doi.org/10.1007/978-3-030-29768-8_16

theranostic applications. Also described are the difficulties imposed to an effective in vivo biodistribution and pharmacokinetic behavior. Current research and preclinical and clinical investigation will be addressed.

Keywords Brain tumor · Gold nanorods · Photoacoustic therapy · Photothermal therapy · Theranostics

Nomenclature

AFM	Atomic force microscopy
anti-EGFR	Monoclonal anti-epidermal growth factor receptor
AuCs	Nanocages
AuNPs	Gold nanoparticles
AuNRs	Nanorods
AuNSs	Nanostars
AuSLs	Nanoshells
AuSs	Nanospheres
BBB	Blood-brain barrier
BSA	Bovine serum albumin
Ce6	Chlorin e6
CPNPs	Colloidal plasmonic nanoparticles
CT	X-ray computed tomography
CTAB	Cetyltrimethylammonium bromide
CTX	Chemotherapy
DR4	Death-4 receptors
FL	Fluorescence
GB	Glioblastoma multiforme
HA	Hyaluronic acid
HP	Hematoporphyrin
HRTEM	High-resolution transmission electron microscopy
ICP-MS	Inductively coupled plasma mass spectrometry
LLS	Laser light scattering
LSPR	Localized surface plasmon resonance
MHDA	Mercaptohexadecanoic acid
MS	Mass spectroscopy
MUA	11-Mercaptoundecanoic acid
NIR	Near-infrared radiation
NPs	Nanoparticles
OCT	Optical coherence tomography
PA	Photoacoustic imaging
PDT	Photodynamic therapy

PEG	Polyethylene glycol
PSS	Poly-styrene sulfonate
PTT	Photothermal therapy
rhTNF	Recombinant human tumor necrosis factor alpha
ROS	Reactive oxygen species
RS	Raman spectroscopy
SAXS	Small angle X-ray scattering
SERS	Surface-enhanced Raman scattering
TCAB	Tetradodecylamonium bromide
TEM	Transmission electron microscopy
Tfr	Transferrin receptor
TNF-α	Tumor necrosis factor-α
TPL	Two-photon luminescence
UV-vis	Ultraviolet-visible spectroscopy
ZnPc	Phthalocyanine

16.1 Introduction

Cancer continues to be the most frequent, aggressive, and lethal set of diseases in the world. In general, the failure of conventional cancer treatment is due to drug resistance at the tumor tissue and/or cellular level, nonspecific therapeutic approaches, distribution failure, and also a late diagnosis, the latter being the main reason for a poor prognosis.

Several nanotechnologies have been developed, in order to solve these problems, including the tendency for a late diagnostic or an inefficient treatment. Advances in nanotechnology with organic or inorganic nanoparticles (NPs) have been described, with progress not only in the real-time tracking or in the engineering of the surface of NPs, but also in the respective biocompatibility and circulation lifetime in the body. As a consequence, there is an increase in the selectivity of drugs to cancer cells and reduction of toxicity. These improvements in nanotechnology have provided a new possibility for the integration of cancer diagnosis and treatment in a single NP. Thus, NPs integrate imaging diagnosis, mechanical therapy (e.g., photothermal therapy), drug delivery, and also a targeting approach on the surface. All these features, together, allow the improvement in treatment, associated to a decrease in side effects, progressing the curative effect and quality of life of patients. Colloidal plasmonic nanoparticles (CPNPs) are described in different compositions, dimensions, and shapes, allowing control over diffusion, reactivity, optical, and magnetic properties, and biodegradability. For these reasons, CPNPs are attractive in almost all fields of engineering, medicine, and science. The respective use in health has grown, due to the wide range of applications that include diagnosis and therapeutics, combined with the delivery of anticancer drugs. Superparamagnetic iron oxide, quantum dots, carbon nanotubes, hybrid nanocomposites, and gold nanoparticles are some examples of CPNPs used for imaging and therapy (Hanske et al. 2018; Liu et al. 2016; Riva et al. 2017; Urries et al. 2014).

Gold nanoparticles (AuNPs) are considered as one of the most convenient INPs, given their high biocompatibility, stability, resistance to oxidation, the diverse optical properties associated to surface plasmons, and the ease of surface functionalization with organic and biological molecules. The electronic structure of AuNPs is central for their clinical applications, specifically in the areas of cancer treatment. AuNPs are inert, innocuous, and nonimmunogenic, with well-established and easy synthetic methods, and control over size and shape (Tian et al. 2018). They can be easily functionalized with multiple targeting and bioactive ligands without compromising their innate characteristics. Thus, colloidal gold is relevant in several clinical applications, including drug and macromolecular delivery, as a contrast agent, in biosensing and bio-detection, catalysis and bioelectronics, bioimaging and photothermal therapy (Alex and Tiwari 2015; Khlebtsov et al. 2013; Koohi et al. 2017; Zhang 2015). This application range is due to its unique characteristics, including high electron density and extinction coefficients, which allows localizing it within cells or in tissues. Depending on shape and functionalities, these particles can interact with biological systems in many ways, depending on the cell type, uptake pathways or targeting different organelles. Also, their marked surface plasmon resonance-enhanced absorption in the NIR range allows obtaining images by different methods including optical coherence tomography (OCT), X-ray computed tomography (CT), two-photon luminescence (TPL), or photoacoustic imaging (PA). All these characteristics are influenced by the particles size, shape, and aggregation state (Fig. 16.1 and Table 16.1) (Zhang et al. 2007).

This chapter focuses on AuNPs properties, their characterization methods, applications, and trends in tumor theranostics, followed by difficulties imposed to an effective biodistribution. Recent advances in brain tumor therapeutics are

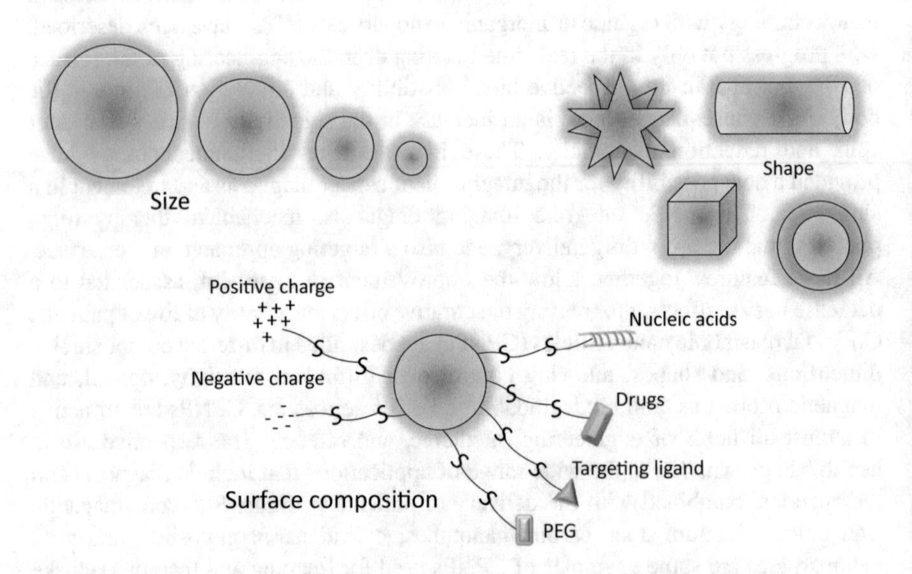

Fig. 16.1 Representation of AuNPs depending on size, shape, and surface composition

Table 16.1 Different types of AuNPs, categorized upon particle shape and tunable optical and electronic properties

AuNPs	Particle size	Plasmon characteristics	Applications
Nanospheres (AuSs)	2–100 nm	~522–539 nm	Cell imaging Photothermal therapy Delivery of therapeutic molecules
Nanorods (AuNRs)	Length: 50–250 Diameter: 5–50 nm	Transversal band ~520 Longitudinal band infrared region dependent of the aspect ratio: 800 nm (aspect ratio = 4.4) or >1200 nm (aspect ratio = 16)	Tumor imaging Photothermal therapy Delivery of therapeutic molecules
Nanoshells (AuSLs)	10–150 nm	510–575 nm	Tumor imaging Photothermal therapy Photodynamic therapy Delivery of therapeutic molecules
Nanocages (AuCs)	Length: 30–70 nm	~532–575 nm	Tumor imaging Photothermal therapy Photodynamic therapy Delivery of therapeutic molecules
Nanostars (AuNSs)	18–90 m	~590–1000 nm	Photothermal therapy Photodynamic therapy

addressed, with special attention on the combination of imaging and therapeutic functions (i.e., theranostics). The use of gold nanoparticles for brain tumor treatment, including current research and preclinical and clinical investigation, will be addressed. The pharmacokinetics and biodistribution of these AuNPs need to be explored so as to understand their fate in vivo. Despite the increasing importance of AuNPs in biomedical research, only limited types of nanoparticles have been approved in clinical practice for simultaneous therapeutics and diagnostics application.

16.2 Optical Properties of Gold Nanoparticles

The gold (Au) atom possesses a high number of electrons per atom ($Z = 79$), and for this reason, AuNPs are capable of absorbing the X-ray radiation energy at a 1000-fold higher probability than any soft tissue (Zhang 2015). Exposure of AuNPs to light induces a collective coherent oscillation of the free electrons of the metal, called localized surface plasmon resonance (LSPR) (Amendola et al. 2017; Kobayashi et al. 2014; Zheng et al. 2012). The electron oscillation around the NPs surface causes a charge separation with respect to the ionic lattice, forming a dipole oscillation along the direction of the electric field of the light. This photophysical phenomenon has two main processes, absorption and scattering, which occur when light passes through metal, resulting in energy loss of the electromagnetic wave. The contribution of these two parts to the total extinction can be calculated by using Mie theory (He et al. 2009). The scattered light has the same frequency as the incident light (Rayleigh scattering) or a shifted frequency (Raman scattering). For plasmonic NPs, like gold or silver, the LSPR band is much stronger than for other metals. The light absorption and scattering are intensely enhanced in AuNPs, being at least 1000 times stronger than the absorption or emission of any organic dye molecules (Jans and Huo 2012). This makes AuNPs the obvious choice in a broad range of applications including drug delivery, photothermal therapy, and imaging of tumor sites. LSPR represents a strong absorption band in the visible region, owing to the collective oscillations of metal conduction band electrons in strong resonance, where light penetration in tissue is optimal. The peak intensity and position frequency of the LSPR are influenced by some parameters, such as the size and shape, surface charges, the dielectric constant of Au and the surrounding medium. For example, the LSPR of spherical AuNPs occurs at ~530 nm, which demonstrate the limitation tunability of the nanosphere LPSR. In the biological field, there are two biological transparency windows, located in 650–950 and 1000–1350 nm with optimal tissue, blood, and water transmission obtained from low scattering and energy absorption (Hemmer et al. 2016). Thus, in order to tune the LSPR to the availability of a suitable laser or to enhance the resonance of the system, an effective approach consists of changing the spherical shape to rod shape. Three main types of interactions between molecules and surface plasmons can be explored: optical, thermal, and mechanical. The resonant excitation of plasmons can severely amplify the electric field near the NPs surface. The large enhancement of the surface electric field on the AuNPs shell layer can be modulated for specific therapeutic purposes, depending on the preclinical or clinical application.

16.3 Gold Nanorods

Gold nanorods (AuNRs) have acquired much attention as a strategy to cancer treatment due to low cytotoxicity, stability, biocompatibility, and suitable physiochemical parameters. Also, the ease of preparation, the large number of synthetic methods available, and the control over the aspect ratio, which is primarily responsible for the change in their optical properties, have attracted the most attention.

The rod shape promotes excellent plasmonic and photothermal activity. They are usually synthesized with a size ranging from 50 to 250 and diameter from 5 to 50 nm and an aspect ratio (the ratio of the longer side to the shorter side) between 1 and 7, and each aspect ratio corresponds to specific longitudinal plasmons. AuNRs have been granted an increased recognition in several biomedical applications, due not only their precisely controlled NIR optical properties but also to their enhanced permeability to tumor vasculature, resulting in higher retention in the tumor tissue. AuNRs also have a higher efficiency of light absorption at their longitudinal plasmon resonance site than any other known nanoparticles. The LSPR spectrum of AuNRs displays two absorption bands, one that represents the longitudinal oscillation of electrons, a stronger long wavelength band in the NIR region, and another in the transverse electronic oscillation, a weaker short-wavelength band in the visible region.

The physicochemical properties of AuNPs could be suitably modified through several strategies during the synthesis phase (Tong et al. 2017, 2018). These properties strongly influence the AuNPs performance and their applicability in cancer treatment, particularly the optical absorption and scattering. Thus, a challenge in nanoscience is the adequate tuning of the size and shape of NPs. Size and shape are crucial for the design and application of NPs, because optoelectronic properties are strongly dependent on them. The influence of AuNRs size was studied, and it was found that larger AuNRs, with the same LSPR wavelength, showed a stronger scattering intensity but a weaker surface-enhanced Raman spectroscopy intensity than small nanorods (Lin et al. 2016). As will be mentioned in a later section, the size is a crucial parameter for the in vitro and in vivo studies. The aspect ratio can markedly change the absorption LSPR spectra, one of the most relevant plasmonic properties of AuNRs, and the LSPR increases with larger aspect ratios. Despite the control of aspect ratio during the synthesis, the mechanisms for the underlying control are not well understood. Compared to AuNSs or nanostars (AuSs), optical properties of AuNRs showed to be 2–3 times stronger absorption efficiency. The main goal is to design AuNRs with the best optical properties and in vivo behavior. Thus, comparisons between shapes were also performed and the SERS property of AuNRs and AuNSs and concluded that AuNRs have a stronger activity.

AuNRs display attractive optical and unique physiochemical properties that play a key role in several biomedical applications. The main limitation of the production of AuNRs is the low yield. However, some strategies and parameters have been considered to overcome this limitation, such as the size of seeds, reaction time, the amount of $AgNO_3$, use of cetyltrimethylammonium bromide (CTAB), pH, and temperature, among others. Several studies have been conducted in order to understand what are the interferents in the AuNRs synthesis, including the CTAB concentration, the influence of pH, and the concentrations of reducing agent and Ag ions (Liu et al. 2017a; Salavatov et al. 2018; Smitha et al. 2013; Xu et al. 2014).

16.3.1 Fabrication Processes

Distinct preparation methods are described for the AuNRs, having in common the use of reduction (e.g., ascorbic acid and sodium borohydride) and stabilization agents (in particular, a cationic surfactant as cetyltrimethylammonium bromide—CTAB). As previously reiterated, the optical properties, size, shape, or aspect ratio are important characteristics with a huge impact on the biomedical application of AuNRs, and thus the selection of fabrication processes could have the impact on their final performance and the yield of the process (Smitha et al. 2013). Thus, all the characteristics can be controlled by using different fabrication processes. The methods to produce AuNRs can be divided into seeded growth methods, seedless methods, template method, and electrochemical. In this section, we will review the various in situ methods and their improvements.

The seeded growth method is the most established, popular, and efficient method (Chhatre et al. 2018). Small gold seeds are added in a controlled growth environment to obtain gold nanorods. The process consists of different steps, including preparation of seeds by reduction of the Au precursor by a strong reducing agent ($NaBH_4$), preparation of growth solution containing $HAuCl_4$, cationic surfactant cetyltrimethylammonium bromide (CTAB), ascorbic acid, and silver ions, and finally the addition of seeds in the solution for AuNRs growth. Ascorbic acid is a weak reducing agent that is crucial for the rod morphology (decrease of length and increase of yield), while the silver ions induce the shape. The growth solution is mixed with gold seeds, and control parameters such as surfactant concentration and combination, pH value or temperature will promote AuNRs with different optical properties (Jiang et al. 2006; Kannadorai et al. 2015; Liu et al. 2017c; Wang et al. 2016b). CTAB is the direct agent that allows producing rods instead of spheres (Becker 2010). The pH is also an important factor for the synthesis of AuNPs, since it affects CTAB micelles stability and reduces their absorption ability on the gold surface. Silver nitrate ($AgNO_3$) was introduced in this method before seed addition to promote rod formation and control the aspect ratio, which is directly proportional to the increase of silver ions concentration. In this method, the mechanisms that lead to a specific shape and the growth of AuNRs are poorly understood. However, there are several limitations associated to seed-mediated methods, including insufficient reproducibility, different variations due to the multi-step procedure, and limited LSPR tunability.

Seedless methods allow overcoming some drawbacks associated with the seed growth method. This method starts with a small-size AuNRs, is easy, simple, and with a simple step. $NaBH_4$ is initially used as reducing agent, and directly added in the growth solution, reducing gold(III) chloride hydrate to Au^0, forming AuNPs as seeds. The nucleation and growth can be properly controlled with a mixture of strong and weak reducing agents, the latter, e.g., ascorbic acid. The strong reducing agent initiates nucleation and the weak reducing agent helps growing the NPs. Moreover, Nikoobakht and El-Sayed (2003) optimized the synthesis scheme by decreasing the pH (<2) of the growth solution, which lowers the reducibility of both $NaBH_4$ and ascorbic acid. The pH decrease leads to a slowdown in the growth of AuNRs, and increases monodispersity. A positive correlation between pH and the reducing ability of ascorbic acid and $NaBH_4$ was found. Another important factor is the concentration of CTAB. This is the main stabilizer of the initial crystalline nuclei and allows decreasing the size of AuNRs. Several studies have selected the important parameters in the seedless synthesis protocol: Au/CTAB ratio, Ag concentration, $NaBH_4$ concentration, and pH value (Liu et al. 2017a; Salavatov et al. 2018; Xu et al. 2014).

Template methods consist of electrochemical deposition of gold within the pores of nanoporous polycarbonate or alumina template membranes (Li and Cao 2012). This approach starts with a small amount of silver (Ag) sputtered onto the alumina template membrane in order to provide a film for electrodeposition, following the electrodeposition of Au within alumina nanopores. The next stage comprises the dissolution of the alumina membrane and the silver film, in the presence of a polymeric stabilizer. Finally, using sonication or agitation, AuNRs can be dispersed in water or organic solvent. Clearly, the control of particle size depends on the diameter of the alumina membrane pores used. The major limitations of this method are the inconsistency in the length of the AuNRs formed, due to the irregular deposition of gold and the yield (Hornyak et al. 1997; Vigderman et al. 2012).

Electrochemical methods are conducted in a two-electrode-type electrochemical cell, in which a gold metal plate is the anode, and a platinum plate the cathode (Li and Cao 2012). The electrolytic solution, containing CTAB, a small amount of tetradodecylammonium bromide (TCAB) as co-surfactant and stabilizer, covers both electrodes and is placed inside an ultrasonic bath at 36 °C. At this point, a certain concentration of acetone and cyclohexane are added, for loosening the micellar framework and enhancing the formation of elongated rod-like CTAB micelles. This method offers a synthetic route for preparing high yields of AuNRs.

There are more fabrication methods for the formation of AuNRs. However, they result in lower amounts, pose problems of reproducibility, and it is difficult to obtain good yields. Finally, after each synthesis, an extra step of AuNRs purification is required. This step will also be important to removing the excess of CTAB, which is recognized as a toxic surfactant for biological applications (He et al. 2018).

16.3.2 Characterization Methods

The quality associated to the fabrication processes needs to be assessed and controlled. This implies that AuNRs need to be characterized in terms of size, aspect ratio, charge, stability, yield, and optical properties. Characterization methods include mass spectroscopy, electroanalytical methods, optical spectroscopy, and various nanoparticle counting techniques, summarized in Table 16.2.

Table 16.2 Description of methods used for the assessment and control of AuNRs

Transmission electron microscopy (TEM)	• Morphology of metallic nanoparticles • Direct information on size, size distribution, and shape of NPs
High-resolution transmission electron microscopy (HRTEM)	• Direct atomic surface image • Interference image • Structural details, individual particles and the surfaces • Distinguish between different shapes: Rods, long rods, and spherical
Small angle X-ray scattering (SAXS)	• Information about the shape, size, and size distributions
Atomic force microscopy (AFM)	• Nanoscale topography by scanning a fine silicon tip across the surface, performing high-resolution imaging under physiological conditions • Morphology of nanoparticles, or visualize the process of cellular uptake of conjugated NPs
Mass spectroscopy (MS)	• Identification and quantification of AuNRs at a sub-nanometer level
Laser light scattering (LLS)	• Detection of particle size and polydispersity index
Ultraviolet-visible spectroscopy (UV-vis)	• Absorbance of ultra violet or visible light by a sample which determine both the size and the concentration of the AuNPs based on the peak position and peak intensity • Evaluate the long term stability of gold NPs colloidal suspensions, indeed unchanged results show the stability of the systems
Raman spectroscopy (RS)	• Molecular vibrations and crystal structures, employing a laser light source (green, red, or near-infrared) to irradiate a sample and generate an infinitesimal amount of Raman scattered light
Inductively coupled plasma mass spectrometry (ICP-MS)	• Information about the elemental (isotopic) composition, their size distribution, and the particle concentration

16.3.3 *Effect of Surface Modification*

Over the years, the use of AuNRs has drawn attention to several biomedical application fields, exploiting their optical properties and/or functional changes. The major disadvantage of AuNRs is the presence of CTAB on the surface. This cationic surfactant is extremely cytotoxic, which, as already stated, limits its application in the biomedical field. Despite the efforts to replace CTAB for another stabilizer, there are difficulties in controlling the growth of AuNRs during the synthesis, and the size or aspect ratio between batches. Some strategies have been put forward, including coating with organic molecules by electrostatic interaction or covalent linkage with thiolic groups or ligand exchange, taking advantage of the replacement of the CTAB by thiol-terminated molecules. Additionally, AuNRs can be modified with different ligands or molecular probes, such as antibodies, oligonucleotides, peptides, carbohydrates, and folic acid, which take advantage of active targeting (Liu et al. 2018a; Peralta et al. 2015; Wang et al. 2014, 2015, 2016a; Xu et al. 2017; Zhang et al. 2016a, c; Zhong et al. 2014). RGD peptides, hyaluronic acid (HA), and transferrin receptor (Tfr) are examples of some protein molecules attached on the AuNRs surface (Xu et al. 2019). Those modifications cannot interfere with the optical properties. The most reported modification relies on the use of polyethylene glycol (PEG), not only due to their abilities to increase the blood circulation, but also allows decreasing the cytotoxicity of CTAB and stabilize nanorods in physiological conditions (Ruff et al. 2016; Schulz et al. 2016). PEG modification was found to be an excellent alternative method to remove CTAB.

Bovine serum albumin (BSA) have been reported due to their strong biocompatibility of the AuNRs, and to perform this modification the method is simple, mixing both together (Li et al. 2018). Thiol-modified chitosan derivatives have been studied due to improving tumor targeting and drug circulation (Duan et al. 2014; Yang et al. 2015). The ligand exchange is used to replace the surface of the AuNRs, including 11-mercaptoundecanoic acid (MUA), poly-styrene sulfonate (PSS), and mercaptohexadecanoic acid (MHDA) (Bhana et al. 2016; He et al. 2018; Mirza 2019; Wan et al. 2015). These molecules can bind on the AuNRs through the Au–S bond, allowing to the carboxyl group or amine groups to conjugate other molecules, increasing their application (Liu et al. 2018a; Peralta et al. 2015; Wang et al. 2014, 2015, 2016a; Zhang et al. 2016a, c; Zhong et al. 2014). Most of the studies indicate not only a great stabilization of AuNRs, but also improvements in biodistribution and viability, as addressed in later sections.

16.4 Gold Nanorods as a Multimodal Nanoparticle

Multimodality is a progress in cancer therapy due to the higher efficacy of combinational treatment and the synchronized monitoring of therapeutic effects. The intrinsic optical properties of AuNPs provide the opportunity to combine in a single NP and in a single treatment, two or more characteristics able to provide therapeutic, diagnostic, and imaging agents. Some characteristics of AuNRs are represented in Table 16.3 and their respective advantages indicated, in the context of a multimodal nanoparticle. It should be recalled that plasmonic nanomaterials, such as AuNRs, are able to perform several therapeutic modalities, including photothermal therapy (PTT) and/or photodynamic therapy (PDT), and in combination with chemotherapeutics may cooperatively suppress cancer cells, developing synergistic effect and reversal of drug resistance (Table 16.4). In this context, AuNRs can be seen as potential NPs for several applications such as delivery of biologically and/or chemically active molecules, imaging detection of diseases at an early stage, and tracking tumor cells proliferation and differentiation. These topics will be discussed in the next sections.

Table 16.3 Representation of the main AuNRs characteristics and their advantages for the in vivo application

Characteristics	Advantages
Tunable absorption band	The absorption peak can be easily tuned to near-infrared region (~600–1200 nm). This is the "transparent window" of biological tissues.
Stability and biocompatibility	The excellent stability and biocompatibility make them suitable for injection in vivo without any systemic or organ toxicity.
Imaging technique	The LSPR, the delivery and distribution of AuNRs within target cells/tissues can be monitored via optical imaging using plasmon-enhanced wavelength selective scattering and CT.
Surface area	The surface area can be easily modified with different functional groups, allowing the incorporation of targeting molecules including drugs, nucleic acids, proteins for specific-receptor delivery, or probes for other imaging modalities for multimodal imaging.
Photo-sensitizer	The AuNRs photothermal phenomenon enables a hyperthermia destruction of the cancer cell or control the release of therapeutic drug at targeted site.

16.5 Therapy

AuNPs possess several advantages including tunable size, high loading capacity, low toxicity, and multifunctional skills, making them attractive NPs for efficient delivery of therapeutic agents to cancer treatment. To obtain a therapeutic effect, chemotherapeutics, functional proteins, DNA, or RNA can be loaded into AuNPs, by either noncovalent interactions or covalent conjugation. Also, AuNPs can convert the light into heat or transfer energy to the substrate, and have been demonstrated to promote an effective therapy. This section describes recent advances in the use of AuNRs as therapeutic NPs for cancer treatment. Therapeutic strategies, photothermal therapy, photodynamic therapy, and the synergistic effects will be addressed. The importance of AuNRs size, shape, and surface coating will also be discussed in each section. The importance of these AuNRs physicochemical characteristics will be mentioned throughout the text, allowing the understanding of these parameters in the in vitro/in vivo performance.

16.5.1 Therapeutic Strategies: Drugs, Nucleic Acids, Proteins

Chemotherapeutics cause cell death or slow the proliferation and spread of cancer cells by directly damaging the genetic code or cellular machinery interfering in cell cycling, inhibiting the cell from dividing uncontrollably. The free drug delivery promotes a lack of accumulation on the right site and high distribution by the normal tissues, affecting not only the cancer cells but also the normal cells. However, the exposure of these drugs administered in free form has a limited use, essentially due to several side effects resulting from nonspecific interactions of drugs with cells and tissues, the low solubility, cellular drug resistance, or unspecific biodistribution.

Table 16.4 Summary of the applications of AuNRs in multimodal approach

AuNRs code	Multimodal approach	Therapeutic approach	Covalent binding/ ligand exchange	Diseases	In vitro	In vivo	References
NR-HS-Ce6-dox	Photodynamic (PDT) Photothermal (PTT) Chemotherapy (CTX)	Doxorubicin	Ligand exchange	Squamous cell carcinoma	– cell uptake – therapeutic efficacy		Yeo et al. (2017a)
GNRs-HA-FA-DOX	Photothermal (PTT) Chemotherapy (CTX)	Doxorubicin	Covalent binding	Breast cancer	– cell cytotoxicity – cellular uptake – targeting ability – therapeutic effect	– pharmacokinetics – biodistribution – anticancer effect – systemic toxicity	Xu et al. (2017)
GNR-PDA	Photodynamic (PDT) Photothermal (PTT) Chemotherapy (CTX)	Methylene blue Doxorubicin	Ligand exchange	HeLa cells	– Photothermal effects – cellular uptake and imaging – intracellular ROS generation – cell cytotoxicity (PDT & PTT and CTX & PTT) – cell apoptosis	– toxicity – Photothermal effect – anticancer effect (PDT & PTT and CTX & PTT)	Wang et al. (2016c)
ISQ@BSA-AuNC@ AuNR@DAC@ DR5	Photodynamic (PDT) Photothermal (PTT) Chemotherapy (CTX)	Dacarbazine	Ligand exchange Covalent binding	Metastatic melanoma	– cell cytotoxicity	– biocompatibility	Sujai et al. (2018)

(continued)

Table 16.4 (continued)

AuNRs code	Multimodal approach	Therapeutic approach	Covalent binding/ ligand exchange	Diseases	In vitro	In vivo	References
Au NRs@ INU-LA-PEG-FA au NRs@ PHEA-EDA-FA	Photodynamic (PDT) Photothermal (PTT) Chemotherapy (CTX)	Nutlin-3	Ligand exchange Covalent binding	Osteosarcoma	– evaluation of hyperthermia – cell cytotoxicity – Thermoablation treatment and hyperthermia-triggered effect – quantitative cell uptake and enhanced hyperthermia-triggered drug internalization – two photons imaging		Li Volsi et al. (2017)
AuNR@CuPDA	Photothermal (PTT) Chemotherapy (CTX)		Ligand exchange	Cancer therapy and diagnosis	– cell cytotoxicity – Photothermal therapy	Pharmacokinetic and safety	Liu et al. (2017b)
AuNR-PTPEGm950	Photodynamic (PDT) Photothermal (PTT)		Ligand exchange	Human umbilical vein endothelial cell line (HUVEC) Human oral epidermoid carcinoma cell line (KB) Human hepatocarcinoma cell line (HepG2)	– cell cytotoxicity – Photothermal therapy	– blood circulation and biodistribution – histology – NIR photothermal assay	Liu et al. (2014)

GNRs-PEI1.8k	Photoacoustic (PA) Photothermal (PTT) Chemotherapy (CTX)	Nucleic acid	Covalent binding	Human cervical carcinoma (HeLa)	– cell cytotoxicity – gene transfection – intracellular uptake – Photothermal therapy	– photoacoustic imaging – Photothermal therapy	Chen et al. (2016)
pH-responsive DNA conjugated gold nanorod	Photothermal (PTT) Chemotherapy (CTX)	Doxorubicin DNA	Covalent binding	Multidrug resistant cancer cells	– cell cytotoxicity – cellular uptake – efflux studies – cellular accumulation under NIR irradiation – apoptosis assay		Zhang et al. (2016c)
RGD-125IPt-PDA@GNRs	Photothermal (PTT) Chemotherapy (CTX)	Cisplatin	Covalent binding	Cancer therapy	– cytotoxicity and chemotherapeutic effect of the probes in vitro – Photothermal and chemo-photothermal therapy effects in vitro	– SPECT/CT imaging – chemo-photothermal combined therapy and histological studies of tumors – photoacoustic imaging of tumor region	Zhang et al. (2016a)
GNR-DSPEI-PEG-RGD/DNA	Photothermal (PTT) Chemotherapy (CTX)	DNA	Covalent binding	Glioblastoma	– cell cytotoxicity – cellular uptake – cellular accumulation under NIR irradiation – apoptosis assay		Wang et al. (2014)
AuNR@MSN	Photodynamic therapy (PDT) Photothermal therapy (PPT)	Indocyanine green	Covalent binding	Cancer therapy	– cellular internalization – detection of multiple enhanced ROS production – cell cytotoxicity – cell apoptosis	– tumor therapy – Photothermal imaging and mice survival	Liu et al. (2018a)

(continued)

Table 16.4 (continued)

AuNRs code	Multimodal approach	Therapeutic approach	Covalent binding/ ligand exchange	Diseases	In vitro	In vivo	References
AuNR-SSPEI-PEG-biotin	Photothermal (PTT) Chemotherapy (CTX) Photodynamic therapy (PDT)	Doxorubicin dsDNA Pyronaridine	Covalent binding	Multidrug resistant (MDR) cancer	– cellular internalization – efflux studies – cell cytotoxicity		Wang et al. (2017)
RGD-PEG-DSPEI-GNR	Photothermal (PTT) Chemotherapy (CTX) Photodynamic therapy (PDT)	Small hairpin RNA	Covalent binding Self-assemble	Glioblastoma	– cell targeting of RGD – intracellular monitoring of the release of shRNA triggered – biocompatibility of the RDG	⇒ Biodistribution. ⇒ RNA interference.	Wang et al. (2015)
GNRs-siRNA nanoplex	Photothermal (PTT) Chemotherapy (CTX)	siRNA oligos BAG3 Doxorubicin	Ligand exchange	Cancer therapy	– heat shock response detection – cell uptake – evaluation of gene silencing efficiency of BAG3 siRNA – cell cytotoxicity and apoptosis	– distribution of GNR-siRNA inside the tumor – Photothermal treatment	Wang et al. (2016a)
cRGD-HN-DOX	Photothermal (PTT) Chemotherapy (CTX)	Doxorubicin	Covalent binding	Glioblastoma	– cellular internalization – cell cytotoxicity	– blood circulation and in vivo imaging – ex vivo imaging and biodistribution – antitumor efficacy – histological analysis	Zhong et al. (2014)

AuNRs@DOX	Photothermal (PTT) Chemotherapy (CTX)	Doxorubicin	Covalent binding	Cancer therapy	– cell uptake – combinational therapy	– combinational therapy	Chen et al. (2018)
DNA-AuNR	Chemotherapy (CTX)	DNA	Covalent binding self-assemble	Cancer therapy	– cytotoxicity of DNA-AuNR		Li et al. (2015)
CS/F-GNRs	Photothermal (PTT) Chemotherapy (CTX)	Doxorubicin	Self-assemble	Cancer therapy	– biocompatibility and hemocompatibility studies – Photothermal conversion efficiency – Photothermal therapy	– Photothermal therapy	Manivasagan et al. (2019)
DOX-DHHC-GNR^{AH}	Photothermal (PTT) Chemotherapy (CTX)	Doxorubicin	Covalent binding	Breast cancer	– stability of DOX-DHHC-GNR^{AH} – Photothermal properties of DOX-DHHC-GNR^{AH} – cellular uptake – Photothermal chemotherapy		Hou et al. (2019)
Ce_6-AuNR@ SiO_2-d-CPP	Photothermal (PTT) Photodynamic therapy (PDT)		Covalent binding	Breast cancer	– cellular uptake – cytotoxicity by PTT/PDT – apoptosis by PTT/PDT	– efficacy	Liu et al. (2018b)
AuNR-Si-ZnPc-HA	Photothermal (PTT) Photodynamic therapy (PDT)	Phthalocyanine (ZnPc)	Covalent binding	Cancer therapy	– cytotoxicity by PTT/PDT		Tham et al. (2016)

(continued)

Table 16.4 (continued)

AuNRs code	Multimodal approach	Therapeutic approach	Covalent binding/ ligand exchange	Diseases	In vitro	In vivo	References
AuNR@CuPDA	Photothermal (PTT) Photodynamic therapy (PDT)	Hematoporphyrin (HP)	Covalent binding	Cancer therapy	– cytotoxicity – Photothermal Therapy in vitro – cellular uptake	– Photothermal therapy	Terentyuk et al. (2014)
AuNRs-LA-COS	Photothermal (PTT)		Covalent binding	Breast cancer	– Photothermal heating – Photothermal ablation	– Photothermal ablation – histology analysis	Manivasagan et al. (2018)
AuNR@res	Photothermal therapy		Ligand exchange	Liver cancer	– Photothermal effect	– tumor accumulation – histology – Photothermal treatment	Wu et al. (2019)
AuNR-ASP-Ce6	Photothermal therapy (PTT) Photodynamic therapy (PDT)	Chlorin e6 (Ce6)	Covalent binding	Leukemia	– study of temperature enhancement in cells bound – cytotoxicity		Wang et al. (2012)

For these reasons, a promising approach comprises the application of NPs which could improve the targeted NPs and overcome drawbacks mentioned before. Thus, linking chemotherapeutics to an AuNRs can improve the functionality of the drug by delivering chemotherapeutic to a designated site of action. Moreover, the high surface area and easy surface modification simplify the loading of not only large therapeutic drugs, nucleic acids, and proteins, but also other biological molecules, including targeting molecules, linkers, or photosensitizers (Chen et al. 2018; Encinas-Basurto et al. 2018; Huang et al. 2011; Shen et al. 2014; Wang et al. 2016a). AuNPs have been extensively studied for application in anticancer therapy. That interest is due to enhance the drug solubility, drug protection from degradation in vivo, increase the bloodstream circulation time, avoid or decrease its removal by the reticuloendothelial system (RES), and consequently, targeted to cancer tissues. These strategies have been explored by several research groups (Table 16.4), in order to improve the therapeutic needs for new treatments in this field. As mentioned before, it is important to decrease the cytotoxicity associated with the AuNRs, and modifying the surface with ligands is a necessary step. Thus, AuNRs have an easy surface modification that allows the conjugation with different chemotherapeutic drugs (e.g., doxorubicin, paclitaxel, temozolomide) by coupling molecules in the surface through covalent bonding or noncovalent attachment. The noncovalent attachment is a good method for adsorption of drugs on AuNRs. However, a premature release can be an issue (Zhang et al. 2015). The covalent bonding is the more commonly used strategy due to their stable delivery. However, the required chemical modification of the drugs and internal/external triggers for active molecule release make this strategy less interesting.

Furthermore, AuNRs offer an attractive carrier for nucleic acids delivery because of its high surface boosting the payload/carrier ratio. There are two methods for coupling nucleic acids on the AuNRs surface by covalent attachment, which in this case is an effective approach. The modification does not inhibit biological activity, taking advantage of the thiol-modified surface or of the electrostatic interaction, between the positive charge of AuNRs and the strong negative charge of DNA/RNA. Both allow protection from enzymatic digestion. These approaches can effectively promote transfection by different cells (Table 16.4).

AuNRs-protein is another strategy that has been explored in cancer therapy, with special attention for targeting peptide sequences or receptor-specific antibodies. These coatings improve internalization of the AuNRs into diverse organelles or targeting specific cancer cells. Bovine serum albumin (BSA), monoclonal anti-epidermal growth factor receptor (anti-EGFR), tumor necrosis factor-α (TNF-α), RGD peptides, death-4 receptors (DR4), and transferrin receptor (Tfr) are examples of several protein molecules attached on the AuNRs surface (Liu et al. 2018a; Peralta et al. 2015; Wang et al. 2014, 2015, 2016a; Xu et al. 2017; Zhang et al. 2016a, c; Zhong et al. 2014). That conjugation provides good biocompatibility and improves the targeting of cancer cells, reducing the nonspecific toxicity. This approach has been implemented as direct conjugation of proteins to AuNPs through thiols or amines on the proteins. However, the preservation of the native structure of the proteins remains a challenge.

The AuNRs straightforward functionalization makes them attractive platforms for the development of drug delivery, nucleic acid delivery, or protein delivery vehicles. The improvement in the stability of conventional drugs and genes in biological fluids, avoidance of enzymatic degradation, and facilitation of their diffusion through biological barriers are some of the favorable attributes of the existing methodology. Despite the advantages and the numerous researches cited in Table 16.4, there are a number of questions to be resolved before AuNRs, as a carrier system, can be applied in clinical trials. The mechanisms of uptake can be used to address the usefulness of targeting moieties. However, it is a challenge to attach the targeting molecules without compromising their functionality. That consideration taken together remains a demand for continued improvement in the design and synthesis of AuNRs to accomplish the ultimate goals of attending as applicable delivery vehicles.

16.5.2 Photothermal Therapy

Photothermal therapy (PTT) has attracted a lot of attention in the cancer treatment, due to their noninvasive nature and low energy radiation, that can intensely penetrate the tissues (Riley and Day 2017). This method allows the application of plasmonic nanoparticles in medicine, where light penetration in tissues is ideal. The plasmonic nanoparticles have the ability to absorb near-infrared radiation (NIR) and effectively convert the light into heat when exposed to an appropriate laser of the right amplitude and frequency, which produces electron excitation and subsequent non-radiative relaxation. Thus, several CPNPs with photothermal characteristics, such as gold nanoparticles, graphene, or carbon nanotubes, have been used for in vitro and in vivo photothermal therapy. However, AuNRs have been described as more appropriate inorganic nanoparticles for effective photothermal therapy, because of their high photothermal conversion efficiency, low toxicity, easy synthesis, functionalization, and easily tuned LSPR-frequency to the NIR region. The intense LSPR-enhanced absorption of AuNRs exposition makes the photothermal conversion process highly efficient. The strong LSPR band supports NIR light absorption and releases large amounts of energy through thermal radiation and cause localized damage to close cells and tissues, usually five orders of magnitude larger than the strongest absorbing dye molecules. The diffused heat can produce a hyperthermia response in near cells, developing cell membrane disruption or protein denaturation inducing apoptosis, and tumor tissue necrosis. Despite site and amount of heat being controllable, the hyperthermia response is nonspecific. The cells heated to a temperature in the range of 41–50 °C begin to show signs of apoptosis, and cellular necrosis takes place for temperatures above 50 °C. That method causes changes in pH, perfusion, and oxygenation of the tumor microenvironment. Tumor cells have been demonstrated to possess more sensitivity to heat than healthy cells, due to their metabolic rate. PTT provides an appealing method for solid tumors treated with a minimally invasive approach and have significant advantages, including rapid recovery, fewer complications, and a deep penetration in tissues up to 4–10 cm. The NIR laser light is weakly absorbed by hemoglobin, melanin, or water, and thus the normal biological tissues are not damaged in this spectral region. The PTT performance strongly depends on the LSPR wavelength, geometries, size, concentration, and assembly state of AuNRs. In Table 16.4, there are several studies described where PTT is applied in distinct types of cancer, and the AuNRs performance is assessed by in vitro and in vivo studies.

Despite the advantages of PTT, the complete eradication of tumor cells using only this strategy is difficult. The heterogeneous distribution of NPs in tumor, the limitation of the penetration depth of NIR light in deep tissues, the direct eradication of metastatic cancer cells or metastatic nodules with PTT alone is a major challenge. Thus, in Sect. 16.5.1 the efficient combination with the current treatments to treat the tumors and metastatic cells in a synergistic approach will be addressed. A high number of studies have described the benefits of the combination of PTT with chemotherapy, photodynamic therapy (Table 16.4).

16.5.3 Photodynamic Therapy

Photodynamic therapy (PDT) involves the administration of a photosensitizer (PS) followed by the irradiation at the corresponding absorbance wavelength and has developed a huge interest as a cancer treatment modality (Luksiene 2003). Thus, the excitation of the PS molecule leads a transference of energy to the substrate, producing singlet oxygen and highly active free radicals that are able to lead the direct death of tumor cells, induce a local inflammatory reaction, and damage the microvasculature. The reactive oxygen species (ROS) generation can be controlled by the specific photoirradiation to the target diseases site and singlet oxygen can diffuse only 10–20 nm in the cell (Nomoto and Nishiyama 2018). Thus, the cytotoxicity can be restricted only to the photoirradiation region with minimal systemic toxicity compared to conventional chemotherapy, leading to unfavorable side effects in normal organs. In Table 16.5 and Fig. 16.2 are represented the four steps involved in the PDT. PS molecules have several disadvantages such as low molar extinction, low water solubility, photobleaching, insufficient selectivity, toxicity to normal cells, and the limitation in tissue penetration depth. For superficial tumors, PS molecules, under light wavelengths ranging from 400 nm to 600 nm, can exert a good penetration. However, for solid or deep-seated tumors a very limited range of PS may be excited by NIR light, as the "optical window" of biological tissue, for efficient ROS generation. Considerable efforts are being made to overcome these weaknesses and improve tumor targeting with minimal side effects. Some NPs have been studied in order to overcome the limitations of conventional PS molecules. The PDT performance of AuNRs has been investigated and exhibits a superior activity over conventional PS. AuNRs have advantages as improvement for targeting cancer cells, good resistance to photobleaching or enzymatic degradation, and the non-limitation in tissue penetration depth. Thus, PS molecules coupling to

Table 16.5 Steps involved in the photodynamic reaction

The absorption of a light photon	The intendency of light lead the excitation of electrons at a low energy molecular orbital to a higher energy molecular orbital or an excited singlet state ($^1PS^*$), causing a promotion of the PS molecule from its ground state to the extremely unstable excited singlet state with a half-life in the range of 10^{-6} to 10^{-9} s.
Fluorescence (FL)	$^1PS^*$ loses its energy by emitting FL through radiative transition. Another situation is undergoing intersystem crossing in which the spin of its excited electron inverts to form a relatively long-lived triplet state ($^3PS^0$).
Type I or II reaction	*I reaction*: $^3PS^0$ reacts with substrates and transfers a proton or an electron to form a radical anion or cation, respectively, which can further react with surrounding oxygen to produce hydrogen peroxides and free radicals. *II reaction*: $^3PS^0$ transfers its energy to molecular oxygen to form excited state singlet oxygen.
ROS	ROS interacts with the tumor cells, resulting in two typical types of cell death: Apoptosis and necrosis.

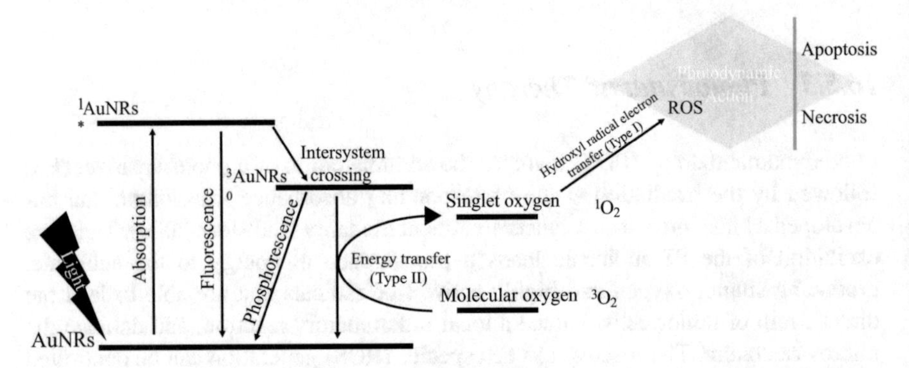

Fig. 16.2 Jablonski diagram showing the mechanisms of PDT using AuNRs. PDT involves tumor accumulation of AuNRs into the target site and subsequent photoactivation of AuNRs to generate cytotoxic ROS (Type I)

AuNRs can be used as an effective contrast agent in biomedical imaging (Ferreira et al. 2017; Freitas et al. 2017; Shi et al. 2016; Tham et al. 2016; Yan et al. 2018b). The AuNRs mechanism of action is correlated with the LSPR capacity. The incidence of light into AuNRs-PS resulted in light-to-heat conversion, as explained in Sect. 16.5.2. Therefore, AuNRs-PS generate hot electrons which can sensitize oxygen to form ROS and further exhibiting photodynamic properties.

Feng et al. (2019) developed a study that compared the PDT performance from different shapes: rods, shells, and cages (Feng et al. 2019). Gold nanoparticles with different shapes were prepared and their LSPR peaks were tuned to 808 nm, which means that different nanostructures absorbed a similar incident light energy and exert their photoactivity. The research demonstrated that AuNPs showed comparable photothermal profiles, but distinctive photodynamic activities. Gold nanocages displayed the most severe photodynamic activity, leading to a cell death and tumor growth inhibition based on the oxidative stress mechanism. Probably the large surface area and abundant corner angles of nanocages are the main reasons for these good results. However, the efficacy of PDT may be additionally boosted by the application of PS molecules in the AuNRs surface, enabling an enhanced excitation of PS molecule and consequent production on ROS. For this purpose, several PS molecules have been attached on the AuNRs surface (Li Volsi et al. 2017; Liu et al. 2014, 2018a, b; Sujai et al. 2018; Terentyuk et al. 2014; Tham et al. 2016; Wang et al. 2016c, 2017; Yeo et al. 2017a).

The PDT approach has been used in the treatment of malignant tumors because of the higher uptake of PS associated to the lower tumor pH, the nontoxicity of PS molecules until light incidence, and the possibility to combine with radiotherapy and chemotherapy for a more effective therapy (Table 16.4).

16.5.4 Synergistic Therapy

Taking into account the physical and chemical properties of AuNPs, including the easy functionalization with therapeutic molecules (chemotherapy), the LSPR conversion of the incident light energy into heat energy (PTT) or conversion of light into energy which produce ROS (PDT), different AuNRs, which combine PTT + PDT or chemotherapeutics + PTT or all in one, have been developed. In Table 16.3 different AuNRs are indicated, which explore the synergistic effect. The synergistic effect between these different strategies is more efficient in killing cancer cells, than each treatment alone. PTT and PDT stimulate some biological and physiological changes in the tumor microenvironment, which enlarge the cytotoxic effects of the different approach and enhance the efficacy of secondary treatment. Thus, the overall effect is crucial for decreasing the required treatment dosages. The hyperthermia generated by AuNRs allows increasing the permeability of tumor vessels and cellular uptake from cancer cell membranes. The increase in permeability leads to other intracellular effects, including DNA damage and protein denaturation.

A co-delivery of anticancer drugs carried by AuNRs displayed the combined photothermal-chemotherapy with enhanced therapeutic benefits (Table 16.4). AuNRs loaded with drugs or nucleic acids offers the possibility to improve therapeutic efficacy and combined with the laser irradiation it is possible to control the release of drugs and nucleic acids that are absorbed or covalently linked. Once again, compared to a single therapeutic treatment, the combined therapy not only avoids the administration of multiple doses of drugs and minimizes the side effects but also can severely improve the therapeutic efficacy. In most cases, beyond the high doses used in chemotherapy, success is conditioned by the resistance and adverse side effects. The synergistic effect between chemotherapy combined with PTT/PDT or both has been explored as to overcome the multidrug-resistant cancer cells and reduce the doses (Bhana et al. 2016; Freitas et al. 2017; Liu et al. 2018b; Tsai et al. 2018; Wang et al. 2017; Yi et al. 2017; Zhang et al. 2016c). Thus, hyperthermia leads to an increase in permeability, in blood vessels, extracellular matrix permeability, and in cancer cell membranes, and consequently improves the chemotherapy efficacy, as mentioned before. Besides that, PTT mediated by AuNRs helps in a localized targeting treatment, with tumor-specific heating, and avoiding the broadside effects. The higher limitation for combined photodynamic and photothermal in cancer therapy is the use of a single near-infrared irradiation.

Moreover, the combination of PTT with conventional treatments has been revealed as an approach to enhance cancer therapy, which has been evidenced to be more effective in reducing tumor, using a lower dosage of X-ray radiation and improving chemotherapy resistance, and inhibiting tumor metastasis than each technique used individually. Less contact and damage to the surrounding normal tissue has been reported as a good consequence of the synergistic effect (Table 16.4). The synergetic effect between strategies has the potential to completely eradicate tumors via noninvasive therapy, preventing tumor reoccurrence and metastasis.

16.6 Diagnosis Approach

An early cancer diagnosis is crucial to understand the localization of cancer cells and metastasis and to improve the therapeutic regimens. Inorganic NPs have been studied as a possible surrogate to conventional molecular probes in biomedical imaging, due to their strong optical properties, and small size (Kim et al. 2018). Besides the better imaging performance, they also provide versatility in drug delivery, colloidal stability, stimuli-responsive ability, nontoxicity, facile synthesis, and surface functionalization. In particular, AuNPs, as plasmon resonant particles, can be used to complement the fluorescent or radioactive labels in bioimaging modalities (Guo et al. 2017). These NPs have very high scattering cross sections, when compared to those of conventionally used labels. The most important feature of the use of plasmonic nanoparticles is the significant reduction in the detection limit of the method. Within the diversity of AuNPs shapes, AuNRs are attracting singular attention due to their scaling-up ability, size-controlled synthesis, and tunable absorption in the NIR region, qualities that make them appropriate to three-dimensional in vivo imaging. Specifically, AuNRs exhibit plasmon resonances four to five orders of magnitude higher than those of conventional dye molecules (Wei et al. 2010). AuNRs have become an effective class of diagnosis agents, with a huge application as photoacoustic tomography (PA), optical coherence tomography (OCT), X-ray computed tomography (CT), and two-photon luminescence (TPL) (Fig. 16.3 and Table 16.6).

In this section, the focus is on biological imaging modalities, considering the NIR absorbing properties of AuNRs, with special emphasis on PA (Yuan and Jiang 2010; Zeiderman et al. 2016). PA combines the advantages of ultrasound and optical imaging in a unique technique, overcoming the individual limitations associated with the isolated methods. Accordingly, the AuNRs are irradiated with light, resulting in localized heating and a small expansion of the tissue. The pressure distribution induced by the tissue expansion prompts acoustic wave propagation

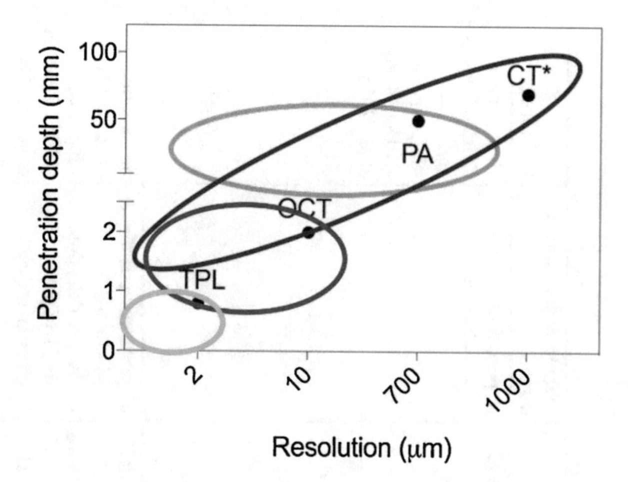

Fig. 16.3 Comparison of several biological imaging modalities according to their penetration depths. The colors represent the range of penetration depth of each bioimaging method: purple—CT; orange—PA; blue—OCT; green—TPL. *Unlimited penetration depth

toward tissue surfaces, which are detected by ultrasonic transducers. This enables the acquisition of real-time information with a high spatial resolution on the anatomical, functional, and molecular content of tumor tissues without ionizing radiation. At the same time, as sound can efficiently penetrate tissues, PA offers a 3D imaging of deeper tissues. The conventional contrast agents are limited in terms of the efficient conversion of the light absorbance and the heat capacity of the optical absorber, consequently, hampering the generation of PA signals, which compromises the sensitivity of the technique. The use of AuNPs as exogenous PA contrast agent allows a high optical absorbance and stability, and improves sensitivity of PA imaging. Additionally, a particularity of this bioimaging method is that it detects only signals from the regions where AuNRs accumulate on the tissue and not from isolated colloidal gold nanoparticles. It can be explained by the fact that the LSPR wavelength of aggregated AuNPs preferentially shifts toward red in detriment of dispersed AuNPs (Sau and Goia 2012). Recent advances in the understanding of the unique optical and electronic properties of AuNRs have supported their theranostic application, to improve the early-stage cancer diagnosis and the control of patients' response to treatment.

As an example, Sujai et al. 2018 developed a targeted three-in-one theranostic AuNRs, gathering photothermal therapy (PTT), photodynamic therapy (PDT), and chemotherapy, for surface-enhanced Raman scattering (SERS) guided photochemotherapy against metastatic melanoma (Sujai et al. 2018). The authors evaluated the apoptotic cell death induced by the theranostic NPs and confirmed the cellular event by molecular fingerprinting. The toxicity of the theranostic NPs displayed excellent biocompatibility in murine models. In another study, Knights and McLaughlan developed AuNRs to be used in PA imaging and plasmonic photothermal therapy as theranostic NPs (Knights and McLaughlan 2018). The study compared four different-sized AuNRs and concluded that the larger-sized AuNRs, presenting a width of <40 nm, showed a stronger PA signal, when compared with <25 nm AuNRs at equivalent NP/mL. However, the smallest-sized AuNRs

Table 16.6 Overview of biological imaging modalities and their characteristics

Imaging modality	Medical imaging procedure	Spatial resolution	Maximum penetration depth	References
Photoacoustic tomography (PA)	The biological tissue is irradiated with light, which results in localized heating, and a small expansion of tissue, causing a sound wave. Use of short laser pulses generates sound waves in the ultrasonic frequency range. ✓ the contrast in photoacoustic images depends primarily on how much light is absorbed by the tissue and/or the contrast agent. ✓ different types of biological tissues have different light absorption coefficients, which make this imaging modality able to discriminate between tissues. ✓ it is a non-ionizing radiation and the use of ultrasound as the output results in higher spatial resolution compared to optical methods due to the lower scattering of ultrasound in tissue.	✓ spatial resolution ~700 μm. ✓ the signal obtained is proportional to the AuNPs concentration.	✓ penetration depth ~10–50 mm.	Hartman et al. (2019), Knights and McLaughlan (2018), Moon et al. (2015)
Optical coherence tomography (OCT)	Optical modality that enables one obtaining cross-sectional or 3D reconstructions of semitransparent tissue by measuring the echo time delay and magnitude of back reflected light. ✓ contrast provided by index of refraction mismatch in tissue. ✓ interferometric depth localization. ✓ non-ionising radiation at biologically safe levels, allowing for long exposure times. Larger AuNRs with greatly enhanced scattering cross-sections could potentially achieve the signal-to-background ratio necessary to realize the advantages of OCT with molecular contrast.	✓ low millimeter range: Axial resolutions on the order of 10 μm, and lateral resolutions in the low micron range. ✓ 10–25 times greater spatial resolution than that produced by ultrasound imaging, and up to 100 times better than MRI or CT ✓ A very high concentration would be required to produce detectable contrast.	✓ penetration depth ~1–2 mm. ✓ AuNRs detection limits in the picomolar range.	de La Zerda et al. (2015), Liba et al. (2016), Popescu et al. (2011), Ratheesh et al. (2016)

X-ray computed tomography (CT)	X-ray imaging is based on the absorption or scattering of photons as a collimated X-ray beam passes through a contrast agent. ✓ high-Z elements are preferred as CT contrast agents. ✓ the blood circulation time of the iodinated contrast agents is very short, preventing their preferential accumulation in a lesion. ✓ AuroVist is a new X-ray contrast agent for research and animal use. X radiation effect.	✓ Spatial resolution: 0.5–1 mm.	✓ Unlimited penetration depth.	Cole et al. (2015), Cormode et al. (2015), Manohar et al. (2016)
Two-photon luminescence (TPL)	TPL and other multiphoton processes can be excited at NIR frequencies between 800 and 1300 nm, the window of greatest transmittivity through biological tissue. TPL imaging provides high 3D spatial resolution and sufficient penetration depth to monitor biological events in vivo. ✓ high resolution. ✓ low photodamage.	✓ spatial resolution 1–2 μm. ✓ TPL signals were approximately three times higher than the background autofluorescence.	✓ penetration depth in ~800 μm.	Chen et al. (2014), Li and Gu (2010), Wang et al. (2009), Zhao et al. (2012)

studied (10 nm) would be the most suitable to induce cell death. Besides the NPs size and the concentration, they concluded that the number of AuNRs is crucial the best performance.

More studies are in progress with AuNRs, including LSPR modulated according to the shape and size, biocompatibility, ability to act as cancer drug carriers, PTT/PDT agents, or alternatively as exogenous contrast agents for image-guided therapy (theranostic application) and the monitoring of therapy outcomes (Liu et al. 2015; Qin et al. 2015). Despite the huge potential to be translated into cancer diagnosis and treatment, the clinical translation of AuNRs will also be impeded by several fundamental limitations (Clancy et al. 2014).

There are already some AuNPs approved for clinical use in diagnosis. However, the scarce number of clinical trials involving AuNPs denotes the difficulty to comply with the regulatory requirements, delaying their approval. Until now, CYT-6091 NCT00356980 and NCT00436410 (Cytimmune 2019), NU-0129 (NCT03020017), and AuroShell® (NCT01679470 and NCT01679470) are the unique Au formulations clinically investigated for cancer treatment. CYT-6091, also known as Aurimune, is used as a carrier to deliver the recombinant human tumor necrosis factor alpha (rhTNF) into tumors, to increase the penetration of chemotherapeutic drugs and the antitumor activity. A thiol-PEG coating allows improving the safe delivery of highly effective doses of rhTNF to tumor cells and immune detection, without reaching and damaging healthy blood vessels and tissues. The company further developed a second generation of Aurimune (CYT-21000) to carry paclitaxel and TNF. This consubstantiates the ability of AuNPs to simultaneously deliver biologic and therapeutic molecules. In turn, NU-0129 is a spherical nucleic acid gold nanoparticle developed for targeting BCL2L12 in recurrent glioblastoma multiforme or gliosarcoma patients, while AuroShell® is used for the refractory and/or recurrent tumors of the head and neck.

However, several AuNRs under development only show basic ideas that are conceptually enthusiastic. Despite the great scientific application, most of them do not yet possess a solid preclinical proof of concept, due to failure in the correlation between the in vitro and the in vivo results, and consequently remain without any predictable clinical impact. A recent review on the preclinical and clinical advances of inorganic nanoparticles for cancer theranostics enlightens the difficulty to implement this kind of NPs in the market (Mendes et al. 2018a).

16.7 Biological Medium

Interaction of NPs with the biological environment has influence on their biological activity, and a systematic understanding of the nature of these interactions is fundamental for suitable designing of the NPs for biological purposes. Gold is a noble metal nontoxic, inert, biocompatible, and with several therapeutic properties. However, AuNPs behavior is very different compared to Au bulk form. Consequently, its safety as NPs form for biomedical applications becomes questionable, and important concerns have been raised in the assessment of risks for humans. The use of AuNPs as a biomedical approach, taking advantages from drug delivery, optical bioimaging for diagnosis, has grown. However, as aforementioned, their translation to the clinic has not been easy. The need to prove their safety in vitro and in vivo has been explored by numerous research groups. Still, the results have been controversial. As if all the contradictions were not enough, the regulation for the NPs is still very tight and their translation into clinical practices is delayed by various factors, which include a poor definition of their biological interactions.

From the fabrication process, physicochemical properties, including size, shape, or surface functionalization, to the biological performance, AuNPs offer an extraordinarily wide range of possibilities. For example, AuNPs exist in a variety of different shapes, such as nanospheres, nanorods, nanocages, nanoshells, and nanostars, among others, and their particle size is also variable. Lastly, the functionalization can be easier using a variety of coating molecules, like proteins, nucleic acids, drugs, surfactants, or polymers. The following parameters need to be analyzed, at the same circumstances, in order to understand the grade of safety associated with AuNRs.

16.7.1 AuNPs-Body Environment, Protein Corona Formation

The colloidal stability of NPs in the biological medium is important to preserve their properties and prompt a desirable biological response. However, NPs in blood circulation interact with the biomolecules leading to the formation of a biocorona (Carnovale et al. 2017; Cheng et al. 2015). This can interfere positively or negatively in the NPs performance, by regulating the NPs-cell interactions, including endocytic pathways and biological responses, with potential ability to limit access to the targeting receptor. It has been reported to cause a substantial inhibition on the cellular uptake of large-sized AuNPs, consequently generating differences in the intracellular location of NPs (Foroozandeh and Aziz 2018).

The biocorona formation also depends on the NPs composition and charge (Mirshafiee et al. 2013).

In order to assess the biological interaction of AuNPs with different charges, Alex et al. investigated the coronation of three biologically relevant proteins (HSA, IgG, and transferrin) and their impact on cellular uptake and viability (Alex et al. 2017b). They described that the formation of IgG biocorona helped to decrease cellular toxicity, and revealed a better cell uptake. Another study showed that after 1 h of exposition, AuNPs exhibited the maximum protein adsorption and the biocorona layer quantified was found to vary based on AuNPs functionalization (CTAB, PEG, and PSS). They concluded that the neutrally charged PEG-AuNPs was the least affected by proteins and suggested that there are more interaction mechanisms apart from electrostatic interactions which play an important role, as hydrophobic interactions, entropy-driven binding, and covalent binding via cysteine group. Other studies described the importance of PEG-coated AuNPs, showing good stability and biocompatibility profiles in comparison with CTAB-capped AuNPs. Additionally, PEG-coated AuNPs manifested a lower reactivity with proteins in comparison to CTAB-capped AuNPs (Alex et al. 2017a, b).

Recently, an in vivo study compared the effect of size and shape on the biocorona formation under in vivo blood flow conditions (García-Álvarez et al. 2018). AuNRs and AuNSs, with 40 and 70 nm, respectively, were intravenously administered in CD-1 mice, and the in vivo protein biocoronas were characterized. The results showed that the total amount of protein adsorbed on NPs and their composition were affected by AuNPs size and shape. The larger NPs adsorbed more proteins on the surface; however, the biocorona complexity did not reflect the total amount of adsorbed protein. The in vivo behavior of AuNRs preincubated with mouse serum or mouse serum albumin before intravenous injection were assessed in order to understand if the biocorona could influence the in vivo metabolic pattern and toxicity of AuNRs (Cai et al. 2016). The preincubation allowed that AuNRs escape from the phagocytosis, having been found in the hepatocytes. At the same time, the biocorona-AuNRs led to long-term retention in the liver. All these aspects need to be in consideration and a complete characterization of the biocorona under realistic in vivo conditions for different NPs is necessary to improve the understanding of clinical performance.

A strategic point of view consists in the design of biocorona-AuNPs, considering the biocorona beneficial properties (Kah et al. 2012). Despite the uncontrolled composition of biocorona in a biological medium, it has the ability to solubilize hydrophobic drugs or charged molecules and stabilize the NPs. Such approach has revealed to improve the AuNRs properties, including drug loading, optical performance, and stability (Yeo et al. 2017b). Thus, it is possible to use biocorona formation as a strategy for cancer treatment, instead of considering it as an undesirable biological artifact.

Summing-up, the biocorona formation on NPs is an active process that is representative of the environment, depends on the patient or the disease, and is far from being completely understood (Nierenberg et al. 2018). Positive aspects that can be pointed out include (1) the ability to avoid aggregation of the AuNRs or, sometimes, (2) the facilitation of cellular uptake; however, there are also negative points, such as (1) obstruction of targeting ligands on the surface of AuNRs, (2) initiation of inflammation or complement activation, resulting in an immunogenic response, and (3) inter- or intraindividual variability (Barbero et al. 2017; Nierenberg et al. 2018).

16.7.2 AuNPs Cell Uptake and Toxicity

Generally, AuNPs toxicity depends on their ability to cross the cell membrane, which, in turn, is influenced by NPs size, shape, and surface coating (Li et al. 2016). The size and the surface coating govern the cellular internalization, due to the type of interaction that is established, encompassing electrostatic, Van der Waals, or hydrophobic forces. A NP hydrophobicity higher than that of cell membrane leads to an improved accessibility to the cell interior. Moreover, larger NPs (>500 nm) enter cells via phagocytosis, while NPs < 100 nm are internalized by receptor-mediated endocytosis.

In what concerns surface coating, the example of AuNRs can be mentioned. AuNRs are typically synthesized considering a high concentration of CTAB solution, which forms a non-covalent CTAB bilayer surrounding NPs (Casas et al. 2013; Garabagiu and Bratu 2013; He et al. 2018; Scaletti et al. 2014). Despite the stabilizing effect of CTAB, this excipient is particularly problematic for biological systems because of its cytotoxicity and cell membrane disrupting ability. On the other hand, the CTAB layer is a barrier to the easy functionalization of AuNRs. Changing the CTAB layer by other biocompatible molecules (e.g., poly-styrene sulfonate—PSS, 11-mercaptoundecanoic acids—MUA, cysteine, PEG-thiol, BSA, PEI) has been employed, having resulted in AuNRs with a largely increased biocompatibility and functionalization ability (Allen et al. 2017; Yan et al. 2018a; Zarska et al. 2018). These different molecules can differently affect, either increasing or hindering, cell uptake (Grabinski et al. 2011).

The effects of different aspect ratios and surface modifications on the cytotoxicity and cellular uptake of AuNRs have also been investigated. There are some studies wherein the aspect ratio did not demonstrate an effect on toxicity, but the surface chemistry influence both the cellular uptake and toxicity (Wan et al. 2015). AuNPs with an aspect ratio of approximated to 1 are reported to perform better than those with an aspect ratio in excess of 1, due to their ability to enter the cell more effectively (Yang et al. 2016). Targeted AuNRs with an aspect ratio between 1 and 2 have a better performance and maximizes the cellular uptake. In contrast, an aspect ratio higher than 4 leads to a lower cellular uptake. Ying Than et al. studied the interaction between A549 cells with a AuNRs. AuNRs with an aspect ratio of 2 was internalized, localized in the lysosomes and membranous vesicles (Tang et al. 2015). The shape or geometry did not show a significant influence on the ability of targeted AuNRs to bind to the cell membrane or engage in a particular uptake pathway. The major part of small AuNRs interacts with cells by endocytosis. This pathway is size-dependent and, for that reason, AuNRs with higher aspect ratio are significantly longer, having a limited internalization (Debrosse et al. 2013).

16.7.3 AuNPs Biodistribution and Excretion

Following the FDA guidelines, NPs should be eliminated via metabolism or excretion processes after their administration. Biodistribution, biokinetics, and excretion are important in vivo parameters that should be evaluated before any AuNPs potential therapeutic applications. AuNRs are easily visualized in biological compartments by TEM and the amount of gold inside the compartments can be measured by ICP-MS, which make them excellent probes to understand the distribution of NPs in the whole body. Particle size, shape, and surface properties (e.g., surface charge, targeting molecules, hydrophilicity, or hydrophobicity) are critical physicochemical characteristics that governed their behavior in vivo. The target organ, the affinity for cell membrane, the specific local accumulation, release from target organs, and toxicological outcomes are controlled not only by surface coating but also by the ability of biocorona formation, as mentioned before. The effective dose reaching the target organ depends not only on the administration route but also on the AuNPs ability to cross physiological barriers to reach their final destination.

The applications of AuNRs as multimodal NPs are summarized in Table 16.4. A total of 26 articles are indicated, but only 16 articles performed in vivo studies addressing biodistribution, pharmacokinetics, and toxicity parameters. This approach demonstrated not only the lack of in vivo studies, but also that simple in vitro experiments may not lead to good predictions regarding in vivo results. A comparative study on the importance of the AuNRs shape and biodistribution profiles was presented (Robinson et al. 2015). The results showed that ca. 20.5% of the original injected dose of AuNRs was still circulating in the blood after 6 h, but they were almost undetectable in the blood by 24 h. The blood distribution of nanocages varied over this time, with 13.8% of the original injected dose circulating at 6 h. The authors have two explanations for these results: differences in particle count or in shape of the NPs promote a different filtration at the organ level.

In what pertains to particle coating, PEG and hyaluronic acid were both efficient in prolonging circulation (Robinson et al. 2015) of the NPs, with the latter also ensuring tumor-targeted delivery (Xu et al. 2017, 2019). The AuNPs biodistribution is mostly affected by the particle volume, rather than by the aspect ratio (Song et al. 2015; Tong et al. 2016). A smaller volume and a high aspect ratio not only allow a high accumulation in the tumor, but also lead to a rapid elimination from the organism. However, in most of the studies, after ca. 1 h the kidney, liver, and spleen (Gao et al. 2016; Xu et al. 2017, 2019) show the highest accumulation of AuNRs. Despite this higher accumulation not being located in the tumor, AuNRs are only toxic when irradiated by light/laser. Thus, if the irradiation is focused only in the tumor, this kind of NPs does not display toxicity for the organs, even in higher amounts (Liu et al. 2014; Manivasagan et al. 2018; Terentyuk et al. 2014; Wang et al. 2015, 2016a; Wu et al. 2019; Zhong et al. 2014).

Some points remain to be addressed, including AuNRs characterization prior to and after mixing with the biological media, determination of the effective therapeutic dosage of these AuNRs and the assessment of the potential toxicity of AuNPs using different administration routes (e.g., inhalation, oral absorption, subcutaneous).

16.8 Gold Nanorod in Glioblastoma

Glioblastoma multiforme (GB) is among the most lethal and aggressive tumors (Seyfried et al. 2015). GB belongs to a class of tumors with high heterogeneity and a high tendency to metastasize, characterized by poor prognosis and low patient survival rate. The conventional treatments do not show progress in reducing the preliminary tumor burden and metastasis in GB. The most challenging problems in the GB are its particularly complex and heterogeneous molecular biology, the presence of different biological barriers as the blood-brain barrier (BBB), multidrug-resistance efflux transporters, and diffuse topography of gliomas (Mendes et al. 2018b). The low therapeutic efficacy, associated with a wide side effects spectrum, involving damage in healthy tissues, the requirement of regular invasive dose regimens, and the late diagnosis are responsible for tumor recurrence and treatment failure. Considering this overview, an early cancer diagnosis, localization, and postoperative imaging inspection are important aspects to improve the beneficial effect of tumor therapy. Thus, AuNRs have become an attractive photothermal therapy, applied as contrast agents for photoacoustic and near-infrared (NIR) imaging suitable to use in GB treatment. This strategy for controlling GB treatment tries to surmount the issues associated with efficacy, lack of specificity, and delivery.

The biofunctionalization of AuNRs with molecules to target cancer cells has proven to be highly effective. There are molecules overexpressed specifically on the cancer cell surface (epidermal growth factor receptors, folic acid, CD44, integrins, angiopoietin 2 or NF-kB), including GB. Thus, biofunctionalization of AuNRs has been explored to specifically target and destroy different cancer cells, rendering the tumor cells susceptible to damage. AuNRs have been combined with antibodies, polymers, or proteins (Fernández-Cabada et al. 2016; Gonçalves et al. 2018; Ruff et al. 2017; Velasco-Aguirre et al. 2017; Verma et al. 2016). The inefficient uniform diffusion of NPs into tumors is the critical point for several therapeutic systems, and the functionalization of NPs has been described as a strategy to overcome these critical points. Additionally, in vitro and in vivo studies have consubstantiated the effectiveness of targeting-AuNRs. The main drawback for NPs delivery into the brain is the presence of BBB, which limits the transport of NPs from blood to brain and vice versa. As such, AuNRs functionalization allows a great cellular uptake by GB cells or cancer stem cells, and penetration abilities (Dixit et al. 2015; Gonçalves et al. 2018; Velasco-Aguirre et al. 2017). Transferrin peptide (Tfpep) exhibited a substantial increase in cellular uptake for targeted conjugates as compared to untargeted AuNPs (Dixit et al. 2015). AuNRs combined with an antibody against anti-epidermal growth factor receptor (EGFR) to eliminate tumor cells were developed and assessed in vitro (Fernández-Cabada et al. 2016). Angiopep-2, a shuttle peptide that can cross the BBB, also demonstrated to improve the in vivo delivery of AuNRs to the central nervous system (Velasco-Aguirre et al. 2017).

On the other hand, studies have reported the development of hydrogels for GB treatment (Basso et al. 2018; Gonçalves et al. 2017; Lin et al. 2017; Zhang et al. 2016b). Zhang et al. developed a hydrogel which combines synergistic strategies, photothermal and localized chemotherapy, for the GB treatment (Zhang et al. 2016b). PEG-AuNRs and paclitaxel were uniformly incorporated in the thermal-gelling matrix and injected intratumorally. In this way, the drug release was sustained by the AuNRs heat. At the same time, the sustained drug release allowed the elimination of the residual tumors cells that have survived to the photothermal treatment. The hydrogel contributed to a delay and even elimination of tumor recurrence. A new strategy considers cell-based delivery of NPs, especially monocytes/macrophages (Mo/Ma) as AuNPs delivery vectors for photothermal therapy of brain tumors (Baek et al. 2011; Hirschberg and Madsen 2017; Madsen et al. 2012, 2015; Trinidad et al. 2014). These types of cells have the capacity to ingest AuNRs and subsequently infiltrate and target the tumor core by an active process. PTT has been addressed but only one in vivo study has considered this strategy (Madsen et al. 2015). PTT mediated by the administration of Ma-AuNPs induced complete tumor inhibition or delayed growth. These cell-delivered AuNPs have shown a high potential for GB treatment.

Several recent studies have explored the effect of AuNRs therapy and imaging on a variety of cancers, including GB. However, few studies have shown in vivo efficacy, so warrant that AuNRs can cross the BBB and effectively target GB. At the same time, most of the in vivo studies use a subcutaneous glioma model, which limits not only the understanding on BBB cross ability and the effect of the local immunological environment, but also if the laser penetrates the skull and soft tissue reaching deep-seated tumors.

16.9 Conclusions

As indicated throughout this chapter, the key properties of AuNRs make them attractive multimodal NPs for cancer treatment. These include biocompatibility, adjustable size, optical properties, and easy functionalization. They also possess a promising potential for imaging, with minimally invasive methods. There are, however, barriers still to be addressed to improve therapeutic efficacy. AuNRs are dynamically employed in photothermal therapy, photodynamic therapy, and chemotherapeutics drug delivery, with promising results. Importantly, in single NPs, different therapeutic approaches can be combined, providing an interesting strategy with a promising outcome. The synergistic effect between optical (PTT or PDT) and chemical stimuli (chemotherapeutic drugs) allows to increase the permeability of blood vessels, and consequently promotes a higher sensitivity of the tumor cells. Strategically, AuNRs can completely eradicate tumors in a noninvasive therapy, with large toxicity effect in the right tissue, and preventing tumor reoccurrence and metastasis. On the other hand, diagnostic properties of AuNRs enable the noninvasive assessment of anatomical and functional information, with image-guided drug delivery sparkling a great deal of attention. However, the selection of the imaging technique requires considering parameters such as target tissue, resolution, sensitivity, depth, contrast, and implementation.

In addition, a new multimodal NPs dual-strategy, diagnostic and therapy, is desired and can have a significant impact on health costs by reducing the number of diagnostic tests and increasing therapy efficacy. AuNRs are a promising theranostic NPs in cancer diseases. Despite the optimal theranostic AuNRs outcomes, the expectations of these reaching the preclinical stage are still far from being applied. There are fundamental issues that need to be addressed before AuNRs can be translated into clinical applications. Questions related to "ideal" size, better surface modification, right biodistribution, and a good excretion from the body need to be clarified. Size and relative dimensions influence optical performance, while surface coating and functionalization determine the in vivo behavior, which includes biocorona formation and biodistribution to the target site or, in alternative, fast recognition as foreign-body. Toxicity, excretion, and elimination from the body must also be taken into consideration. Most of the studies have controversial results, which hinders acceptance of AuNRs in the market.

AuNRs as a multifunctional theranostic NPs present great potential for cancer therapy and exhibit promising in vitro and in vivo results. The main challenges to the translation of AuNRs rely on establishing the required therapeutic and toxic doses, the biological monitoring, and the long-term fate of NPs in the body.

Acknowledgements The authors acknowledge Fundação para a Ciência e a Tecnologia (FCT), Portuguese Agency for Scientific Research, for financial support through the Research Project No. 016648 (Ref. POCI-01-0145-FEDER-016648), the project Pest UID/NEU/04539/2013, COMPETE (Ref. POCI-01-0145-FEDER-007440), the project POCI-01-0145-FEDER-007440 and UID/NEU/04539/2013 (CNC.IBILI Consortium strategic project). Maria Mendes acknowledges the PhD research grant SFRH/BD/133996/2017 also assigned by FCT.

References

Alex S, Tiwari A. Functionalized gold nanoparticles: synthesis, properties and applications—a review. J Nanosci Nanotechnol. 2015;15(3):1869–94. ISSN: 1533-4880.

Alex SA, Rajiv S, et al. Significance of surface functionalization of gold Nanorods for reduced effect on IgG stability and minimization of cytotoxicity. Mater Sci Eng C. 2017a;71:744–54. ISSN: 0928-4931.

Alex SA, Chandrasekaran N, Mukherjee A. Impact of gold nanorod functionalization on biocorona formation and their biological implication. J Mol Liq. 2017b;248:703–12. ISSN: 0167-7322.

Allen JM, et al. Synthesis of less toxic gold nanorods by using dodecylethyldimethylammonium bromide as an alternative growth-directing surfactant. J Colloid Interface Sci. 2017;505:1172–6. ISSN: 0021-9797.

Amendola V, et al. Surface plasmon resonance in gold nanoparticles: a review. J Phys Condens Matter. 2017;29(20):203002. ISSN: 0953-8984.

Baek S-K, et al. Photothermal treatment of glioma; an in vitro study of macrophage-mediated delivery of gold nanoshells. J Neurooncol. 2011;104(2):439–48. ISSN: 0167-594X.

Barbero F, et al. Formation of the protein corona: the interface between nanoparticles and the immune system. Semin Immunol. 2017;34:52–60. ISBN: 1044-5323.

Basso J, et al. Hydrogel-based drug delivery nanosystems for the treatment of brain tumors. Gels. 2018;4(3):62.

Becker R. CTAB promoted synthesis of au nanorods–temperature effects and stability considerations. J Colloid Interface Sci. 2010;343(1):25–30. ISSN: 0021-9797.

Bhana S, et al. Photosensitizer-loaded gold nanorods for near infrared photodynamic and photothermal cancer therapy. J Colloid Interface Sci. 2016;469:8–16. https://doi.org/10.1016/j.jcis.2016.02.012. ISSN: 0021-9797.

Cai H, et al. Protein corona influences liver accumulation and hepatotoxicity of gold nanorods. NanoImpact. 2016;3:40–6. ISSN: 2452-0748.

Carnovale C, et al. Gold nanoparticle biodistribution and toxicity: role of biological corona in relation with nanoparticle characteristics. In: Metal nanoparticles in pharma. Cham: Springer; 2017. p. 419–36.

Casas J, et al. Replacement of cetyltrimethylammoniumbromide bilayer on gold nanorod by alkanethiol crosslinker for enhanced plasmon resonance sensitivity. Biosensor Bioelectron. 2013;49:525–30. ISSN: 0956-5663.

Chen N-T, et al. Enhanced plasmonic resonance energy transfer in mesoporous silica-encased gold nanorod for two-photon-activated photodynamic therapy. Theranostics. 2014;4(8):798.

Chen J, et al. Gold-nanorods-based gene carriers with the capability of photoacoustic imaging and photothermal therapy. ACS Appl Mater Interfaces. 2016;8(46):31558–66. ISSN: 1944-8244.

Chen J, et al. Doxorubicin-conjugated pH-responsive gold nanorods for combined photothermal therapy and chemotherapy of cancer. Bioact Mater. 2018;3(3):347–54. ISSN: 2452-199X.

Cheng X, et al. Protein corona influences cellular uptake of gold nanoparticles by phagocytic and nonphagocytic cells in a size-dependent manner. ACS Appl Mater Interfaces. 2015;7(37):20568–75. ISSN: 1944-8244.

Chhatre A, et al. Formation of gold nanorods by seeded growth: mechanisms and modeling. Cryst Growth Des. 2018;18(6):3269–82. ISSN: 1528-7483.

Clancy MK, et al. Clinical translation and regulations of theranostics. In: Cancer theranostics. Amsterdam: Elsevier Inc.; 2014. ISBN: 9780124077225.

Cole LE, et al. Gold nanoparticles as contrast agents in X-ray imaging and computed tomography. Nanomedicine. 2015;10(2):321–41. ISSN: 1743-5889.

Cormode DP, et al. Nanoparticle contrast agents for computed tomography: a focus on micelles. HHS Public Access. 2015;9(1):37–52. https://doi.org/10.1002/cmmi.1551.Nanoparticle.

Cytimmune. Aurimune: a nanomedicine platform—[Em linha] [Consult. 27 Mar 2019]. Assessed on http://cytimmune.com/#pipeline.

de La Zerda A, et al. Optical coherence contrast imaging using gold nanorods in living mice eyes. Clin Exp Ophthalmol. 2015;43(4):358–66. ISSN: 1442-6404.

Debrosse MC, et al. High aspect ratio gold nanorods displayed augmented cellular internalization and surface chemistry mediated cytotoxicity. Mater Sci Eng C. 2013;33(7):4094–100. ISSN: 0928-4931.

Dixit S, et al. Transferrin receptor-targeted theranostic gold nanoparticles for photosensitizer delivery in brain tumors. Nanoscale. 2015;7(5):1782–90.

Duan R, et al. Chitosan-coated gold nanorods for cancer therapy combining chemical and photothermal effects. Macromol Biosci. 2014;14(8):1160–9. ISSN: 1616-5187.

Encinas-Basurto D, et al. Hybrid folic acid-conjugated gold nanorods-loaded human serum albumin nanoparticles for simultaneous photothermal and chemotherapeutic therapy. Mater Sci Eng C. 2018;91:669–78. ISSN: 0928-4931.

Feng Y, et al. Differential photothermal and photodynamic performance behaviors of gold nanorods, nanoshells and nanocages under identical energy conditions. Biomater Sci. 2019;7(4):1448–62. https://doi.org/10.1039/C8BM01122B.

Fernández-Cabada T, et al. Optical hyperthermia using anti-epidermal growth factor receptor-conjugated gold nanorods to induce cell death in glioblastoma cell lines. J Nanosci Nanotechnol. 2016;16(7):7689–95. ISSN: 1533-4880.

Ferreira DC, et al. Hybrid systems based on gold nanostructures and porphyrins as promising photosensitizers for photodynamic therapy. Colloids Surf B: Biointerfaces. 2017;150:297–307. https://doi.org/10.1016/j.colsurfb.2016.10.042. ISSN: 0927-7765.

Foroozandeh P, Aziz AA. Insight into cellular uptake and intracellular trafficking of nanoparticles. Nanoscale Res Lett. 2018;13(1):339. ISSN: 1931-7573.

Freitas LF, et al. Zinc phthalocyanines attached to gold nanorods for simultaneous hyperthermic and photodynamic therapies against melanoma in vitro. J Photochem Photobiol B Biol. 2017;173:181–6. https://doi.org/10.1016/j.jphotobiol.2017.05.037. ISSN: 1011-1344.

Gao B, et al. Cellular uptake and intra-organ biodistribution of functionalized silica-coated gold nanorods. Mol Imaging Biol. 2016;18(5):667–76. ISSN: 1536-1632.

Garabagiu S, Bratu I. Thiol containing carboxylic acids remove the CTAB surfactant onto the surface of gold nanorods: an FTIR spectroscopic study. Appl Surf Sci. 2013;284:780–3. ISSN: 0169-4332.

García-Álvarez R, et al. In vivo formation of protein corona on gold nanoparticles. The effect of their size and shape. Nanoscale. 2018;10(3):1256–64.

Gonçalves DPN, et al. Enhanced targeting of invasive glioblastoma cells by peptide-functionalized gold nanorods in hydrogel-based 3D cultures. Acta Biomater. 2017;58:12–25. https://doi.org/10.1016/j.actbio.2017.05.054. ISSN: 18787568.

Gonçalves DPN, et al. Modular peptide-functionalized gold nanorods for effective glioblastoma multicellular tumor spheroid targeting. Biomater Sci. 2018;6(5):1140–6.

Grabinski C, et al. Effect of gold nanorod surface chemistry on cellular response. ACS Nano. 2011;5(4):2870–9. ISSN: 1936-0851.

Guo J, et al. Gold nanoparticles enlighten the future of cancer theranostics. Int J Nanomed. 2017;12:6131.

Hanske C, et al. Silica-coated plasmonic metal nanoparticles in action. Adv Mater. 2018;30(27):1707003. ISSN: 0935-9648.

Hartman RK, et al. Photoacoustic imaging of gold nanorods in the brain delivered via microbubble-assisted focused ultrasound: a tool for in vivo molecular neuroimaging. Laser Phys Lett. 2019;16(2):25603. ISSN: 1612-202X.

He GS, et al. Rayleigh, Mie, and Tyndall scatterings of polystyrene microspheres in water: wavelength, size, and angle dependences. J Appl Phys. 2009;105(2):23110. ISSN: 0021-8979.

He J, et al. The facile removal of CTAB from the surface of gold nanorods. Colloids Surf B Biointerfaces. 2018;163:140–5. ISSN: 0927-7765.

Hemmer E, et al. Exploiting the biological windows: current perspectives on fluorescent bioprobes emitting above 1000 nm. Nanoscale Horizons. 2016;1(3):168–84.

Hirschberg H, Madsen SJ. Cell mediated photothermal therapy of brain tumors. J Neuroimmune Pharmacol. 2017;12(1):99–106. ISSN: 1557-1890.

Hornyak GL, et al. Fabrication, characterization, and optical properties of gold nanoparticle/porous alumina composites: the nonscattering Maxwell–Garnett limit. J Phys Chem B. 1997;101(9):1548–55. ISSN: 1520-6106.

Hou G, et al. A novel pH-sensitive targeting polysaccharide-gold nanorod conjugate for combined photothermal-chemotherapy of breast cancer. Carbohydr Polym. 2019;212:334–44. ISSN: 0144-8617.

Huang P, et al. Folic acid-conjugated silica-modified gold nanorods for X-ray/CT imaging-guided dual-mode radiation and photo-thermal therapy. Biomaterials. 2011;32(36):9796–809. ISSN: 0142-9612.

Jans H, Huo Q. Gold nanoparticle-enabled biological and chemical detection and analysis. Chem Soc Rev. 2012;41(7):2849–66.

Jiang XC, et al. Gold nanorods: limitations on their synthesis and optical properties. Colloids Surf A Physicochem Eng Asp. 2006;277(1–3):201–6. ISSN: 0927-7757.

Kah JCY, et al. Exploiting the protein corona around gold nanorods for loading and triggered release. ACS Nano. 2012;6(8):6730–40. ISSN: 1936-0851.

Kannadorai RK, et al. Dual functions of gold nanorods as photothermal agent and autofluorescence enhancer to track cell death during plasmonic photothermal therapy. Cancer Lett. 2015;357(1):152–9. ISSN: 0304-3835.

Khlebtsov N, et al. Analytical and theranostic applications of gold nanoparticles and multifunctional nanocomposites. Theranostics. 2013;3(3):167. https://doi.org/10.7150/thno.8382.

Kim D, et al. Recent development of inorganic nanoparticles for biomedical imaging. ACS Cent Sci. 2018;4:324–36. https://doi.org/10.1021/acscentsci.7b00574.

Knights O, McLaughlan JR. Gold nanorods for light-based lung cancer theranostics. Int J Mol Sci. 2018;19(11):3318.

Kobayashi K, et al. Surface engineering of nanoparticles for therapeutic applications. Polym J. 2014;46(8):460–8. https://doi.org/10.1038/pj.2014.40. ISSN: 0032-3896.

Koohi SR, et al. Plasmonic photothermal therapy of colon cancer cells utilising gold nanoshells: an in vitro study. IET Nanobiotechnol. 2017;12(2):196–200. ISSN: 1751-875X.

Li Q, Cao Y. Preparation and characterization of gold nanorods. In: Nanorods. Rijeka: IntechOpen; 2012.

Li J-L, Gu M. Surface plasmonic gold nanorods for enhanced two-photon microscopic imaging and apoptosis induction of cancer cells. Biomaterials. 2010;31(36):9492–8. ISSN: 0142-9612.

Li Volsi A, et al. Near-infrared light responsive folate targeted gold nanorods for combined photothermal-chemotherapy of osteosarcoma. ACS Appl Mater Interfaces. 2017;9(16):14453–69. ISSN: 1944-8244.

Li J, et al. Simple and rapid functionalization of gold nanorods with oligonucleotides using an mPEG-SH/tween 20-assisted approach. Langmuir. 2015;31(28):7869–76. ISSN: 0743-7463.

Li Z, et al. Small gold nanorods laden macrophages for enhanced tumor coverage in photothermal therapy. Biomaterials. 2016;74:144–54. ISSN: 0142-9612.

Li D, et al. Biomimetic albumin-modified gold nanorods for photothermo-chemotherapy and macrophage polarization modulation. Acta Pharm Sin B. 2018;8(1):74–84. ISSN: 2211-3835.

Liba O, et al. Contrast-enhanced optical coherence tomography with picomolar sensitivity for functional in vivo imaging. Sci Rep. 2016;6:23337. ISSN: 2045-2322.

Lin K-Q, et al. Size effect on SERS of gold nanorods demonstrated via single nanoparticle spectroscopy. J Phys Chem C. 2016;120(37):20806–13. ISSN: 1932-7447.

Lin F-W, et al. Rapid in situ MRI traceable gel-forming dual-drug delivery for synergistic therapy of brain tumor. Theranostics. 2017;7(9):2524.

Liu X, et al. Multidentate polyethylene glycol modified gold nanorods for in vivo near-infrared photothermal cancer therapy. ACS Appl Mater Interfaces. 2014;6(8):5657–68. ISSN: 1944-8244.

Liu Y, et al. Gold nanorods/mesoporous silica-based nanocomposite as theranostic agents for targeting near-infrared imaging and photothermal therapy induced with laser. Int J Nanomed. 2015;10:4747.

Liu K, et al. Theoretical comparison of optical properties of near-infrared colloidal plasmonic nanoparticles. Sci Rep. 2016;6:34189. ISSN: 2045-2322.

Liu K, et al. Seedless synthesis of monodispersed gold nanorods with remarkably high yield: synergistic effect of template modification and growth kinetics regulation. Chem Eur J. 2017a;23(14):3291–9. ISSN: 0947-6539.

Liu S, et al. Cu (II)-doped polydopamine-coated gold nanorods for tumor theranostics. ACS Appl Mater Interfaces. 2017b;9(51):44293–306. ISSN: 1944-8244.

Liu X, et al. Effect of growth temperature on tailoring the size and aspect ratio of gold nanorods. Langmuir. 2017c;33(30):7479–85. ISSN: 0743-7463.

Liu J, et al. Tumor acidity activating multifunctional nanoplatform for NIR-mediated multiple enhanced photodynamic and photothermal tumor therapy. Biomaterials. 2018a;157:107–24. ISSN: 0142-9612.

Liu L, et al. Functional chlorin gold nanorods enable to treat breast cancer by photothermal/photodynamic therapy. Int J Nanomed. 2018b;13:8119.

Luksiene Z. Photodynamic therapy: mechanism of action and ways to improve the efficiency of treatment. Medicina. 2003;39(12):1137–50. ISSN: 1010-660X.

Madsen SJ, et al. Macrophages as cell-based delivery systems for nanoshells in photothermal therapy. Ann Biomed Eng. 2012;40(2):507–15. ISSN: 0090-6964.

Madsen SJ, et al. Nanoparticle-loaded macrophage-mediated photothermal therapy: potential for glioma treatment. Lasers Med Sci. 2015;30(4):1357–65. ISSN: 0268-8921.

Manivasagan P, et al. Biocompatible chitosan oligosaccharide modified gold nanorods as highly effective photothermal agents for ablation of breast cancer cells. Polymers. 2018;10(3):232.

Manivasagan P, et al. Chitosan/fucoidan multilayer coating of gold nanorods as highly efficient near-infrared photothermal agents for cancer therapy. Carbohydr Polym. 2019;211:360–9. ISSN: 0144-8617.

Manohar N, et al. Quantitative imaging of gold nanoparticle distribution in a tumor-bearing mouse using benchtop X-ray fluorescence computed tomography. Sci Rep. 2016;6:22079. ISSN: 2045-2322.

Mendes M, et al. Targeted theranostic nanoparticles for brain tumor treatment. Pharmaceutics. 2018a;10(4):181.

Mendes M, et al. Clinical applications of nanostructured drug delivery systems: from basic research to translational medicine. In: Core-shell nanostructures for drug delivery and theranostics. San Diego, CA: Elsevier; 2018b. p. 43–116.

Mirshafiee V, et al. Protein corona significantly reduces active targeting yield. Chem Commun. 2013;49(25):2557–9.

Mirza AZ. Fabrication and characterization of doxorubicin functionalized PSS coated gold nanorod. Arab J Chem. 2019;12(1):146–50. https://doi.org/10.1016/j.arabjc.2014.08.009. ISSN: 1878-5352.

Moon H, et al. Amplified photoacoustic performance and enhanced photothermal stability of reduced graphene oxide coated gold nanorods for sensitive photoacoustic imaging. ACS Nano. 2015;9(3):2711–9. ISSN: 1936-0851.

Nierenberg D, et al. Formation of a protein corona influences the biological identity of nanomaterials. Rep Pract Oncol Radiother. 2018;23(4):300–8. ISSN: 1507-1367.

Nikoobakht B, El-Sayed MA. Preparation and growth mechanism of gold nanorods (NRs) using seed-mediated growth method. Chem Mater. 2003;15(10):1957–62. ISSN: 0897-4756.

Nomoto T, Nishiyama N. Photodynamic therapy. In: Photochemistry for biomedical applications. Singapore: Springer; 2018. p. 301–13.

Peralta DV, et al. Hybrid paclitaxel and gold nanorod-loaded human serum albumin nanoparticles for simultaneous chemotherapeutic and photothermal therapy on 4T1 breast cancer cells. ACS Appl Mater Interfaces. 2015;7(13):7101–11. ISSN: 1944-8244.

Popescu DP, et al. Optical coherence tomography: fundamental principles, instrumental designs and biomedical applications. Biophys Rev. 2011;3(3):155. ISSN: 1867-2450.

Qin J, et al. Gold nanorods as a theranostic platform for in vitro and in vivo imaging and photothermal therapy of inflammatory macrophages. Nanoscale. 2015;7(33):13991–4001.

Ratheesh KM, et al. Gold nanorods with higher aspect ratio as potential contrast agent in optical coherence tomography and for photothermal applications around 1300 nm imaging window. Biomed Phys Eng Exp. 2016;2(5):55005. ISSN: 2057-1976.

Riley RS, Day ES. Gold nanoparticle-mediated photothermal therapy: applications and opportunities for multimodal cancer treatment. Wiley Interdiscip Rev Nanomed Nanobiotechnol. 2017;9(4):e1449. ISSN: 1939-5116.

Riva ER, et al. Plasmonic/magnetic nanocomposites: gold nanorods-functionalized silica coated magnetic nanoparticles. J Colloid Interface Sci. 2017;502:201–9. ISSN: 0021-9797.

Robinson R, et al. Comparative effect of gold nanorods and nanocages for prostate tumor hyperthermia. J Control Release. 2015;220:245–52. ISSN: 0168-3659.

Ruff J, et al. Multivalency of PEG-thiol ligands affects the stability of NIR-absorbing hollow gold nanospheres and gold nanorods. J Mater Chem B. 2016;4(16):2828–41.

Ruff J, et al. The effects of gold nanoparticles functionalized with ss-amyloid specific peptides on an in vitro model of blood–brain barrier. Nanomedicine. 2017;13(5):1645–52. ISSN: 1549-9634.

Salavatov NA, et al. Some aspects of seedless synthesis of gold nanorods. Colloid J. 2018;80(5):541–9. ISSN: 1061-933X.

Sau TK, Goia DV. Biomedical applications of gold nanoparticles. In: Fine particles in medicine and pharmacy. New York: Springer; 2012. p. 101–45.

Scaletti F, et al. Rapid purification of gold nanorods for biomedical applications. Methods X. 2014;1:118–23. ISSN: 2215-0161.

Schulz F, et al. Effective PEGylation of gold nanorods. Nanoscale. 2016;8(13):7296–308.

Seyfried TN, et al. Metabolic therapy: a new paradigm for managing malignant brain cancer. Cancer Lett. 2015;356(2):289–300. ISSN: 0304-3835.

Shen J, et al. Multifunctional gold nanorods for siRNA gene silencing and photothermal therapy. Adv Healthcare Mater. 2014;3(10):1629–37. ISSN: 2192-2640.

Shi Z, et al. Gold nanorods for biomedical imaging and therapy in cancer. Advances in nanotheranostics I. Springer, Berlin, Heidelberg, 2016. 103-136.2016. ISBN: 9783662485446.

Smitha SL, et al. Size-dependent optical properties of au nanorods. Prog Nat Sci Mater Int. 2013;23(1):36–43. ISSN: 1002-0071.

Song J, et al. Ultrasmall gold nanorod vesicles with enhanced tumor accumulation and fast excretion from the body for cancer therapy. Adv Mater. 2015;27(33):4910–7. ISSN: 0935-9648.

Sujai PT, et al. Biogenic cluster-encased gold nanorods as a targeted three-in-one theranostic nanoenvelope for SERS-guided photochemotherapy against metastatic melanoma. ACS Appl Biomater. 2018;2(1):588–600. ISSN: 2576-6422.

Tang Y, et al. In vitro cytotoxicity of gold nanorods in A549 cells. Environ Toxicol Pharmacol. 2015;39(2):871–8. ISSN: 1382-6689.

Terentyuk G, et al. Gold nanorods with a hematoporphyrin-loaded silica shell for dual-modality photodynamic and photothermal treatment of tumors in vivo. Nano Res. 2014;7(3):325–37. ISSN: 1998-0124.

Tham HP, et al. Photosensitizer anchored gold nanorods for targeted combinational photothermal and photodynamic therapy. Chem Commun. 2016;52(57):8854–7.

Tian X, et al. Biofunctional magnetic hybrid nanomaterials for theranostic applications. Nanotechnology. 2018;30(3):32002. ISSN: 0957-4484.

Tong X, et al. Size dependent kinetics of gold nanorods in EPR mediated tumor delivery. Theranostics. 2016;6(12):2039.

Tong W, et al. Control of symmetry breaking size and aspect ratio in gold nanorods: underlying role of silver nitrate. J Phys Chem C. 2017;121(6):3549–59. ISSN: 1932-7447.

Tong W, et al. The evolution of size, shape, and surface morphology of gold nanorods. Chem Commun. 2018;54(24):3022–5.

Trinidad AJ, et al. Combined concurrent photodynamic and gold nanoshell loaded macrophage-mediated photothermal therapies: an in vitro study on squamous cell head and neck carcinoma. Lasers Surg Med. 2014;46(4):310–8. ISSN: 0196-8092.

Tsai DP, et al. Single 808 nm laser treatment comprising photothermal and photodynamic therapies by using gold nanorods hybrid upconversion particles. J Phys Chem. 2018;122:2402–12. https://doi.org/10.1021/acs.jpcc.7b10976.

Urries I, et al. Magneto-plasmonic nanoparticles as theranostic platforms for magnetic resonance imaging, drug delivery and NIR hyperthermia applications. Nanoscale. 2014;6(15):9230–40.

Velasco-Aguirre C, et al. Improving gold nanorod delivery to the central nervous system by conjugation to the shuttle Angiopep-2. Nanomedicine. 2017;12(20):2503–17. ISSN: 1743-5889.

Verma J, et al. Delivery and cytotoxicity of doxorubicin and temozolomide to primary glioblastoma cells using gold nanospheres and gold nanorods. Eur J Nanomed. 2016;8(1):49–60. ISSN: 1662-596X.

Vigderman L, et al. Functional gold nanorods: synthesis, self-assembly, and sensing applications. Adv Mater. 2012;24(36):4811–41. ISSN: 0935-9648.

Wan J, et al. Surface chemistry but not aspect ratio mediates the biological toxicity of gold nanorods in vitro and in vivo. Sci Rep. 2015;5:11398. ISSN: 2045-2322.

Wang D-S, et al. Surface plasmon effects on two photon luminescence of gold nanorods. Opt Express. 2009;17(14):11350–9. ISSN: 1094-4087.

Wang J, et al. Assembly of aptamer switch probes and photosensitizer on gold nanorods for targeted photothermal and photodynamic cancer therapy. ACS Nano. 2012;6(6):5070–7. ISSN: 1936-0851.

Wang F, et al. Efficient, dual-stimuli responsive cytosolic gene delivery using a RGD modified disulfide-linked polyethylenimine functionalized gold nanorod. J Control Release. 2014;196:37–51. ISSN: 0168-3659.

Wang F, et al. Efficient RNA delivery by integrin-targeted glutathione responsive polyethyleneimine capped gold nanorods. Acta Biomater. 2015;23:136–46. ISSN: 1742-7061.

Wang B, et al. Biomaterials gold-nanorods-siRNA nanoplex for improved photothermal therapy by gene silencing. Biomaterials. 2016a;78:27–39. https://doi.org/10.1016/j.biomaterials.2015.11.025. ISSN: 0142-9612.

Wang J, et al. Localized surface plasmon resonance of gold nanorods and assemblies in the view of biomedical analysis. Trends Anal Chem. 2016b;80:429–43. https://doi.org/10.1016/j.trac.2016.03.015. ISSN: 0165-9936.

Wang S, et al. Biologically inspired polydopamine capped gold nanorods for drug delivery and light-mediated cancer therapy. ACS Appl Mater Interfaces. 2016c;8(37):24368–84. ISSN: 1944-8244.

Wang Y, et al. pH, redox and photothermal tri-responsive DNA/polyethylenimine conjugated gold nanorods as nanocarriers for specific intracellular co-release of doxorubicin and chemosensitizer pyronaridine to combat multidrug resistant cancer. Nanomedicine. 2017;13(5):1785–95. ISSN: 1549-9634.

Wei A, et al. Gold nanorods: multifunctional agents for cancer imaging and therapy. In: Cancer nanotechnology. Berlin: Springer; 2010. p. 119–30.

Wu L, et al. Enzyme-responsive multifunctional peptide coating of gold nanorods improves tumor targeting and photothermal therapy efficacy. Acta Biomater. 2019;86:363–72. ISSN: 1742-7061.

Xu X, et al. Seedless synthesis of high aspect ratio gold nanorods with high yield. J Mater Chem A. 2014;2(10):3528–35.

Xu W, et al. Hyaluronic acid-functionalized gold nanorods with pH/NIR dual-responsive drug release for synergetic targeted photothermal chemotherapy of breast cancer. ACS Appl Mater Interfaces. 2017;9(42):36533–47. ISSN: 1944-8244.

Xu W, et al. A dual-targeted hyaluronic acid-gold nanorod platform with triple-stimuli responsiveness for photodynamic/photothermal therapy of breast cancer. Acta Biomater. 2019;83:400–13. https://doi.org/10.1016/j.actbio.2018.11.026. ISSN: 1742-7061.

Yan C, et al. Concentration effect on large scale synthesis of high quality small gold nanorods and their potential role in cancer theranostics. Mater Sci Eng C. 2018a;87:120–7. ISSN: 0928-4931.

Yan J, et al. A theranostic plaster combining photothermal therapy and photodynamic therapy based on chlorin e6/gold nanorods (Ce6/au nrs) composite. Colloids Surf A. 2018b;537:460–6. https://doi.org/10.1016/j.colsurfa.2017.10.051. ISSN: 0927-7757.

Yang Z, et al. Chitosan layered gold nanorods as synergistic therapeutics for photothermal ablation and gene silencing in triple-negative breast cancer. Acta Biomater. 2015;25:194–204. ISSN: 1742-7061.

Yang H, et al. Mechanism for the cellular uptake of targeted gold nanorods of defined aspect ratios. Small. 2016;12(37):5178–89. ISSN: 1613-6810.

Yeo ELL, Cheah JU-J, et al. Protein corona around gold nanorods as a drug carrier for multimodal cancer therapy. ACS Biomater Sci Eng. 2017a;3(6):1039–50. ISSN: 2373-9878.

Yeo ELL, Joshua U, et al. Exploiting the protein corona around gold nanorods for low-dose combined photothermal and photodynamic therapy. J Mater Chem B. 2017b;5(2):254–68.

Yi Y, et al. A smart, photocontrollable drug release nanosystem for multifunctional synergistic cancer therapy. ACS Appl Mater Interfaces. 2017;9(7):5847–54. ISSN: 1944-8244.

Yuan Z, Jiang H. Photoacoustic tomography for imaging nanoparticles. In: Cancer nanotechnology. Berlin: Springer; 2010. p. 309–24.

Zarska M, et al. Biological safety and tissue distribution of (16-mercaptohexadecyl) trimethylammonium bromide-modified cationic gold nanorods. Biomaterials. 2018;154:275–90. ISSN: 0142-9612.

Zeiderman MR, et al. Acidic pH-targeted chitosan-capped mesoporous silica coated gold nanorods facilitate detection of pancreatic tumors via multispectral optoacoustic tomography. ACS Biomater Sci Eng. 2016;2(7):1108–20. ISSN: 2373-9878.

Zhang X. Gold nanoparticles: recent advances in the biomedical applications. Cell Biochem Biophys. 2015;72:771–5. https://doi.org/10.1007/s12013-015-0529-4. ISSN: 1085-9195.

Zhang J, et al. Shape-selective synthesis of gold nanoparticles with controlled sizes, shapes, and plasmon resonances. Adv Funct Mater. 2007;17(16):3295–303. ISSN: 1616-301X.

Zhang Z, et al. Silver nanoparticle gated, mesoporous silica coated gold nanorods (AuNR@MS@AgNPs): low premature release and multifunctional cancer theranostic platform. ACS Appl Mater Interfaces. 2015;7(11):6211–9. ISSN: 1944-8244.

Zhang L, et al. A multifunctional platform for tumor angiogenesis-targeted chemo-thermal therapy using polydopamine-coated gold nanorods. ACS Nano. 2016a;10(11):10404–17. ISSN: 1936-0851.

Zhang N, et al. Nanocomposite hydrogel incorporating gold nanorods and paclitaxel-loaded chitosan micelles for combination photothermal–chemotherapy. Int J Pharm. 2016b;497(1–2):210–21. ISSN: 0378-5173.

Zhang W, et al. pH and near-infrared light dual-stimuli responsive drug delivery using DNA-conjugated gold nanorods for effective treatment of multidrug resistant cancer cells. J Control Release. 2016c;232:9–19. ISSN: 0168-3659.

Zhao T, et al. Gold nanorods as dual photo-sensitizing and imaging agents for two-photon photodynamic therapy. Nanoscale. 2012;4(24):7712–9.

Zheng YB, et al. Molecular plasmonics for biology and nanomedicine. Nanomedicine. 2012;7(5):751–70. ISSN: 1743-5889.

Zhong Y, et al. cRGD-directed, NIR-responsive and robust AuNR/PEG–PCL hybrid nanoparticles for targeted chemotherapy of glioblastoma in vivo. J Control Release. 2014;195:63–71. ISSN: 0168-3659.

Index

A
Absorption, distribution, metabolism, and excretion (ADME), 45
Acidic tumor microenvironment, 235
Adenosine-5'-triphosphate (ATP), 10
Aggregation induced emission (AIE) molecules, 72, 73
Alternative method, 299–329
Amine group, 348
Aminolevulinic acid (ALA), 338–356
Amplified nucleic acid detector, 81, 82, 110
Angiogenesis, 338
Animal model, 300, 303
Animals, 349, 351, 352
Anti-angiogenic, 236
Antibacterial and tissue regeneration agent, 105, 106
Antibacterial nanocarrier, 102, 103
Antibiofilm susceptibility testing (ABST), 46
Antibodies, 198–203, 205, 207, 208
Antimicrobial cluster bombs, 45, 46
Antimicrobial peptides, 121, 128
Antimicrobial resistance (AMR), 42
Antimicrobial susceptibility testing (AST), 48
Antitoxoplasmic agent, 100
Antivirals, 138–145, 147, 148, 150, 153–158, 161–164
Apoptosis, 299, 306, 308, 316, 323, 329, 353
Aptamer-based sandwich assay (ABSA), 206
Aptamers, 10, 14, 71, 72
Aptasensors, 204, 206, 208
Atheroma, 343
Atherosclerosis, 338, 351, 352, 355, 356
Attractive interactions, 253–255
Avian influenza viruses (AIV), 140, 150–152

B
Bacteremia, 65
β-Glucan biosensor, 92
Biocompatible, 21, 23, 26, 28, 29, 33, 34, 231, 232, 236, 299, 301, 302, 305–317, 321
Bioencapsulated, 302
Biofunctionalized, 300, 305, 319, 321
Biogenic, 46
Biogenic nanomaterials, 237, 238
Bioimaging, 26, 29
Biological medium, 390–394
Biomarkers, 4–14, 50
Biomimetic, 46
Bioresponsive, 6, 9–11, 15
Biosensors, 44
Blood, 338, 339, 344, 349, 351, 352, 356
Body environment interaction, 391–392
Breast cancer, 339
Brownian relaxations, 257, 258

C
Cancer, 214–223, 338, 339, 341–345, 349, 352, 353, 355, 356
Cancer cell imaging, 86, 87, 111
Cancer cells, 300, 305–308, 310, 311, 315, 320, 321, 323, 326
Cancer stem cells (CSCs), 279, 283
Cancer therapy, 26, 31, 34, 299–327, 329, 365–397
Capping agents, 239
Carbon dots (CD), 49
Carbon quantum dots (CQDs), 72
Carboxyl group, 348
Cardiovascular, 338, 344
Cell, 342, 344, 345, 352–355

M. Rai, B. Jamil (eds.), *Nanotheranostics*,
https://doi.org/10.1007/978-3-030-29768-8

Cellular nucleases, 207
Characterization, 338, 339, 346–356
Characterization methods, 366, 371
Chemotherapy, 198–208, 214–216, 265–267, 301, 319, 321, 323, 329
Chitosan, 74
Chitosan-gold nanoparticle, 81–99, 110
Chitosan-silver nanoparticle, 99–106, 111
Combined cancer theranostics, 247, 248, 253, 255–258, 261, 264–269
Computed tomography (CT) imaging, 7, 11, 14, 22–24, 29–31, 33
Contrast agents, 247, 248, 253, 255–258, 260–262, 264–267
Controlled drug release, 125
Cyclodextrins (CDs), 304
Cytotoxicity, 338, 344, 352

D

Delivery of STAT3 siRNA, 90
Dendrimers, 24, 117
Dengue viruses (DENV), 139, 140, 161–162
Deoxyribonucleic acid (DNA), 26
Diagnose, 214, 216, 217, 220, 221, 338, 356
Diagnosis approach, 386–394
Diagnostics, 278, 281–282, 285–287, 292
DNA damage, 184, 185, 187–189
Drug, 340, 342–344, 349, 356
Drug delivery, 217, 220, 223, 278, 280–288, 290, 292, 365, 368, 374, 382, 386, 391, 396

E

Ebola viruses (EBOVs), 140, 159, 160
Eco-friendly, 236, 239
Effect of surface modification, 372–373
Electromagnetic radiation, 3
Electron transport chain, 184
Endogenous, 339, 341, 352
Enhanced permeability and retention (EPR), 4, 278, 282–283, 288
Enzyme-linked aptamer sorbent assay (ELASA), 206
Enzyme-linked immunosorbent assay (ELISA), 3, 13, 201
Enzyme-responsive nanotheranostics, 9
Epigenetic toxicity, 53
Erlotinib delivery, 85, 86

F

Feces, 338, 349, 351, 352, 356
Ferrimagnetic nanoparticles (FiMNPs), 301, 304, 306
Ferrofluids, 247, 248, 252–258, 261, 264, 267, 269
Ferromagnetism, 27
Fluorescence, 338, 339, 341–343, 349, 354, 356
Fluorescence imaging, 265, 266, 269
Fluorescence resonance energy transfer (FRET)-based drug delivery system, 51
Fluorescent inorganic nanoparticles (FINs), 43
Fluorescent organic nanoparticles (FONs), 43
Fluorophore, 341
Foodborne pathogens, 118
Food poisoning, 66
Functional coatings, 47, 48

G

Gadolinium nanoparticles, 29
Gallium based formulations, 26
Gene therapy, 22, 26, 30
Genome, 198
Genotoxicity, 52
Glioblastoma (GB), 377, 378, 390, 394–396
Global incidence of cancer, 230
Gold nanomaterials (GNMs), 219
Gold nanoparticles (AuNPs), 23–25, 181–187, 190, 191, 345–356
Gold nanoparticles platform (AuNPs), 48
Gold nanorods (AuNRs), 29, 365–397
Gold nanospheres, 23
Graphene-family nanomaterials (GFNs), 53

H

Healthcare-associated infections (HAI), 47
Health cells, 300
Heavy metal ion sensor, 94
Heme, 340, 341, 352
Hepatitis B viruses (HBV), 139, 140, 157–159
Hepatitis C viruses (HCV), 140, 157–159
Hepatocellular carcinoma, 83, 84
Herpes simplex virus (HSV), 139–141, 154–156
High-throughput screening (HTS), 51
Human immunodeficiency virus (HIV), 139–146, 157
Hydrolytic synthetic routes, 247–250, 253, 255–258, 261, 264, 267

I

Image-guided drug delivery, 21, 23
Image-guided therapy, 4, 5, 7, 8, 10, 11, 13–14, 26
Imaging, 301, 319–321, 323, 325, 329
Imaging modalities, 374, 386–389
Imaging probes, 4–8, 10–13
Immunotherapy, 198, 265, 268
Immunotoxicity, 53
In vitro, 300, 303, 305–317, 321, 323, 329, 342, 352, 355
In vivo, 52
Inductively coupled plasma optical emission spectroscopy (ICP-OES), 239
Infectious diseases, 198
Inflammatory, 338
Influenza A viruses (subtypes H3N2 and H1N1), 140, 146–150, 152, 162
Intracellular metal ions, 239
Intracellular pathogens, 116–131
Intradermal applications, 140
Intramuscular, 140, 146, 150–152, 158, 159
Intranasal, 140, 143, 145, 147, 149–153, 160, 161
Intrarectal, 140, 143
Intravaginal, 140, 143, 154
Intravenous, 140, 142, 150
Invasive, 351
Iron, 340
Iron oxide, 302, 305, 306, 308, 312, 314, 316, 317, 321
Iron oxide nanoparticles (IONPs), 27–29, 31, 32

L

Lab-on-Chip (LoC), 44
Light, 339, 342–344, 346
Liposomes (Ls), 4, 50, 116, 117, 121–123, 125, 126, 128, 217, 221, 222, 302, 321, 323–325
Liquid metal nanoparticles, 26, 34
Listeria monocytogenes, 116, 118
Localized surface plasmon resonance (LSPR), 368
Locked nucleic acid (LNA), 207
Lymphatic drainage, 4

M

Macrophages, 117, 118, 120–122, 126–128, 352–354
Magnetic field, 304–307, 314–318, 321, 323
Magnetic fluid hyperthermia (MFH), 246, 255–257, 259, 260, 265–269
Magnetic hyperthermia (MHT), 299, 301, 302, 305, 308, 310, 315, 319, 321, 323–328
Magnetic nanoparticles (MNPs), 21, 26–27, 299–303, 305–317, 319–321, 323, 327, 329
Magnetic nanotheranostics, 217, 219
Magnetic resonance imaging (MRI), 7, 11, 12, 14, 22, 24, 27–31, 33, 52, 246, 255, 260–263, 265–267, 269
Magnetic targeting (MT), 262–264
Magnetite, 246–249, 258, 259, 266
Magnetofection, 27
Malignant tumors, 299–329
Mannitol fermentation-positive test, 64
Marburg viruses, 140, 159, 160
Marker, 351
Mass spectrometry imaging (MSI), 51
Medicinal plants, 232, 236, 239
Membrane integrity, 184, 186, 187
Membrane potential, 183–186, 189
Metabolomics, 46
Metabolomics-on-a-Chip (MoC), 52
Metacaspase, 184
Metal-free silicon nanoparticles, 73
Metallic, 338
Metallic nanoparticles (MNPs), 21–27, 230–240
Metal oxide, 139
Metals, 138–140, 145, 146, 154, 162
Methicillin-resistant *Staphylococcus aureus* (MRSA), 49, 64, 66–70, 72
Methyl-ALA (MALA), 338, 345–348, 352–356
Micelles, 30, 31, 116, 117, 121, 126
Microbial enzymes, 233
Microbial infection, 180, 182–190
Microneedle (MN), 50
Mitochondria, 183, 184, 187, 188
Molecular imaging, 4, 5, 7, 8, 10–13
Molecular recognition elements (MRE), 200
Monoclonal antibodies (mAbs), 13, 199–201
Mucosal vaccine delivery, 97
Multimodal nanoparticles, 373
Multi-parameter HTS (MPHTS), 51
Mycobacterium tuberculosis, 116–118, 121, 123

N

Nanoantibiotics, 46
Nano-assemblies, 47
Nanobiotechnology, 239, 300, 305, 318, 329

Nanocapsules, 124
Nanocarrier, 4
Nanoclusters, 44, 306, 308
Nanodrugs, 69, 71
Nanoformulations, 138–164
Nanomaterials (NMs), 43, 116–131, 217, 219, 221, 222
Nanomedicines, 14, 281–283
Nano-oncology, 278, 283, 290–292
Nanoparticle-based vaccines, 139, 153
Nanoparticles mediated toxicity, 234
Nanoparticles (NPs), 3–6, 9, 12–14, 116, 121–130, 138–150, 152–163, 180–182, 186, 190, 191, 214, 217, 220–222, 278, 280–292, 348, 353, 354
Nanopyramids (DPs), 44
Nanorisk, 52
Nanospheres, 124
Nanostructured lipid carriers (NLCs), 49
Nanotechnology, 2, 3, 14, 15
Nanotheranostic agents, 79–111
Nanotheranostics, 42–54, 215–223, 300, 301, 318–327, 329
Near-infrared (NIR), 3, 11, 12
Necrosis, 344, 353
Neel's relaxation, 257, 258
Newcastle disease virus (NDV), 140, 151, 160, 161
New Zealand rabbits, 351
Non-hydrolytic synthetic routes, 248, 251
Noninvasive, 299–329
Non-toxicity, 302, 329
Nose-to-brain delivery of thymoquinone, 108, 109, 111
Nosocomial infections, 42
Novobiocin-sensitive test, 64
Nucleic acids, 9, 10

O
Optoelectronics, 46
Oral, 140, 144, 150, 158
Osteomyelitis (OM), 66
Oxidation, 340, 346
Oxidative stress, 338, 353

P
Pathogens, 198–201, 203
Peptide–drug conjugates (PDCs), 49
Personalized medicine, 20, 21, 23, 26, 27
Personalized treatment, 6, 14
Pharmacogenomics, 50
Pharmacokinetic (PK), 45
pH-dependent Ag release, 236
Phosphate modification, 207
Photoacoustic imaging (PA), 23, 24, 26, 28
Photoactivatable compounds, 47
Photo-antimicrobial, 43
Photodynamic therapy (PDT), 3, 22, 23, 29, 30, 33, 338, 342–345, 352, 354–356, 367, 373, 374, 377–380, 383–385, 387, 396
Photodynamic vaccination (PDV), 50
Photoreduction, 345, 346
Photosensitizers (PS), 3, 25, 29, 33, 341, 343, 345
Photothermal therapy (PTT), 3, 22–25, 29–31, 33, 46, 365–368, 373, 374, 376, 377, 379, 380, 382–383, 387, 395, 396
Phytochemicals, 237
Phytonanotechnology, 232
PI3K/Akt pathway, 236
Plaques, 338, 341, 344, 345, 352
Pneumonia, 64, 66
Polyethylene glycol (PEG), 284, 286, 288
Polyethylenimine, 74, 75
Polymerase chain reaction (PCR), 281
Polymeric nanoparticles, 21, 30
Polymeric nanosystems, 117
Polymers, 217, 218, 222, 223
Polymersomes, 30–32
Porphyrin, 339, 342, 351, 352
Positron emission tomography (PET), 7, 11, 13, 24, 27, 33, 52
Precursor, 341, 343
Programmed cell death, 185, 186
Prostate cancer, 342, 349
Proteome, 198
Protoporphyrin IX (PpIX), 338–343, 345, 349–352, 354, 356
Pseudorabies virus (PRV), 140, 162, 163

Q
Quantum dots (QDs), 4, 12, 21, 25, 32–34, 49, 281–284, 287

R
Radiolabeling, 107–109, 111
Radiotherapy (RT), 23, 25, 33, 198, 300, 301, 317, 319, 329
Reactive oxygen species (ROS), 3, 8, 182–188, 234, 236, 343–345, 353
Reduction, 340, 344, 346, 355
Repulsive interactions, 253–255
Resistance, 198, 202, 207

Resistant mechanism, 67
Respiratory syncytial virus (RSV), 139–141, 152, 153
RNA, 22, 31, 33
RNA-induced silencing complex (RISC), 288

S
Salmonella enterica serovar Typhi, 118
Selective accumulation, 352
Self-assembled TPIP-FONs Nanoprobes, 70
Semiconductor nanoparticles, 32
Serum therapy, 199
Silica nanoparticles (Si NPs), 4
Silver nanoparticles (AgNPs), 32–33, 181, 186–191
Single-photon emission computed tomography (SPECT), 27, 30
Single-stranded DNA, 200
SiO_2-Cy-Van nanoprobes, 70, 71
Skin and soft tissue infections, 65
Skin infections, 180, 181, 190
Solid lipid nanoparticles (SLNPs), 117, 126, 127
Sonodynamic therapies (SDT), 338, 342–345, 352, 355–356
SOS repair, 185, 189
Spectra, 348, 354
Staphylococcal enterotoxins (SEs), 203
Staphylococcus aureus, 64–75
Subcutaneous tissue, 180
Sub-G1 phase arrest, 235
Superparamagnetic hyperthermia (SPMHT), 299–329
Superparamagnetic iron oxide nanoparticles (SPIONs), 47, 246–269
Superparamagnetic iron oxide (SPIO), 12
Superparamagnetic nanoparticles (SPMNPs), 27, 300, 321
Superparamagnetism, 27
Surface-enhanced Raman spectroscopy (SERS), 24, 48
Surface functionalization, 23, 26, 27
Surface plasmon resonance (SPR), 3, 21, 23, 25, 48, 345, 346
Surfactants, 246, 250–256, 262
Swine influenza viruses (SwIV), 140, 150–152
Synergistic therapy, 385–386
Systematic evolution of ligands by exponential enrichment (SELEX), 204–206, 208

T
Target, 342, 344, 345, 351
Targeted DNA delivery, 99
Targeted drug delivery, 3, 6
Technetium radiolabeled chitosan nanoparticle, 111
Theranostic approach, 365–397
Theranostics, 64, 68–75, 128–130, 278–292, 338
Therapy, 214–217, 219, 220, 223, 278, 280, 281, 283–287, 290, 291
Thermal decomposition method, 251
Thermoacoustic imaging, 24
Tissues, 338–342, 344, 345, 351–353
Toxicity, 182, 191, 278, 280, 283, 287, 288, 291
Toxoplasma gondii, 120
Transcriptome, 198
Transmission electron microscopy, 347
Tumor microenvironments, 4, 6, 8, 9
Tumors, 278, 279, 282, 283, 287, 288, 291, 292, 338, 339, 341–344, 349

U
Ultrasound, 343, 344, 355

V
Vascular endothelial growth factor (VEGF), 234, 236
Virulent, 202

X
Xenon (Xe), 346

Z
Zeta potential, 347
Zika viruses (ZIKV), 140, 161–162
Zwitterionic surface, 47

Printed in the United States
By Bookmasters